Hansjörg Kielhöfer

Bifurcation Theory

An Introduction with Applications to Partial Differential Equations

Second Edition

Springer

Hansjörg Kielhöfer
Institut für Mathematik
Universität Augsburg
D-86135 Augsburg
Germany
Kielhoefer@math.uni-augsburg.de

ISSN 0066-5452   Applied Mathematical Sciences
ISBN 978-1-4939-0140-1   ISBN 978-1-4614-0502-3 (eBook)
DOI 10.1007/978-1-4614-0502-3
Springer New York Dordrecht Heidelberg London

British Library Cataloguing in Publication Data
A catalogue record for this book is available from the British Library

Mathematics Subject Classification (2010): 34C23, 35B32, 35P30, 37G15, 37J20, 47H11, 47J15, 47J30, 47N20

© Springer Science+Business Media, LLC 2012
Softcover reprint of the hardcover 2nd edition 2012

Printed on acid-free paper

Springer is part of Springer Science+Business Media (www.springer.com)

Contents

Preface to the Second Edition

All misprints that were known to us have been corrected, and many arguments have been made more transparent by additional comments.

Although it is modeled by an ODE, we discuss now in Section III.2.1 the buckling of the Euler rod in detail, since it is an important historical paradigm for bifurcation. This discussion requires some comments on one-dimensional elliptic operators, so-called Sturm–Liouville operators, which play a special role among elliptic operators in general dimensions.

In Remark III.2.4 we discuss the appearance of Taylor vortices in the Couette–Taylor model. From a mathematical point of view it is one of the best known examples of a symmetry-breaking bifurcation. Nonetheless, there remain open mathematical problems, some of which are purely technical, some of which are really deep. We comment also on other pattern formations of this prominent model.

We include in Remark III.6.4 a detailed proof of the singular limit process of the Cahn–Hilliard model when the interfacial energy tends to zero. In Remark III.6.5 we explain how this method is used for more complicated nonconvex variational problems.

Two sections are new: In Section I.19.1 we prove bifurcation with a two-dimensional kernel, and in Section III.2.2 we apply this method to nonlinear elliptic systems.

Section III.7.5 is completely revised.

January 2011   Hansjörg Kielhöfer

Chapter 0
Introduction

Bifurcation Theory attempts to explain various phenomena that have been discovered and described in the natural sciences over the centuries. The buckling of the Euler rod, the appearance of Taylor vortices, and the onset of oscillations in an electric circuit, for instance, all have a common cause: A specific physical parameter crosses a threshold, and that event forces the system to the organization of a new state that differs considerably from that observed before.

Mathematically speaking, the following occurs: The observed states of a system correspond to solutions of nonlinear equations that model the physical system. A state can be observed if it is stable, an intuitive notion that is made precise for a mathematical solution. One expects that a slight change of a parameter in a system should not have a big influence, but rather that stable solutions change continuously in a unique way. That expectation is verified by the Implicit Function Theorem. Consequently, as long as a continuous branch of solutions preserves its stability, no dramatic change is observed when the parameter is varied. However, if that "ground state" loses its stability when the parameter reaches a critical value, then the state is no longer observed, and the system itself organizes a new stable state that "bifurcates" from the ground state.

Bifurcation is a paradigm for nonuniqueness in Nonlinear Analysis.

We sketch that scenario in Figure 1, which is referred to as a "pitchfork bifurcation." The solutions bifurcate in pairs that typically describe one state in two possible representations. Also typically, the bifurcating state has less symmetry than the ground state (also called a "trivial solution"), in which case one calls it a "symmetry-breaking bifurcation." In Figure 1 we show the solution set of the odd "bifurcation equation" $\lambda x - x^3 = 0$, where $x \in \mathbb{R}$ represents the state and $\lambda \in \mathbb{R}$ is the parameter.

In the case in which solutions correspond to critical points of a parameter-dependent functional, Figure 2 shows how a slight change of the potential turns a stable equilibrium into an unstable one and creates at the same time two new stable equilibria. That exchange of stability, however, is not restricted to variational problems, but is typical of all "generic" bifurcations.

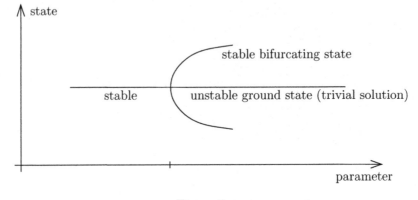

Figure 1

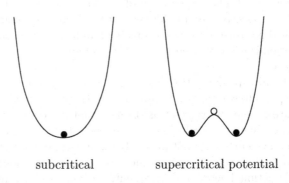

Figure 2

Bifurcation Theory provides the mathematical existence of bifurcation scenarios observed in various systems and experiments. A necessary condition is obviously the failure of the Implicit Function Theorem.

In this book we present some sufficient conditions for "one-parameter bifurcation," which means that the bifurcation parameter is a real scalar. We do not treat "multiparameter bifurcation theory."

We distinguish a local theory, which describes the bifurcation diagram in a neighborhood of the bifurcation point, and a global theory, where the continuation of local solution branches beyond that neighborhood is investigated. In applications we also prove specific qualitative properties of solutions on global branches, which, in turn, help to separate global branches, to decide on their unboundedness, and, in special cases, to establish their smoothness and asymptotic behavior.

As mentioned before, bifurcation is often related to a breaking of symmetry. We sometimes make use of symmetry in the applications in investigating the qualitative properties of solutions on global branches. However, we typically exploit symmetry in an ad hoc manner. For a systematic treatment of symmetry and bifurcation, we refer to the monographs [18], [58], [59], [164]. Symmetry ideas do not play a dominant role in this book.

We present the results of Chapter I and Chapter II in an abstract way, and we apply these abstract results to concrete problems for partial differential equations only in Chapter III. The theory is separated from applications for the following reasons: It is our opinion that mathematical understanding can be reached only via abstraction and not by examples or applications. Moreover, only an abstract result is suitable to be adapted to a new problem. Therefore, we resisted mixing the general theory with our personal selection of applications.

The general theory of Chapters I and II is formulated for operators acting in infinite-dimensional spaces. This lays the groundwork for Chapter III, where detailed applications to concrete partial differential equations are provided. The abstract versions of the Hopf Bifurcation Theorem in Chapter I are directly applicable to ODEs, RFDEs, and Hamiltonian or reversible systems. For stability considerations we employ throughout the principle of linearized stability, which means, in turn, that stability is determined by the perturbation of the critical eigenvalue or Floquet exponent.

The motivation to write this book came from many questions of students and colleagues about bifurcation theorems. Most of the results contained herein are not new. But many are apparently known only to a few experts, and a unified presentation was not available. Indeed, while there exist many good books treating various aspects of bifurcation theory, e.g., [11], [18], [19], [35], [58], [59], [60], [64], [81], [164], [171], there is precious little analysis of problems governed by partial differential equations available in textbook form. This work addresses that gap. We apologize to all who have obtained similar or better results that are not mentioned here. During the last thirty years a vast literature on bifurcation theory has been published, and we have not been able to write a survey. A reason for this limitation is that we feel competent only in fields where we have worked ourselves.

In many of the above-mentioned books we find the "basic" or "generic" bifurcations in simple settings illustrating the geometric ideas behind them, mostly from a dynamical viewpoint; cf.[64]. In view of that excellent heuristic literature, we think that there is no need to repeat these ideas but that it is necessary to give the calculations in a most general setting. This might be hard for beginners, but we hope that it is useful to advanced students.

Apart from the Cahn–Hilliard model (serving as a paradigm), our applications to partial differential equations are motivated only by, but are not directly related to, mathematical physics. The formulation of a specific problem of physics and the verification of all hypotheses are typically quite involved, and such an expenditure might disguise the essence of Bifurcation Theory.

For these reasons we believe that a detailed presentation of the cascade of bifurcations appearing in the Taylor model, for instance, is not appropriate here; rather, we refer to the literature, [17], for example. On the other hand, we hope that our choice of mathematical applications offers a broad selection of techniques illustrating the use of the abstract theory without getting lost in too many technicalities. Finally, if necessary, the analysis can be completed by numerical analysis as expounded in [4], [87], and [159].

I am indebted to Rita Moeller for having typed the entire text in LaTeX. And in particular, I thank my friend Tim Healey for his encouragement and help in writing this book: Many of the results obtained in a fruitful collaboration with him are presented here.

Chapter I
Local Theory

I.1 The Implicit Function Theorem

One of the most important analytic tools for the solution of a nonlinear problem

(I.1.1) $$F(x, y) = 0,$$

where F is a mapping $F : U \times V \to Z$ with open sets $U \subset X, V \subset Y$, and where X, Y, Z are (real) Banach spaces, is the following **Implicit Function Theorem**:

Theorem I.1.1 *Let (I.1.1) have a solution $(x_0, y_0) \in U \times V$ such that the Fréchet derivative of F with respect to x at (x_0, y_0) is bijective:*

(I.1.2)
$$F(x_0, y_0) = 0,$$
$$D_x F(x_0, y_0) : X \to Z \quad \text{is bounded (continuous)}$$
$$\text{with a bounded inverse (Banach's Theorem).}$$

Assume also that F and $D_x F$ are continuous:

(I.1.3)
$$F \in C(U \times V, Z),$$
$$D_x F \in C(U \times V, L(X, Z)), \text{ where } L(X, Z)$$
$$\text{denotes the Banach space of bounded linear operators}$$
$$\text{from } X \text{ into } Z \text{ endowed with the operator norm.}$$

Then there exist a neighborhood $U_1 \times V_1$ in $U \times V$ of (x_0, y_0) and a mapping $f : V_1 \to U_1 \subset X$ such that

(I.1.4)
$$f(y_0) = x_0,$$
$$F(f(y), y) = 0 \text{ for all } y \in V_1.$$

Furthermore, f is continuous on V_1:

(I.1.5) $f \in C(V_1, X)$.

Finally, every solution of (I.1.1) in $U_1 \times V_1$ is of the form $(f(y), y)$.

For a proof we refer to [38]. For the prerequisites to this book we recommend also [19], [10], which present sections on analysis in Banach spaces.

Let us consider Y as a space of parameters and X as a space of configurations (a phase space, for example). Then the Implicit Function Theorem allows the following interpretation: The configuration described by problem (I.1.1) persists for perturbed parameters if it exists for some particular parameter, and it depends smoothly and in a unique way on the parameters. In other words, this theorem describes what one expects: A small change of parameters entails a unique small change of configuration (without any "surprise"). Thus "dramatic" changes in configurations for specific parameters can happen only if the assumptions of Theorem I.1.1 are violated, in particular, if

(I.1.6) $D_x F(x_0, y_0) : X \to Z$ is not bijective.

Bifurcation Theory can be briefly described by the investigation of problem (I.1.1) in a neighborhood of (x_0, y_0) where (I.1.6) holds.

For later use we need the following addition to Theorem I.1.1:

(I.1.7)
> If the mapping F in (I.1.1) is k-times
> continuously differentiable on $U \times V$, i.e.,
> $F \in C^k(U \times V, Z)$, then the mapping f
> in (I.1.4) is also k-times continuously
> differentiable on V_1; i.e., $f \in C^k(V_1, X), k \geq 1$.
> If the mapping F is analytic, then the
> mapping f is also analytic.

For a proof we refer again to [38].

I.2 The Method of Lyapunov–Schmidt

The method of Lyapunov and Schmidt describes the reduction of problem (I.1.1) (which is high- or infinite-dimensional) to a problem having only as many dimensions as the defect (I.1.6). To be more precise, we need the following definition:

Definition I.2.1 *A continuous mapping $F : U \to Z$, where $U \subset X$ is open and where X, Z are Banach spaces, is a nonlinear Fredholm operator if it is Fréchet differentiable on U and if $DF(x)$ fulfills the following:*
(i) $\dim N(DF(x)) < \infty$ ($N = $ null space or kernel),

(ii) $\operatorname{codim} R(DF(x)) < \infty$ *(R = range),*

(iii) $R(DF(x))$ *is closed in* Z.

The integer $\dim N(DF(x)) - \operatorname{codim} R(DF(x))$ *is called the Fredholm index* of $DF(x)$.

Remark I.2.2 *As remarked in [86], p.230, assumption (iii) is redundant. If* DF *depends continuously on* x *and possibly on a parameter* y, *in the sense of* (I.1.3), *and if* U *or* $U \times V$ *is connected in* X *or also in* $X \times Y$, *respectively, then it can be shown that the Fredholm index of* $DF(x)$ *is independent of* x; *cf.[86], IV. 5.*

We consider now $F : U \times V \to Z$, $U \subset X$, $V \subset Y$, where

(I.2.1)
$$F(x_0, y_0) = 0 \quad \text{for some} \quad (x_0, y_0) \in U \times V,$$
$$F \in C(U \times V, Z),$$
$$D_x F \in C(U \times V, L(X, Z)) \quad (\text{see (I.1.3)}).$$

We assume that for $y = y_0$ the mapping F is a nonlinear Fredholm operator with respect to x; i.e., $F(\cdot, y_0) : U \to Z$ satisfies Definition I.2.1. In particular, observe that the spaces N and Z_0 defined below are finite-dimensional.

Thus there exist closed complements in the Banach spaces X and Z such that

(I.2.2)
$$X = N(D_x F(x_0, y_0)) \oplus X_0,$$
$$Z = R(D_x F(x_0, y_0)) \oplus Z_0$$

(see [39], p.553). These decompositions, in turn, define projections

(I.2.3)
$$\begin{aligned} P : X \to N \quad &\text{along} \quad X_0 \quad (N = N(D_x F(x_0, y_0))), \\ Q : Z \to Z_0 \quad &\text{along} \quad R \quad (R = R(D_x F(x_0, y_0))), \end{aligned}$$

in a natural way. By the Closed Graph Theorem (see [170]) these projections are continuous.

Then the following **Reduction Method of Lyapunov–Schmidt** holds:

Theorem I.2.3 *There is a neighborhood* $U_2 \times V_2$ *of* (x_0, y_0) *in* $U \times V \subset X \times Y$ *such that the problem*

(I.2.4)
$$F(x, y) = 0 \quad \text{for} \quad (x, y) \in U_2 \times V_2$$

is equivalent to a finite-dimensional problem

(I.2.5)
$$\begin{aligned} &\Phi(v, y) = 0 \text{ for } (v, y) \in \tilde{U}_2 \times V_2 \subset N \times Y, \text{ where} \\ &\Phi : \tilde{U}_2 \times V_2 \to Z_0 \text{ is continuous} \\ &\text{and } \Phi(v_0, y_0) = 0, \ (v_0, y_0) \in \tilde{U}_2 \times V_2. \end{aligned}$$

The function Φ, *called a bifurcation function, is given in* (I.2.9) *below.*

(If the parameter space Y is finite-dimensional, then (I.2.5) is indeed a purely finite-dimensional problem.)

Proof. Problem (I.2.4) is obviously equivalent to the system

$$\text{(I.2.6)} \qquad \begin{aligned} QF(Px + (I - P)x, y) &= 0, \\ (I - Q)F(Px + (I - P)x, y) &= 0, \end{aligned}$$

where we set $Px = v \in N$ and $(I - P)x = w \in X_0$. Next we define

$$\text{(I.2.7)} \quad \begin{aligned} &G : \tilde{U}_2 \times W_2 \times V_2 \to R \text{ via} \\ &G(v, w, y) \equiv (I - Q)F(v + w, y), \text{ where} \\ &v_0 = Px_0 \in \tilde{U}_2 \subset N, \\ &w_0 = (I - P)x_0 \in W_2 \subset X_0, \\ &\text{and } \tilde{U}_2, W_2 \text{ are neighborhoods such that } \tilde{U}_2 + W_2 \subset U \subset X. \end{aligned}$$

We have $G(v_0, w_0, y_0) = 0$, and by our choice of the spaces, $D_w G(v_0, w_0, y_0) = (I-Q)D_x F(x_0, y_0) : X_0 \to R$ is bijective. Application of the Implicit Function Theorem then yields

$$\text{(I.2.8)} \quad \begin{aligned} &G(v, w, y) = 0 \text{ for } (v, w, y) \in \tilde{U}_2 \times W_2 \times V_2 \text{ is equivalent to} \\ &w = \psi(v, y) \text{ for some } \psi : \tilde{U}_2 \times V_2 \to W_2 \subset X_0 \text{ such that} \\ &\psi(v_0, y_0) = w_0. \end{aligned}$$

Insertion of the function ψ into (I.2.6)$_1$ yields

$$\text{(I.2.9)} \qquad \Phi(v, y) \equiv QF(v + \psi(v, y), y) = 0.$$

The Implicit Function Theorem also gives the continuity of ψ. □

Corollary I.2.4 *In the notation of Theorem I.2.3, if $F \in C^1(U \times V, Z)$, we also obtain $\psi \in C^1(\tilde{U}_2 \times V_2, X_0)$, $\Phi \in C^1(\tilde{U}_2 \times V_2, Z_0)$, and*

$$\text{(I.2.10)} \qquad \begin{aligned} \psi(v_0, y_0) &= w_0, \quad D_v \psi(v_0, y_0) = 0 \in L(N, X_0), \\ D_v \Phi(v_0, y_0) &= 0 \in L(N, Z_0). \end{aligned}$$

Proof. The regularity of ψ and Φ follows from (I.1.7). Differentiating $(I - Q)F(v + \psi(v, y), y) = 0$ for all $(v, y) \in \tilde{U}_2 \times V$ with respect to v yields

$$\text{(I.2.11)} \qquad (I - Q)D_x F(v + \psi(v, y), y)(I_N + D_v \psi(v, y)) = 0,$$

where I_N denotes the identity in N. Since N is the kernel of $D_x F(x_0, y_0)$, we obtain at (v_0, y_0),

$$\text{(I.2.12)} \qquad (I - Q)D_x F(x_0, y_0)D_v \psi(v_0, y_0) = 0.$$

Since $D_v \psi(v_0, y_0)$ maps into X_0, which is complementary to N, we necessarily have $D_v \psi(v_0, y_0) = 0$. By virtue of (I.2.9) we then get

(I.2.13) $$D_v \Phi(v_0, y_0) = Q D_x F(x_0, y_0) I_N = 0.$$

$\square$

I.3 The Lyapunov–Schmidt Reduction for Potential Operators

In applications, the following situation often occurs: $G : U \to Z$ is a mapping, where U is an open subset of a real Banach space X, and X is continuously embedded into Z. Furthermore, a scalar product can be defined on the real Banach space Z such that

(I.3.1) $(\ ,\ ) : Z \times Z \to \mathbb{R}$ is bilinear, symmetric, continuous, and definite; i.e., $(z, z) \geq 0$, and $(z, z) = 0$ if and only if $z = 0$.

Definition I.3.1 *A continuous mapping* $G : U \to Z$, *where* $U \subset X$, X *is continuously embedded into* Z, *and* Z *is endowed with a scalar product* $(\ ,\ )$ *satisfying (I.3.1), is called a potential operator (with respect to that scalar product) if there exists a continuously differentiable mapping* $g : U \to \mathbb{R}$ *such that*

(I.3.2) $$Dg(x)h = (G(x), h) \quad \text{for all} \quad x \in U, h \in X.$$

The function g *is called the potential of* G. *We use also the notation* $G = \nabla g$.

Proposition I.3.2 *If* $G : U \to Z$ *is a potential operator and differentiable, then the derivative* $DG(x) \in L(X, Z)$ *is symmetric with respect to* $(\ ,\ )$; *i.e.,*

(I.3.3) $(DG(x)h_1, h_2) = (h_1, DG(x)h_2) = (DG(x)h_2, h_1)$
for all $x \in U$, $h_1, h_2 \in X$.

Proof. The potential g is twice differentiable, and its second derivative is a continuous bilinear mapping from $X \times X$ into $\mathbb{R}$ is given by

(I.3.4) $$D^2 g(x)[h_1, h_2] = (DG(x)h_1, h_2).$$

A well-known result is that this bilinear mapping $D^2 g(x)$ is symmetric; i.e., $D^2 g(x)[h_1, h_2] = D^2 g(x)[h_2, h_1]$ (see [38], [19], [10]). $\square$

Proposition I.3.3 *Let* $G : U \to Z$ *be continuously differentiable and assume that the open set* $U \subset X$ *is star-shaped with respect to the origin* $0 \in X$. *If* $DG(x)$ *is symmetric with respect to* $(\ ,\ )$ *in the sense of (I.3.3) for all* $x \in U$, *then* G *is a potential operator with respect to* $(\ ,\ )$.

Proof. We define

$$(I.3.5) \qquad g(x) = \int_0^1 (G(tx), x) dt \qquad \text{for} \quad x \in U.$$

Then

$$(I.3.6) \qquad \begin{aligned} \frac{d}{ds} g(x + sh)\big|_{s=0} &= \int_0^1 [(DG(tx)th, x) + (G(tx), h)] dt \\ &= \int_0^1 \frac{d}{dt} (G(tx), th) dt = (G(x), h) \text{ for all } x \in U, h \in X. \end{aligned}$$

This proves that the Gâteaux derivative of g at x in the direction h is linear and continuous in h for all $x \in U$, and furthermore, that the Gâteaux derivative of g is continuous in x (with respect to the norm in $L(X, \mathbb{R}) = X'$, the dual space). Accordingly, the Gâteaux derivative is actually the Fréchet derivative (see [38], [19], [10]). □

If $G : U \to Z$, $U \subset X$, is a differentiable potential operator (see Definition I.3.1) and a nonlinear Fredholm operator of index zero in the sense of Definition I.2.1, then the kernel of $DG(x)$ and its range have equal finite dimension and codimension, respectively. By the symmetry as stated in Proposition I.3.2, the following assumption is reasonable:

$$(I.3.7) \qquad \begin{aligned} &Z = R(DG(x)) \oplus N(DG(x)), \\ &\text{where } R \text{ and } N \text{ are orthogonal with respect to} \\ &\text{the scalar product } (\quad , \quad) \text{ on } Z. \end{aligned}$$

We recall that $N(DG(x)) \subset X \subset Z$ (with continuous embedding) in this section.

Next we consider $F : U \times V \to Z$, $U \subset X \subset Z$, $V \subset Y$, where (I.2.1) is satisfied. Furthermore, we assume that F is a potential operator and a nonlinear Fredholm operator of index zero with respect to x; i.e., $F(\cdot, y)$ satisfies Definitions I.2.1 and I.3.1 for all $y \in V$. Finally, we assume the orthogonal decomposition

$$(I.3.8) \qquad Z = R(D_x F(x_0, y_0)) \oplus N(D_x F(x_0, y_0)),$$

cf. (I.2.2); i.e., $Z_0 = N$. This decomposition defines an orthogonal projection

$$(I.3.9) \qquad Q : Z \to N \quad \text{along } R \text{ (as in (I.2.3))}$$

that is continuous on Z. By the continuous embedding $X \subset Z$, its restriction

$$(I.3.10) \qquad Q|_X : X \to N \subset X$$

is continuous as well, and will be denoted by P. This projection, in turn, defines the decomposition

(I.3.11) $$X = N \oplus (R \cap X),$$

where $R \cap X$ is closed in X.

Using these projections, the Lyapunov–Schmidt reduction as stated in Theorem I.2.3 has the following additional property:

Theorem I.3.4 *If F is a potential operator with respect to x, then the finite-dimensional mapping Φ obtained by the orthogonal Lyapunov–Schmidt reduction (cf.(I.2.9)) is also a potential operator with respect to v. (The scalar product on Z induces a scalar product on $N \subset X \subset Z$, and this same scalar product is employed in the definition of a potential operator in both cases.)*

Proof. We use the same notation as in the proof of Theorem I.2.3. Let $f(x, y)$ be the potential for $F(x, y)$; i.e.,

(I.3.12) $$\begin{aligned} &D_x f(x,y)h = (F(x,y), h) \\ &\text{for all } (x, y) \in U \times V \subset X \times Y \text{ and for all } h \in X. \end{aligned}$$

Then we claim that $f(v + \psi(v, y), y)$ (see (I.2.8)) is a potential for $\Phi(v, y) = QF(v + \psi(v, y), y)$. For every $(v, y) \in \tilde{U}_2 \times V_2 \subset N \times Y$ and $h \in N$ we get by differentiation of f with respect to v,

(I.3.13) $$\begin{aligned} &D_v f(v + \psi(v, y), y)h \\ &= D_x f(v + \psi(v, y), y)(I_N + D_v \psi(v, y))h \\ &= (F(v + \psi(v, y), y), h + D_v \psi(v, y)h) \\ &= (QF(v + \psi(v, y), y), h) \\ &\quad + ((I - Q)F(v + \psi(v, y), y), D_v \psi(v, y)h) \text{ (by orthogonality)} \\ &= (\Phi(v, y), h), \end{aligned}$$

where we have employed $(I - Q)F(v + \psi(v, y), y) = 0$; cf. (I.2.7), (I.2.8). □

Corollary I.3.5 $D_v \Phi(v, y) = QD_x F(v + \psi(v, y), y)(I_N + D_v \psi(v, y))$ *is a symmetric operator in $L(N, N)$ with respect to the scalar product $(\ ,\ )$.*

Proof. The proof is the same as that for Proposition I.3.2. □

I.4 An Implicit Function Theorem for One-Dimensional Kernels: Turning Points

In this section we consider mappings $F : U \times V \to Z$ with open sets $U \subset X, V \subset Y$, where X and Z are Banach spaces, but where this time $Y = \mathbb{R}$.

Following a long tradition, we change the notation and denote parameters in $\mathbb{R}$ by λ. We assume

$$(I.4.1) \qquad \begin{aligned} &F(x_0, \lambda_0) = 0 \text{ for some } (x_0, \lambda_0) \in U \times V, \\ &\dim N(D_x F(x_0, \lambda_0)) = 1. \end{aligned}$$

Obviously, the Implicit Function Theorem, Theorem I.1.1, is not directly applicable. We assume now the hypotheses of the Lyapunov–Schmidt reduction (Theorem I.2.3) for F with the additional assumption that

$$(I.4.2) \qquad \begin{aligned} &\text{the Fredholm index of } D_x F(x_0, \lambda_0) \text{ is zero;} \\ &\text{i.e., by (I.4.1), codim} R(D_x F(x_0, \lambda_0)) = 1. \end{aligned}$$

Since $Y = \mathbb{R}$, we can identify the Fréchet derivative $D_\lambda F(x, \lambda)$ with an element of Z, namely, by

$$(I.4.3) \qquad D_\lambda F(x, \lambda)1 = D_\lambda F(x, \lambda) \in Z, \quad 1 \in \mathbb{R}.$$

Theorem I.4.1 *Assume that $F : U \times V \to Z$ is continuously differentiable on $U \times V \subset X \times \mathbb{R}$, i.e.,*

$$(I.4.4) \qquad F \in C^1(U \times V, Z),$$

and (I.4.1), (I.4.2), (I.4.3), and that

$$(I.4.5) \qquad D_\lambda F(x_0, \lambda_0) \notin R(D_x F(x_0, \lambda_0)).$$

Then there is a continuously differentiable curve through (x_0, λ_0); that is, there exists

$$(I.4.6) \qquad \{(x(s), \lambda(s))|s \in (-\delta, \delta), \ (x(0), \lambda(0)) = (x_0, \lambda_0)\}$$

such that
$$(I.4.7) \qquad F(x(s), \lambda(s)) = 0 \text{ for } s \in (-\delta, \delta),$$

and all solutions of $F(x, \lambda) = 0$ in a neighborhood of (x_0, λ_0) belong to the curve (I.4.6).

Proof. We apply Theorem I.2.3, and we know that all solutions of $F(x, \lambda) = 0$ near (x_0, λ_0) can be found by solving $\Phi(v, \lambda)$ near (v_0, λ_0). Using the terminology of the proof of that Theorem, assumption (I.4.4) together with (I.1.7) for $k = 1$ gives the continuous differentiability of Φ with respect to λ, and in particular,

$$(I.4.8) \qquad D_\lambda \Phi(v_0, \lambda_0) = Q D_\lambda F(v_0 + \psi(v_0, \lambda_0), \lambda_0) = Q D_\lambda F(x_0, \lambda_0) \neq 0,$$

by assumption (I.4.5). Now, by (I.4.1), (I.4.2) the spaces N and Z_0 are one-dimensional, and also $Y = \mathbb{R}$ is one-dimensional. Since

$$\Phi : \tilde{U}_2 \times V_2 \to Z_0, \quad \tilde{U}_2 \times V_2 \subset N \times \mathbb{R},$$
(I.4.9)
$$\Phi(v_0, \lambda_0) = 0, \quad D_\lambda \Phi(v_0, \lambda_0) \neq 0,$$

the Implicit Function Theorem implies the existence of a continuously differentiable mapping

(I.4.10)
$$\varphi : \tilde{U}_2 \to V_2 \subset \mathbb{R} \quad \text{such that} \quad \varphi(v_0) = \lambda_0,$$
$$\Phi(v, \varphi(v)) = 0 \quad \text{for all} \quad v \in \tilde{U}_2 \subset N.$$

(In fact, it may be necessary to shrink the neighborhood $\tilde{U}_2$, but for simplicity we use the same notation.)

Let

(I.4.11) $N(D_x F(x_0, \lambda_0)) = \text{span}[\hat{v}_0], \ \hat{v}_0 \in X, \ \|\hat{v}_0\| = 1.$

Then $v = v_0 + s\hat{v}_0 \in \tilde{U}_2$ for $s \in (-\delta, \delta)$, and

(I.4.12)
$$x(s) = v_0 + s\hat{v}_0 + \psi(v_0 + s\hat{v}_0, \varphi(v_0 + s\hat{v}_0)),$$
$$\lambda(s) = \varphi(v_0 + s\hat{v}_0),$$

gives the curve (I.4.6), having all properties claimed in Theorem I.4.1. □

Corollary I.4.2 *The tangent vector of the solution curve (I.4.6) at (x_0, λ_0) is given by*

(I.4.13) $(\hat{v}_0, 0) \in X \times \mathbb{R};$

i.e., (I.4.6) is tangent at (x_0, λ_0) to the one-dimensional kernel of $D_x F(x_0, \lambda_0)$.

Proof. Since $\Phi(v, \varphi(v)) = 0$ for all $v \in \tilde{U}_2$, and $D_v \Phi(v_0, \lambda_0) = 0$ by Corollary I.2.4, we get

(I.4.14) $D_\lambda \Phi(v_0, \lambda_0) D_v \varphi(v_0) = 0 \quad (\varphi(v_0) = \lambda_0).$

By (I.4.9), $D_\lambda \Phi(v_0, \lambda_0) \neq 0$, and thus $D_v \varphi(v_0) = 0$.
Now, by (I.4.12),

$$\frac{d}{ds} x(s)|_{s=0} = \hat{v}_0 + D_v \psi(v_0, \lambda_0)\hat{v}_0 + D_\lambda \psi(v_0, \lambda_0) D_v \varphi(v_0)\hat{v}_0,$$
(I.4.15)
$$= \hat{v}_0 \text{ by Corollary I.2.4 and } D_v \varphi(v_0) = 0,$$

$$\frac{d}{ds} \lambda(s)|_{s=0} = D_v \varphi(v_0)\hat{v}_0 = 0. \qquad \square$$

Let us assume more differentiability on F, namely, $F \in C^2(U \times V, Z)$. Then differentiation of (I.4.7) with respect to s gives, in view of (I.4.15),

$$\frac{d}{ds}F(x(s), \lambda(s))\big|_{s=0}$$

$$= D_x F(x_0, \lambda_0)\dot{x}(0) + D_\lambda F(x_0, \lambda_0)\dot{\lambda}(0) \quad \left(\cdot = \frac{d}{ds}\right)$$

$$= D_x F(x_0, \lambda_0)\hat{v}_0 = 0,$$

(I.4.16)

$$\frac{d^2}{ds^2}F(x(s), \lambda(s))\big|_{s=0}$$

$$= D_{xx}^2 F(x_0, \lambda_0)[\hat{v}_0, \hat{v}_0] + D_x F(x_0, \lambda_0)\ddot{x}(0) + D_\lambda F(x_0, \lambda_0)\ddot{\lambda}(0) = 0$$

(observe that $\dot{\lambda}(0) = 0$).

Application of the projection Q (see (I.2.3)) yields

(I.4.17) $QD_{xx}^2 F(x_0, \lambda_0)[\hat{v}_0, \hat{v}_0] + QD_\lambda F(x_0, \lambda_0)\ddot{\lambda}(0) = 0.$

Since $QD_\lambda F(x_0, \lambda_0) \neq 0$ by virtue of (I.4.5), the additional assumption

(I.4.18) $D_{xx}^2 F(x_0, \lambda_0)[\hat{v}_0, \hat{v}_0] \notin R(D_x F(x_0, \lambda_0))$

guarantees (according to (I.4.17), which is an equation in the one-dimensional space Z_0)

(I.4.19) $\ddot{\lambda}(0) > 0 \quad \text{or} \quad \ddot{\lambda}(0) < 0.$

This means that schematically, the curve (I.4.6) through $(x_0, \lambda_0) \in X \times \mathbb{R}$ has one of the shapes sketched in Figure I.4.1.

In the literature, this is commonly called a **saddle-node bifurcation**, a nomenclature that makes sense only if the vector fields $F(\cdot, \lambda) : X \to Z$ generate a flow, which, in turn, requires $X \subset Z$. Since that is not always true in our general setting, we prefer the terminology **turning point** or **fold**.

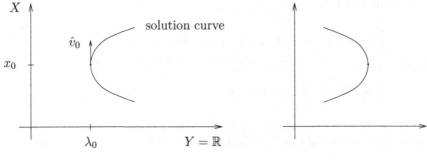

Figure I.4.1

In order to replace the nonzero quantities in (I.4.17) by real numbers, we introduce the following explicit representation of the projection Q in (I.2.3). Recall that the complement Z_0 of $R(D_x F(x_0, \lambda_0))$ is one-dimensional:

(I.4.20) $\qquad\qquad Z_0 = \text{span}[\hat{v}_0^*], \quad \hat{v}_0^* \in Z, \quad \|\hat{v}_0^*\| = 1.$

By the Hahn–Banach Theorem (see [170]), there exists a vector

$\qquad\qquad\qquad \hat{v}_0' \in Z'$ (the dual space) such that

(I.4.21) $\qquad\qquad \langle \hat{v}_0^*, \hat{v}_0' \rangle = 1 \quad$ and

$\qquad\qquad\qquad \langle z, \hat{v}_0' \rangle = 0 \quad$ for all $z \in R(D_x F(x_0, \lambda_0)).$

Here $\langle \ , \ \rangle$ denotes the duality between Z and Z'.

Then the projection Q in (I.2.3) is given by

(I.4.22) $\qquad\qquad Qz = \langle z, \hat{v}_0' \rangle \hat{v}_0^* \quad$ for all $z \in Z,$

and (I.4.17), (I.4.18) imply

(I.4.23) $\qquad\qquad \ddot{\lambda}(0) = -\dfrac{\langle D_{xx}^2 F(x_0, \lambda_0)[\hat{v}_0, \hat{v}_0], \hat{v}_0' \rangle}{\langle D_\lambda F(x_0, \lambda_0), \hat{v}_0' \rangle},$

and the sign of $\ddot{\lambda}(0)$ determines the appropriate diagram in Figure I.4.1. If $\ddot{\lambda}(0) = 0$, however, the shape of the curve (I.4.6) is determined by higher derivatives of $\lambda(s)$ at $s = 0$.

Remark I.4.3 *There is also an Implicit Function Theorem for higher-dimensional kernels if the parameter space Y is higher-dimensional, too. To be more precise, if $\dim N(D_x F(x_0, \lambda_0)) = n$ for some $(x_0, \lambda_0) \in U \times V \subset X \times \mathbb{R}^n$ and if a complement of $R(D_x F(x_0, \lambda_0))$ is spanned by $D_{\lambda_i} F(x_0, \lambda_0)$, $i = 1, \dots, n$, then the analogous proof yields an n-dimensional manifold of the form $\{(x(s), \lambda(s)))|s \in \tilde{U}_3 \subset \mathbb{R}^n\} \subset X \times \mathbb{R}^n$ through $(x(0), \lambda(0)) = (x_0, \lambda_0)$ such that $F(x(s), \lambda(s)) = 0$ for all $s \in \tilde{U}_3$ (which is a neighborhood of $0 \in \mathbb{R}^n$). Moreover, the manifold is tangent to $N(D_x F(x_0, \lambda_0)) \times \{0\}$ in $X \times \mathbb{R}^n$.*

I.5 Bifurcation with a One-Dimensional Kernel

We assume the existence of a solution curve of $F(x, \lambda) = 0$ through (x_0, λ_0) and prove the intersection of a second solution curve at (x_0, λ_0), a situation that is rightly called bifurcation. A necessary condition for this is again (I.1.6), which excludes the application of the Implicit Function Theorem near (x_0, λ_0).

As in Section I.4, we assume again that the parameter space Y is one-dimensional, i.e., $Y = \mathbb{R}$, and we normalize the first curve of solutions to the so-called trivial solution line $\{(0, \lambda) | \lambda \in \mathbb{R}\}$. This is done as follows: If $F(x(s), \lambda(s)) = 0$, then we set $\hat{F}(x, s) = F(x(s) + x, \lambda(s))$, and obviously, $\hat{F}(0, s) = 0$ for all parameters s. Returning to our original notation, this leads to the following assumptions:

(I.5.1)
$$\begin{aligned}
&F(0, \lambda) = 0 \text{ for all } \lambda \in \mathbb{R}, \\
&\dim N(D_x F(0, \lambda_0)) = \operatorname{codim} R(D_x F(0, \lambda_0)) = 1, \\
&\text{i.e., } F(\cdot, \lambda_0) \text{ is a Fredholm operator of index zero} \\
&(\text{cf. Definition I.2.1}).
\end{aligned}$$

The assumed regularity of F is as follows:

(I.5.2)
$$\begin{aligned}
&F \in C^2(U \times V, Z), \\
&\text{where } 0 \in U \subset X, \quad \lambda_0 \in V \subset \mathbb{R}, \\
&\text{are open neighborhoods,}
\end{aligned}$$

where we identify again the derivative $D^2_{x\lambda} F(x, \lambda)$ with an element in $L(X, Z)$; cf.(I.4.3). By assumption (I.5.2) we have $D^2_{x\lambda} = D^2_{\lambda x}$ (see [38], [10]).

The **Crandall–Rabinowitz Theorem** then reads as follows:

Theorem I.5.1 *Assume (I.5.1), (I.5.2), and that*

(I.5.3)
$$N(D_x F(0, \lambda_0)) = \operatorname{span}[\hat{v}_0], \quad \hat{v}_0 \in X, \quad \|\hat{v}_0\| = 1,$$
$$D^2_{x\lambda} F(0, \lambda_0) \hat{v}_0 \notin R(D_x F(0, \lambda_0)).$$

Then there is a nontrivial continuously differentiable curve through $(0, \lambda_0)$,

(I.5.4) $$\{(x(s), \lambda(s)) \big| s \in (-\delta, \delta), (x(0), \lambda(0)) = (0, \lambda_0)\},$$

such that

(I.5.5) $$F(x(s), \lambda(s)) = 0 \quad \text{for} \quad s \in (-\delta, \delta),$$

and all solutions of $F(x, \lambda)$ in a neighborhood of $(0, \lambda_0)$ are on the trivial solution line or on the nontrivial curve (I.5.4). The intersection $(0, \lambda_0)$ is called a bifurcation point.

Proof. The Lyapunov–Schmidt reduction (Theorem I.2.3) reduces $F(x, \lambda) = 0$ near $(0, \lambda_0)$ equivalently to a one-dimensional problem, the so-called **Bifurcation Equation**; that is,

(I.5.6)
$$\begin{aligned}
&\Phi(v, \lambda) = 0 \text{ near } (0, \lambda_0) \in \tilde{U}_2 \times V_2 \subset N \times \mathbb{R}, \text{ where} \\
&\Phi : \tilde{U}_2 \times V_2 \to Z_0 \text{ with } \dim Z_0 = 1,
\end{aligned}$$

and $\Phi \in C^2(\tilde{U}_2 \times V_2, Z_0)$, by assumption (I.5.2) and (I.1.7). By $F(0, \lambda) = 0$ for all $\lambda \in \mathbb{R}$ (cf. (I.5.1)$_1$) we get, when using the notation of Theorem I.2.3

and Corollary I.2.4,

(I.5.7)
$$\psi(0, \lambda) = 0 \quad \text{for all } \lambda \in V_2, \text{ whence}$$
$$D_\lambda \psi(0, \lambda) = 0 \quad \text{for all } \quad \lambda \in V_2.$$

Inserting $(0, \lambda)$ into the definition (I.2.9) of Φ yields

(I.5.8)
$$\Phi(0, \lambda) = 0 \text{ for all } \lambda \in V_2,$$

which gives the trivial solution line. By (I.5.8), $\Phi(v, \lambda) = \int_0^1 \frac{d}{dt} \Phi(tv, \lambda) dt$, or

(I.5.9)
$$\Phi(v, \lambda) = \int_0^1 D_v \Phi(tv, \lambda) v \, dt \quad \text{for} \quad (v, \lambda) \in \tilde{U}_2 \times V_2.$$

Setting $v = s\hat{v}_0$, $s \in (-\delta, \delta)$, for $v \in \tilde{U}_2 \subset N$, we get nontrivial solutions $(s \neq 0)$ of (I.5.6) by solving

(I.5.10) $\quad \tilde{\Phi}(s, \lambda) \equiv \int_0^1 D_v \Phi(st\hat{v}_0, \lambda) \hat{v}_0 dt = 0 \quad$ for nontrivial $s \in (-\delta, \delta)$.

By assumption (I.5.2), $\tilde{\Phi} \in C^1((-\delta, \delta) \times V_2, Z_0)$, and by Corollary I.2.4 (see $(I.2.10)_2$),

(I.5.11)
$$\tilde{\Phi}(0, \lambda_0) = 0.$$

The following computation leads to $D_\lambda \tilde{\Phi}(0, \lambda_0)$:

(I.5.12)
$$D_\lambda(D_v \Phi(v, \lambda) \hat{v}_0)$$
$$= D_\lambda(Q D_x F(v + \psi(v, \lambda), \lambda)(\hat{v}_0 + D_v \psi(v, \lambda) \hat{v}_0)$$
$$= Q D_{xx}^2 F(v + \psi(v, \lambda), \lambda)[\hat{v}_0 + D_v \psi(v, \lambda) \hat{v}_0, D_\lambda \psi(v, \lambda)]$$
$$+ Q D_x F(v + \psi(v, \lambda), \lambda) D_{\lambda v}^2 \psi(v, \lambda) \hat{v}_0$$
$$+ Q D_{x\lambda}^2 F(v + \psi(v, \lambda), \lambda)(\hat{v}_0 + D_v \psi(v, \lambda) \hat{v}_0).$$

Inserting $(v, \lambda) = (0, \lambda_0)$ into (I.5.12), we find that the first term vanishes in view of (I.5.7), the second term vanishes by the definition (I.2.3) of the projection Q, and Corollary I.2.4 together with assumption $(I.5.3)_2$ finally yields

(I.5.13)
$$D_\lambda \tilde{\Phi}(0, \lambda_0) = Q D_{x\lambda}^2 F(0, \lambda_0) \hat{v}_0 \neq 0 \in Z_0.$$

The Implicit Function Theorem for (I.5.10) gives a continuously differentiable function

(I.5.14)
$$\varphi : (-\delta, \delta) \to V_2 \text{ such that } \varphi(0) = \lambda_0,$$
$$\tilde{\Phi}(s, \varphi(s)) = 0 \text{ for all } s \in (-\delta, \delta).$$

(Again, the interval $(-\delta, \delta)$ is shrunk if necessary.)

Then

(I.5.15) $\Phi(s\hat{v}_0, \varphi(s)) = s\tilde{\Phi}(s, \varphi(s)) = 0$ for $s \in (-\delta, \delta)$,

and

(I.5.16) $\begin{aligned} x(s) &= s\hat{v}_0 + \psi(s\hat{v}_0, \varphi(s)), \\ \lambda(s) &= \varphi(s), \end{aligned}$

is the curve (I.5.4) having all desired properties.   □

Corollary I.5.2 *The tangent vector of the nontrivial solution curve (I.5.4) at the bifurcation point $(0, \lambda_0)$ is given by*

(I.5.17) $(\hat{v}_0, \dot{\lambda}(0)) \in X \times \mathbb{R}.$

Proof. By (I.5.16),

(I.5.18) $\begin{aligned} \frac{d}{ds}(x(s))\Big|_{s=0} &= \hat{v}_0 + D_v\psi(0, \lambda_0)\hat{v}_0 + D_\lambda\psi(0, \lambda_0)\dot{\lambda}(0) \\ &= \hat{v}_0 \quad \text{by Corollary I.2.4 and (I.5.7).} \end{aligned}$

Figure I.5.1 depicts the schematic bifurcation diagram.   □

Under the general assumptions of this section, it is not clear whether the component $\dot{\lambda}(0)$ of the tangent vector (I.5.17) vanishes. Therefore, for now, we cannot decide on sub-, super-, or transcritical bifurcation. These notions will be made precise in the next section.

Remark I.5.3 *The generalization of Theorem I.5.1 to higher-dimensional kernels is given by Theorem I.19.2, provided that the parameter space is higher-dimensional, too. To be more precise, we need as many parameters as the codimension of the range amounts to.*

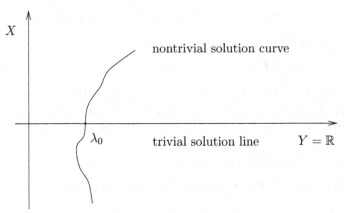

Figure I.5.1

I.6 Bifurcation Formulas (Stationary Case)

In this section we give formulas to compute $\dot{\lambda}(0) = \dot{\varphi}(0)$ in the tangent (I.5.17) or $\ddot{\lambda}(0)$ if $\dot{\lambda}(0) = 0$. For this purpose we assume that the mapping F is in $C^3(U \times V, Z)$. Using $\tilde{\Phi}(s, \lambda(s)) = 0$ for all $s \in (-\delta, \delta)$ (recall $\lambda(s) = \varphi(s)$, by (I.5.16)), we obtain

(I.6.1) $\qquad \dfrac{d}{ds}\tilde{\Phi}(s, \lambda(s))\big|_{s=0} = D_s\tilde{\Phi}(0, \lambda_0) + D_\lambda\tilde{\Phi}(0, \lambda_0)\dot{\lambda}(0) = 0.$

By (I.5.13), $D_\lambda\tilde{\Phi}(0, \lambda_0) \neq 0$, and thus $\dot{\lambda}(0)$ is determined by $D_s\tilde{\Phi}(0, \lambda_0)$. By definition (I.5.10) and (I.2.13),

(I.6.2)
$$D_s\tilde{\Phi}(0, \lambda_0) = \int_0^1 D_{vv}^2\Phi(0, \lambda_0)[\hat{v}_0, t\hat{v}_0]dt$$
$$= \frac{1}{2}QD_{xx}^2F(0, \lambda_0)[\hat{v}_0, \hat{v}_0],$$

where again we have used the definition (I.2.3) of the projection Q, yielding $QD_xF(0, \lambda_0)x = 0$ for all $x \in X$. If (I.6.2) is nonzero, we can easily derive our first formula. Using the representation (I.4.22) of the projection Q, (I.6.1) yields

(I.6.3) $\qquad \dot{\lambda}(0) = -\dfrac{1}{2}\dfrac{\langle D_{xx}^2F(0, \lambda_0)[\hat{v}_0, \hat{v}_0], \hat{v}_0'\rangle}{\langle D_{x\lambda}^2F(0, \lambda_0)\hat{v}_0, \hat{v}_0'\rangle}.$

If $D_{xx}^2F(0, \lambda_0)[\hat{v}_0, \hat{v}_0] \notin R(D_xF(0, \lambda_0))$, the number $\dot{\lambda}(0)$ is nonzero. Since this represents the component in $\mathbb{R}$ of the tangent vector of the curve (I.5.4), the bifurcation is called **transcritical** in this case (see Figure I.6.1).

However, if $D_{xx}^2F(0, \lambda_0)[\hat{v}_0, \hat{v}_0] \in R(D_xF(0, \lambda_0))$, then $\dot{\lambda}(0) = 0$, and the local shape of the curve (I.5.4) is determined by $\ddot{\lambda}(0)$. Differentiating $\tilde{\Phi}(s, \lambda(s)) = 0$ twice with respect to s gives

(I.6.4)
$$\dfrac{d^2}{ds^2}\tilde{\Phi}(s, \lambda(s))\big|_{s=0} = D_{ss}^2\tilde{\Phi}(0, \lambda_0) + D_\lambda\tilde{\Phi}(0, \lambda_0)\ddot{\lambda}(0) = 0$$
when $\dot{\lambda}(0) = 0$.

We now compute $D_{ss}^2\tilde{\Phi}(0, \lambda_0)$. By definition (I.5.10), this amounts to computing $D_{vvv}^3\Phi(0, \lambda_0)[\hat{v}_0, \hat{v}_0, \hat{v}_0]$. Using (I.2.13) we get

$$D^2_{vv}\Phi(v,\lambda_0)[\hat{v}_0,\hat{v}_0]$$

$$= QD^2_{xx}F(v+\psi(v,\lambda),\lambda)[\hat{v}_0+D_v\psi(v,\lambda)\hat{v}_0,\hat{v}_0+D_v\psi(v,\lambda)\hat{v}_0]$$

$$+ QD_xF(v+\psi(v,\lambda),\lambda)D^2_{vv}\psi(v,\lambda)[\hat{v}_0,\hat{v}_0],$$

(I.6.5)　　$$D^3_{vvv}\Phi(0,\lambda_0)[\hat{v}_0,\hat{v}_0,\hat{v}_0]$$

$$= QD^3_{xxx}F(0,\lambda_0)[\hat{v}_0,\hat{v}_0,\hat{v}_0]$$

$$+ 2QD^2_{xx}F(0,\lambda_0)[\hat{v}_0,D^2_{vv}\psi(0,\lambda_0)[\hat{v}_0,\hat{v}_0]]$$

$$+ QD^2_{xx}F(0,\lambda_0)[\hat{v}_0,D^2_{vv}\psi(0,\lambda_0)[\hat{v}_0,\hat{v}_0]],$$

where we have used $QD_xF(0,\lambda_0)x = 0$ for all $x \in X$, and also $(I.2.10)_1$. To compute $D^2_{vv}\psi(0,\lambda_0)[\hat{v}_0,\hat{v}_0]$ we use

(I.6.6)　　$$(I-Q)F(v+\psi(v,\lambda),\lambda) = 0 \quad \text{for all} \quad (v,\lambda) \in \tilde{U}_2 \times V_2$$

(cf. (I.2.7), (I.2.8)). This gives by differentiation

$$(I-Q)D_xF(v+\psi(v,\lambda),\lambda)(\hat{v}_0+D_v\psi(v,\lambda)\hat{v}_0) = 0,$$

(I.6.7)　　$$(I-Q)D^2_{xx}F(0,\lambda_0)[\hat{v}_0,\hat{v}_0]$$

$$+(I-Q)D_xF(0,\lambda_0)D^2_{vv}\psi(0,\lambda_0)[\hat{v}_0,\hat{v}_0] = 0.$$

Taking into account that $D_xF(0,\lambda_0) : X_0 \to R = (I-Q)Z$ is an isomorphism (see (I.2.3)), we get

(I.6.8)
$$D^2_{vv}\psi(0,\lambda_0)[\hat{v}_0,\hat{v}_0]$$
$$= -(D_xF(0,\lambda_0))^{-1}(I-Q)D^2_{xx}F(0,\lambda_0)[\hat{v}_0,\hat{v}_0]$$
$$\in X_0 \text{ for } \hat{v}_0 \in N.$$

In order to emphasize that the preimage (I.6.8) is in $X_0 = (I-P)X$, we insert the projection $(I-P)$, and combining (I.6.5) with (I.6.8) gives

(I.6.9)
$$D^3_{vvv}\Phi(0,\lambda_0)[\hat{v}_0,\hat{v}_0,\hat{v}_0]$$
$$= QD^3_{xxx}F(0,\lambda_0)[\hat{v}_0,\hat{v}_0,\hat{v}_0]$$
$$-3QD^2_{xx}F(0,\lambda_0)[\hat{v}_0,(I-P)(D_xF(0,\lambda_0))^{-1}(I-Q)D^2_{xx}F(0,\lambda_0)[\hat{v}_0,\hat{v}_0]].$$

Definition (I.5.10) of $\tilde{\Phi}$ implies

(I.6.10)　　　　$$D^2_{ss}\tilde{\Phi}(0,\lambda_0) = \frac{1}{3}D^3_{vvv}\Phi(0,\lambda_0)[\hat{v}_0,\hat{v}_0,\hat{v}_0].$$

Relation (I.6.4), the representation (I.4.22) of the projection Q, and (I.6.10) give our second bifurcation formula for the case $\dot{\lambda}(0) = 0$:

(I.6.11) $\ddot{\lambda}(0) = -\dfrac{1}{3} \dfrac{\langle D^3_{vvv}\Phi(0,\lambda_0)[\hat{v}_0, \hat{v}_0, \hat{v}_0], \hat{v}'_0\rangle}{\langle D^2_{x\lambda}F(0,\lambda_0)\hat{v}_0, \hat{v}'_0\rangle}$ (cf. (I.6.9)).

If $\ddot{\lambda}(0) < 0$, the bifurcation is **subcritical**, and if $\ddot{\lambda}(0) > 0$, it is **supercritical**. In both cases the diagram is referred to as a **pitchfork bifurcation** (see Figure I.6.1).

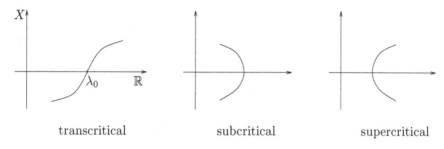

transcritical   subcritical   supercritical

Figure I.6.1

I.7 The Principle of Exchange of Stability (Stationary Case)

Stability is a property of solutions of evolution equations, in particular of equilibria or stationary solutions. Here we consider formally

(I.7.1) $$\frac{dx}{dt} = F(x, \lambda),$$

where F is a mapping as considered in Sections I.4–I.6. Such an evolution equation, however, makes sense only if $X \subset Z$, and as in Section I.3 we assume that the Banach space X is continuously embedded in the Banach space Z.

Let $F(x_0, \lambda_0) = 0$; i.e., $x_0 \in X$ is an equilibrium of (I.7.1) for the parameter $\lambda_0 \in \mathbb{R}$. According to the **Principle of Linearized Stability** we call

(I.7.2)   the equilibrium x_0 stable (linearly stable)
if the spectrum of $D_x F(x_0, \lambda_0)$ is in the left complex half-plane.

Of course, (I.7.2) implies true nonlinear stability of x_0 if one has rigorous dynamics for (I.7.1), e.g., when (I.7.1) represents a system of ordinary differential equations ($X = Z = \mathbb{R}^n$) or a parabolic partial differential equation;

i.e., $F(x, \lambda)$ is a semilinear elliptic partial differential operator over a bounded domain; cf. Section III.4.

We cannot go into a detailed discussion about the general validity of this principle, but we refer to [89] and [91].

The stability criterion (I.7.2) is also viable for (Lagrangian) evolution equations of the form

$$(\text{I.7.3}) \qquad\qquad \frac{d^2x}{dt^2} = F(x, \lambda),$$

where $F(x, \lambda)$ is a potential operator with respect to x with potential $f(x, \lambda)$; cf. Definition I.3.1. Proposition I.3.2 then gives that the linear map $D_x F(x_0, \lambda_0) = D_x \nabla_x f(x_0, \lambda_0)$ is formally self-adjoint, in particular, its spectrum is real. Accordingly, (I.7.2) ensures that $D_x \nabla_x f(x_0, \lambda_0)$ is negative definite. Now it is easy to show that the *total energy*

$$E = \frac{1}{2} \left(\frac{dx}{dt}, \frac{dx}{dt} \right) - f(x, \lambda_0)$$

is constant along all classical solutions of (I.7.3) at $\lambda = \lambda_0$. Thus, E defines a Lyapunov function, and again (I.7.2) implies nonlinear stability if one has rigorous dynamics for (I.7.3).

Remark I.7.1 *In view of various approaches to bifurcation theory via a "Center Manifold Reduction," we give the following warning: Do not mix the problem of existence of equilibria with the problem of their stability.*

Solutions of $F(x, \lambda) = 0$ are equilibria of (I.7.1) and of (I.7.3), for example, but their dynamics are obviously different. The perturbation of an equilibrium $F(x_0, \lambda_0) = 0$ depends only on the spectral properties of the number zero for the linear operator $D_x F(x_0, \lambda_0)$. The stability properties of perturbed equilibria, however, depend on the entire spectrum of $D_x F(x_0, \lambda_0)$.

In this section we study the perturbation of the critical eigenvalue zero along the perturbed equilibria, and this eigenvalue perturbation determines the stability of the perturbed equilibria if the rest of the spectrum is in the left complex half-plane. This condition on the rest of the spectrum, however, is required neither for the existence of perturbed equilibria nor for their critical eigenvalue perturbation.

A center manifold for (I.7.1), provided that it exists, depends on the spectrum of $D_x F(x_0, \lambda_0)$ on the imaginary axis. One finds bifurcation theorems for hyperbolic equilibria of (I.7.1) on perturbed one-dimensional center manifolds, where accordingly nonzero eigenvalues on the imaginary axis are excluded. If the existence is not separated from the stability analysis, one might get the wrong impression that all purely imaginary eigenvalues have an influence on the bifurcation of equilibria.

Under the assumptions of the previous sections, the Principle of Linearized Stability does not apply to the equilibrium x_0 when $D_x F(x_0, \lambda_0)$ has a one-dimensional kernel, i.e., if zero is an eigenvalue of $D_x F(x_0, \lambda_0)$. But we can apply the principle to solution curves through $(x_0, \lambda_0) \in X \times \mathbb{R}$ (apart from (x_0, λ_0)) under certain nondegeneracy conditions. In fact, the same calculations from Section I.6 leading to the shape of the solution curves help us to study the perturbation of the so-called critical zero eigenvalue of $D_x F(x_0, \lambda_0)$ to an eigenvalue of $D_x F(x(s), \lambda(s))$, where $\{(x(s), \lambda(s)) | s \in (-\delta, \delta)\}$ is a curve through (x_0, λ_0) established in Sections I.4–I.6 .

First we need to be sure that such a perturbation of the zero eigenvalue exists in a suitable way. Accordingly, we assume that

(I.7.4)   0 is a simple eigenvalue of $D_x F(x_0, \lambda_0)$; i.e.,
            if $N(D_x F(x_0, \lambda_0)) = \text{span}[\hat{v}_0]$, then $\hat{v}_0 \notin R(D_x F(x_0, \lambda_0))$.

Recall that $X \subset Z$. This definition is the generalization of the algebraic simplicity of an eigenvalue of a matrix. *Note that a simple eigenvalue means throughout an algebraically simple eigenvalue in the sense of (I.7.4).* In particular, (I.7.4) implies that we have a decomposition

(I.7.5) $$Z = R(D_x F(x_0, \lambda_0)) \oplus N(D_x F(x_0, \lambda_0)),$$

which induces a decomposition

(I.7.6) $$X = N \oplus (R \cap X)$$

for the continuously embedded space $X \subset Z$. Thus the projections are

(I.7.7) $\begin{aligned} Q : Z \to N \quad &\text{along } R, \text{ and} \\ P = Q|_X : X \to N \quad &\text{along} \quad R \cap X. \end{aligned}$

We shall use these projections for the Lyapunov–Schmidt reduction as well as the representation (I.4.22) for Q:

(I.7.8) $\begin{aligned} &Qz = \langle z, \hat{v}_0' \rangle \hat{v}_0 \text{ for all } z \in Z, \text{ where} \\ &\langle \hat{v}_0, \hat{v}_0' \rangle = 1, \ \langle z, \hat{v}_0' \rangle = 0 \text{ for all } z \in R, \\ &\text{and } \hat{v}_0' \in Z', \text{ the dual space.} \end{aligned}$

We now consider a continuously differentiable curve of solutions through (x_0, λ_0):

(I.7.9) $\begin{aligned} &\{(x(s), \lambda(s)) | s \in (-\delta, \delta), (x(0), \lambda(0)) = (x_0, \lambda_0)\} \subset X \times \mathbb{R} \\ &\text{such that } F(x(s), \lambda(s)) = 0 \text{ for } s \in (-\delta, \delta). \end{aligned}$

The assumed regularity of F is that $F \in C^2(U \times V, Z)$, where $(x_0, \lambda_0) \in U \times V \subset X \times \mathbb{R}$.

Proposition I.7.2 *There is a continuously differentiable curve of perturbed eigenvalues $\{\mu(s) | s \in (-\delta, \delta), \mu(0) = 0\}$ in $\mathbb{R}$ such that*

(I.7.10) $$D_x F(x(s), \lambda(s))(\hat{v}_0 + w(s)) = \mu(s)(\hat{v}_0 + w(s)),$$

where $\{w(s)|s \in (-\delta, \delta), w(0) = 0\} \subset R \cap X$ is continuously differentiable. (The interval $(-\delta, \delta)$ is not necessarily the same as in (I.7.9) but possibly shrunk.) In this sense, $\mu(s)$ is the perturbation of the critical zero eigenvalue of $D_x F(x_0, \lambda_0)$.

Proof. Define a mapping

(I.7.11) $\quad \begin{aligned} &G : U \times V \times (R \cap X) \times \mathbb{R} \to Z, \quad x_0 \in U \subset X, \quad \lambda_0 \in V \subset \mathbb{R} \text{ by} \\ &G(x, \lambda, w, \mu) = D_x F(x, \lambda)(\hat{v}_0 + w) - \mu(\hat{v}_0 + w). \end{aligned}$

Then $G(x_0, \lambda_0, 0, 0) = 0$ and

(I.7.12) $\qquad \begin{aligned} D_w G(x_0, \lambda_0, 0, 0) &= D_x F(x_0, \lambda_0), \\ D_\mu G(x_0, \lambda_0, 0, 0) &= -\hat{v}_0, \end{aligned}$

so that assumption (I.7.4) implies that

(I.7.13) $\qquad D_{(w,\mu)} G(x_0, \lambda_0, 0, 0) : (R \cap X) \times \mathbb{R} \to Z$

is an isomorphism. The Implicit Function Theorem then gives continuously differentiable functions $w : U_1 \times V_1 \to R \cap X$, $\mu : U_1 \times V_1 \to \mathbb{R}$ such that $x_0 \in U_1 \subset U_2 \subset X$, $\lambda_0 \in V_1 \subset V_2 \subset \mathbb{R}$, $w(x_0, \lambda_0) = 0$, $\mu(x_0, \lambda_0) = 0$, and $G(x, \lambda, w(x, \lambda), \mu(x, \lambda)) = 0$ for all $(x, \lambda) \in U_1 \times V_1$. Inserting the curve (I.7.9) into w and μ, we obtain

(I.7.14) $\quad \mu(s) = \mu(x(s), \lambda(s)), \; w(s) = w(x(s), \lambda(s)), \quad s \in (-\delta, \delta),$

having all required properties. $\qquad \qquad \qquad \qquad \qquad \qquad \qquad \qquad \square$

Assuming that the spectrum of $D_x F(x_0, \lambda_0)$ is in the left complex half-plane apart from the simple eigenvalue zero, the linearized stability of the curve is then determined by the sign of the perturbed eigenvalue $\mu(s)$, at least for small values of $s \in (-\delta, \delta)$.

Recall that the solution curve (I.7.9) is found by the method of Lyapunov–Schmidt:

(I.7.15) $\quad \begin{aligned} &\{(v(s), \lambda(s))|s \in (-\delta, \delta), (v(0), \lambda(0)) = (v_0, \lambda_0)\} \text{ satisfies} \\ &(\text{see } (I.2.9), y = \lambda) \\ &\Phi(v(s), \lambda(s)) = 0, \text{ where } v(s) = Px(s) \in N. \end{aligned}$

In order to transform this reduced problem into a problem in $\mathbb{R}^2$ we set

(I.7.16) $\quad \begin{aligned} &v = v_0 + y\hat{v}_0 \in N, \quad y \in \mathbb{R}, \\ &\Psi(y, \lambda) \equiv \langle \Phi(v_0 + y\hat{v}_0, \lambda), \hat{v}_0' \rangle, \\ &\Psi : \tilde{U}_2 \times V_2 \to \mathbb{R}, \quad (0, \lambda_0) \in \tilde{U}_2 \times V_2 \subset \mathbb{R}^2. \end{aligned}$

Setting $\langle v(s) - v_0, \hat{v}_0' \rangle = y(s)$, we have a local solution curve of Ψ through $(0, \lambda_0)$:

(I.7.17) $\Psi(y(s), \lambda(s)) = 0$, $(y(0), \lambda(0)) = (0, \lambda_0)$.

For the subsequent analysis we require more differentiability of the solution curve $\{(y(s), \lambda(s))\}$: We assume that F is in $C^3(U \times V, Z)$ (cf. Section I.6).

Proposition I.7.3 *Under all assumptions of this section,*

(I.7.18) $\dfrac{d}{ds} D_y\Psi(y(s), \lambda(s))\big|_{s=0} = \dfrac{d}{ds}\mu(s)\big|_{s=0}$,

and if $\frac{d}{ds}\mu(s)\big|_{s=0} = 0$, then

(I.7.19) $\dfrac{d^2}{ds^2} D_y\Psi(y(s), \lambda(s))\big|_{s=0} = \dfrac{d^2}{ds^2}\mu(s)\big|_{s=0}$.

Proof. By definition (I.7.16),

(I.7.20) $D_y\Psi(y, \lambda) = \langle D_v\Phi(v_0 + y\hat{v}_0, \lambda)\hat{v}_0, \hat{v}_0' \rangle$.

Then $(\dot{} = \frac{d}{ds})$

(I.7.21) $\begin{aligned}\dfrac{d}{ds} &\langle D_v\Phi(v_0 + y(s)\hat{v}_0, \lambda(s))\hat{v}_0, \hat{v}_0'\rangle\big|_{s=0} \\ &= \langle D_{vv}^2\Phi(v_0, \lambda_0)[\dot{y}(0)\hat{v}_0, \hat{v}_0], \hat{v}_0'\rangle + \langle D_{v\lambda}^2\Phi(v_0, \lambda_0)\hat{v}_0, \hat{v}_0'\rangle\dot{\lambda}(0).\end{aligned}$

Differentiating equation (I.7.10) with respect to s at $s = 0$ yields

(I.7.22)
$$D_{xx}^2 F(x_0, \lambda_0)[\dot{x}(0), \hat{v}_0] + D_{x\lambda}^2 F(x_0, \lambda_0)\hat{v}_0\dot{\lambda}(0)$$
$$+ D_x F(x_0, \lambda_0)\dot{w}(0) = \dot{\mu}(0)\hat{v}_0,   \text{and by (I.7.8)},$$
$$\dot{\mu}(0) = \langle D_{xx}^2 F(x_0, \lambda_0)[\dot{x}(0), \hat{v}_0], \hat{v}_0'\rangle + \langle D_{x\lambda}^2 F(x_0, \lambda_0)\hat{v}_0, \hat{v}_0'\rangle\dot{\lambda}(0).$$

Using $x(s) = Px(s) + \psi(v(s), \lambda(s)) = v(s) + \psi(v(s), \lambda(s))$, we get by Corollary I.2.4,

(I.7.23) $\dot{x}(0) = \dot{y}(0)\hat{v}_0 + D_\lambda\psi(x_0, \lambda_0)\dot{\lambda}(0)$.

On the other hand (see (I.6.5) and again (I.2.10)$_1$),

(I.7.24)
$$\langle D_{vv}^2\Phi(v_0, \lambda_0)[\dot{y}(0)\hat{v}_0, \hat{v}_0], \hat{v}_0'\rangle = \langle D_{xx}^2 F(x_0, \lambda_0)[\dot{y}(0)\hat{v}_0, \hat{v}_0], \hat{v}_0'\rangle,$$
$$\langle D_{v\lambda}^2\Phi(v_0, \lambda_0)\hat{v}_0, \hat{v}_0'\rangle$$
$$= \langle D_{xx}^2 F(v_0, \lambda_0)[\hat{v}_0, D_\lambda\psi(x_0, \lambda_0)], \hat{v}_0'\rangle + \langle D_{x\lambda}^2 F(v_0, \lambda_0)\hat{v}_0, \hat{v}_0'\rangle.$$

Combining (I.7.21), (I.7.23), (I.7.24) with (I.7.22) gives (I.7.18).

We prove (I.7.19) for the special case in which we are mainly interested: We assume $x(s) = s\hat{v}_0 + \psi(s\hat{v}_0, \lambda(s))$, i.e., $x(0) = 0$, $\dot{x}(0) = \hat{v}_0$, $\dot{\lambda}(0) = 0$, and also $F(0, \lambda) = 0$ for all $\lambda \in V \subset \mathbb{R}$. A second differentiation of (I.7.10) with respect to s (see (I.7.22)) gives

$$D_{xxx}^3 F(0, \lambda_0)[\hat{v}_0, \hat{v}_0, \hat{v}_0] + 2D_{xx}^2 F(0, \lambda_0)[\hat{v}_0, \dot{w}(0)]$$

(I.7.25)
$$+ D_{xx}^2 F(0, \lambda_0)[\ddot{x}(0), \hat{v}_0] + D_{x\lambda}^2 F(0, \lambda_0)\hat{v}_0 \ddot{\lambda}(0)$$

$$+ D_x F(0, \lambda_0)\ddot{w}(0) = \ddot{\mu}(0)\hat{v}_0,$$

where we also assumed that $\dot{\mu}(0) = 0$.

Next we compute $\ddot{x}(0)$ and $\dot{w}(0)$. By our assumptions on $x(s)$ and by $\dot{\lambda}(0) = 0$ and from $(I.5.7)_2$, we get

(I.7.26)
$$\ddot{x}(0) = D_{vv}^2 \psi(0, \lambda_0)[\hat{v}_0, \hat{v}_0],$$

which is given by (I.6.8). Equation $(I.7.22)_1$ for $\dot{\mu}(0) = 0$ can be solved for $\dot{w}(0) \in R \cap X = (I - P)X$; that is,

(I.7.27)
$$\dot{w}(0) = -(I - P)(D_x F(0, \lambda_0))^{-1}(I - Q)D_{xx}^2 F(0, \lambda_0)[\hat{v}_0, \hat{v}_0]$$

$$= \ddot{x}(0), \text{ by (I.7.26) and (I.6.8)}.$$

Returning now to (I.7.25), observe that $D_x F(0, \lambda_0)\ddot{w}(0) \in R = (I - Q)Z$.

Applying the functional $\hat{v}_0' \in Z'$ after inserting (I.7.26), (I.7.27) into (I.7.25) gives us (see (I.7.8))

(I.7.28)
$$\ddot{\mu}(0) = \langle D_{xxx}^3 F^0[\hat{v}_0, \hat{v}_0, \hat{v}_0], \hat{v}_0' \rangle$$

$$-3\langle D_{xx}^2 F^0[\hat{v}_0, (I - P)(D_x F^0)^{-1}(I - Q)D_{xx}^2 F^0[\hat{v}_0, \hat{v}_0]], \hat{v}_0' \rangle$$

$$+\langle D_{x\lambda}^2 F^0 \hat{v}_0, \hat{v}_0' \rangle \ddot{\lambda}(0),$$

where "0" denotes evaluation at $(0, \lambda_0)$. On the other hand, one more differentiation of (I.7.21) with respect to s yields (using $\dot{\lambda}(0) = 0, y(s) = s$)

(I.7.29)
$$\frac{d^2}{ds^2} D_y \Psi(s, \lambda(s))\big|_{s=0}$$

$$= \langle D_{vvv}^3 \Phi(0, \lambda_0)[\hat{v}_0, \hat{v}_0, \hat{v}_0], \hat{v}_0' \rangle + \langle D_{v\lambda}^2 \Phi(0, \lambda_0)\hat{v}_0, \hat{v}_0' \rangle \ddot{\lambda}(0).$$

Formulas (I.6.9) (I.5.12), and (I.5.13) (replace v by $v_0 = 0$) together with $(I.5.7)_2$ prove the equality of (I.7.28) and (I.7.29).

The general case is reduced to a special case as follows: Define $\hat{F}(x, s) \equiv F(x(s) + x, \lambda(s))$ for x in a neighborhood of 0 in X. Then $\hat{F}(0, s) = 0$ for $s \in (-\delta, \delta)$ and $D_x \hat{F}(0, s) = D_x F(x(s), \lambda(s))$, yielding for $s = 0$ the same projections for the method of Lyapunov–Schmidt as before. The function $\hat{\Phi}$ of (I.2.9) to solve $\hat{F}(x, s) = 0$ near $(x, s) = (0, 0)$ is therefore given by $\hat{\Phi}(v, s) = \Phi(v(s) + v, \lambda(s))$, and the application of formula (I.7.19) for the trivial solution line $\{(0, s)|s \in (-\delta, \delta)\}$ of $\hat{F}(x, s) = 0$ proves (I.7.19) for the solution curve $\{(x(s), \lambda(s))|s \in (-\delta, \delta)\}$ of $F(x, \lambda) = 0$. (More details of this argument can be found in the proof of Theorem I.16.6, which generalizes Proposition I.7.3 considerably; see, in particular, (I.16.35)–(I.16.39).) We remark that formula

(I.16.36) for $m = 2$ is valid also under the regularity condition of Proposition I.7.3. We recommend proving (I.7.19) for the trivial solution line directly and comparing it with the coefficients μ_2 and c_{20} given in (I.16.9) and (I.16.23), respectively.   □

We now apply Proposition I.7.3 to determine the linearized stability of the solution curve $\{(x(s), \lambda(s))|s \in (-\delta, \delta)\}\backslash\{(x_0, \lambda_0)\}$; cf.(I.7.9). As stated previously, if we assume that the critical zero eigenvalue of $D_x F(x_0, \lambda_0)$ has the largest real part of all points of the spectrum of $D_x F(x_0, \lambda_0)$, then stability is determined by the sign of the perturbed eigenvalue $\mu(s)$ as given by Proposition I.7.2. We now carry out this program for the cases studied in Sections I.4–I.6.

1. Turning Point or Saddle-Node Bifurcation

This is described in Theorem I.4.1 and Corollary I.4.2: Under assumption (I.4.5) there is a unique curve of solutions through (x_0, λ_0), and its tangent vector at (x_0, λ_0) is $(\dot{x}(0), \dot{\lambda}(0)) = (\hat{v}_0, 0)$. If in addition, (I.4.18) is satisfied, then $\ddot{\lambda}(0) \neq 0$, so that the curve has one of the shapes sketched in Figure I.4.1. Formula (I.7.22) gives

$$(I.7.30) \qquad \dot{\mu}(0) = \langle D^2_{xx} F(x_0, \lambda_0)[\hat{v}_0, \hat{v}_0], \hat{v}'_0 \rangle \neq 0 \quad (cf.(I.4.23)).$$

Together with the bifurcation formula (I.4.23) we obtain

$$(I.7.31) \qquad \dot{\mu}(0) = -\langle D_\lambda F(x_0, \lambda_0), \hat{v}'_0 \rangle \ddot{\lambda}(0),$$

and assumption (I.4.5) is precisely that $\langle D_\lambda F(x_0, \lambda_0), \hat{v}'_0 \rangle \neq 0$. Depending on the signs of $\langle D_\lambda F(x_0, \lambda_0), \hat{v}'_0 \rangle$ and $\langle D^2_{xx} F(x_0, \lambda_0)[\hat{v}_0, \hat{v}_0], \hat{v}'_0 \rangle$, the signs of $\dot{\mu}(0)$ and $\ddot{\lambda}(0)$ are determined. In any case, in view of $\mu(0) = 0$, $\dot{\mu}(0) \neq 0$, the sign of $\mu(s)$ changes at $s = 0$, which implies that the stability of the curve $\{x(s), \lambda(s))\}$ changes at the turning point (x_0, λ_0). The possibilities are sketched in Figure I.7.1.

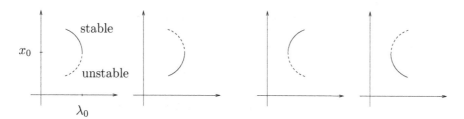

Figure I.7.1

That exchange of stability is also true at degenerate turning points, i.e., where $\dot{\lambda}(0) = \cdots = \lambda^{(k-1)}(0) = 0$ but $\lambda^{(k)}(0) \neq 0$ for some even $k \geq 2$ and where the mapping F is analytic. For details we refer to Section III.7.5, where formula (I.7.31) is generalized to (III.7.129).

2. The Transcritical Bifurcation

Here we have two curves intersecting at $(x_0, \lambda_0) = (0, \lambda_0)$: the trivial solution line $\{(0, \lambda)\}$ and the nontrivial solution curve $\{(x(s), \lambda(s))\}$. Although formula (I.7.22) applies to the trivial solution line as well, we give a new argument for the eigenvalue perturbation (I.7.10), which we parameterize by λ (near λ_0):

$$(I.7.32) \qquad D_x F(0, \lambda)(\hat{v}_0 + w(\lambda)) = \mu(\lambda)(\hat{v}_0 + w(\lambda)).$$

Differentiation of (I.7.32) with respect to λ yields by $\mu(\lambda_0) = 0, w(\lambda_0) = 0$,

$$(I.7.33) \qquad D_{x\lambda}^2 F(0, \lambda_0)\hat{v}_0 + D_x F(0, \lambda_0)w'(\lambda_0) = \mu'(\lambda_0)\hat{v}_0,$$

where $' = \frac{d}{d\lambda}$. By the choice of $\hat{v}_0'$ (cf. (I.7.8) and in particular, $\langle z, \hat{v}_0' \rangle = 0$ for all $z \in R$), (I.7.33) implies

$$(I.7.34) \qquad \mu'(\lambda_0) = \langle D_{x\lambda}^2 F(0, \lambda_0)v_0, \hat{v}_0' \rangle.$$

Here we recognize the nondegeneracy (I.5.3): Given (I.7.8), we now see that the hypothesis

$$(I.7.35) \qquad D_{x\lambda}^2 F(0, \lambda_0)\hat{v}_0 \notin R(D_x F(0, \lambda_0))$$

of the Crandall–Rabinowitz Theorem, Theorem I.5.1, is equivalent to the assumption
$$(I.7.36) \qquad \mu'(\lambda_0) \neq 0,$$

which can be stated as follows: The real eigenvalue $\mu(\lambda)$ of $D_x F(0, \lambda)$ crosses the imaginary axis at $\mu(\lambda_0) = 0$ "with nonvanishing speed." If the spectrum of $D_x F(x_0, \lambda_0)$ is in the left complex half-plane apart from the simple eigenvalue $\mu(\lambda_0) = 0$, then $\mu'(\lambda_0) > 0$ describes a loss of stability of the trivial solution: $(0, \lambda)$ (as a solution of $F(x, \lambda) = 0$) is stable for $\lambda < \lambda_0$ and unstable for $\lambda > \lambda_0$ (locally).

For the bifurcating solution curve $\{(x(s), \lambda(s))\}$ as described by Theorem I.5.1, we assume $\dot{\lambda}(0) \neq 0$, which guarantees a transcritical bifurcation (see Figure I.6.1). By the bifurcation formula (I.6.3), this is equivalent to

$$(I.7.37) \qquad \langle D_{xx}^2 F(x_0, \lambda_0)[\hat{v}_0, \hat{v}_0], \hat{v}_0' \rangle \neq 0,$$

and we rewrite (I.6.3) as

$$(I.7.38) \qquad \langle D_{xx}^2 F(0, \lambda_0)[\hat{v}_0, \hat{v}_0], \hat{v}_0' \rangle + 2\mu'(\lambda_0)\dot{\lambda}(0) = 0,$$

where we have used (I.7.34). On the other hand, (I.7.22) and (I.7.34) give for the eigenvalue perturbation $\hat{\mu}(s)$ along the bifurcating curve (we change the notation in order to distinguish it from $\mu(\lambda)$)

$$(I.7.39) \qquad \langle D_{xx}^2 F(0, \lambda_0)[\hat{v}_0, \hat{v}_0], \hat{v}_0' \rangle + \mu'(\lambda_0)\dot{\lambda}(0) = \dot{\hat{\mu}}(0),$$

where we have used $\dot{x}(0) = \hat{v}_0$; cf.(I.5.18). Combining (I.7.38) with (I.7.39) yields the crucial formula that locks the eigenvalue perturbations to the bifurcation direction, namely,

(I.7.40)
$$\mu'(\lambda_0)\dot{\lambda}(0) = -\dot{\hat{\mu}}(0).$$

Now assume $\mu'(\lambda_0) > 0$, which means a loss of stability of the trivial solution $x = 0$ at $\lambda = \lambda_0$. Then by $\lambda(0) = \lambda_0$, $\hat{\mu}(0) = 0$, $\dot{\lambda}(0) \neq 0$, and $\dot{\hat{\mu}}(0) \neq 0$ (cf.(I.7.37)),

(I.7.41)
$$\text{sign}(\lambda(s) - \lambda_0) = \text{sign}\hat{\mu}(s) \quad \text{for} \quad s \in (-\delta, \delta),$$

which proves the stability of $x(s)$ for $\lambda(s) > \lambda_0$ as well as its instability for $\lambda(s) < \lambda_0$. If $\mu'(\lambda_0) < 0$, the stability properties of all solution curves are reversed, which is sketched in Figure I.7.2.

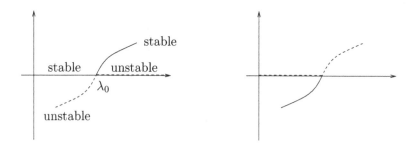

Figure I.7.2

3. The Pitchfork Bifurcation

Again, we have the trivial solution line $\{(0, \lambda)\}$, and bifurcation is caused by (I.7.36), which means a nondegenerate loss or gain of stability of $x = 0$ at $\lambda = \lambda_0$. For the bifurcating solution curve $\{(x(s), \lambda(s))\}$ we have $\dot{\lambda}(0) = 0$, and we assume that $\ddot{\lambda}(0) \neq 0$ (see Figure I.6.1). By the bifurcation formula (I.6.11), this is equivalent to

(I.7.42)
$$\langle D^3_{vvv}\Phi(0, \lambda_0)[\hat{v}_0, \hat{v}_0, \hat{v}_0], \hat{v}_0' \rangle \neq 0,$$

and we rewrite (I.6.11) as

(I.7.43)
$$\langle D^3_{vvv}\Phi(0, \lambda_0)[\hat{v}_0, \hat{v}_0, \hat{v}_0], \hat{v}_0' \rangle + 3\mu'(\lambda_0)\ddot{\lambda}(0) = 0,$$

where we have used (I.7.34). By (I.7.40) we see that assumption $\dot{\lambda}(0) = 0$ is equivalent to $\dot{\hat{\mu}}(0) = 0$, where again $\hat{\mu}(s)$ denotes the eigenvalue perturbation along the bifurcating curve. Formula (I.7.28) is then valid, which is rewritten, using (I.6.9), as

(I.7.44)
$$\langle D^3_{vvv}\Phi(0, \lambda_0)[\hat{v}_0, \hat{v}_0, \hat{v}_0], \hat{v}_0' \rangle + \mu'(\lambda_0)\ddot{\lambda}(0) = \ddot{\hat{\mu}}(0).$$

Together with (I.7.43), this implies the crucial formula

(I.7.45) $$2\mu'(\lambda_0)\ddot{\lambda}(0) = -\dddot{\hat{\mu}}(0).$$

Assume $\mu'(\lambda_0) > 0$. Then by $\lambda(0) = \lambda_0$, $\dot{\lambda}(0) = 0$, $\hat{\mu}(0) = 0$, $\dot{\hat{\mu}}(0) = 0$, and $\ddot{\lambda}(0) \neq 0$ (cf. (I.7.42)),

(I.7.46) $$\text{sign}(\lambda(s) - \lambda_0) = -\text{sign}\hat{\mu}(s) \quad \text{for} \quad s \in (-\delta, \delta),$$

which proves an exchange of stability as sketched in Figure I.7.3. In that standard situation, in which the trivial solution $x = 0$ loses stability at $\lambda = \lambda_0$, a supercritical bifurcation is stable, whereas a subcritical bifurcation is unstable. If $\mu'(\lambda_0) < 0$, the stability properties of all solution curves are reversed.

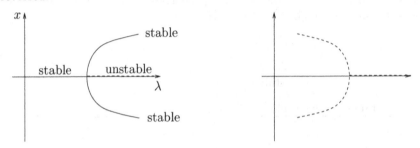

Figure I.7.3

That exchange of stability at transcritical or pitchfork bifurcations holds true also in degenerate cases when $\dot{\lambda}(0) = \cdots = \lambda^{(k-1)}(0) = 0$ but $\lambda^{(k)}(0) \neq 0$ for some $k \geq 2$; cf. (I.16.30) and (I.16.51), generalizing formulas (I.7.40) and (I.7.45).

Each of the cases 1 through 3 above illustrates what is typically referred as the **Principle of Exchange of Stability**. In Figures I.7.1 through I.7.3 we fix a typical value of λ and consider adjacent solution curves. In each case, note that the curves have alternating stability properties.

Before stating the principle as a theorem, we assume, in keeping with cases 1 through 3, that for the eigenvalue perturbation $\mu(s)$ of any solution curve of $F(x, \lambda) = 0$ through (x_0, y_0) in the sense of (I.7.10),

(I.7.47) $$\begin{aligned} &\dot{\mu}(0) \neq 0 \text{ or} \\ &\dot{\mu}(0) = 0 \quad \text{and} \quad \ddot{\mu}(0) \neq 0. \end{aligned}$$

Proposition I.7.3 then yields

(I.7.48) $$\begin{aligned} &\text{sign}\mu(s) = \text{sign}D_y\Psi(y(s), \lambda(s)) \neq 0 \\ &\text{for } s \in (-\delta, \delta)\backslash\{0\}. \end{aligned}$$

A simple observation from one-dimensional calculus gives the following **Principle of Exchange of Stability**.

Theorem I.7.4 *Assume (I.7.47) for the eigenvalue perturbation $\mu(s)$ of all solution curves of $F(x, \lambda) = 0$ through (x_0, λ_0). Assume that there are two solution curves $\{(x_i(s), \lambda_i(s))\}$, $i = 1, 2$, of $F(x, \lambda) = 0$ through (x_0, λ_0) that are adjacent in the following sense: If $Px_i(s) = v_0 + y_i(s)\hat{v}_0$, $i = 1, 2$ (cf. (I.7.16)), then there are parameters s_1 and s_2 such that $y_1(s_1)$ and $y_2(s_2)$ are consecutive zeros of the function $\Psi(\cdot, \lambda)$ at $\lambda = \lambda(s_1) = \lambda(s_2)$ on the y-axis. Then $x_1(s_1)$ and $x_2(s_2)$ have opposite stability properties; i.e., $\mu_1(s_1)\mu_2(s_2) < 0$ for the perturbed eigenvalues $\mu_i(s)$ of $D_xF(x_i(s), \lambda_i(x))$, $i = 1, 2$, near zero.*

Proof. Since a real differentiable function on the real line has derivatives of opposite sign at consecutive zeros with nonzero derivatives, the claim follows from (I.7.48). □

We sketch the situation of Theorem I.7.4 in Figure I.7.4. From (I.7.48) it follows also that the lowest and uppermost curves in the $N \times \mathbb{R}$ plane have the same stability properties on both sides of the bifurcation point (v_0, λ_0), respectively.

Later, in Section I.16, we generalize Theorem I.7.4 to degenerate cases in which (I.7.47) does not hold; cf. Theorem I.16.8.

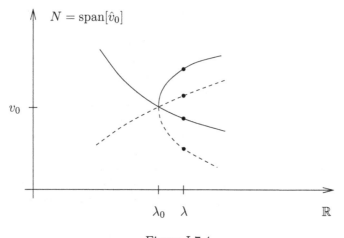

Figure I.7.4

Remark I.7.5 *Note that in all cases discussed in this section, zero is a simple eigenvalue of $D_xF(x_0, \lambda_0)$ in its algebraic sense, whereas the existence results in Sections I.4 and I.5 are proved if $\dim N(D_xF(x_0, \lambda_0)) = 1$. If $X \subset Z$ (which is not necessary in Sections I.4–I.6), this means that zero is a simple eigenvalue only in its geometric sense, and the eigenvalue perturbation of the eigenvalue zero can be complicated if its algebraic multiplicity is larger than one. We discuss this in Sections II.3, II.4. A general condition for bifurcation at $(0, \lambda_0)$ in terms of the eigenvalue perturbation in the spirit of*

(I.7.36) is given in Theorem II.4.4. It is not at all obvious how the nondege-neracy (I.5.3) is related to the eigenvalue perturbation, and we give the result in Case 1 of Theorem II.4.4. That knowledge, however, is not needed in the proof of Theorem I.5.1.

I.8 Hopf Bifurcation

Here as in Section I.5 we assume a trivial solution line $\{(0, \lambda)|\lambda \in \mathbb{R}\} \subset X \times \mathbb{R}$ for the parameter-dependent evolution equation

$$(I.8.1) \qquad\qquad \frac{dx}{dt} = F(x, \lambda);$$

i.e., $F(0, \lambda) = 0$ for all $\lambda \in \mathbb{R}$. Recall that a bifurcation of nontrivial stationary solutions of (I.8.1) (i.e., of $F(x, \lambda) = 0$) can be caused by a loss of stability of the trivial solution at $\lambda = \lambda_0$. To be more precise, that loss of stability is described by a simple real eigenvalue of $D_x F(0, \lambda)$ leaving the "stable" left complex half-plane through 0 at the critical value $\lambda = \lambda_0$ "with nonvanishing speed." This was proven in the previous section, see (I.7.36), and this scenario is resumed in Section I.16, where the loss of stability of the trivial solution is "slow" or "degenerate." (Observe, however, Remark I.7.5.)

In this section we describe the effect of a loss of stability of the trivial solution of (I.8.1) via a pair of complex conjugate eigenvalues of $D_x F(0, \lambda)$ leaving the left complex half-plane through complex conjugate points on the imaginary axis at some critical value $\lambda = \lambda_0$. If 0 is not an eigenvalue of $D_x F(0, \lambda_0)$, then by the Implicit Function Theorem, stationary solutions of (I.8.1) cannot bifurcate from the trivial solution line at $(0, \lambda_0)$. The Hopf Bifurcation Theorem, however, states that (time-) periodic solutions of (I.8.1) bifurcate at $(0, \lambda_0)$. This type of bifurcation is explained and proved in this section, and in Section I.12 we generalize the Principle of Exchange of Stability to this setting.

As before,

$$(I.8.2) \quad \begin{array}{l} F : U \times V \to Z, \quad \text{where} \\ 0 \in U \subset X \quad \text{and } \lambda_0 \in V \subset \mathbb{R} \text{ are open neighborhoods.} \end{array}$$

The function F is sufficiently smooth (see our assumptions (I.8.13) below), and in particular,

$$(I.8.3) \quad \begin{array}{l} F(0, \lambda) = 0 \quad \text{and} \\ D_x F(0, \lambda) \quad \text{exists in } L(X, Z) \text{ for all } \lambda \in V. \end{array}$$

In order to define the evolution equation (I.8.1), we assume for the real Banach spaces X and Z that

(I.8.4) $X \subset Z$ is continuously embedded,

and the derivative of x with respect to t is taken to be an element of Z.

Under assumption (I.8.4), a spectral theory for $D_x F(0, \lambda)$ is possible, and introducing complex eigenvalues of the linear operator $D_x F(0, \lambda)$ requires a natural complexification of the real Banach spaces X and Z: This can be done by a formal sum $X_c = X + iX$ (or by a pair $X \times X$), where we define $(\alpha + i\beta)(x + iy) = \alpha x - \beta y + i(\beta x + \alpha y)$ for every complex number $\alpha + i\beta$. In particular, a real and imaginary part of any vector in X_c is well defined, and a real linear operator A in $L(X, Z)$ is extended in a natural way to a complex linear operator A_c in $L(X_c, Z_c)$. If $\mu \in \mathbb{C}$ is an eigenvalue of A_c with eigenvector φ, then $\overline{\mu} \in \mathbb{C}$ is also an eigenvalue of A_c with eigenvector $\overline{\varphi}$. Here, the bar denotes complex conjugation. In the subsequent analysis we omit, for simplicity, the subscript c, but we keep in mind that our given operators are real and that we are interested in real solutions of (I.8.1).

In this section we assume

(I.8.5)
$i\kappa_0 (\neq 0)$ is a simple eigenvalue of $D_x F(0, \lambda_0)$
with eigenvector $\varphi_0 \notin R(i\kappa_0 I - D_x F(0, \lambda_0))$ (cf.(I.7.4)),
$\pm i\kappa_0 I - D_x F(0, \lambda_0)$ are Fredholm operators
of index zero.

As mentioned before, $-i\kappa_0$ is a simple eigenvalue of $D_x F(0, \lambda_0)$ with eigenvector $\overline{\varphi}_0$.

We can apply Proposition I.7.2 in order to guarantee perturbed eigenvalues $\mu(\lambda)$ of $D_x F(0, \lambda)$:

(I.8.6)
$$D_x F(0, \lambda)\varphi(\lambda) = \mu(\lambda)\varphi(\lambda) \quad \text{such that}$$
$$\mu(\lambda_0) = i\kappa_0, \quad \varphi(\lambda_0) = \varphi_0.$$

(Consider the mapping (I.7.11) with $R = R(i\kappa_0 I - D_x F(0, \lambda_0))$, $\hat{v}_0 = \varphi_0$ at $(0, \lambda_0, 0, i\kappa_0)$.) These eigenvalues $\mu(\lambda)$ are continuously differentiable with respect to λ near λ_0, and following E. Hopf, we assume that

(I.8.7) $\text{Re}\mu'(\lambda_0) \neq 0, \quad \text{where} \quad ' = \dfrac{d}{d\lambda},$

and Re denotes "real part." In this sense the eigenvalue $\mu(\lambda)$ crosses the imaginary axis with "nonvanishing speed," or the exchange of stability of the trivial solution $\{(0, \lambda)\}$ is "nondegenerate." As we will show later, (I.8.7) can be expressed via (I.8.43) below, and it is therefore similar to the nondegeneracy (I.5.3), which is equivalent to (I.7.36) in the case of a simple eigenvalue 0.

Apart from the spectral properties (I.8.5) and (I.8.7), we need more assumptions in order to give the evolution equation (I.8.1) a meaning in the (possibly) infinite-dimensional Banach space Z. The following condition on the linearization serves this purpose:

(I.8.8) $A_0 = D_x F(0, \lambda_0)$ as a mapping in Z, with dense domain of definition $D(A_0) = X$, generates an analytic (holomorphic) semigroup $e^{A_0 t}$, $t \geq 0$, on Z that is compact for $t > 0$.

For a definition of analytic semigroups we refer, for example, to [86], [142], or [170]. The compactness of $e^{A_0 t}$ for $t > 0$ is true if the embedding (I.8.4) is compact. (Assumption (I.8.8) is used only in the proof of Proposition I.8.1 below. The Fredholm property of J_0 is crucial, and if in applications this property can be proved under a weaker assumption, condition (I.8.8) can be weakened accordingly; cf. Remark I.9.2 and Remark III.4.2.)

We look for periodic solutions of (I.8.1) of small amplitude for λ near λ_0 where the period is a priori unknown. A simple but crucial step in proving the Hopf bifurcation theorem is based upon the following observation:

(I.8.9) $x = x(t)$ is a 2π-periodic solution of $\kappa \dfrac{dx}{dt} = F(x, \lambda)$ if and only if $\tilde{x}(t) = x(\kappa t)$ is a $2\pi/\kappa$-periodic solution of (I.8.1).

In other words, we may rescale "time" $= t$ and focus on 2π-periodic solutions of

(I.8.10) $$G(x, \kappa, \lambda) \equiv \kappa \frac{dx}{dt} - F(x, \lambda) = 0.$$

We can then give problem (I.8.10) a functional-analytic setting, in which the method of Lyapunov–Schmidt is applicable. Accordingly, we introduce the following Banach spaces of 2π-periodic Hölder continuous functions having values in X or Z:

$$E \equiv C_{2\pi}^{\alpha}(\mathbb{R}, X) \equiv \left\{ x : \mathbb{R} \to X \,|\, x(t + 2\pi) = x(t),\ t \in \mathbb{R}, \right.$$

$$\left. \|x\|_E = \|x\|_{X,\alpha} \equiv \max_{t \in \mathbb{R}} \|x(t)\|_X + \sup_{s \neq t} \frac{\|x(t) - x(s)\|_X}{|t - s|^\alpha} < \infty \right\},$$

(I.8.11) $W \equiv C_{2\pi}^{\alpha}(\mathbb{R}, Z)$ analogously,

$$Y \equiv C_{2\pi}^{1+\alpha}(\mathbb{R}, Z) \equiv \left\{ x : \mathbb{R} \to Z \,\Big|\, x, \frac{dx}{dt} \text{ (exists) } \in C_{2\pi}^{\alpha}(\mathbb{R}, Z), \right.$$

$$\left. \|x\|_Y = \|x\|_{Z,1+\alpha} \equiv \|x\|_{Z,\alpha} + \left\| \frac{dx}{dt} \right\|_{Z,\alpha} \right\}.$$

The Hölder exponent α is in the interval $(0, 1]$. Obviously, $Y \cap E = C_{2\pi}^{1+\alpha}(\mathbb{R}, Z) \cap C_{2\pi}^{\alpha}(\mathbb{R}, X)$ is a Banach space with norm $\|x\|_{X,\alpha} + \|\frac{dx}{dt}\|_{Z,\alpha}$ (cf.(I.8.4)).

Next we define G, given in (I.8.10), via

$$G : \tilde{U} \times \tilde{V} \to W, \text{ where}$$
(I.8.12) $\quad 0 \in \tilde{U} \subset E \cap Y$ and
$$(\kappa_0, \lambda_0) \in \tilde{V} \subset \mathbb{R}^2, \text{with } \tilde{U} \text{ and } \tilde{V} \text{ open neighborhoods.}$$

We leave the following statements as an exercise:

(I.8.13) $\qquad\qquad \begin{array}{c} G \in C^2(\tilde{U} \times \tilde{V}, W) \text{ if} \\ F \in C^3(U \times V, Z). \end{array}$

(Note that the meaning of x is different in the contexts of $D_x F$ and $D_x G$: x is a vector in X in the first case, while x denotes a function in $Y \cap E = C_{2\pi}^{1+\alpha}(\mathbb{R}, Z) \cap C_{2\pi}^{\alpha}(\mathbb{R}, X)$ in the second case.)

Returning to (I.8.10), we obviously have $G(0, \kappa_0, \lambda_0) = 0$. Observe that $D_x G(0, \kappa_0, \lambda_0) = \kappa_0 \frac{d}{dt} - D_x F(0, \lambda_0)$ and recall that $A_0 = D_x F(0, \lambda_0)$. In order to apply the Lyapunov–Schmidt reduction described in Theorem I.2.3 to (I.8.10), the following proposition is crucial.

Proposition I.8.1 *Assume (I.8.5), (I.8.8), and the following nonresonance condition:*

(I.8.14) $\qquad\qquad \begin{array}{l} \text{for all } n \in \mathbb{Z} \backslash \{1, -1\}, \ in\kappa_0 \text{ is not} \\ \text{an eigenvalue of } A_0 = D_x F(0, \lambda_0). \end{array}$

Then the linear operator

(I.8.15) $$J_0 \equiv \kappa_0 \frac{d}{dt} - A_0 : Y \cap E \to W$$

is continuous and is a Fredholm operator of index zero, with $\dim N(J_0) = 2$.

Proof. It is clear that J_0 is continuous when the intersection $Y \cap E = C_{2\pi}^{1+\alpha}(\mathbb{R}, Z) \cap C_{2\pi}^{\alpha}(\mathbb{R}, X)$ is given the norm $\|x\|_{X,\alpha} + \|\frac{dx}{dt}\|_{Z,\alpha}$. We compute the kernel $N(J_0)$:

$$J_0 x = \kappa_0 \frac{dx}{dt} - A_0 x = 0, \ x(0) = x(2\pi) \Leftrightarrow$$

(I.8.16) $\qquad x(t) = e^{A_0 t/\kappa_0} x(0), \quad (I - e^{A_0 2\pi/\kappa_0}) x(0) = 0 \Leftrightarrow$

$$x(0) \in N(I - e^{A_0 2\pi/\kappa_0}).$$

(By the regularizing property of $e^{A_0 t}$ for $t > 0$, we have $N(I - e^{A_0 2\pi/\kappa_0}) \subset X$.) On the other hand, a Fourier analysis of the real continuous 2π-periodic function $x = x(t)$ yields for any $z' \in Z'$,

$$\langle x(t), z' \rangle = \sum_{n \in \mathbb{Z}} \langle a_n, z' \rangle e^{int}, \quad a_{-n} = \overline{a}_n \in X,$$

$$a_n = \frac{1}{2\pi} \int_0^{2\pi} x(t) e^{-int} dt, \quad x(t) = e^{A_0 t/\kappa_0} x(0),$$

(I.8.17)
$$in a_n = \frac{1}{\kappa_0} A_0 a_n \qquad \text{(integration by parts)},$$

$$a_n = 0 \text{ for } n \in \mathbb{Z} \backslash \{1, -1\} \quad \text{(by (I.8.14))},$$
$$a_1 = c\varphi_0, \quad c \in \mathbb{C} \text{ (by (I.8.5))}, \text{ and}$$
$$x(0) \in \{c\varphi_0 + \overline{c}\,\overline{\varphi}_0 | c \in \mathbb{C}\} = N(I - e^{A_0 2\pi/\kappa_0}),$$

$$x \in \{c\varphi_0 e^{it} + \overline{c}\,\overline{\varphi}_0 e^{-it} | c \in \mathbb{C}\} = N(J_0).$$

(Note that $e^{A_0 t/\kappa_0} \varphi_0 = \varphi_0 e^{it}$, which follows from $A_0 \varphi_0 = i\kappa_0 \varphi_0$.)

By assumption (I.8.8), $A_0 : Z \to Z$ is densely defined, and thus its dual operator $A_0' : Z' \to Z'$ exists. The simplicity of the eigenvalue $i\kappa_0$ (cf.(I.8.5)) implies that the eigenvector φ_0' of A_0' with eigenvalue $i\kappa_0$ can be chosen so that

$$\langle \varphi_0, \varphi_0' \rangle = 1$$

(where $\langle \ , \ \rangle$ denotes the bilinear pairing of Z and Z'),

$$R(i\kappa_0 I - A_0) = \{z \in Z | \langle z, \varphi_0' \rangle = 0\}, \quad \text{(Closed Range Theorem)},$$

(I.8.18)   and the eigenprojection $Q_0 \in L(Z, Z)$ onto
$N(i\kappa_0 I - A_0) \oplus N(-i\kappa_0 I - A_0)$ is given by

$$Q_0 z = \langle z, \varphi_0' \rangle \varphi_0 + \langle z, \overline{\varphi}_0' \rangle \overline{\varphi}_0.$$

(By $(A_0 - i\kappa_0 I)\overline{\varphi}_0 = -2i\kappa_0 \overline{\varphi}_0$ we have $\langle \overline{\varphi}_0, \varphi_0' \rangle = 0$, and $\langle \varphi_0, \overline{\varphi}_0' \rangle = 0$ follows in the same way.)

Note that $A_0 Q_0 x = Q_0 A_0 x$ for all $x \in X = D(A_0)$. Hence, $R(Q_0)$ as well as $N(Q_0)$ are invariant spaces under A_0. As shown in (I.8.17),

(I.8.19) $$N(I - e^{A_0 2\pi/\kappa_0}) = R(Q_0) \subset Z,$$

when Q_0 as given by (I.8.18) is restricted to the real space Z. The invariance of $R(Q_0)$ and of $N(Q_0)$ under A_0 implies their invariance under $e^{A_0 2\pi/\kappa_0}$, and the compactness of $e^{A_0 2\pi/\kappa_0}$ (cf. (I.8.8)) finally proves that

(I.8.20) $$I - e^{A_0 2\pi/\kappa_0} \in L(N(Q_0), N(Q_0)), \quad N(Q_0) \subset Z,$$
is an isomorphism.

(Injectivity in (I.8.20) is clear; surjectivity then follows from the Riesz–Schauder Theory for compact operators; cf. [170], for example.) Next, we introduce a projection $Q \in L(W, W)$ as follows:

$$(Qz)(t) = \frac{1}{2\pi} \int_0^{2\pi} \langle z(t), \psi_0'(t) \rangle dt \psi_0(t)$$

(I.8.21)
$$+ \frac{1}{2\pi} \int_0^{2\pi} \langle z(t), \overline{\psi}_0'(t) \rangle dt \overline{\psi}_0(t),$$
$$\text{where } \psi_0(t) = \varphi_0 e^{it}, \quad \psi_0'(t) = \varphi_0' e^{-it}.$$

(Although Q is defined by complex functions, the restriction to the real space gives a real operator. Recall that the pairing $\langle \ , \ \rangle$ is bilinear also in the complex case.) The result of (I.8.17) can be restated as

(I.8.22)
$$N(J_0) = R(Q), \text{ or in other words,}$$
$$Q \text{ projects onto } N(J_0); \text{ cf. (I.8.17).}$$

The proof of Proposition I.8.1 will be complete when we have shown that $R(J_0) = N(Q)$.

Let $f \in W = C_{2\pi}^\alpha(\mathbb{R}, Z)$. As proved in [86], IX, 1.7, a solution of

(I.8.23)
$$J_0 x = \kappa_0 \frac{dx}{dt} - A_0 x = f \quad \text{is given by}$$
$$x(t) = e^{A_0 t/\kappa_0} x(0) + \frac{1}{\kappa_0} \int_0^t e^{A_0(t-s)/\kappa_0} f(s) ds$$

for any $x(0) \in Z$. Due to the regularity of x as proved in [86], the function x is in $Y \cap E$ if it is 2π-periodic. That, in turn, is proved if

(I.8.24)
$$(I - e^{A_0 2\pi/\kappa_0}) x(0) = \frac{1}{\kappa_0} \int_0^{2\pi} e^{A_0(2\pi-s)/\kappa_0} f(s) ds$$
$$\text{has a solution } x(0) \in X = D(A_0).$$

For every $f \in W$, we have

(I.8.25)
$$\left\langle \int_0^{2\pi} e^{A_0(2\pi-s)/\kappa_0} f(s) ds, \varphi_0' \right\rangle = \int_0^{2\pi} \langle f(s), e^{A_0'(2\pi-s)/\kappa_0} \varphi_0' \rangle ds$$
$$= \int_0^{2\pi} \langle f(s), e^{i(2\pi-s)} \varphi_0' \rangle ds.$$

(We use that $(e^{A_0 t/\kappa_0})' \varphi_0' = e^{A_0' t/\kappa_0} \varphi_0' = e^{it} \varphi_0'$, since $A_0' \varphi_0' = i \kappa_0 \varphi_0'$; in general $(e^{A_0 t})' = e^{A_0' t}$ on $D(A_0') \subset Z'$; cf. [170], IX, 13.) Since (I.8.25) is true for $\overline{\varphi}_0'$ as well, we have shown (cf. (I.8.18), (I.8.21)) that

(I.8.26)
$$f \in N(Q) \Leftrightarrow \int_0^{2\pi} e^{A_0(2\pi-s)/\kappa_0} f(s) ds \in N(Q_0).$$

Thus, by (I.8.20), there is a solution $x(0) \in N(Q_0) \subset Z$ of (I.8.24) if and only if $f \in N(Q)$. Finally,

$$(I.8.27) \quad x(0) = e^{A_0 2\pi/\kappa_0} x(0) + \frac{1}{\kappa_0} \int_0^{2\pi} e^{A_0(2\pi-s)/\kappa_0} f(s)ds \in N(Q_0)$$

proves that $x(0) \in D(A_0) = X$, since both terms on the right-hand side of (I.8.27) are in $D(A_0)$ (cf. [86]).

We have proved that

(I.8.28)
$$R(J_0) = N(Q), \text{ and, in view of (I.8.22)},$$
$$W = C_{2\pi}^\alpha(\mathbb{R}, Z) = R(J_0) \oplus N(J_0),$$

which completes the proof of Proposition I.8.1. $\qquad\qquad\qquad\qquad$ □

For the problem

(I.8.29)
$$G(x, \kappa, \lambda) \equiv \kappa \frac{dx}{dt} - F(x, \lambda) = 0$$
$$\text{near } (0, \kappa_0, \lambda_0) \quad (\text{cf. (I.8.12)}),$$

the method of Lyapunov–Schmidt is now applicable, since all hypotheses are satisfied. For that reduction as described in Section I.2 we use the decomposition (I.8.28) and the projection Q. Since $Y \cap E = C_{2\pi}^{1+\alpha}(\mathbb{R}, Z) \cap C_{2\pi}^\alpha(\mathbb{R}, X)$ is continuously embedded into $W = C_{2\pi}^\alpha(\mathbb{R}, Z)$, the projection $Q|_{Y \cap E}$ projects $Y \cap E$ onto $N(J_0)$ along $R(J_0) \cap (Y \cap E)$. We set $Q|_{Y \cap E} = P$. Thus, by Theorem I.2.3, (I.8.29) is equivalent to

(I.8.30)
$$QG(Px + \psi(Px, \kappa, \lambda), \kappa, \lambda) = 0,$$
$$\text{where } P : Y \cap E \to N(J_0) \text{ along } R(J_0) \cap (Y \cap E),$$
$$\text{and } Q : W \to N(J_0) \text{ along } R(J_0).$$

The efficient analysis of this two-dimensional system (I.8.30)₁ makes use of another crucial property of (I.8.29), namely, its equivariance. Equivariance here means that every phase-shifted solution of (I.8.1) is again a solution, and for 2π-periodic solutions of (I.8.10) or (I.8.29), this is expressed as S^1-equivariance: Let

(I.8.31)
$$(S_\theta x)(t) = x(t + \theta), \quad \theta \in \mathbb{R}(\text{mod } 2\pi). \text{ Then}$$
$$G(S_\theta x, \kappa, \lambda) = S_\theta G(x, \kappa, \lambda)$$
$$\text{for all } x \in \tilde{U} \subset \mathbb{R}^2.$$

Since the projection Q commutes with S_θ (i.e., $QS_\theta = S_\theta Q$) the Lyapunov–Schmidt reduction preserves the equivariance by the uniqueness of the solution given via the Implicit Function Theorem:

$$(I.8.32) \qquad \psi(PS_\theta x, \kappa, \lambda) = \psi(S_\theta Px, \kappa, \lambda) = S_\theta \psi(Px, \kappa, \lambda).$$

A real function $Px \in N(J_0)$ is of the following form:

$$(Px)(t) = c\varphi_0 e^{it} + \bar{c}\,\bar{\varphi}_0 e^{-it}, \quad c \in \mathbb{C}, \text{ and}$$

(I.8.33)

$$(PS_\theta x)(t) = (S_\theta Px)(t) = ce^{i\theta}\varphi_0 e^{it} + \bar{c}e^{-i\theta}\bar{\varphi}_0 e^{-it}.$$

By the definition of Q (cf. (I.8.21), for a *real* function Px, the bifurcation equation $(I.8.30)_1$ is equivalent to the following one-dimensional complex equation:

(I.8.34) $$\frac{1}{2\pi}\int_0^{2\pi}\langle G(Px + \psi(Px, \kappa, \lambda), \kappa, \lambda), \psi_0'\rangle dt = 0.$$

Indeed, the second complex equation in order to satisfy $QG(./.) = 0$ is complex conjugate to (I.8.34). We insert (I.8.33) into (I.8.34) and write it as

(I.8.35) $$\begin{aligned}&\hat{\Phi}(c, \kappa, \lambda) = 0, \quad \text{where}\\ &\hat{\Phi} : \tilde{U}_2 \times \tilde{V}_2 \to \mathbb{C}, \quad 0 \in \tilde{U}_2 \subset \mathbb{C} \quad (\kappa_0, \lambda_0) \in \tilde{V}_2 \subset \mathbb{R}^2.\end{aligned}$$

The S^1-equivariance for G (I.8.31) or for QG (I.8.30) is expressed for (I.8.35) as follows:

(I.8.36) $$\hat{\Phi}(e^{i\theta}c, \kappa, \lambda) = e^{i\theta}\hat{\Phi}(c, \kappa, \lambda), \quad \theta \in [0, 2\pi).$$

That property has the following consequences:

(I.8.37) $$\begin{aligned}&\hat{\Phi}(c, \kappa, \lambda) = 0 \Leftrightarrow \hat{\Phi}(|c|, \kappa, \lambda) = 0,\\ &\hat{\Phi}(-c, \kappa, \lambda) = -\hat{\Phi}(c, \kappa, \lambda); \quad \text{i.e., } \hat{\Phi} \text{ is odd in } c \in \mathbb{C}.\end{aligned}$$

In particular, $\hat{\Phi}(0, \kappa, \lambda) = 0$ for all $(\kappa, \lambda) \in \tilde{V}_2$, which reflects the trivial solution. In view of $(I.8.37)_1$, it suffices to solve $\hat{\Phi}$ for real c; i.e., we consider henceforth

(I.8.38) $$\hat{\Phi}(r, \kappa, \lambda) = 0, \quad r \in (-\delta, \delta) \subset \mathbb{R}, \quad (\kappa, \lambda) \in \tilde{V}_2 \subset \mathbb{R}^2,$$

which is the **Bifurcation Equation for Hopf Bifurcation**. In order to eliminate the trivial solution, we proceed in the same way as in Section I.5 when we proved Theorem I.5.1: For $r \neq 0$ we set

(I.8.39) $$\begin{aligned}&\tilde{\Phi}(r, \kappa, \lambda) = \hat{\Phi}(r, \kappa, \lambda)/r, \quad \text{which we rewrite as}\\ &\tilde{\Phi}(r, \kappa, \lambda) = \int_0^1 D_r\hat{\Phi}(\tau r, \kappa, \lambda)d\tau = 0, \quad r \in (-\delta, \delta).\end{aligned}$$

By the definition of $\hat{\Phi}$, $(Px)(t) = r(\varphi_0 e^{it} + \bar{\varphi}_0 e^{-it})$, and using Corollary I.2.4 we have

(I.8.40) $$\tilde{\Phi}(0, \kappa_0, \lambda_0) = \frac{1}{2\pi}\int_0^{2\pi}\langle D_x G(0, \kappa_0, \lambda_0)\psi_0, \psi_0'\rangle dt = 0,$$

since $\psi_0 \in N(J_0)$. The computations of $D_\kappa\tilde{\Phi}(0, \kappa_0, \lambda_0)$ and $D_\lambda\tilde{\Phi}(0, \kappa_0, \lambda_0)$ follow precisely the lines of the proof of Theorem I.5.1, in particular, of (I.5.12)

((I.5.7) holds analogously). Therefore, we give only the result:

$$D_\kappa \tilde{\Phi}(0, \kappa_0, \lambda_0) = \frac{1}{2\pi} \int_0^{2\pi} \langle D_{x\kappa}^2 G(0, \kappa_0, \lambda_0)\psi_0, \psi_0' \rangle dt$$

$$= \frac{1}{2\pi} \int_0^{2\pi} \langle \frac{d}{dt}\psi_0, \psi_0' \rangle dt = i,$$

(I.8.41)

$$D_\lambda \tilde{\Phi}(0, \kappa_0, \lambda_0) = \frac{1}{2\pi} \int_0^{2\pi} \langle D_{x\lambda}^2 G(0, \kappa_0, \lambda_0)\psi_0, \psi_0' \rangle dt$$

$$= -\langle D_{x\lambda}^2 F(0, \lambda_0)\varphi_0, \varphi_0' \rangle.$$

Decomposing $\tilde{\Phi}$ into real and imaginary parts, we finally get (without changing the notation)

$$\tilde{\Phi} : (-\delta, \delta) \times \tilde{V}_2 \to \mathbb{R}^2 \text{ and}$$

(I.8.42)

$$D_{(\kappa,\lambda)}\tilde{\Phi}(0, \kappa_0, \lambda_0) = \begin{pmatrix} 0 & -\mathrm{Re}\langle D_{x\lambda}^2 F(0, \lambda_0)\varphi_0, \varphi_0' \rangle \\ 1 & -\mathrm{Im}\langle D_{x\lambda}^2 F(0, \lambda_0)\varphi_0, \varphi_0' \rangle \end{pmatrix}.$$

The last condition to solve $\tilde{\Phi}(r, \kappa, \lambda) = 0$ for $r \in (-\delta, \delta)$ (which is possibly shrunk) by the Implicit Function Theorem is

(I.8.43) $\mathrm{Re}\langle D_{x\lambda}^2 F(0, \lambda_0)\varphi_0, \varphi_0' \rangle \neq 0,$

and we obtain continuously differentiable functions $(\kappa, \lambda) : (-\delta, \delta) \to \tilde{V}_2 \subset \mathbb{R}^2$, $(\kappa(0), \lambda(0)) = (\kappa_0, \lambda_0)$, such that $\tilde{\Phi}(r, \kappa(r), \lambda(r)) = 0$ for all $r \in (-\delta, \delta)$. Obviously, $\hat{\Phi}(r, \kappa(r), \lambda(r)) = r\tilde{\Phi}(r, \kappa(r), \lambda(r)) = 0$, and the bifurcation equation (I.8.38) or (I.8.35) is solved nontrivially.

(Instead of solving the system $\tilde{\Phi}(r, \kappa, \lambda) = 0$ in one step, we could do it in two steps: By (I.8.42) and the Implicit Function Theorem, the solution of the imaginary part defines κ as a function of (r, λ), and when it is inserted into the real part, we obtain the **Reduced Bifurcation Equation** $\tilde{\Phi}_{re}(r, \lambda) = 0$. By (I.8.42), (I.8.43), and the Implicit Function Theorem, that scalar equation gives λ in terms of r. That reduction eliminates the period, and the bifurcation is reduced to the (r, λ)-plane.)

At the end we identify (I.8.43) with assumption (I.8.7). Differentiation of (I.8.6) with respect to λ at $\lambda = \lambda_0$ yields

$$D_{x\lambda}^2 F(0, \lambda_0)\varphi_0 + D_x F(0, \lambda_0)\varphi'(\lambda_0) = \mu'(\lambda_0)\varphi_0 + \mu(\lambda_0)\varphi'(\lambda_0),$$

$$\langle D_{x\lambda}^2 F(0, \lambda_0)\varphi_0, \varphi_0' \rangle + \langle \varphi'(\lambda_0), A_0'\varphi_0' \rangle$$

(I.8.44)

$$= \mu'(\lambda_0) + i\kappa_0 \langle \varphi'(\lambda_0), \varphi_0' \rangle,$$

$$\langle D_{x\lambda}^2 F(0, \lambda_0)\varphi_0, \varphi_0' \rangle = \mu'(\lambda_0),$$

since $A_0'\varphi_0' = i\kappa_0\varphi_0'$ and $\langle \ , \ \rangle$ is bilinear.

We summarize, and we obtain the **Hopf Bifurcation Theorem**.

Theorem I.8.2 *For the parameter-dependent evolution equation*

$$\frac{dx}{dt} = F(x, \lambda)$$

in a Banach space Z we make the regularity assumptions (I.8.2), (I.8.3), (I.8.4), (I.8.13), on the mapping F. We make the spectral assumptions (I.8.5), (I.8.6), (I.8.7) on the linearization $D_x F(0, \lambda)$ along the trivial solutions:

> $D_x F(0, \lambda)\varphi(\lambda) = \mu(\lambda)\varphi(\lambda)\, with\ \mu(\lambda_0) = i\kappa_0 \neq 0$,
> $\mu(\lambda)$ *are simple eigenvalues, and we assume the nondegeneracy*
> $\mathrm{Re}\mu'(\lambda_0) \neq 0$.

We impose the nonresonance condition (I.8.14):

> *For all $n \in \mathbb{Z}\backslash\{1, -1\}$, $in\kappa_0$ is not an eigenvalue of $A_0 = D_x F(0, \lambda_0)$.*

We assume that the operator $A_0 = D_x F(0, \lambda_0)$ generates a holomorphic semigroup according to (I.8.8):

$$e^{A_0 t} \in L(Z, Z) \text{ for } t \geq 0, \text{ which is compact for } t > 0.$$

Then there exists a continuously differentiable curve $\{(x(r), \lambda(r))\}$ of (real) $2\pi/\kappa(r)$-periodic solutions of (I.8.1) through $(x(0), \lambda(0)) = (0, \lambda_0)$ with $2\pi/\kappa(0)$
$= 2\pi/\kappa_0$ in $(C^{1+\alpha}_{2\pi/\kappa(r)}(\mathbb{R}, Z) \cap C^{\alpha}_{2\pi/\kappa(r)}(\mathbb{R}, X)) \times \mathbb{R}$. Every other periodic solution of (I.8.1) in a neighborhood of $(0, \lambda_0)$ (in that topology) is obtained from $(x(r), \lambda(r))$ by a phase shift $S_\theta x(r)$.
In particular, $x(-r) = S_{\pi/\kappa(r)}x(r)$, $\kappa(-r) = \kappa(r)$, and $\lambda(-r) = \lambda(r)$ for all $r \in (-\delta, \delta)$.

We give the arguments for the last statements: By the oddness of $\hat{\Phi}$ (cf. (I.8.37)), $\hat{\Phi}(-r, \kappa, \lambda) = 0 \Leftrightarrow \hat{\Phi}(r, \kappa, \lambda)$, whence $\kappa(-r) = \kappa(r)$ and $\lambda(-r) = \lambda(r)$ by the uniqueness of the solutions $(r, \kappa(r), \lambda(r))$ of $\hat{\Phi} = 0$ near $(0, \kappa_0, \lambda_0)$. By (I.8.33), $S_\pi Px = -Px$, so that by (I.8.32), $\psi(-Px, \kappa, \lambda) = S_\pi \psi(Px, \kappa, \lambda)$. Therefore, in the space of 2π-periodic functions, $x(-r) = S_\pi x(r)$. After substituting $\kappa(r)t$ for t (cf.(I.8.9)), we get $x(-r) = S_{\pi/\kappa(r)}x(r)$ for the $2\pi/\kappa(r)$-periodic solution (see also (I.8.47) below).

Corollary I.8.3 *The tangent vector of the nontrivial solution curve $(x(r), \lambda(r))$ at the bifurcation point $(0, \lambda_0)$ is given by*

(I.8.45) $(2\mathrm{Re}(\varphi_0 e^{i\kappa_0 t}), 0) \in (C^{1+\alpha}_{2\pi/\kappa_0}(\mathbb{R}, Z) \cap C^{\alpha}_{2\pi/\kappa_0}(\mathbb{R}, X)) \times \mathbb{R}$.

Proof. By $\lambda(-r) = \lambda(r)$ and $\kappa(-r) = \kappa(r)$ we have clearly

$$(I.8.46) \qquad \frac{d}{dr}\lambda(r)\Big|_{r=0} = \frac{d}{dr}\kappa(r)\Big|_{r=0} = 0.$$

After substituting $\kappa(r)t$ for t (cf. (I.8.9)), the bifurcating $2\pi/\kappa(r)$-periodic solutions of (I.8.1) are by construction

$$(I.8.47) \qquad \begin{aligned} x(r)(t) &= r(\varphi_0 e^{i\kappa(r)t} + \overline{\varphi}_0 e^{-i\kappa(r)t}) \\ &\quad + \psi(r(\varphi_0 e^{i\kappa(r)t} + \overline{\varphi}_0 e^{-i\kappa(r)t}), \kappa(r), \lambda(r)). \end{aligned}$$

The same arguments as for (I.5.18) give (I.8.45).   □

Remark I.8.4 *If (I.8.1) is an ODE, i.e., if $X = Z = \mathbb{R}^n$, then we can choose $E = W = C_{2\pi}(\mathbb{R}, \mathbb{R}^n) = \{x : \mathbb{R} \to \mathbb{R}^n | x \text{ is continuous and } 2\pi\text{-periodic}\}$ and $Y = C_{2\pi}^1(\mathbb{R}, \mathbb{R}^n) = \{x : \mathbb{R} \to \mathbb{R}^n | x, \frac{dx}{dt} \in C_{2\pi}(\mathbb{R}, \mathbb{R}^n)\}$ with norms $\|x\|_E = \max_{t \in \mathbb{R}} \|x(t)\|$ and $\|x\|_Y = \|x\|_E + \|\frac{dx}{dt}\|_E$. The crucial Proposition I.8.1 holds for these spaces, too. In particular, for the validity of (I.8.23) Hölder continuity is not needed. For an ODE we can therefore save one order of differentiability in (I.8.13): $F \in C^2(U \times V, \mathbb{R}^n)$ implies $G \in C^2(\tilde{U} \times \tilde{V}, W)$.*

I.9 Bifurcation Formulas for Hopf Bifurcation

Since $\dot{\lambda}(0) = 0$ (where $\dot{} = \frac{d}{dr}$), the sign of $\lambda(r) = \lambda(-r)$ is not yet determined, and in order to sketch the bifurcation diagram in $(C_{2\pi/\kappa(r)}^{1+\alpha}(\mathbb{R}, Z) \cap C_{2\pi/\kappa(r)}^{\alpha}(\mathbb{R}, X)) \times \mathbb{R}$ in lowest order, we give a formula for how to compute $\ddot{\lambda}(0)$ (and also $\ddot{\kappa}(0)$). We follow precisely the procedure of Section I.6, and we make use of the formulas derived there, in particular, (I.6.9). In order to have enough differentiability of G as given by (I.8.12), we assume that $F \in C^4(U \times V, Z)$. If (I.8.1) is an ODE, then it suffices that $F \in C^3(U \times V, Z)$; cf. Remark I.8.4. Our starting point is $\tilde{\Phi}(r, \kappa(r), \lambda(r)) = 0$ for all $r \in (-\delta, \delta)$, whence

$$(I.9.1) \qquad \begin{aligned} &\frac{d^2}{dr^2}\tilde{\Phi}(r, \kappa(r), \lambda(r))|_{r=0} \\ &= D_{rr}^2\tilde{\Phi}(0, \kappa_0, \lambda_0) + D_\kappa\tilde{\Phi}(0, \kappa_0, \lambda_0)\ddot{\kappa}(0) + D_\lambda\tilde{\Phi}^0\ddot{\lambda}(0) = 0, \end{aligned}$$

since $\dot{\kappa}(0) = \dot{\lambda}(0) = 0$ (cf. I.8.46).

Here "0" means evaluation at $(0, \kappa_0, \lambda_0)$. Decomposing $\tilde{\Phi}$ into real and imaginary parts as in (I.8.42), we solve (I.9.1) for $(\ddot{\kappa}(0), \ddot{\lambda}(0))$ using the inverse matrix of (I.8.42) and (I.8.44):

$$\begin{pmatrix} \ddot{\kappa}(0) \\ \ddot{\lambda}(0) \end{pmatrix} = \frac{1}{\mathrm{Re}\mu'(\lambda_0)} \begin{pmatrix} \mathrm{Im}\mu'(\lambda_0) & -\mathrm{Re}\mu'(\lambda_0) \\ 1 & 0 \end{pmatrix} \begin{pmatrix} \mathrm{Re}D^2_{rr}\tilde{\Phi}^0 \\ \mathrm{Im}D^2_{rr}\tilde{\Phi}^0 \end{pmatrix},$$

(I.9.2)

$$\mu'(\lambda_0) = \langle D^2_{x\lambda}F(0,\lambda_0)\varphi_0, \varphi'_0\rangle.$$

By definitions (I.8.39) of $\tilde{\Phi}$ and (I.8.34) of $\hat{\Phi}$,

$$D^2_{rr}\tilde{\Phi}(0,\kappa_0,\lambda_0) = \frac{1}{3}D^3_{rrr}\hat{\Phi}(0,\kappa_0,\lambda_0), \quad \text{where}$$

(I.9.3) $\quad \hat{\Phi}(r,\kappa,\lambda)$

$$= \frac{1}{2\pi}\int_0^{2\pi} \langle G(r(\psi_0 + \overline{\psi}_0) + \psi(r(\psi_0 + \overline{\psi}_0),\kappa,\lambda),\kappa,\lambda),\psi'_0\rangle dt.$$

In order to apply directly the computations of Section I.6, we make the following definitions:

(I.9.4)
$$\Phi(v,\kappa,\lambda) = QG(v + \psi(v,\kappa,\lambda),\kappa,\lambda),$$
$$v = r(\psi_0 + \overline{\psi}_0), \quad \hat{v}_0 = \psi_0 + \overline{\psi}_0.$$

Observe that Φ coincides with definition (I.2.9) used in Section I.6. Then, in view of the definition of Q (cf.(I.8.21)),

(I.9.5)
$$\hat{\Phi}(r,\kappa,\lambda) = \frac{1}{2\pi}\int_0^{2\pi} \langle \Phi(v,\kappa,\lambda),\psi'_0\rangle dt \quad \text{and}$$
$$D^3_{rrr}\hat{\Phi}(0,\kappa_0,\lambda_0) = \frac{1}{2\pi}\int_0^{2\pi} \langle D^3_{vvv}\Phi(0,\kappa_0,\lambda_0)[\hat{v}_0,\hat{v}_0,\hat{v}_0],\psi'_0\rangle dt.$$

From (I.6.9) we can read off

$$D^3_{vvv}\Phi(0,\kappa_0,\lambda_0)[\hat{v}_0,\hat{v}_0,\hat{v}_0]$$

(I.9.6)
$$= QD^3_{xxx}G(0,\kappa_0,\lambda_0)[\hat{v}_0,\hat{v}_0,\hat{v}_0] - 3QD^2_{xx}G(0,\kappa_0,\lambda_0)[./.],$$

where $[./.] =$
$$[\hat{v}_0,(I-P)(D_xG(0,\kappa_0,\lambda_0))^{-1}(I-Q)D^2_{xx}G(0,\kappa_0,\lambda_0)[\hat{v}_0,\hat{v}_0]].$$

We compute these terms using the definitions of G (I.8.10), (I.9.4), and (I.9.5). The first term is simple:

(I.9.7)
$$\frac{1}{2\pi}\int_0^{2\pi} \langle QD^3_{xxx}G(0,\kappa_0,\lambda_0)[\hat{v}_0,\hat{v}_0,\hat{v}_0],\psi'_0\rangle dt$$
$$= -3\langle D^3_{xxx}F(0,\lambda_0)[\varphi_0,\varphi_0,\overline{\varphi}_0],\varphi'_0\rangle.$$

For the second term, we compute

$$D^2_{xx} G(0, \kappa_0, \lambda_0)[\hat{v}_0, \hat{v}_0]$$

(I.9.8)
$$= -D^2_{xx}F(0, \lambda_0)[\varphi_0, \varphi_0]e^{2it} - 2D^2_{xx}F(0, \lambda_0)[\varphi_0, \overline{\varphi}_0]$$
$$\quad - D^2_{xx}F(0, \lambda_0)[\overline{\varphi}_0, \overline{\varphi}_0]e^{-2it}$$
$$= (I - Q)D^2_{xx}G(0, \kappa_0, \lambda_0)[\hat{v}_0, \hat{v}_0].$$

The last equation follows from $QD^2_{xx}G(0, \kappa_0, \lambda_0)[\hat{v}_0, \hat{v}_0] = 0$.

Recall that $D_x G(0, \kappa_0, \lambda_0) = \kappa_0 \frac{d}{dt} - A_0 = J_0$. Then the unique solutions of $J_0 x = z_0 e^{\pm 2it}$ or $J_0 x = z_0$ for $z_0 \in Z$ can be given explicitly. We consider J_0 as a mapping from $(I - P)(E \cap Y)$ onto $(I - Q)W = R(J_0)$; i.e., we consider J_0 in the complement of its kernel $N(J_0) = R(Q)$; cf. (I.6.8).

Now,

(I.9.9)
$$x(t) = (\pm 2i\kappa_0 - A_0)^{-1} z_0 e^{\pm 2it} \text{ solves}$$
$$J_0 x = z_0 e^{\pm 2it},$$
$$x(t) = (-A_0)^{-1} z_0 \text{ solves}$$
$$J_0 x = z_0,$$

and both solutions are in $(I - P)(E \cap Y)$. Here we use again the nonresonance condition (I.8.14). We insert these preimages $J_0^{-1}(I - Q)D^2_{xx}G(0, \kappa_0, \lambda_0)[\hat{v}_0, \hat{v}_0]$ into (I.9.6), and we get

(I.9.10)
$$\frac{1}{2\pi} \int_0^{2\pi} \langle QD^2_{xx}G(0, \kappa_0, \lambda_0)[\varphi_0 e^{it} + \overline{\varphi}_0 e^{-it},$$
$$\quad - (2i\kappa_0 - A_0)^{-1}D^2_{xx}F(0, \lambda_0)[\varphi_0, \varphi_0]e^{2it}$$
$$\quad + 2A_0^{-1}D^2_{xx}F(0, \lambda_0)[\varphi_0, \overline{\varphi}_0]$$
$$\quad - (-2i\kappa_0 - A_0)^{-1}D^2_{xx}F(0, \lambda_0)[\overline{\varphi}_0, \overline{\varphi}_0]e^{-2it}], \varphi_0' e^{-it}\rangle dt$$
$$= \langle D^2_{xx}F(0, \lambda_0)[\overline{\varphi}_0, (2i\kappa_0 - A_0)^{-1}D^2_{xx}F(0, \lambda_0)[\varphi_0, \varphi_0]], \varphi_0'\rangle$$
$$\quad - 2\langle D^2_{xx}F(0, \lambda_0)[\varphi_0, A_0^{-1}D^2_{xx}F(0, \lambda_0)[\varphi_0, \overline{\varphi}_0]], \varphi_0'\rangle.$$

This gives finally, using (I.9.3), (I.9.5), and (I.9.6),

(I.9.11)
$$D^2_{rr}\tilde{\Phi}(0, \kappa_0, \lambda_0) = \frac{1}{3}D^3_{rrr}\hat{\Phi}(0, \kappa_0, \lambda_0)$$
$$= -\langle D^3_{xxx}F(0, \lambda_0)[\varphi_0, \varphi_0, \overline{\varphi}_0], \varphi_0'\rangle$$
$$\quad - \langle D^2_{xx}F(0, \lambda_0)[\overline{\varphi}_0, (2i\kappa_0 - A_0)^{-1}D^2_{xx}F(0, \lambda_0)[\varphi_0, \varphi_0]], \varphi_0'\rangle$$
$$\quad + 2\langle D^2_{xx}F(0, \lambda_0)[\varphi_0, A_0^{-1}D^2_{xx}F(0, \lambda_0)[\varphi_0, \overline{\varphi}_0]], \varphi_0'\rangle.$$

Inserting this complex number (I.9.11) into (I.9.2) provides the **Bifurcation Formulas** for the Hopf bifurcation, where we make use also of Corollary I.8.3:

Theorem I.9.1 *Let $\{(x(r), \lambda(r))\}$ be the curve of $2\pi/\kappa(r)$-periodic solutions of (I.8.1) according to Theorem I.8.2. Then*

(I.9.12)
$$\frac{d}{dr}x(r)\Big|_{r=0}(t) = 2\mathrm{Re}(\varphi_0 e^{i\kappa_0 t}),$$

$$\frac{d}{dr}\kappa(r)\Big|_{r=0} = \frac{d}{dr}\lambda(r)\Big|_{r=0} = 0,$$

$$\frac{d^2}{dr^2}\kappa(r)\Big|_{r=0} = \frac{\mathrm{Im}\mu'(\lambda_0)}{\mathrm{Re}\mu'(\lambda_0)}\mathrm{Re}D^2_{rr}\tilde{\Phi}^0 - \mathrm{Im}D^2_{rr}\tilde{\Phi}^0,$$

$$\frac{d^2}{dr^2}\lambda(r)\Big|_{r=0} = \frac{1}{\mathrm{Re}\mu'(\lambda_0)}\mathrm{Re}D^2_{rr}\tilde{\Phi}^0,$$

where $D^2_{rr}\tilde{\Phi}^0$ is given by (I.9.11), and $\mathrm{Re}\mu'(\lambda_0) \neq 0$ is the nondegeneracy condition for $\mu'(\lambda_0) = \langle D^2_{x\lambda}F(0,\lambda_0)\varphi_0, \varphi'_0\rangle$ (cf. (I.8.44)). (We recall that the pairing $\langle \; , \; \rangle$ is bilinear also in the complex case.)

If $\mathrm{Re}D^2_{rr}\tilde{\Phi}^0 \neq 0$, we have a sub- or supercritical "pitchfork" bifurcation of periodic solutions sketched in Figure I.9.1.

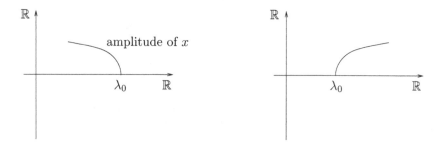

Figure I.9.1

In contrast to Figure I.5.1, we sketch only one branch in Figure I.9.1, which represents the amplitude $\max_{t\in\mathbb{R}} \|x(t)\|$ (with norm in X) of the bifurcating periodic solution. By (I.9.12)$_1$ that amplitude is of order $|r|$. Recall that the curve $\{(x(r), \lambda(r))\}$ for negative r is obtained from that for positive r by a phase shift of half the period (and all other periodic solutions are obtained by phase shifts).

Remark I.9.2 *One might ask why the linearization $A_0 = D_xF(0,\lambda_0)$ has to generate a holomorphic semigroup $e^{A_0 t}$ that is compact for $t > 0$ (cf.(I.8.8)). As a matter of fact, this strong assumption was used only for the proof of Proposition I.8.1. (The compactness was used only for (I.8.20).)*

The assumption (I.8.8) is obviously satisfied if:

(I.8.1) is an ODE; i.e., $X = Z = \mathbb{R}^n$;

(I.8.1) is a parabolic PDE; i.e., if A_0 is an elliptic partial differential

operator over a bounded domain; cf. Section III.4.

There is a large class of evolution equations, however, for which assumption (I.8.8) is not satisfied, namely, the RFDEs (retarded functional differential equations). We briefly comment on these.

For a function $x : \mathbb{R} \to \mathbb{R}^n$ we define $x_t : [-h, 0] \to \mathbb{R}^n$ by $x_t(\theta) = x(t + \theta)$ for $-h \le \theta \le 0$, and a parameter-dependent RFDE is of the form

$$(I.9.13) \qquad \frac{dx}{dt}(t) = F(x_t, \lambda).$$

Here

$$(I.9.14) \qquad \begin{aligned} &F : U \times V \to Z = \mathbb{R}^n, \text{ where} \\ &0 \in U \subset X = C([-h, 0], \mathbb{R}^n) \text{ and } \lambda_0 \in V \subset \mathbb{R}. \\ &U \text{ and } V \text{ are open and } F \in C^2(U \times V, Z). \end{aligned}$$

Let $F(0, \lambda) = 0$ (such that there is the trivial solution line $\{(0, \lambda)\} \subset X \times \mathbb{R}$), and let $A(\lambda) = D_x F(0, \lambda)$ as before. Then (I.9.13) is written as

$$(I.9.15) \qquad \begin{aligned} &\frac{dx}{dt}(t) = A(\lambda)x_t + R(x_t, \lambda) \text{ with a remainder} \\ &R(x_t, \lambda) = F(x_t, \lambda) - A(\lambda)x_t. \end{aligned}$$

As expounded in [36], (I.9.15) can be considered as an abstract evolution equation

$$(I.9.16) \qquad \begin{aligned} &\frac{du}{dt} = A^{\odot *}(\lambda)u + R^{\odot *}(u, \lambda) \text{ in some} \\ &\text{infinite-dimensional phase space } X^{\odot *} \end{aligned}$$

that is related to X. (We cannot go into the details here.) The linearization $A_0^{\odot *} = A^{\odot *}(\lambda_0)$, unfortunately, does not satisfy assumptions (I.8.8). It generates a strongly continuous semigroup that is eventually compact (for $t \ge h$) and that eventually regularizes (i.e., it maps into $D(A_0^{\odot *})$ for $t \ge h$). To summarize, the proof of Proposition I.8.1 cannot be modified for this situation. That is the reason why for a proof of the Hopf Bifurcation Theorem for RFDEs the setting (I.9.16) is not useful. For the original formulation (I.9.13) or (I.9.15), however, we can follow the lines of the proof of Section I.8.

By a substitution κt for t we normalize the unknown period $2\pi/\kappa$ to 2π, but in contrast to an ODE, the parameter κ appears also on the right-hand side of (I.9.13); that is,

$$(I.9.17) \qquad \begin{aligned} &\kappa \frac{dx}{dt}(t) = \tilde{F}(x_t, \kappa, \lambda), \text{ where} \\ &\tilde{F}(x_t, \kappa, \lambda) = F(x_t(\kappa \cdot), \lambda); \end{aligned}$$

i.e., the mapping F is evaluated at the function $x_t(\kappa\theta) = x(t + \kappa\theta)$ for $-h \le \theta \le 0$. Setting

$$\tilde{G}(x, \kappa, \lambda)(t) \equiv \kappa \frac{dx}{dt}(t) - \tilde{F}(x_t, \kappa, \lambda),$$

(I.9.18)
$$\tilde{G} : \tilde{U} \times \tilde{V} \to C_{2\pi}(\mathbb{R}, Z), \quad Z = \mathbb{R}^n,$$

$$0 \in \tilde{U} \subset C^1_{2\pi}(\mathbb{R}, Z), \ (\kappa_0, \lambda_0) \in \tilde{V} \subset \mathbb{R}^2,$$

we obtain

(I.9.19)
$$D_x \tilde{G}(0, \kappa_0, \lambda_0) = \kappa_0 \frac{d}{dt} - \tilde{A}(\kappa_0, \lambda_0) \equiv \tilde{J}_0, \ \text{where}$$

$$(\tilde{A}(\kappa_0, \lambda_0)x)(t) = D_x F(0, \lambda_0)x_t(\kappa_0 \cdot) = A_0 x_t(\kappa_0 \cdot).$$

As shown in [63], Chapter 9, or in [36], Section VI.4, the operator

(I.9.20)
$$\tilde{J}_0 : C^1_{2\pi}(\mathbb{R}, Z) \to C_{2\pi}(\mathbb{R}, Z)$$
is a Fredholm operator of index zero.

Therefore, the crucial property for the application of the method of Lyapunov–Schmidt for $\tilde{G}(x, \kappa, \lambda) = 0$ *near* $(0, \kappa_0, \lambda_0)$ *is satisfied. Observe also that the equivariance (I.8.31) holds as well for autonomous RFDEs.*

Let $i\kappa_0$ *be a simple eigenvalue of* A_0 *(in the sense that* $A_0 x_t = i\kappa_0 x(t)$ *holds for some function* $x(t) = e^{i\kappa_0 t}\varphi_0$ *with* $\varphi_0 \in \mathbb{C}^n$*). According to the theory expounded in [36], Chapter IV,* $i\kappa_0$ *is a simple zero of the determinant of the characteristic matrix* $\Delta(z, \lambda_0) \in L(\mathbb{C}^n, \mathbb{C}^n)$ *of* $A_0 = A(\lambda_0) \in L(X, Z)$ *(see (I.9.21) below), and* 0 *is a simple eigenvalue of* $\Delta(i\kappa_0, \lambda_0)$*; i.e.,* $\Delta(i\kappa_0, \lambda_0)\varphi_0 = 0$ *and* $\Delta^T(i\kappa_0, \lambda_0)\varphi_0' = 0$ *for some* $\varphi_0, \varphi_0' \in \mathbb{C}^n$*, where we can normalize* $\langle \varphi_0, \varphi_0' \rangle = 1$*. The pairing* $\langle \ , \ \rangle$ *between* $Z = \mathbb{C}^n = Z'$ *is the bilinear complexification of the Euclidean scalar product in* $\mathbb{R}^n$*. The nonresonance condition means that* $\Delta(im\kappa_0, \lambda_0)$ *is regular for all* $m \in \mathbb{Z} \backslash \{1, -1\}$*.*

Setting $\psi_0(t) = e^{it}\varphi_0$*, then* $N(\tilde{J}_0) = \text{span}[\psi_0, \overline{\psi}_0]$*, and for the definition of the projection* Q *(cf. (I.8.21)) we take* $\psi_0'(t) = e^{-it}\varphi_0'$*. Observe that* $\psi_0'(-t), \overline{\psi}_0'(-t)$ *are* 2π*-periodic solutions of the so-called transposed RFDE; that is,* $(\kappa_0 \frac{d}{dt} - \tilde{A}^T(\kappa_0, \lambda_0))z = 0$*, and property (I.9.20) is proved by showing that* $R(\tilde{J}_0) = N(Q)$*.*

The method of Lyapunov–Schmidt provides a one-dimensional complex equation of the form (I.8.34) or, by equivariance, one complex equation $\hat{\Phi}(r, \kappa, \lambda) = 0$ *for the real variables* (r, κ, λ) *near* $(0, \kappa_0, \lambda_0)$*; cf. (I.8.37). However, the computations (I.8.41) are different now. We use the representations*

(I.9.21)
$$A(\lambda)x_t = \int_0^h d\zeta(\theta, \lambda)x(t - \theta),$$

$$\Delta(z, \lambda) = zE - \int_0^h d\zeta(\theta, \lambda)e^{-z\theta}; \ \text{cf.[36], Chapter IV.}$$

Then for $\tilde{\Phi}$ *as defined in (I.8.39) we obtain in view of (I.8.41), (I.9.19), (I.9.21), and* $\langle \varphi_0, \varphi_0' \rangle = 1,$

$$D_\kappa \tilde{\Phi}(0, \kappa_0, \lambda_0)$$

(I.9.22)
$$= i(1 + \int_0^h \langle d\zeta(\theta, \lambda_0)\theta e^{-i\kappa_0\theta}\varphi_0, \varphi_0'\rangle)$$
$$= i\langle D_z\Delta(i\kappa_0, \lambda_0)\varphi_0, \varphi_0'\rangle,$$

$$D_\lambda\tilde{\Phi}(0, \kappa_0, \lambda_0) = \langle D_\lambda\Delta(i\kappa_0, \lambda_0)\varphi_0, \varphi_0'\rangle.$$

Let $\mu(\lambda)$ be the simple eigenvalue perturbation of $A(\lambda) = D_x F(0, \lambda)$ through $\mu(\lambda_0) = i\kappa_0$. Then $\mu(\lambda)$ satisfies the characteristic equation $\det \Delta(\mu(\lambda), \lambda) = 0$ or $\Delta(\mu(\lambda), \lambda)\varphi(\lambda) = 0$, where $\varphi(\lambda_0) = \varphi_0$. By differentiating we obtain, in view of $\langle \Delta(i\kappa_0, \lambda_0)\varphi'(\lambda_0), \varphi_0'\rangle = \langle\varphi'(\lambda_0), \Delta^T(i\kappa_0, \lambda_0)\varphi_0'\rangle = 0$,

(I.9.23) $\langle D_z\Delta(i\kappa_0, \lambda_0)\varphi_0, \varphi_0'\rangle\mu'(\lambda_0) + \langle D_\lambda\Delta(i\kappa_0, \lambda_0)\varphi_0, \varphi_0'\rangle = 0.$

Decomposing $\tilde{\Phi}$ into real and imaginary parts, we get as in (I.8.42),

$$D_{(\kappa,\lambda)}\tilde{\Phi}(0, \kappa_0, \lambda_0) = \begin{pmatrix} \mathrm{Re}D_\kappa\tilde{\Phi}(0, \kappa_0, \lambda_0) & \mathrm{Re}D_\lambda\tilde{\Phi}(0, \kappa_0, \lambda_0) \\ \mathrm{Im}D_\kappa\tilde{\Phi}(0, \kappa_0, \lambda_0) & \mathrm{Im}D_\lambda\tilde{\Phi}(0, \kappa_0, \lambda_0) \end{pmatrix},$$

(I.9.24)   and by (I.9.22), (I.9.23),

$$\det D_{(\kappa,\lambda)}\tilde{\Phi}(0, \kappa_0, \lambda_0) = \mathrm{Re}\mu'(\lambda_0)|\langle D_z\Delta(i\kappa_0, \lambda_0)\varphi_0, \varphi_0'\rangle|^2$$
$$\neq 0 \Leftrightarrow \mathrm{Re}\mu'(\lambda_0) \neq 0,$$

since $\langle D_z\Delta(i\kappa_0, \lambda_0)\varphi_0, \varphi_0'\rangle \neq 0$ by the simplicity of the eigenvalue $i\kappa_0$ and the results of [36], Section IV.5. (Note that we normalize $\langle\varphi_0, \varphi_0'\rangle = 1$ but not $\langle D_z\Delta(i\kappa_0, \lambda_0)\varphi_0, \varphi_0'\rangle$.)

The conclusion from (I.9.24) is the same as that from (I.8.42)–(I.8.44), namely, the **Hopf Bifurcation Theorem for RFDEs**:

Under formally the same assumptions as for Theorem I.8.2, summarized and explained above in this remark, there exists a unique (up to phase shifts) smooth curve $\{(x(r), \lambda(r))\}$ of (real) $2\pi/\kappa(r)$-periodic solutions of (I.9.13) through $(x(0), \lambda(0)) = (0, \lambda_0)$ with $2\pi/\kappa(0) = 2\pi/\kappa_0$ in $C^1_{2\pi/\kappa(r)}(\mathbb{R}, Z) \times \mathbb{R}$. Furthermore, $x(-r) = S_{\pi/\kappa(r)}x(r)$, $\kappa(-r) = \kappa(r)$, $\lambda(-r) = \lambda(r)$ for all $r \in (-\delta, \delta)$, and $\frac{d}{dr}x(r)|_{r=0}(t) = 2\mathrm{Re}(\varphi_0 e^{i\kappa_0 t})$.

The **Bifurcation Formulas** are derived as we did before in this section: The inversion of the matrix $(I.9.24)_1$ yields via (I.9.1) the formulas for $\ddot{\kappa}(0)$ and $\ddot{\lambda}(0)$ (where clearly, we use (I.9.22), (I.9.23)). This yields, for instance,

(I.9.25) $\ddot{\lambda}(0) = \dfrac{\mathrm{Re}\langle D_z\Delta(i\kappa_0, \lambda_0)\varphi_0, \varphi_0'\rangle\mathrm{Re}D^2_{rr}\tilde{\Phi}^0 + \mathrm{Im}\langle./.\rangle\mathrm{Im}D^2_{rr}\tilde{\Phi}^0}{\mathrm{Re}\mu'(\lambda_0)|\langle D_z\Delta(i\kappa_0, \lambda_0)\varphi_0, \varphi_0'\rangle|^2}.$

The crucial quantity is $D^2_{rr}\tilde{\Phi}^0$ given in (I.9.3)–(I.9.6), where now G is replaced by $\tilde{G}$ from (I.9.18). This means that in (I.9.7), the vector φ_0 has to be replaced by the function $e^{i\kappa_0\theta}\varphi_0$. (Recall that $F(\cdot, \lambda_0)$ acts on $X = C([-h, 0], \mathbb{C}^n)$.) The same holds true in (I.9.8), and for the inversion of $\tilde{J}_0$

from (I.9.19) in $(I - Q)W$ we make use of the following:

$$\tilde{J}_0 x = z_0 e^{imt} \Leftrightarrow x(t) = w_0 e^{imt}, \ \ where$$

(I.9.26) $\Delta(im\kappa_0, \lambda_0) w_0 = z_0 \ \ \ for \ w_0, z_0 \in \mathbb{C}^n$

$$\Leftrightarrow w_0 = \Delta^{-1}(im\kappa_0, \lambda_0) z_0 \ \ \ for \ m \in \mathbb{Z} \backslash \{1, -1\}.$$

(By the nonresonance condition, $\Delta(im\kappa_0, \lambda_0)$ is regular for all $m \in \mathbb{Z} \backslash \{1, -1\}$.)
This yields for $D_{rr}\tilde{\Phi}^0$ the expression (I.9.11), where now the resolvents
$(2i\kappa_0 - A_0)^{-1}$ and A_0^{-1} are replaced by $e^{2i\kappa_0\theta}\Delta^{-1}(2i\kappa_0, \lambda_0)$ and $-\Delta^{-1}(0, \lambda_0)$,
respectively, and where, as before, φ_0 is replaced by $e^{i\kappa_0\theta}\varphi_0$. (Recall that
$F(\cdot, \lambda_0)$ maps into $Z = \mathbb{C}^n$.) We leave formula for $\ddot{\kappa}(0)$ as a simple con-
sequence to the reader.

The preceding proof of the Hopf Bifurcation Theorem for RFDEs follows the proof for infinite-dimensional evolution equations (I.8.1). The alternative suggested in [36] is the following: Reduce the RFDE to a finite-
dimensional center manifold and afterwards apply the Hopf Bifurcation Theorem for ODEs, whose proof is the same as for infinite-dimensional evolution
equations. The proof in [63], however, is in the spirit of that presented here.

I.10 A Lyapunov Center Theorem

In this section we prove a bifurcation of periodic solutions of an evolution
equation that does not depend on a real parameter. We show the existence
of periodic solutions with small amplitude of a Hamiltonian system

$$\text{(I.10.1)} \qquad\qquad \frac{dx}{dt} = F(x)$$

bifurcating from the stationary solution $x = 0$ (i.e., $F(0) = 0$). Although
there is no explicit parameter involved in (I.10.1), the notion of a bifurcation is justified: We learned in Section I.8 that for autonomous systems the
period can be considered as a second hidden parameter, and we show that
the existence of small-amplitude periodic solutions of (I.10.1) is due to a one-
parameter bifurcation with the period as a parameter. Recall that the Hopf
bifurcation is a two-parameter bifurcation, and the release of one parameter obviously requires more restrictions on (I.10.1), which is its Hamiltonian
structure.

In the following proof it will be nowhere used that the spaces $X \subset Z$ are
finite-dimensional. But since we know reasonable applications only if $X = Z$
is finite-dimensional (see Remark I.10.1 below), we assume that

$$F : U \to Z = \mathbb{R}^n, \text{ where}$$

(I.10.2) $0 \in U \subset X = \mathbb{R}^n$, and U is an open neighborhood, and

$$F(0) = 0, \quad F \in C^2(U, Z).$$

We identify F with $F(\cdot, \lambda_0)$ and impose the same hypotheses on F as we did in Section I.8 on $F(\cdot, \lambda_0)$. In particular, we assume (I.8.5) and (I.8.14), and we realize that (I.8.8) is automatically satisfied for $A_0 = DF(0) \in L(\mathbb{R}^n, \mathbb{R}^n)$. We will return to the assumption (I.8.5) (cf. (I.10.10)) when we explain the Hamiltonian structure of (I.10.1).

The space $X = Z = \mathbb{R}^n$ is a Hilbert space with a scalar product $(\quad , \quad)$. Let

(I.10.3) $H : U \to \mathbb{R}$ be in $C^3(U, \mathbb{R})$, $U \subset X = \mathbb{R}^n$;

we call such a mapping Hamiltonian.

This Hamiltonian H defines a potential operator ∇H in the sense of Definition I.3.1 as follows:

(I.10.4) $DH(x)h = (\nabla H(x), h)$ for all $x \in U, h \in X = \mathbb{R}^n$,

since all linear functionals $DH(x) : X \to \mathbb{R}$ are represented as in (I.10.4) via a scalar product. Obviously, $\nabla H : U \to Z = \mathbb{R}^n$ is in $C^2(U, Z)$.

Next we choose a particular element J in $L(\mathbb{R}^n, \mathbb{R}^n)$:

(I.10.5)   Let $J \in L(\mathbb{R}^n, \mathbb{R}^n)$ satisfy

$$J^2 = -I, \quad (Jx, x) = 0 \text{ for all } x \in X = \mathbb{R}^n.$$

Then (I.10.1) is called a **Hamiltonian system** if

(I.10.6)
$$\frac{dx}{dt} = F(x) = J\nabla H(x).$$

Typically, when $X = Z = \mathbb{R}^n$ is endowed with the canonical orthonormal basis with respect to the Euclidean scalar product, such a mapping J as required in (I.10.5) exists if $n = 2m$ and J is represented by the matrix

(I.10.7)
$$J = \begin{pmatrix} 0 & -E \\ E & 0 \end{pmatrix}, \text{ where } E \text{ is the}$$

m-dimensional identity matrix.

It is easily verified that for any solution $x(t)$ of (I.10.6), the Hamiltonian H gives a first conservation law in the following sense:

(I.10.8)
$$H(x(t)) = \text{ const.}$$

We define $A_0 = DF(0)$ and by (I.10.6),

(I.10.9)
$$A_0 = JD\nabla H(0) \equiv JB_0.$$

When expressed in coordinates, $\nabla H(x)$ is a column called the "gradient of H at x," and $D\nabla H(x)$ is the "Hessian matrix of H at x," which is obviously symmetric (cf. Proposition I.3.3). In particular, B_0 is symmetric; i.e., $(B_0 x_1, x_2) = (x_1, B_0 x_2)$ for all $x_1, x_2 \in X = \mathbb{R}^n$.

Remark I.10.1 *Infinite-dimensional Hamiltonian systems are typically hyperbolic PDEs, and the infinitely many (discrete) negative eigenvalues of B_0 become infinitely many (discrete) eigenvalues of A_0 on the imaginary axis (cf. Remark III.3.1). In this case, however, A_0 does not generate a holomorphic semigroup, and assumption (I.8.8) is not satisfied. On the other hand, from an abstract point of view, one might consider (I.10.6) as an infinite-dimensional system, where ∇H is the gradient of a potential H in the sense of Section I.3 and the mapping $J \in L(Z, Z)$ satisfies (I.10.5) with the scalar product (I.3.1). But then the spectral property (I.11.32) is also an obstruction to the generation of a holomorphic semigroup unless $A_0 \in L(Z, Z)$ is bounded. For that situation, however, we do not know an application.*

We assume for A_0 as given by (I.10.9) the special (but typical) situation that

(I.10.10)
$\quad i\kappa_0 (\neq 0)$ is an eigenvalue of $A_0 = JB_0$ with eigenvector $\varphi_0 \in \mathbb{C}^n$ (= complexification of $\mathbb{R}^n$) such that $J\varphi_0 = i\varphi_0$, and $\dim N(\pm i\kappa_0 I - A_0) = 1$.

Clearly, $Z = \mathbb{R}^n$ is complexified to $Z = Z_c = \mathbb{C}^n$. In order to stay with our notation of Section I.8, the pairing $\langle \ , \ \rangle$ between $Z = \mathbb{C}^n$ and $Z' = \mathbb{C}^n$ is the bilinear complexification of the real scalar product $(\ , \ )$ on $\mathbb{R}^n$. Therefore, $J \in L(\mathbb{R}^n, \mathbb{R}^n)$ is complexified to $J \in L(\mathbb{C}^n, \mathbb{C}^n)$ satisfying $\langle Jz, z \rangle = 0$ or $\langle Jz_1, z_2 \rangle = -\langle z_1, Jz_2 \rangle$ for all $z, z_1, z_2 \in Z = \mathbb{C}^n$. Also, $A_0' = (JB_0)' = B_0'J' = -B_0 J$ has the eigenvalue $i\kappa_0$ with eigenvector $\overline{\varphi}_0$, since $JB_0\varphi_0 = B_0 J\varphi_0$ by assumption $J\varphi_0 = i\varphi_0$. Thus $\varphi_0' = \overline{\varphi}_0$ and $\langle \varphi_0, \varphi_0' \rangle = \langle \varphi_0, \overline{\varphi}_0 \rangle \neq 0$ for $\varphi_0 \neq 0$. From (I.10.10) it follows that

(I.10.11)
$\quad \dim N(\pm i\kappa_0 I - A_0') = 1$, whence
$\quad N(i\kappa_0 I - A_0') = \text{span}[\overline{\varphi}_0]$.

Finally, by $R(i\kappa_0 I - A_0) = \{ z \in Z | \langle z, \overline{\varphi}_0 \rangle = 0 \}$ and $\langle \varphi_0, \overline{\varphi}_0 \rangle \neq 0$, assumption (I.10.10) implies that

(I.10.12)
$\quad i\kappa_0$ is a simple eigenvalue of A_0
$\quad$ in the sense of (I.8.5).

The assumption (I.10.10) is satisfied if B_0 has a double real eigenvalue κ_0 ($\neq 0$) with eigenvectors x_1 and $x_2 = Jx_1$, and $-\kappa_0$ is not an eigenvalue of B_0. Then (I.10.10) holds with $\varphi_0 = -x_1 + ix_2$. Note that $\langle \varphi_0, \varphi_0 \rangle = 0 = \langle B_0\varphi_0, \varphi_0 \rangle$ in this case.

In view of (I.10.12), we can proceed as in Section I.8 using also its notation. Here we have in particular

(I.10.13)
$$\varphi_0' = \overline{\varphi}_0, \ \psi_0' = \overline{\psi}_0,$$
$$\text{and we normalize } \langle \varphi_0, \overline{\varphi}_0 \rangle = 1.$$

Since (I.8.8) is trivially satisfied, we can use Proposition I.8.1 for a Lyapunov–Schmidt reduction of

(I.10.14)
$$G(x, \kappa) \equiv \kappa \frac{dx}{dt} - F(x) = 0, \ \text{where}$$
$$G : \tilde{U} \times \tilde{V} \longrightarrow C_{2\pi}(\mathbb{R}, Z), \ \text{and}$$
$$0 \in \tilde{U} \subset C_{2\pi}^1(\mathbb{R}, Z), \ \kappa_0 \in \tilde{V} \subset \mathbb{R} \ \text{are neighborhoods.}$$

(Observe that in our finite-dimensional setting we can choose $\alpha = 0$ in (I.8.11), and Proposition I.8.1 holds. This saves us one order of differentiability of F in (I.8.13): $G \in C^k(\tilde{U} \times \tilde{V}, C_{2\pi}(\mathbb{R}, Z))$, provided that $F \in C^k(U, Z)$.)

By Theorem I.2.3, equation (I.10.14) is equivalent to the one-dimensional complex equation (cf. (I.8.34))

(I.10.15)
$$\frac{1}{2\pi} \int_0^{2\pi} \langle G(Px + \psi(Px, \kappa), \kappa), \psi_0' \rangle dt = 0,$$
$$\text{or, by definition, to}$$
$$\hat{\Phi}(r, \kappa) = 0 \quad \text{if } Px = r(\psi_0 + \overline{\psi}_0) \ \text{(cf. (I.8.34), (I.8.38)).}$$

Proposition I.10.2 *Under the assumptions of this section,*

(I.10.16) $\text{Re}\hat{\Phi}(r, \kappa) = 0 \quad \text{for all } r \in (-\delta, \delta), \quad \kappa \in \tilde{V}_2 \subset \mathbb{R};$

i.e., the bifurcation equation (I.10.15) is in fact a one-dimensional real equation.

Proof. By the definition of the function ψ (cf. the proof of Theorem I.2.3, in particular (I.2.8)),

(I.10.17)
$$(I - Q)G(Px + \psi(Px, \kappa), \kappa) = 0, \quad \text{or}$$
$$(I - Q)\left(\kappa \frac{d}{dt}(Px + \psi(Px, \kappa)) - F(Px + \psi(Px, \kappa))\right) = 0.$$

Since $Q\frac{d}{dt} = \frac{d}{dt}P$ and $\psi(Px, \kappa) \in N(P)$, this implies

(I.10.18) $\kappa \dfrac{d}{dt}\psi(Px, \kappa) = (I - Q)F(Px + \psi(Px, \kappa)).$

Now,

$$(I.10.19) \quad \hat{\Phi}(r, \kappa) = \frac{1}{2\pi} \int_0^{2\pi} \kappa r \left\langle \frac{d}{dt}(\psi_0 + \overline{\psi}_0), \psi_0' \right\rangle dt$$
$$- \frac{1}{2\pi} \int_0^{2\pi} \langle F(Px + \psi(Px, \kappa)), \psi_0' \rangle dt,$$

since $\frac{d}{dt}\psi(Px, \kappa) \in N(P)$ and $Px = r(\psi_0 + \overline{\psi}_0)$. The first term in (I.10.19) equals $i\kappa r$, so that for real $F(Px + \psi(Px, \kappa))$,

$$(I.10.20) \quad \mathrm{Re}\hat{\Phi}(r, \kappa) = -\frac{1}{2\pi} \int_0^{2\pi} \langle F(Px + \psi(Px, \kappa)), \mathrm{Re}\psi_0' \rangle dt.$$

We show that (I.10.20) vanishes identically. By $J\varphi_0 = i\varphi_0$ and $\psi_0' = \overline{\psi}_0$, we have for $Px = r(\psi_0 + \overline{\psi}_0)$,

$$(I.10.21) \quad \frac{d}{dt}Px = r\mathrm{Re}J\psi_0'.$$

Using $F = J\nabla H$, $J' = -J$, and (I.10.21), we obtain for $r \neq 0$,

$$\mathrm{Re}\hat{\Phi}(r, \kappa) = \frac{1}{2\pi} \int_0^{2\pi} \left\langle \nabla H(Px + \psi(Px, \kappa)), \frac{1}{r}\frac{d}{dt}Px \right\rangle dt$$

$$= \frac{1}{2\pi r} \int_0^{2\pi} \frac{d}{dt} H(Px + \psi(Px, \kappa)) dt$$

$$(I.10.22)$$

$$- \frac{1}{2\pi r} \int_0^{2\pi} \left\langle \nabla H(Px + \psi(Px, \kappa)), \frac{d}{dt}\psi(Px, \kappa) \right\rangle dt$$

$$= \frac{1}{2\pi r \kappa} \int_0^{2\pi} \langle \nabla H(Px + \psi(Px, \kappa)), (I - Q)J\nabla H(./.) \rangle dt$$

by 2π-periodicity and (I.10.18). Next we use $JQ = QJ$ and the following symmetry of the projection Q:

$$(I.10.23) \quad \frac{1}{2\pi} \int_0^{2\pi} \langle Qz_1(t), z_2(t) \rangle dt = \frac{1}{2\pi} \int_0^{2\pi} \langle z_1(t), Qz_2(t) \rangle dt$$
for all $z_1, z_2 \in C_{2\pi}(\mathbb{R}, Z)$.

This gives finally, in view of $(I - Q)^2 = I - Q$,

$$\mathrm{Re}\hat{\Phi}(r, \kappa)$$

$$(I.10.24) \quad = \frac{1}{2\pi r \kappa} \int_0^{2\pi} \langle (I - Q)\nabla H(Px + \psi(Px, \kappa)), J(I - Q)\nabla H(./.) \rangle dt$$

$$= 0 \text{ by } \langle z, Jz \rangle = 0 \text{ for all } z \in Z.$$

Since $\hat{\Phi}(0, \kappa) = 0$, we have shown that $\mathrm{Re}\hat{\Phi}(r, \kappa) = 0$ for all $r \in (-\delta, \delta)$ and for all $\kappa(\neq 0) \in \tilde{V}_2$. $\square$

Proposition I.10.2 clearly implies that $\mathrm{Re}\tilde{\Phi}(r,\kappa) = \mathrm{Re}\hat{\Phi}(r,\kappa)/r \equiv 0$ and nontrivial solutions $(r \neq 0)$ of (I.10.15) are given by

$$(\text{I.10.25}) \qquad\qquad \mathrm{Im}\tilde{\Phi}(r,\kappa) = 0.$$

The computation (I.8.41) gives $D_\kappa(\mathrm{Im}\tilde{\Phi}(0,\kappa_0)) = 1$, and by the Implicit Function Theorem, (I.10.25) is solved by a continuously differentiable function $\kappa : (-\delta,\delta) \to \tilde{V}_2$, $\kappa(0) = \kappa_0$, such that $\mathrm{Im}\tilde{\Phi}(r,\kappa(r)) = 0$ for all $r \in (-\delta,\delta)$. Furthermore, by the oddness of $\hat{\Phi}$ with respect to r (cf. (I.8.37)), we have $\kappa(-r) = \kappa(r)$, and Corollary I.8.3 and Theorem I.9.1 hold as well in a simplified form. We summarize:

Theorem I.10.3 *For a Hamiltonian system*

$$\frac{dx}{dt} = J\nabla H(x)$$

in $X = \mathbb{R}^n$ we assume that $H \in C^3(U,\mathbb{R})$ in a neighborhood U of 0 and $\nabla H(0) = 0$. We make the spectral assumption (I.10.10), which implies that

> $i\kappa_0(\neq 0)$ *is a simple eigenvalue of $A_0 = JD\nabla H(0)$*
> $(D\nabla H = \text{the Hessian of } H)$ *with eigenvector $\varphi_0 \in \mathbb{C}^n$*
> *such that $J\varphi_0 = i\varphi_0$.*

We impose the nonresonance condition:

> *For all $m \in \mathbb{Z}\backslash\{1,-1\}$, $im\kappa_0$ is not an eigenvalue of A_0.*

Then there exists a continuously differentiable curve $\{x(r)|r \in (-\delta,\delta)\}$ of (real) $2\pi/\kappa(r)$-periodic solutions of (I.10.6) through $x(0) = 0$ and $\kappa(0) = \kappa_0$ in $C^1_{2\pi/\kappa(r)}(\mathbb{R},X)$. Furthermore, $\kappa(-r) = \kappa(r)$ and

$$(\text{I.10.26}) \quad x(r)(t) = 2r\mathrm{Re}(\varphi_0 e^{i\kappa(r)t}) + \psi(2r\mathrm{Re}(\varphi_0 e^{i\kappa(r)t}),\kappa(r))$$

such that

$$(\text{I.10.27}) \qquad \begin{aligned} &\frac{d}{dr}x(r)\Big|_{r=0}(t) = 2\mathrm{Re}(\varphi_0 e^{i\kappa_0 t}), \\ &\frac{d}{dr}\kappa(r)\Big|_{r=0} = 0, \quad \frac{d^2}{dr^2}\kappa(r)\Big|_{r=0} = -\mathrm{Im}D^2_{rr}\tilde{\Phi}^0, \end{aligned}$$

where $D^2_{rr}\tilde{\Phi}^0$ is given by (I.9.11) with $F(0,\lambda_0) = J\nabla H(0)$ and $D_x = D$. (For the last formula we need $H \in C^4(U,\mathbb{R})$ and by Proposition I.10.2, the number $D^2_{rr}\tilde{\Phi}^0$ is purely imaginary.)

In Section I.11 we generalize Theorem I.10.3 to the case that $i\kappa_0$ is a simple eigenvalue of A_0 without the restriction $J\varphi_0 = i\varphi_0$ for the eigenvector φ_0; cf. Theorem I.11.4. In that case, Proposition I.10.2 is not necessarily true but is replaced by (I.11.25). The parameter λ in Theorem I.11.4 can be frozen such that the statement is precisely the same as that of Theorem I.10.3.

I.11 Constrained Hopf Bifurcation for Hamiltonian, Reversible, and Conservative Systems

The Hopf Bifurcation Theorem, Theorem I.8.2, for a parameter-dependent evolution equation

$$\text{(I.11.1)} \qquad\qquad \frac{dx}{dt} = F(x, \lambda)$$

in a Banach space Z is proved under the following spectral hypotheses on the Fréchet derivative $D_x F(0, \lambda)$ along the trivial solutions $F(0, \lambda) = 0$:

$$\text{(I.11.2)}$$
$i\kappa_0 (\neq 0)$ is a simple eigenvalue of $D_x F(0, \lambda_0)$, and for the simple eigenvalue perturbation $D_x F(0, \lambda)\varphi(\lambda) = \mu(\lambda)\varphi(\lambda)$ with $\mu(\lambda_0) = i\kappa_0$, the nondegeneracy $\text{Re}\mu'(\lambda_0) \neq 0$ holds.

The algebraically simple eigenvalue $\mu(\lambda)$ crosses the imaginary axis in a nondegenerate way such that by the Principle of Linearized Stability, the trivial solution $x = 0$ loses stability when the parameter λ crosses the critical value λ_0; cf. Section I.7. For physical reasons one expects that a loss of stability "creates" new solutions, and by the absence of bifurcating stationary solutions (zero is not an eigenvalue of $D_x F(0, \lambda_0)$), those are the periodic solutions guaranteed by the Hopf Bifurcation Theorem. (In Section I.12 we show that the periodic solutions actually gain stability.) This scenario, however, is not restricted to the situation described in (I.11.2): Any crossing of eigenvalues of arbitrary multiplicity causes a loss of stability, and the physical expectation about the creation of new solutions is indeed satisfied under quite general assumptions; see Remark I.11.13 below. In this section we do not consider Hopf Bifurcation in its greatest generality, but we admit two slight generalizations of (I.11.2):

(1) $i\kappa_0$ is an algebraically simple eigenvalue of $D_x F(0, \lambda_0)$, but we give up the nondegeneracy $\text{Re}\mu'(\lambda_0) \neq 0$,

(2) $i\kappa_0$ is a geometrically simple eigenvalue of $D_x F(0, \lambda_0)$, but its algebraic multiplicity is arbitrary.

We comment on these generalizations:

(1) In Section I.17 we prove Hopf Bifurcation under the degenerate assumption $\text{Re}\mu'(\lambda_0) = \cdots = \text{Re}\mu^{(m-1)}(\lambda_0) = 0$ but $\text{Re}\mu^{(m)}(\lambda_0) \neq 0$ for some odd m.

That loss of stability of the trivial solution causes the bifurcation of at least one and at most m curves of periodic solutions. We prove also a general Principle of Exchange of Stability for the bunch of periodic orbits including the line of trivial solutions; cf. Section I.17.

As we see below in this section, there are systems (I.11.1) with a structural constraint such that $\mathrm{Re}\mu(\lambda) \equiv 0$ for all λ near λ_0; in other words, the perturbed eigenvalues $\mu(\lambda)$ in (I.11.2) stay necessarily on the imaginary axis. A Constrained Hopf Bifurcation Theorem guarantees bifurcation of periodic solutions in this case, too; see Theorem I.11.2 below. Since the spectral properties are the same for all λ near λ_0, the parameter λ plays no role, and when λ is frozen, Theorem I.11.2 is a Center Theorem as in Section I.10 with the period as a hidden parameter.

(2) The generalization from algebraic to geometric simplicity of the eigenvalue $i\kappa_0$ seems to be quite natural in view of analogous results in stationary bifurcation theory: In Theorem I.5.1, if $X \subset Z$, the eigenvalue 0 of $D_x F(0, \lambda_0)$ is only geometrically simple, and the assumed nondegeneracy (I.5.3) is not related to the eigenvalue perturbation, which might be complicated if the algebraic multiplicity of the eigenvalue zero is large. As discussed in Case 1 of Theorem II.4.4, the nondegeneracy (I.5.3) implies an "odd crossing number"; cf. Definition II.4.1.

If the eigenvalue $i\kappa_0$ is only geometrically simple, we give an analogous nondegeneracy that guarantees bifurcation of periodic solutions. However, in contrast to the stationary case, only systems (I.11.1) with a structural constraint are admitted; see Theorem I.11.2. It turns out that the period can be frozen, and in this case, Theorem I.11.2 describes again a one-parameter bifurcation.

The peculiarity of the Constrained Hopf Bifurcation Theorem proved in this section consists on the one hand in extensions of the hypotheses of the (classical) Hopf Bifurcation Theorem, and on the other hand in restrictions to systems with structural constraints. These constraints are satisfied by Hamiltonian, reversible, and conservative systems.

Accordingly, we replace the assumptions (I.11.2) by

(I.11.3)
$$\dim N(\pm i\kappa_0 I - D_x F(0, \lambda_0)) = 1,$$
$$\pm i\kappa_0 I - D_x F(0, \lambda_0) \text{ are Fredholm operators}$$
$$\text{of index zero.}$$

Note that (I.11.3) replaces only assumption (I.8.5); all other assumptions on F and the notation of Section I.8 are kept. (Assumptions (I.8.6), (I.8.7), however, make no sense under (I.11.3).) In a first step we prove the possibility of a Lyapunov–Schmidt reduction for

$$G(x, \kappa, \lambda) \equiv \kappa \frac{dx}{dt} - F(x, \lambda) = 0$$

(I.11.4)
$$\text{for } (x, \kappa, \lambda) \in \tilde{U} \times \tilde{V},$$
$$0 \in \tilde{U} \subset E \cap Y, \ (\kappa_0, \lambda_0) \in \tilde{V} \subset \mathbb{R}^2;$$

cf. (I.8.12). Note that the proof of Proposition I.8.1 uses the simplicity (I.8.5) of the eigenvalue $i\kappa_0$. As before, we set $A_0 = D_x F(0, \lambda_0)$, and the Fredholm property (I.11.3) implies

(I.11.5)
$$X = N(i\kappa_0 I - A_0) \oplus X_0,$$
$$Z = R(i\kappa_0 I - A_0) \oplus Z_0, \text{ cf. (I.2.2)},$$

where $\dim Z_0 = 1$. We choose

(I.11.6)
$$N(-i\kappa_0 I - A_0) \subset X_0 \text{ and}$$
$$Z_0 \subset R(-i\kappa_0 I - A_0).$$

While the possibility of the first choice is obvious, the second choice is justified as follows. By the Closed Range Theorem,

(I.11.7)
$$\dim N(\pm i\kappa_0 I - A_0') = 1 \text{ and}$$
$$R(i\kappa_0 I - A_0) = \{z \in Z | \langle z, \varphi_0' \rangle = 0\}$$
$$\text{if } N(i\kappa_0 I - A_0') = \text{span}[\varphi_0'].$$

Let $x_0 \in X \backslash R(i\kappa_0 I - A_0)$, which exists, since $X \subset Z$ is dense. Then, by (I.11.7)$_2$, $\langle (-i\kappa_0 I - A_0)x_0, \varphi_0' \rangle = -2i\kappa_0 \langle x_0, \varphi_0' \rangle \neq 0$ and $(-i\kappa_0 I - A_0)x_0 \in R(-i\kappa_0 I - A_0) \backslash R(i\kappa_0 I - A_0)$. Let

(I.11.8)
$$N(i\kappa_0 I - A_0) = \text{span}[\varphi_0],$$
$$Z_0 = \text{span}[\varphi_1], \ \varphi_1 \in R(-i\kappa_0 I - A_0).$$

According to the Hahn–Banach Theorem, there are

(I.11.9)
$$\varphi_1' \in X' \text{ with } \langle \varphi_0, \varphi_1' \rangle = 1 \text{ and}$$
$$\langle x, \varphi_1' \rangle = 0 \text{ for all } x \in X_0,$$
$$\varphi_0' \in Z' \text{ with } \langle \varphi_1, \varphi_0' \rangle = 1 \text{ and}$$
$$\langle z, \varphi_0' \rangle = 0 \text{ for all } z \in R(i\kappa_0 I - A_0).$$

Again by the Closed Range Theorem, $\text{span}[\varphi_0'] = N(i\kappa_0 I - A_0')$, and by the choices of X_0 and Z_0 in (I.11.6),

(I.11.10)
$$\langle \overline{\varphi}_0, \varphi_1' \rangle = 0, \text{ since } \text{span}[\overline{\varphi}_0] = N(-i\kappa_0 I - A_0),$$
$$\langle \varphi_1, \overline{\varphi}_0' \rangle = 0, \text{ since } \text{span}[\overline{\varphi}_0'] = N(-i\kappa_0 I - A_0').$$

The relations (I.11.9), (I.11.10) clearly imply

(I.11.11)
$$\langle \varphi_0, \overline{\varphi}_1' \rangle = 0, \quad \langle \overline{\varphi}_0', \overline{\varphi}_1' \rangle = 1,$$
$$\langle \varphi_1, \overline{\varphi}_0' \rangle = 0, \quad \langle \overline{\varphi}_1, \overline{\varphi}_0' \rangle = 1,$$

so that

$$P_0 x = \langle x, \varphi_1' \rangle \varphi_0 + \langle x, \overline{\varphi}_1' \rangle \overline{\varphi}_0$$
is a projection $P_0 \in L(X, X)$

(I.11.12)   onto $N(i\kappa_0 I - A_0) \oplus N(-i\kappa_0 I - A_0)$, and

$$Q_0 z = \langle z, \varphi_0' \rangle \varphi_1 + \langle z, \overline{\varphi}_0' \rangle \overline{\varphi}_1$$
is a projection $Q_0 \in L(Z, Z)$ onto $Z_0 \oplus \overline{Z}_0$.

Defining

$$\psi_0(t) = \varphi_0 e^{it}, \quad \psi_1'(t) = \varphi_1' e^{-it},$$
$$\psi_1(t) = \varphi_1 e^{it}, \quad \psi_0'(t) = \varphi_0' e^{-it},$$

$$(Px)(t) = \frac{1}{2\pi} \int_0^{2\pi} \langle x(t), \psi_1'(t) \rangle dt \psi_0(t)$$

(I.11.13)

$$+ \frac{1}{2\pi} \int_0^{2\pi} \langle x(t), \overline{\psi}_1'(t) \rangle dt \overline{\psi}_0(t),$$

$$(Qz)(t) = \frac{1}{2\pi} \int_0^{2\pi} \langle z(t), \psi_0'(t) \rangle dt \psi_1(t)$$

$$+ \frac{1}{2\pi} \int_0^{2\pi} \langle z(t), \overline{\psi}_0'(t) \rangle dt \overline{\psi}_1(t),$$

we end up with projections $P \in L(Y \cap E, Y \cap E)$ and $Q \in L(W, W)$. A similar proof to that of Proposition I.8.1 then yields the following result:

Proposition I.11.1 *Assume (I.11.3), (I.8.8), and the nonresonance condition (I.8.14). Then*

(I.11.14) $$J_0 \equiv \kappa_0 \frac{d}{dt} - A_0 : Y \cap E \to W$$

is a (continuous) Fredholm operator of index zero with $\dim N(J_0) = 2$.

Proof. As in (I.8.17) we obtain

(I.11.15) $$N(J_0) = R(P); \quad \text{cf. (I.8.22).}$$

We show that $R(J_0) = N(Q)$. By compactness of $e^{A_0 2\pi/\kappa_0} \in L(Z, Z)$ the operator $I - e^{A_0 2\pi/\kappa_0}$ is a Fredholm operator of index zero. As in (I.8.19),

(I.11.16)

$$N(I - e^{A_0 2\pi/\kappa_0}) = R(P_0) = \text{span}[\varphi_0, \overline{\varphi}_0],$$

$$N(I - (e^{A_0 2\pi/\kappa_0})') = \text{span}[\varphi_0', \overline{\varphi}_0'],$$

cf. Section I.8, after (I.8.25), and by the Closed Range Theorem,

(I.11.17) $$R(I - e^{A_0 2\pi/\kappa_0}) = \{z \in Z \,|\, \langle z, \varphi_0' \rangle = \langle z, \overline{\varphi}_0' \rangle = 0\}$$
$$= N(Q_0).$$

The proof that $R(J_0) = N(Q)$ is then the same as in (I.8.25)–(I.8.27).   □

Proposition I.11.1 gives a suitable Lyapunov–Schmidt decomposition

$$\text{(I.11.18)} \qquad \begin{aligned} Y \cap E &= N(J_0) \oplus N(P), \\ W &= R(J_0) \oplus R(Q), \end{aligned}$$

with projections $P : Y \cap E \to N(J_0)$ and $Q : W \to R(Q)$ as given in (I.11.13). Thus, as in Section I.8, the problem

$$\text{(I.11.19)} \qquad G(x, \kappa, \lambda) \equiv \kappa \frac{dx}{dt} - F(x, \lambda) = 0 \in W \quad \text{near } (0, \kappa_0, \lambda_0)$$

in $(Y \cap E) \times \mathbb{R} \times \mathbb{R}$ is equivalent to

$$\text{(I.11.20)} \qquad QG(Px + \psi(Px, \kappa, \lambda), \kappa, \lambda) = 0.$$

For a real function (I.8.33) the bifurcation equation (I.11.20), in turn, is equivalent to the complex equation

$$\text{(I.11.21)} \qquad \begin{aligned} &\frac{1}{2\pi} \int_0^{2\pi} \langle G(Px + \psi(Px, \kappa, \lambda), \kappa, \lambda), \psi_0' \rangle dt = 0, \quad \text{or} \\ &\hat{\Phi}(c, \kappa, \lambda) = 0 \quad \text{if } (Px)(t) = c\varphi_0 e^{it} + \bar{c}\,\overline{\varphi}_0 e^{-it}, \ c \in \mathbb{C}; \end{aligned}$$

cf. (I.8.35). Since both projections P and Q commute with the phase shift S_θ, the equivariance of G with respect to S_θ implies

$$\text{(I.11.22)} \qquad \hat{\Phi}(e^{i\theta}c, \kappa, \lambda) = e^{i\theta}\hat{\Phi}(c, \kappa, \lambda) \quad \text{for } \theta \in [0, 2\pi);$$

see (I.8.31)–(I.8.36). Therefore, the remaining calculations (I.8.37)–(I.8.41) hold literally also in the general case of this section, except, however, that

$$\text{(I.11.23)} \qquad D_\kappa \tilde{\Phi}(0, \kappa_0, \lambda_0) = i\langle \varphi_0, \varphi_0' \rangle$$

might be zero ($\tilde{\Phi}(r, \kappa, \lambda) = \hat{\Phi}(r, \kappa, \lambda)/r$; see (I.8.39)). As a matter of fact,

$$\text{(I.11.24)} \qquad \begin{aligned} &\langle \varphi_0, \varphi_0' \rangle \neq 0 \ (= 1 \text{ w.l.o.g.}) \Leftrightarrow \\ &i\kappa_0 \text{ is a simple eigenvalue of } A_0; \text{ cf. (I.8.5).} \end{aligned}$$

Therefore, in case of an algebraically nonsimple eigenvalue $i\kappa_0$ of A_0, the derivative $D_{(\kappa,\lambda)}\tilde{\Phi}(0, \kappa_0, \lambda_0)$ (cf. (I.8.42)) is singular.

On the other hand, if $i\kappa_0$ is an algebraically simple eigenvalue of A_0 but $\mathrm{Re}\mu'(\lambda_0) = 0$ for the eigenvalue perturbation (I.11.2), then, in view of (I.8.44), the derivative $D_{(\kappa,\lambda)}\tilde{\Phi}(0, \kappa_0, \lambda_0)$ is singular, too (cf. (I.8.42)).

Therefore, assumptions (I.11.2) are indispensable in treating the Hopf Bifurcation in the spirit of Section I.8. (They are clearly dispensable for the general Hopf Bifurcation as expounded in Remark I.11.13 below.) Nonetheless, we can follow our path and solve (I.11.21) nontrivially by the Implicit Function Theorem under additional constraints on F.

In Section I.10 we learn that a Hamiltonian structure reduces the complex equation (I.11.21) to a real bifurcation equation. This allows us to release one parameter, and the period is used as a hidden parameter. It turns out that the special case considered in Section I.10 allows a fruitful generalization to systems that play an important role in applications. Before we give the special classes of systems to which our method applies, we give the general **Constrained Hopf Bifurcation Theorem:**

Theorem I.11.2 *For the system (I.11.1) we make the general assumptions (I.8.2), (I.8.3), (I.8.4), (I.11.3), (I.8.8), (I.8.13), (I.8.14), so that the problem of finding real periodic solutions with small amplitude and periods near $2\pi/\kappa_0$ is reduced via (I.11.4) to the complex equation $\hat{\Phi}(r, \kappa, \lambda) = 0$ for (r, κ, λ) near $(0, \kappa_0, \lambda_0)$ in $\mathbb{R}^3$; cf. (I.11.21).*

If $i\kappa_0$ is an algebraically simple eigenvalue of $A_0 = D_x F(0, \lambda_0)$, assume that a constraint on F implies

$$(I.11.25) \qquad \begin{aligned} &\mathrm{Im}\hat{\Phi}(r, \kappa, \lambda) = 0 \ \Rightarrow \ \mathrm{Re}\hat{\Phi}(r, \kappa, \lambda) = 0 \\ &\text{for all } (r, \kappa, \lambda) \text{ near } (0, \kappa_0, \lambda_0). \end{aligned}$$

Then without any further assumption there is a continuously differentiable surface $\{(x(r, \lambda), \lambda) | r \in (-\delta, \delta), \lambda \in (\lambda_0 - \delta, \lambda_0 + \delta)\}$ of nontrivial (real) $2\pi/\kappa(r, \lambda)$-periodic solutions of (I.11.1) through $(x(0, \lambda), \lambda) = (0, \lambda)$ and $\kappa(0, \lambda_0) = \kappa_0$ in $(C^{1+\alpha}_{2\pi/\kappa(r,\lambda)}(\mathbb{R}, Z) \cap C^{\alpha}_{2\pi/\kappa(r,\lambda)}(\mathbb{R}, X)) \times \mathbb{R}$. Furthermore, $\kappa(-r, \lambda) = \kappa(r, \lambda)$, and $x(-r, \lambda)$ is obtained from $x(r, \lambda)$ by a phase shift of half the period $\pi/\kappa(r, \lambda)$.

If $i\kappa_0$ is a geometrically but not necessarily algebraically simple eigenvalue of A_0, assume that a constraint on F implies

$$(I.11.26) \qquad \begin{aligned} &\mathrm{Im}\hat{\Phi}(r, \kappa, \lambda) = 0 \Rightarrow \mathrm{Re}\hat{\Phi}(r, \kappa, \lambda) = 0, \text{ or} \\ &\mathrm{Re}\hat{\Phi}(r, \kappa, \lambda) = 0 \Rightarrow \mathrm{Im}\hat{\Phi}(r, \kappa, \lambda) = 0 \\ &\text{for all } (r, \kappa, \lambda) \text{ near } (0, \kappa_0, \lambda_0). \end{aligned}$$

According to the two cases (I.11.26), assume a nondegeneracy

$$(I.11.27) \qquad \begin{aligned} &\mathrm{Im}\langle D^2_{x\lambda} F(0, \lambda_0)\varphi_0, \varphi'_0 \rangle \neq 0 \text{ or} \\ &\mathrm{Re}\langle D^2_{x\lambda} F(0, \lambda_0)\varphi_0, \varphi'_0 \rangle \neq 0. \end{aligned}$$

Then there exists a continuously differentiable surface $\{(x(r, \kappa), \lambda(r, \kappa)) | r \in (-\delta, \delta), \kappa \in (\kappa_0 - \delta, \kappa_0 + \delta)\}$ of nontrivial (real) $2\pi/\kappa$-periodic solutions of (I.11.1) through $(x(0, \kappa), \lambda(0, \kappa)) = (0, \lambda(0, \kappa))$ and $\lambda(0, \kappa_0) = \lambda_0$ in the space $(C^{1+\alpha}_{2\pi/\kappa}(\mathbb{R}, Z) \cap C^{\alpha}_{2\pi/\kappa}(\mathbb{R}, X)) \times \mathbb{R}$. Furthermore, $\lambda(-r, \kappa) = \lambda(r, \kappa)$ and $x(-r, \lambda)$ is obtained from $x(r, \lambda)$ by a phase shift of half the period π/κ.

Proof. As in Section I.8 we set $\tilde{\Phi}(r, \kappa, \lambda) = \hat{\Phi}(r, \kappa, \lambda)/r$ for $r \neq 0$; cf. (I.8.39). By assumptions (I.11.25), (I.11.26), the real bifurcation equation

$$(I.11.28) \qquad \mathrm{Im}\tilde{\Phi}(r, \kappa, \lambda) = 0 \quad \text{or} \quad \mathrm{Re}\tilde{\Phi}(r, \kappa, \lambda) = 0$$

can be solved by the Implicit Function Theorem, since

(I.11.29)
$$D_\kappa(\mathrm{Im}\tilde{\Phi}(0,\kappa_0,\lambda_0)) = 1 \text{ in the first case and}$$
$$D_\lambda\tilde{\Phi}(0,\kappa_0,\lambda_0) = -\langle D^2_{x\lambda}F(0,\lambda_0)\varphi_0,\varphi_0'\rangle$$

in the second case; cf. (I.8.41). The oddness of $\hat{\Phi}$ with respect to r implies the evenness of κ and λ with respect to r, and it is explained after Theorem I.8.2 that the transition from r to $-r$ corresponds to a phase shift of half the period. □

More bifurcation formulas are given when Theorem I.11.2 is applied to special classes below. Here we make two remarks on the respective surfaces of periodic solutions:

In the first case, the parameter λ can be frozen in $(\lambda_0 - \delta, \lambda_0 + \delta)$, and the vertical Hopf Bifurcation describes a **Center Theorem** for systems (I.11.1) for each fixed λ near λ_0.

In the second case, the parameter κ can be frozen in $(\kappa_0 - \delta, \kappa_0 + \delta)$, and each point on the "bifurcation curve" $\{(\kappa, \lambda(0, \kappa))\}$ in the (κ, λ)-plane gives rise to a bifurcation of $2\pi/\kappa$-periodic solutions of system (I.11.1) with $\lambda = \lambda(r, \kappa)$.

I.11.1 Hamiltonian Systems: Lyapunov Center Theorem and Hamiltonian Hopf Bifurcation

We return to the Hamiltonian systems introduced in Section I.10, but this time the Hamiltonian depends on a real parameter λ. For the reason explained in Remark I.10.1 we confine ourselves to $X = Z = \mathbb{R}^n$, and we assume that

(I.11.30)
$$H : U \times V \to \mathbb{R} \quad \text{is in } C^4(U \times V, \mathbb{R}) \text{ for}$$
$$0 \in U \subset \mathbb{R}^n, \lambda_0 \in V \subset \mathbb{R},$$
$$F(x,\lambda) = J\nabla_x H(x,\lambda) \quad \text{for } (x,\lambda) \in U \times V,$$
$$F(0,\lambda) = 0 \text{ for all } \lambda \in V,$$

where the gradient "∇_x" refers to the variable $x \in U$. (By the remark after (I.10.14), $F \in C^3(U \times V, \mathbb{R}^n)$ is enough for our analysis.) We define

(I.11.31)
$$A(\lambda) = D_x F(0,\lambda) = JD_x\nabla_x H(0,\lambda) = JB(\lambda),$$
$$A(\lambda_0) = A_0 = JB_0,$$

where $B(\lambda), B_0 \in L(\mathbb{R}^n, \mathbb{R}^n)$ are the "Hessians" of $H(0,\lambda)$, $H(0,\lambda_0)$, respectively, and are symmetric with respect to the chosen scalar product $(\ ,\ )$.

Before proving (I.11.25), (I.11.26)$_1$ we make an observation on the spectrum $\sigma(A)$ of any $A = JB$ of the form (I.11.31):

$$(I.11.32) \qquad \begin{aligned} &\mu = \alpha + i\beta \in \sigma(A) \Leftrightarrow \\ &\bar{\mu} = \alpha - i\beta, \ -\mu = -\alpha - i\beta \in \sigma(A), \end{aligned}$$

so that all eigenvalues that are not real and not purely imaginary appear as quadruplets.

For a proof, consider the dual $A' = -BJ$, and clearly, $\mu \in \sigma(A')$. Then $-BJ\varphi' = \mu\varphi'$ and $JBJ\varphi' = -\mu J\varphi'$, so that $-\mu$ is an eigenvalue of $A = JB$ with eigenvector $J\varphi'$. By the reality of A, we clearly have $A\varphi = \mu\varphi$ and $A\bar{\varphi} = \bar{\mu}\bar{\varphi}$.

Remark I.11.3 *Property (I.11.32) has the consequence that an algebraically simple eigenvalue $\mu(\lambda_0) = i\kappa_0$ of $A(\lambda_0)$ is perturbed to a simple eigenvalue $\mu(\lambda)$ of $A(\lambda)$ according to (I.8.6) that stays necessarily on the imaginary axis for all $\lambda \in (\lambda_0 - \delta, \lambda_0 + \delta)$. Therefore, the transversality $\operatorname{Re}\mu'(\lambda_0) \neq 0$ required in (I.8.7) is excluded, which means that a Hopf Bifurcation in the sense of Section I.8 (or in a degenerate sense of Section I.17) cannot occur to a Hamiltonian system.*

The bifurcation function (I.11.20) or in its reduced form (I.11.21) for real Px depends on the projections P and Q, which, in turn, are defined by the vectors φ, φ_0', φ_1, φ_1'; cf. (I.11.13). Properties (I.11.25), (I.11.26)$_1$ depend, therefore, on a suitable choice of these vectors.

Choose φ_0 according to (I.11.8). Then

$$(I.11.33) \qquad \begin{aligned} &A_0' J\bar{\varphi}_0 = -B_0 JJ\bar{\varphi}_0 = B_0\bar{\varphi}_0 = i\kappa_0 J\bar{\varphi}_0 \text{ by} \\ &A_0\bar{\varphi}_0 = JB_0\bar{\varphi}_0 = -i\kappa_0\bar{\varphi}_0. \end{aligned}$$

Therefore, $J\bar{\varphi}_0$ is an eigenvector of A_0' with eigenvalue $i\kappa_0$.

Let $i\kappa_0$ be an algebraically simple eigenvalue of A_0.

In this case, we have to normalize $\langle \varphi_0, \varphi_0' \rangle = 1$. By $\langle \varphi_0, J\bar{\varphi}_0 \rangle = -\langle \bar{\varphi}_0, J\varphi_0 \rangle$ this product is purely imaginary, and we choose

$$(I.11.34) \qquad \begin{aligned} &\varphi_0' = i\beta J\bar{\varphi}_0 \quad \text{with} \ \beta = -i\langle \varphi_0, J\bar{\varphi}_0 \rangle^{-1} \in \mathbb{R}, \\ &\varphi_1 = \varphi_0, \ \varphi_1' = \varphi_0'. \end{aligned}$$

(Recall that $\langle \ , \ \rangle$ is the bilinear complexification of $(\ , \ )$.)

Let $i\kappa_0$ not be an algebraiclly simple eigenvalue of A_0.

This means that $\langle \varphi_0, \varphi_0' \rangle = 0$, and we can choose $\varphi_0' = cJ\bar{\varphi}_0$ with any $c \in \mathbb{C} \backslash \{0\}$. We take

$$(I.11.35) \qquad \varphi_0' = iJ\bar{\varphi}_0, \ \varphi_1, \varphi_1' \text{ as in (I.11.8), (I.11.9).}$$

The infinite-dimensional equation of the Lyapunov–Schmidt decomposition (cf.(I.2.7), (I.2.8)),

$$(I.11.36) \qquad (I - Q)G(Px + \psi(Px, \kappa, \lambda), \kappa, \lambda) = 0,$$

yields the identity

$$\kappa \frac{d}{dt}\left(Px + \psi(Px, \kappa, \lambda)\right)$$

(I.11.37)
$$= \kappa Q \frac{d}{dt}(Px + \psi(Px, \kappa, \lambda)) + (I - Q)F(Px + \psi(Px, \kappa, \lambda), \lambda).$$

By 2π-periodicity,

$$\int_0^{2\pi} \frac{d}{dt}H(Px + \psi(Px, \kappa, \lambda), \lambda)dt = 0$$

(I.11.38)
$$= \int_0^{2\pi} \left\langle \nabla_x H(Px + \psi(Px, \kappa, \lambda), \lambda), \frac{d}{dt}(Px + \psi(Px, \kappa, \lambda))\right\rangle dt.$$

Inserting (I.11.37) into (I.11.38) gives by (I.11.31),

$$\int_0^{2\pi} \langle \nabla_x H(Px + \psi(Px, \kappa, \lambda), \lambda), QG(Px + \psi(Px, \kappa, \lambda), \kappa, \lambda)\rangle dt$$

(I.11.39)
$$+ \int_0^{2\pi} \langle \nabla_x H(Px + \psi(Px, \kappa, \lambda), \lambda), J\nabla_x H(./.)\rangle dt = 0.$$

In view of $\langle z, Jz \rangle = 0$ for all $z \in Z$, the second term in (I.11.39) vanishes, and the first term, involving the bifurcation function $QG(Px + \psi(Px, \kappa, \lambda), \kappa, \lambda)$, cf. (I.2.9), is of the following form when the projection Q is used explicitly, cf. (I.11.13):

$$\frac{1}{2\pi} \int_0^{2\pi} \langle \nabla_x H(Px + \psi(Px, \kappa, \lambda), \lambda), \psi_1\rangle dt$$

(I.11.40)
$$\times \int_0^{2\pi} \langle G(Px + \psi(Px, \kappa, \lambda), \kappa, \lambda), \psi_0'\rangle dt$$

$$+ \frac{1}{2\pi} \int_0^{2\pi} \langle \nabla_x H(./.), \overline{\psi}_1\rangle dt \int_0^{2\pi} \langle G(./.), \overline{\psi}_0'\rangle dt = 0.$$

For real $Px = r(\psi_0 + \overline{\psi}_0)$ the terms $\nabla_x H(Px + \psi(Px, \kappa, \lambda), \lambda)$ and $G(Px + \psi(Px, \kappa, \lambda), \kappa, \lambda)$ are real, so that the second summand in (I.11.40) is the complex conjugate to the first one. If we define

(I.11.41)
$$h(r, \kappa, \lambda) \equiv \frac{1}{2\pi} \int_0^{2\pi} \langle \nabla_x H(Px + \psi(Px, \kappa, \lambda), \lambda), \psi_1\rangle dt$$
$$\text{for } Px = r(\psi_0 + \overline{\psi}_0), \ r \in (-\delta, \delta), (\kappa, \lambda) \in \tilde{V}_2 \subset \mathbb{R}^2,$$

where $\tilde{V}_2$ is a neighborhood of (κ_0, λ_0), the identity (I.11.40) implies, in view of (I.11.21),

$$\text{Re}[h(r, \kappa, \lambda)\hat{\Phi}(r, \kappa, \lambda)] = 0 \text{ or}$$

(I.11.42)
$$\text{Re}h(r, \kappa, \lambda)\text{Re}\hat{\Phi}(r, \kappa, \lambda) = \text{Im}h(r, \kappa, \lambda)\text{Im}\hat{\Phi}(r, \kappa, \lambda)$$
$$\text{for all } r \in (-\delta, \delta), (\kappa, \lambda) \in \tilde{V}_2 \subset \mathbb{R}^2.$$

By definition (I.11.41) we have $h(0, \kappa, \lambda) = 0$ and

$$
\begin{aligned}
D_r h(0, \kappa_0, \lambda_0) &= \frac{1}{2\pi} \int_0^{2\pi} \langle B_0(\psi_0 + \overline{\psi}_0), \psi_1 \rangle dt \\
&= \langle B_0 \overline{\varphi}_0, \varphi_1 \rangle = i\kappa_0 \langle J\overline{\varphi}_0, \varphi_1 \rangle \text{ by (I.11.33)} \\
&= \kappa_0 \beta^{-1} \langle \varphi_1, \varphi_0' \rangle \text{ by (I.11.34) or (I.11.35) with } \beta = 1 \\
&= \kappa_0 \beta^{-1} \neq 0 \text{ and real by } \langle \varphi_1, \varphi_0' \rangle = 1.
\end{aligned}
$$

(I.11.43)

(For (I.11.43)$_1$ observe that $\psi(0, \kappa, \lambda) = 0$, $D_v \psi(0, \kappa_0, \lambda_0) = 0$ for $v = Px$; cf. Corollary I.2.4, and $B_0 = D_x \nabla_x H(0, \lambda_0)$.) Therefore, $\mathrm{Re} h(r, \kappa, \lambda) \neq 0$ for $r \in (-\delta, \delta) \backslash \{0\}$ and $(\kappa, \lambda) \in \tilde{V}_2$, which proves (I.11.25), (I.11.26)$_1$ in view of (I.11.42).

For the nondegeneracy (I.11.27) we compute

$$
\begin{aligned}
\langle D_{x\lambda}^2 F(0, \lambda_0)\varphi_0, \varphi_0' \rangle &\\
= \langle JD_{x\lambda}^2 \nabla_x H(0, \lambda_0)\varphi_0, \varphi_0' \rangle &= \left\langle J\frac{d}{d\lambda}B(\lambda_0)\varphi_0, \varphi_0' \right\rangle \\
= i\beta\left\langle J\frac{d}{d\lambda}B(\lambda_0)\varphi_0, J\overline{\varphi}_0 \right\rangle &\text{ by (I.11.34) or (I.11.35)} \\
= i\beta\left\langle \frac{d}{d\lambda}B(\lambda_0)\varphi_0, \overline{\varphi}_0 \right\rangle, &\text{ which is purely imaginary}
\end{aligned}
$$

(I.11.44)

by the symmetry of $B(\lambda)$ and therefore of $\frac{d}{d\lambda}B(\lambda_0)$.

The first case of Theorem I.11.2 then implies the **Lyapunov Center Theorem**, generalizing Theorem I.10.3:

Theorem I.11.4 *For a parameter-dependent Hamiltonian system*

$$
\text{(I.11.45)} \qquad \frac{dx}{dt} = J\nabla_x H(x, \lambda) \quad \text{in } \mathbb{R}^n,
$$

we assume that $\nabla_x H(0, \lambda) = 0$, that $i\kappa_0(\neq 0)$ is an algebraically simple eigenvalue of $A_0 = JD_x\nabla_x H(0, \lambda_0)$, and that for all $m \in \mathbb{Z}\backslash\{1, -1\}$, $im\kappa_0$ is not an eigenvalue of A_0. Then there exists a continuously differentiable surface $\{(x(r, \lambda), \lambda)|r \in (-\delta, \delta), \lambda \in (\lambda_0 - \delta, \lambda_0 + \delta)\}$ of nontrivial (real) $2\pi/\kappa(r, \lambda)$-periodic solutions of (I.11.45) through $(x(0, \lambda), \lambda) = (0, \lambda)$ and $\kappa(0, \lambda_0) = \kappa_0$ in $C_{2\pi/\kappa(r,\lambda)}^1(\mathbb{R}, \mathbb{R}^n) \times \mathbb{R}$ with the properties stated in Theorem I.11.2.

The surface is fibered into vertically bifurcating curves of $2\pi/\kappa(r, \lambda)$-periodic solutions for each fixed λ near λ_0. As stated in Remark I.11.3, the simple eigenvalue perturbation $\mu(\lambda)$ of $A(\lambda) = JD_x\nabla_x H(0, \lambda)$ consists of purely imaginary eigenvalues near $i\kappa_0$ given by $i\kappa(0, \lambda)$. Therefore, clearly, $\frac{d}{d\lambda}\kappa(0, \lambda_0) = \mathrm{Im}\mu'(\lambda_0)$; cf. (I.11.53) below. For the parameter-dependent Hamiltonian systems it is not necessary to distinguish between λ_0 and $\lambda \in (\lambda_0 - \delta, \lambda_0 + \delta)$, since the spectral properties of $A(\lambda)$ are identical for all

$\lambda \in (\lambda_0 - \delta, \lambda_0 + \delta)$. Therefore, Lyapunov's Center Theorem is usually stated without any parameter.

The second case of Theorem I.11.2 implies the **Hamiltonian Hopf Bifurcation:**

Theorem I.11.5 *If under the same hypotheses as stated in Theorem I.11.4 the eigenvalue $i\kappa_0$ of A_0 is geometrically but not necessarily algebraically simple, and if the nondegeneracy of the Hessian*

$$(I.11.46) \qquad \left\langle \frac{d}{d\lambda} D_x \nabla_x H(0, \lambda_0) \varphi_0, \overline{\varphi}_0 \right\rangle \neq 0$$

holds for the eigenvector φ_0 of A_0 with eigenvalue $i\kappa_0$, then there exists a continuously differentiable surface $\{(x(r, \kappa), \lambda(r, \kappa)) \mid r \in (-\delta, \delta), \ \kappa \in (\kappa_0 - \delta, \ \kappa_0 + \delta)\}$ of nontrivial (real) $2\pi/\kappa$-periodic solutions of (I.11.45) through $(x(0, \kappa), \lambda(0, \kappa)) = (0, \lambda(0, \kappa))$ with $\lambda(0, \kappa_0) = \lambda_0$ in $C^1_{2\pi/\kappa}(\mathbb{R}, \mathbb{R}^n) \times \mathbb{R}$ with the properties stated in Theorem I.11.2. Note that the algebraic multiplicity of the eigenvalue $i\kappa_0$ is arbitrary.

As mentioned after Theorem I.11.2, for any fixed $\kappa \in (\kappa_0 - \delta, \kappa_0 + \delta)$ each point on the bifurcation curve $\{(\kappa, \lambda(0, \kappa))\}$ in the (κ, λ)-plane gives rise to a curve of $2\pi/\kappa$-periodic solutions of (I.11.45) with $\lambda = \lambda(r, \kappa)$, and the surface stated in Theorem I.11.5 is fibered into these curves. The shape of the bifurcation curve follows from **Bifurcation Formulas.**

Theorem I.11.6 *For the surface of the Lyapunov Center Theorem, Theorem I.11.4, the following formulas hold (if the Hamiltonian H is in $C^4(U \times V, \mathbb{R})$; cf. (I.11.30)):*

$$\frac{d}{dr} x(r, \lambda)\big|_{(r,\lambda)=(0,\lambda)}(t) = 2\mathrm{Re}(\varphi_0 e^{i\kappa(0,\lambda)t}),$$

$$(I.11.47) \quad \frac{d}{dr} \kappa(r, \lambda)\big|_{(r,\lambda)=(0,\lambda)} = 0, \quad \frac{d^2}{dr^2} \kappa(r, \lambda)\big|_{(r,\lambda)=(0,\lambda_0)} = -\mathrm{Im} D^2_{rr} \tilde{\Phi}^0,$$

$$\frac{d}{d\lambda} \kappa(r, \lambda)\big|_{(r,\lambda)=(0,\lambda_0)} = -\frac{\langle \frac{d}{d\lambda} D_x \nabla_x H(0, \lambda_0) \varphi_0, \overline{\varphi}_0 \rangle}{\mathrm{Im}\langle \varphi_0, J\overline{\varphi}_0 \rangle},$$

where the quantity $D^2_{rr} \tilde{\Phi}^0$ is given by (I.9.11) with φ'_0 as defined in (I.11.34).

Proof. Formula (I.11.47)$_1$ follows from

$$(I.11.48) \quad x(r, \lambda)(t) = 2r\mathrm{Re}(\varphi_0 e^{i\kappa(r,\lambda)t}) + \psi(2r\mathrm{Re}(\varphi_0 e^{i\kappa(r,\lambda)t}), \kappa(r, \lambda), \lambda)$$

by its construction (cf. (I.8.47)), where $\kappa(r, \lambda)$ satisfies the bifurcation equation

$$(I.11.49) \quad \mathrm{Im}\tilde{\Phi}(r, \kappa(r, \lambda), \lambda) = 0 \quad \text{for all } r \in (-\delta, \delta), \lambda \in (\lambda_0 - \delta, \lambda_0 + \delta).$$

Since $\kappa(-r, \lambda) = \kappa(r, \lambda)$, we obtain as in (I.9.1),

(I.11.50)
$$\frac{d^2}{dr^2}\text{Im}\tilde{\Phi}(r, \kappa(r, \lambda), \lambda)\big|_{(r,\lambda)=(0,\lambda_0)}$$
$$= \text{Im}D^2_{rr}\tilde{\Phi}^0 + D_\kappa(\text{Im}\tilde{\Phi}(0, \kappa_0, \lambda_0))\frac{d^2}{dr^2}\kappa(0, \lambda_0) = 0,$$

which implies $(I.11.47)_2$ by $(I.11.29)_1$. Differentiating (I.11.49) with respect to λ gives

(I.11.51)
$$\frac{d}{d\lambda}\kappa(0, \lambda_0) + D_\lambda(\text{Im}\tilde{\Phi}(0, \kappa_0, \lambda_0)) = 0,$$

proving $(I.11.47)_3$ by $(I.11.29)_2$ and (I.11.44) with (I.11.34). $\square$

Remark I.11.7 *As stated in Remark I.11.3, the simple eigenvalue perturbation $\mu(\lambda)$ of $A(\lambda) = JD_x\nabla_x H(0, \lambda)$ with $\mu(\lambda_0) = i\kappa_0$ is purely imaginary. By (I.8.44) and (I.11.44),*

(I.11.52)
$$\mu'(\lambda_0) = i\beta\Big\langle \frac{d}{d\lambda}D_x\nabla_x H(0, \lambda_0)\varphi_0, \overline{\varphi}_0\Big\rangle \in i\mathbb{R},$$

so that formula $(I.11.47)_3$ can be restated as

(I.11.53)
$$\frac{d}{d\lambda}\kappa(r, \lambda)\big|_{(r,\lambda)=(0,\lambda_0)} = \text{Im}\frac{d}{d\lambda}\mu(\lambda)\big|_{\lambda=\lambda_0}.$$

Theorem I.11.8 *For the surface of the Hamiltonian Hopf Bifurcation Theorem, Theorem I.11.5, the following formulas hold (in case $H \in C^4(U \times V, \mathbb{R})$):*

(I.11.54)
$$\frac{d}{dr}x(r, \kappa)\big|_{(r,\kappa)=(0,\kappa)}(t) = 2\text{Re}(\varphi_0 e^{i\kappa t}),$$
$$\frac{d}{dr}\lambda(r, \kappa)\big|_{(r,\kappa)=(0,\kappa)} = 0,$$
$$\frac{d^2}{dr^2}\lambda(r, \kappa)\big|_{(r,\kappa)=(0,\kappa_0)} = \frac{\text{Im}D^2_{rr}\tilde{\Phi}^0}{\langle\frac{d}{d\lambda}D_x\nabla_x H(0, \lambda_0)\varphi_0, \overline{\varphi}_0\rangle},$$
$$\frac{d}{d\kappa}\lambda(r, \kappa)\big|_{(r,\kappa)=(0,\kappa_0)} = 0,$$
$$\frac{d^2}{d\kappa^2}\lambda(r, \kappa)\big|_{(r,\kappa)=(0,\kappa_0)} = \frac{2\langle\varphi_0^1, J\overline{\varphi}_0\rangle}{\langle\frac{d}{d\lambda}D_x\nabla_x H(0, \lambda_0)\varphi_0, \overline{\varphi}_0\rangle},$$

where $(i\kappa_0 I - A_0)\varphi_0^1 = \varphi_0$. If $i\kappa_0$ is an algebraically double eigenvalue of A_0, then the last second derivative with respect to κ is nonzero. The quantity $D^2_{rr}\tilde{\Phi}^0$ is given in (I.9.11) with φ_0' as in (I.11.35).

Proof. By its construction,

(I.11.55) $\text{Im}\tilde{\Phi}(r, \kappa, \lambda(r, \kappa)) = 0$ for all $r \in (-\delta, \delta), \kappa \in (\kappa_0 - \delta, \kappa_0 + \delta).$

Since $\lambda(-r, \kappa) = \lambda(r, \kappa)$, we obtain as in (I.9.1),

$$\text{(I.11.56)} \qquad \operatorname{Im}D^2_{rr}\tilde{\Phi}^0 + \operatorname{Im}D_\lambda\tilde{\Phi}^0 \frac{d^2}{dr^2}\lambda(0,\kappa_0) = 0,$$

which proves (I.11.54)$_3$ by (I.11.29)$_2$, (I.11.44), (I.11.35).

If $i\kappa_0$ is not an algebraically simple eigenvalue of A_0, then $\operatorname{Im}D_\kappa\tilde{\Phi}^0 = 0$, cf. (I.11.23), (I.11.24), which implies (I.11.54)$_4$. Therefore,

$$\text{(I.11.57)} \qquad \operatorname{Im}D^2_{\kappa\kappa}\tilde{\Phi}^0 + \operatorname{Im}D_\lambda\tilde{\Phi}^0 \frac{d^2}{d\kappa^2}\lambda(0,\kappa_0) = 0.$$

For the computation of $D^2_{\kappa\kappa}\tilde{\Phi}^0$ we proceed as in Sections I.6, I.9. By the definition $\tilde{\Phi}(r,\kappa,\lambda) = \hat{\Phi}(r,\kappa,\lambda)/r$ we have $D^2_{\kappa\kappa}\tilde{\Phi}^0 = D^3_{r\kappa\kappa}\hat{\Phi}^0$. Following (I.9.4),

$$\text{(I.11.58)} \qquad \begin{aligned} \Phi(v,\kappa,\lambda) &= QG(v + \psi(v,\kappa,\lambda),\kappa,\lambda), \\ v &= r(\psi_0 + \overline{\psi}_0)\,, \ \hat{v}_0 = \psi_0 + \overline{\psi}_0, \end{aligned}$$

we get by definition (I.11.21),

$$\text{(I.11.59)} \qquad \begin{aligned} \hat{\Phi}(r,\kappa,\lambda) &= \frac{1}{2\pi}\int_0^{2\pi}\langle\Phi(v,\kappa,\lambda),\psi'_0\rangle dt, \ \text{whence} \\ D^3_{r\kappa\kappa}\hat{\Phi}^0 &= \frac{1}{2\pi}\int_0^{2\pi}\langle D_{v\kappa\kappa}\Phi(0,\kappa_0,\lambda_0)\hat{v}_0,\psi'_0\rangle dt. \end{aligned}$$

By $G(x,\kappa,\lambda) = \kappa\frac{d}{dt}x - F(x,\lambda)$ we have $D^3_{x\kappa\kappa}G(x,\kappa,\lambda) = 0$, and therefore (use (I.5.12) replacing λ by κ),

$$\text{(I.11.60)} \quad D^3_{v\kappa\kappa}\Phi(0,\kappa_0,\lambda_0)\hat{v}_0 = 2QD^2_{x\kappa}G(0,\kappa_0,\lambda_0)D^2_{v\kappa}\psi(0,\kappa_0,\lambda_0)\hat{v}_0.$$

Since $\psi(v,\kappa,\lambda)$ solves $(I-Q)G(v+\psi(v,\kappa,\lambda),\kappa,\lambda) = 0$, we obtain as in (I.6.6)–(I.6.8),

$$\text{(I.11.61)} \quad \begin{aligned} &D^2_{v\kappa}\psi(0,\kappa_0,\lambda_0)\hat{v}_0 \\ &= -(I-P)(D_xG(0,\kappa_0,\lambda_0))^{-1}(I-Q)D^2_{x\kappa}G(0,\kappa_0,\lambda_0)\hat{v}_0. \end{aligned}$$

Inserting (I.11.61) into (I.11.60), we get

$$\text{(I.11.62)} \quad \begin{aligned} &D^3_{v\kappa\kappa}\Phi(0,\kappa_0,\lambda_0)\hat{v}_0 \\ &= -2QD^2_{x\kappa}G^0(I-P)(D_xG^0)^{-1}(I-Q)D^2_{x\kappa}G^0\hat{v}_0, \end{aligned}$$

where, as usual, "0" denotes evaluation at $(0,\kappa_0,\lambda_0)$. In our particular case we have $D^2_{x\kappa}G^0 = \frac{d}{dt}$ and $D_xG^0 = \kappa_0\frac{d}{dt} - A_0 = J_0$.

By assumption, $i\kappa_0$ is not an algebraically simple eigenvalue of A_0. This means that $\langle\varphi_0,\overline{\varphi}'_0\rangle = \langle\overline{\varphi}_0,\varphi'_0\rangle = 0$, and as in (I.8.18) we have also $\langle\varphi_0,\overline{\varphi}'_0\rangle = \langle\overline{\varphi}_0,\varphi'_0\rangle = 0$, so that the projections (I.11.13) satisfy $QP = 0$ or $R(P) \subset N(Q)$. Since $\hat{v}_0 = \psi_0 + \overline{\psi}_0 \in R(P)$, also $\frac{d}{dt}\hat{v}_0 \in R(P)$, so that $QD^2_{x\kappa}G^0\hat{v}_0 = 0$ and $(I-Q)D^2_{x\kappa}G^0\hat{v}_0 = i\psi_0 - i\overline{\psi}_0$. Choose φ^1_0 as a generalized eigenvector solving $(i\kappa_0 I - A_0)\varphi_0^1 = \varphi_0$. Then

$$(I.11.63) \quad (I - P)J_0^{-1}(i\psi_0 - i\overline{\psi}_0) = (I - P)(i\varphi_0^1 e^{it} - i\overline{\varphi}_0^1 e^{-it}),$$

and by $\frac{d}{dt}P = P\frac{d}{dt}$ and $QP = 0$ we obtain for (I.11.62),

$$(I.11.64) \qquad \begin{aligned} D_{v\kappa\kappa}^3 \Phi^0 \hat{v}_0 &= 2Q(\varphi_0^1 e^{it} - \overline{\varphi}_0^1 e^{-it}), \text{ whence} \\ D_{r\kappa\kappa}^3 \hat{\Phi}^0 &= 2\langle \varphi_0^1, \varphi_0' \rangle = D_{\kappa\kappa}^2 \tilde{\Phi}^0. \end{aligned}$$

Formula (I.11.54)$_5$ then follows from (I.11.57), (I.11.44), (I.11.35), since $\langle \varphi_0^1, J\overline{\varphi}_0 \rangle$ is real.   □

If the algebraic multiplicity of the eigenvalue $i\kappa_0$ of A_0 is larger than two, then $\langle \varphi_0^1, \varphi_0' \rangle = \langle \varphi_0^1, J\overline{\varphi}_0 \rangle = 0$ and $\frac{d^2}{d\kappa^2}\lambda(0, \kappa_0) = 0$. The computation of higher derivatives follows the same lines but is tedious.

We discuss the case in which $i\kappa_0$ is a *geometrically simple* and an *algebraically double* eigenvalue of A_0. By (I.11.54)$_5$, the bifurcation curve $\{(\kappa, \lambda(0, \kappa))\}$ through (κ_0, λ_0) is of the form sketched in Figure I.11.1.

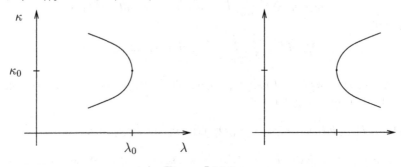

Figure I.11.1

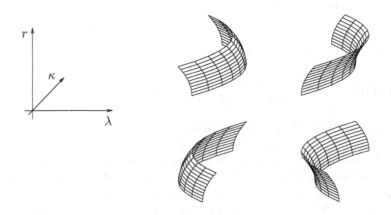

Figure I.11.2

By formula (I.11.54)$_3$, each point $(\kappa, \lambda(0, \kappa))$ on that bifurcation curve gives rise to a standard, when $D^2_{rr}\tilde{\Phi}^0 \neq 0$, or degenerate pitchfork of $2\pi/\kappa$-periodic solutions $x(r, \kappa)$ of (I.11.45) with $\lambda = \lambda(r, \kappa)$. The direction of the pitchfork is given by the sign of $D^2_{rr}\tilde{\Phi}^0$; cf. Figure I.9.1. The surface $\{(r, \kappa, \lambda(r, \kappa))\}$ in $\mathbb{R}^3$, or $\{(x(r, \kappa), \kappa, \lambda(r, \kappa))\}$ in $C^1_{2\pi/\kappa}(\mathbb{R}, \mathbb{R}^n) \times \mathbb{R}^2$, is fibered by all bifurcating pitchforks. A sketch of these surfaces is given in Figure I.11.2.

A necessary condition for bifurcation of $2\pi/\kappa$-periodic solutions of (I.11.1) at $(0, \lambda)$ is that $A(\lambda) = D_xF(0, \lambda)$ have the purely imaginary eigenvalue $i\kappa$. This means that for $\lambda < \lambda_0$ (or $\lambda > \lambda_0$) the two values of κ on the bifurcation curve, giving rise to $2\pi/\kappa$-periodic solutions, correspond to two simple eigenvalues $i\kappa$ of $A(\lambda)$ for $\lambda = \lambda(0, \kappa)$. For $\lambda > \lambda_0$ (or $\lambda < \lambda_0$), however, there is no bifurcation of $2\pi/\kappa$-periodic solutions (for κ near κ_0), so that the eigenvalues of $A(\lambda)$ are no longer on the imaginary axis. We sketch the eigenvalue perturbation for the two cases of Figure I.11.1 in Figure I.11.3, where the arrows point in the direction of increasing λ when λ passes through λ_0.

Two pairs of simple purely imaginary eigenvalues of $A(\lambda)$ collide at $\pm i\kappa_0$ and form a quadruplet of complex eigenvalues when λ passes through λ_0, where we take account of (I.11.32). This spectral scenario is referred to as **Hamiltonian Hopf Bifurcation** in the literature. Note that we give only heuristic arguments for the eigenvalue perturbation sketched in Figure I.11.3. In order to apply Theorem I.11.5, the eigenvalue perturbation does not have to be verified. The only hypotheses are the geometric simplicity of the eigenvalues $\pm i\kappa_0$ and the nondegeneracy (I.11.46).

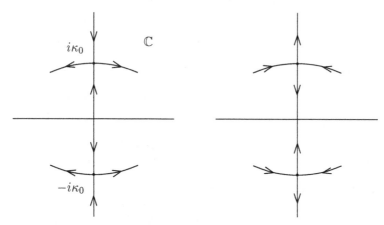

Figure I.11.3

In the general case of Theorem I.11.5, however, when the algebraic multiplicity of $i\kappa_0$ is arbitrary, the eigenvalue perturbation of $A(\lambda)$ near $\pm i\kappa_0$ can be much more complicated. There are two restrictions: Depending on the shape of the bifurcation curve $\{(\kappa, \lambda(0, \kappa))\}$, at least one pair of eigenvalues of $A(\lambda)$ stays on the imaginary axis near $\pm i\kappa_0$ for $\lambda < \lambda_0$ and/or $\lambda > \lambda_0$, and those eigenvalues that leave the imaginary axis do so in quadruplets to the right and left half-planes; cf. (I.11.32). We do not know whether the nondegeneracy (I.11.46) has more implications for the eigenvalue perturbation of $A(\lambda) = D_x F(0, \lambda) = D_x \nabla_x H(0, \lambda)$.

I.11.2 Reversible Systems

We consider (I.11.1) under a structural constraint called reversibility. The mapping $F : U \times V \to Z$, where $0 \in U \subset X, \lambda_0 \in V \subset \mathbb{R}$, is called **reversible** if

(I.11.65)
$$F(Rx, \lambda) = -RF(x, \lambda) \text{ for all } x \in U, \lambda \in V,$$
and for a reflection $R \in L(X, X) \cap L(Z, Z)$ such that $R^2 = I$.

For a function $z : \mathbb{R} \to Z$ we define the time reversal

(I.11.66) $$(Sz)(t) = z(-t).$$

Then the evolution equation (I.11.1) for a reversible mapping F has the following property:

(I.11.67)
$$\text{For } x \in C^1(\mathbb{R}, Z) \cap C(\mathbb{R}, X),$$
$$RS\left(\frac{dx}{dt} - F(x, \lambda)\right) = -\left(\frac{d}{dt}RSx - F(RSx, \lambda)\right).$$

If F is differentiable, we obtain from (I.11.65) by the chain rule

(I.11.68) $$A(\lambda)R = -RA(\lambda) \quad \text{for } A(\lambda) = D_x F(0, \lambda) \in L(X, Z).$$

For any real $A \in L(X, Z)$ satisfying (I.11.68) the point spectrum $\sigma_p(A)$ has the symmetry (I.11.32) in $\mathbb{C}$. For this reason, there is a great similarity between Hamiltonian and reversible systems. This fact was discovered by many people in KAM Theory (see [158] for a survey and references) and in Equivariant Bifurcation Theory (see [57] and the references therein).

Remark I.11.9 *The spectral symmetry (I.11.32) of A_0 is an obstruction to the generation of a holomorphic semigroup if the (point) spectrum is unbounded in $\mathbb{C}$. Therefore, reversibility requires $X = Z$ and $A_0 \in L(X, X)$, and the compactness of the semigroup might have reasonable applications only if $X = Z = \mathbb{R}^n$; cf. Remark I.10.1. Nonetheless, we stay with our infinite-dimensional setting, and we keep the notation of the general Theorem I.11.2. Remark I.11.3 holds accordingly also for reversible systems: a simple eigenvalue of $A(\lambda) = D_x F(0, \lambda)$ cannot cross the imaginary axis, so that a Hopf*

Bifurcation in the sense of Section I.8 (or even in its degenerate version of Section I.17) is not possible for a reversible system.

In order to apply Theorem I.11.2, we make its general assumptions on the reversible mapping F. Before proving (I.11.25), (I.11.26)$_1$, we have to take care of the choice of the vectors $\varphi_0, \varphi_0', \varphi_1, \varphi_1'$ for the projections P and Q in (I.11.13). Extend the reflection (I.11.65) to the complexified spaces in keeping $R^2 = I$. By (I.11.68), if φ_0 is an eigenvector of A_0 with eigenvalue $i\kappa_0$, then $R\varphi_0$ is an eigenvector of A_0 for $-i\kappa_0$. Therefore, by (I.11.3), $R\varphi_0 = c\overline{\varphi}_0$ for some $c \in \mathbb{C}$ with $|c| = 1$. Replacing φ_0 by $e^{-i\theta}\varphi_0$ if $c = e^{2i\theta}$, then $R\varphi_0 = \overline{\varphi}_0$. By $A_0' R' = -R'A_0'$ and $(R')^2 = I$ for the dual operators, the same arguments hold for an eigenvector φ_0' of A_0' for $i\kappa_0$. We summarize:

$$(\text{I}.11.69) \qquad R\varphi_0 = \overline{\varphi}_0, \ R'\varphi_0' = \overline{\varphi}_0'.$$

For the choices of φ_1 and φ_1', note that the ranges $R(\pm i\kappa_0 I - A_0)$ are invariant under the involution $z \mapsto R\overline{z}$. Therefore, if for some $\varphi_1 \in R(-i\kappa_0 I - A_0)$ spanning Z_0 we have $\langle \varphi_1, \varphi_0' \rangle = 1$, cf. (I.11.8), (I.11.9), then also $R\overline{\varphi}_1 \in R(-i\kappa_0 I - A_0)$ and $\langle R\overline{\varphi}_1, \varphi_0' \rangle = 1$ by (I.11.69). Replacing φ_1 by $\frac{1}{2}(\varphi_1 + R\overline{\varphi}_1)$, we have a vector φ_1 satisfying (I.11.8), (I.11.9), and $R\varphi_1 = \overline{\varphi}_1$.

For the choice of φ_1' we have to choose the complement X_0 satisfying (I.11.5) and (I.11.6). Let $X = N(i\kappa_0 I - A_0) \oplus N(-i\kappa_0 I - A_0) \oplus X_1$ with projection $P_1 : X \to X_1$ along $N(i\kappa_0 I - A_0) \oplus N(-i\kappa_0 I - A_0)$. Since the kernels $N(\pm i\kappa_0 I - A_0)$ are invariant under the involution $z \mapsto R\overline{z}$, the space $\hat{X}_1 = \{P_1 x + R\overline{P_1 R\overline{x}} \,|\, x \in X\}$ is a complement of $N(i\kappa_0 I - A_0) \oplus N(-i\kappa_0 I - A_0)$ in X, too: From $x = (I - P_1)x + P_1 x$ and $R\overline{x} = (I - P_1)R\overline{x} + P_1 R\overline{x}$ we conclude that $x = R\overline{(I - P_1)R\overline{x}} + R\overline{P_1 R\overline{x}}$, and the first summand is in $N(i\kappa_0 I - A_0) \oplus N(-i\kappa_0 I - A_0)$. Choosing $X_0 = N(-i\kappa_0 I - A_0) \oplus \hat{X}_1$, we have a complement satisfying (I.11.5), (I.11.6), and X_0 is invariant under the involution $z \mapsto R\overline{z}$. If $\varphi_1' \in X'$ is chosen according to (I.11.9), then $1 = \langle \varphi_0, \varphi_1' \rangle = \langle \overline{\varphi}_0, \overline{\varphi}_1' \rangle = \langle R\varphi_0, \overline{\varphi}_1' \rangle = \langle \varphi_0, R'\overline{\varphi}_1' \rangle$, and $\langle \varphi_0, R'\overline{\varphi}_1' - \varphi_1' \rangle = 0$. For $x \in X_0$ we have $\langle x, \varphi_1' \rangle = 0$, and since $R\overline{x} \in X_0$, it follows that $\langle R\overline{x}, \varphi_1' \rangle = 0$. Therefore, $\langle x, R'\overline{\varphi}_1' - \varphi_1' \rangle = 0$ for all $x \in X_0$, which implies $R'\overline{\varphi}_1' - \varphi_1' = 0$. We summarize:

$$(\text{I}.11.70) \qquad R\varphi_1 = \overline{\varphi}_1, \ R'\varphi_1' = \overline{\varphi}_1'.$$

Inserting the vectors $\varphi_0, \varphi_0', \varphi_1, \varphi_1'$ satisfying (I.11.69), (I.11.70) into the projections P and Q given in (I.11.13), we obtain the equivariance

$$(\text{I}.11.71) \qquad PRS = RSP \quad \text{and} \quad QRS = RSQ$$

for the time reversion S given in (I.11.66). For the function G defined in (I.11.4), the property (I.11.67) implies the "skew-equivariance"

$$(\text{I}.11.72) \qquad \begin{aligned} &G(RSx, \kappa, \lambda) = -RSG(x, \kappa, \lambda) \\ &\text{for all } (x, \kappa, \lambda) \in \tilde{U} \times \tilde{V} \subset (E \cap Y) \times \mathbb{R}^2. \end{aligned}$$

By the uniqueness of the function ψ solving $(I-Q)G(Px+\psi(Px,\kappa,\lambda),\kappa,\lambda) = 0$ (cf. (I.2.7), (I.2.8)), the equivariances (I.11.71), (I.11.72) imply

$$(I.11.73) \qquad \psi(RSPx,\kappa,\lambda) = RS\psi(Px,\kappa,\lambda)$$

for all (Px,κ,λ) near $(0,\kappa_0,\lambda_0)$.

A real function $Px = r(\psi_0 + \overline{\psi}_0)$ has the "isotropy"

$$(I.11.74) \qquad RSPx = Px$$

by (I.11.66), (I.11.69). This has the following consequences for the complex bifurcation function given in (I.11.21):

$$(I.11.75)$$
$$\begin{aligned}
\text{Re}\hat{\Phi}(r,\kappa,\lambda) &\\
&= \frac{1}{2\pi}\int_0^{2\pi}\langle G(Px+\psi(Px,\kappa,\lambda),\kappa,\lambda),\text{Re}\psi_0'\rangle dt \\
&= \frac{1}{2\pi}\int_0^{2\pi}\langle G(RS(Px+\psi(Px,\kappa,\lambda),\kappa,\lambda),\kappa,\lambda),\text{Re}\psi_0'\rangle dt \\
&= -\frac{1}{2\pi}\int_0^{2\pi}\langle RSG(Px+\psi(Px,\kappa,\lambda),\kappa,\lambda),\text{Re}\psi_0'\rangle dt \\
&= -\frac{1}{2\pi}\int_0^{2\pi}\langle SG(Px+\psi(Px,\kappa,\lambda),\kappa,\lambda),S\text{Re}\psi_0'\rangle dt \\
&= -\text{Re}\hat{\Phi}(r,\kappa,\lambda)
\end{aligned}$$

by (I.11.74), (I.11.72), and $R'\psi_0' = S\overline{\psi}_0'$; cf. (I.11.69). This proves that

$$(I.11.76) \quad \text{Re}\hat{\Phi}(r,\kappa,\lambda) = 0 \quad \text{for all} \quad r \in (-\delta,\delta), (\kappa,\lambda) \in \tilde{V}_2 \subset \mathbb{R}^2,$$

which clearly implies (I.11.25) and (I.11.26)$_1$.

Therefore, the **Constrained Hopf Bifurcation Theorem, Theorem I.11.2**, applies to reversible systems (I.11.1):

If $i\kappa_0$ is an algebraically simple eigenvalue of $A_0 = D_x F(0,\lambda_0)$, then a **Center Theorem for Reversible Systems** holds.

If $i\kappa_0$ is not necessarily an algebraically simple eigenvalue of A_0 and if the nondegeneracy
$$(I.11.77) \qquad \langle D_{x\lambda}^2 F(0,\lambda_0)\varphi_0,\varphi_0'\rangle \neq 0$$

is valid, then a **Hamiltonian Hopf Bifurcation Theorem for Reversible Systems** in the sense of Theorem I.11.5 holds. Observe that we choose the eigenvectors φ_0,φ_0' according to (I.11.69), and by

$$(I.11.78)$$
$$\begin{aligned}
\langle D_{x\lambda}^2 F(0,\lambda_0)\varphi_0,\varphi_0'\rangle &= \langle RD_{x\lambda}^2 F(0,\lambda_0)\varphi_0, R'\varphi_0'\rangle \\
&= -\langle D_{x\lambda}^2 F(0,\lambda_0)R\varphi_0, R'\varphi_0'\rangle = -\langle D_{x\lambda}^2 F(0,\lambda_0)\overline{\varphi}_0,\overline{\varphi}_0'\rangle,
\end{aligned}$$

the nondegeneracy (I.11.77) is purely imaginary.

The **Bifurcation Formulas** are the following:

For the surface of the Center Theorem the formulas (I.11.47) hold accordingly with the following modifications: By (I.11.76), the quantity $D_{rr}^2 \tilde{\Phi}^0$ given in (I.9.11) is purely imaginary. Formula (I.11.47)$_3$ reads in this case

(I.11.79)
$$\frac{d}{d\lambda}\kappa(r,\lambda)|_{(r,\lambda)=(0,\lambda_0)} = \mathrm{Im}\langle D_{x\lambda}^2 F(0,\lambda_0)\varphi_0, \varphi_0' \rangle$$
$$= \mathrm{Im}\frac{d}{d\lambda}\mu(\lambda)|_{\lambda=\lambda_0};$$

cf. Remark I.11.7. (Note the normalization $\langle \varphi_0, \varphi_0' \rangle = 1$.)

For the surface of the Hamiltonian Hopf Bifurcation we have again the formulas (I.11.54), which are modified as follows:

(I.11.80)
$$\frac{d^2}{dr^2}\lambda(r,\kappa)|_{(r,\kappa)=(0,\kappa_0)} = \frac{D_{rr}^2\tilde{\Phi}^0}{\langle D_{x\lambda}^2 F(0,\lambda_0)\varphi_0, \varphi_0' \rangle},$$
$$\frac{d^2}{d\kappa^2}\lambda(r,\kappa)|_{(r,\kappa)=(0,\kappa_0)} = \frac{2\langle \varphi_0^1, \varphi_0' \rangle}{\langle D_{x\lambda}^2 F(0,\lambda_0)\varphi_0, \varphi_0' \rangle},$$

where both quotients are real, since numerator and denominator are purely imaginary. The vector φ_0^1 is a generalized eigenvector satisfying the equation $(i\kappa_0 I - A_0)\varphi_0^1 = \varphi_0$.

If the eigenvalue $i\kappa_0$ of A_0 is algebraically double, then the last second derivative with respect to κ is nonzero, yielding bifurcation curves as sketched in Figure I.11.1. The eigenvalue perturbation $\mu(\lambda)$ of $A(\lambda)$ near $\pm i\kappa_0$ is sketched in Figure I.11.3.

Remark I.11.10 *In case of the Hamiltonian Hopf Bifurcation the eigenvalue $i\kappa_0$ of A_0 is not necessarily algebraically simple, which means that possibly $\langle \varphi_0, \varphi_0' \rangle = 0$. These eigenvectors have to satisfy $\langle \varphi_0, \varphi_1' \rangle = 1$ and $\langle \varphi_1, \varphi_0' \rangle = 1$; cf. (I.11.9). In the Bifurcation Formulas (I.11.54) and (I.11.80), only the vectors φ_0, φ_0' (or $\overline{\varphi}_0$ via (I.11.35)) and the generalized eigenvector φ_0^1 appear. Whereas formula (I.11.54)$_5$ (or (I.11.80)$_2$) is invariant for any choice of φ_0, φ_0', formula (I.11.54)$_3$ (or (I.11.80)$_1$) does not have this invariance with respect to the choice of φ_0. Note, however, that $x = Px + \psi(Px, \kappa, \lambda)$ is represented in terms of $\psi_0(t) = \varphi_0 e^{it}$, so that formula (I.11.54)$_1$ also depends explicitly on the vector φ_0. Therefore, the second derivative of λ with respect to r depends on the choice of φ_0, too; cf. (I.11.54)$_3$. The same holds also for formulas (I.11.47)$_1$ and (I.11.47)$_2$ when $\langle \varphi_0, \varphi_0' \rangle = 1$.*

I.11.3 Nonlinear Oscillations

We apply the results for reversible systems to nonlinear oscillations

$$(I.11.81) \qquad \ddot{x} = f(x, \dot{x}, \lambda),$$

where

$$(I.11.82) \qquad \begin{aligned} &f : U \times V \to \mathbb{R}^n, \ (0,0) \in U \subset \mathbb{R}^n \times \mathbb{R}^n, \ \lambda_0 \in V \subset \mathbb{R} \\ &\text{and } f(0,0,\lambda) = 0 \text{ for all } \lambda \in V. \end{aligned}$$

Clearly, $f \in C^3(U \times V, \mathbb{R}^n)$ is enough for the subsequent analysis; cf. the remarks after (I.10.14). We write (I.11.81) as a first-order system

$$(I.11.83) \qquad \begin{aligned} \dot{x} &= y, \\ \dot{y} &= f(x,y,\lambda), \end{aligned} \quad \text{or} \quad \frac{d}{dt}\begin{pmatrix} x \\ y \end{pmatrix} = F(x,y,\lambda),$$

and if F is *not* reversible, then a Hopf Bifurcation (in a nondegenerate or degenerate sense) is clearly possible for the system (I.11.83); cf. the example (I.17.56). In particular, one-dimensional oscillations (I.11.81) with a linear part $\ddot{x} = -x + d(\lambda)\dot{x}$ such that $d(\lambda)$ changes sign at $\lambda = \lambda_0$ give rise to Hopf Bifurcations: At $\lambda = \lambda_0$ the damping $d(\lambda)$ switches to a forcing, which creates nontrivial oscillations, as for the example (I.17.56), where $d(\lambda) = \lambda^7$. (This is also true for higher-dimensional nonlinear oscillations; cf. Remark I.11.13.) For reversible systems, however, Hopf Bifurcations are excluded; cf. Remark I.11.3.

The system (I.11.83) is **reversible** under the following conditions:

$$(I.11.84) \qquad \begin{aligned} &\text{For } R = \begin{pmatrix} -E & 0 \\ 0 & E \end{pmatrix} \quad \text{if} \quad f(-x,y,\lambda) = -f(x,y,\lambda), \\ &\text{for } R = \begin{pmatrix} E & 0 \\ 0 & -E \end{pmatrix} \quad \text{if} \quad f(x,-y,\lambda) = f(x,y,\lambda), \end{aligned}$$

where E denotes the n-dimensional identity matrix. In both cases in (I.11.84), $D_y f(0,0,\lambda) = 0$, and for $A_0 = D_{(x,y)} F(0,0,\lambda_0)$ we obtain

$$(I.11.85) \qquad A_0 = \begin{pmatrix} 0 & E \\ D_x f(0,0,\lambda_0) & 0 \end{pmatrix} \in L(\mathbb{R}^n \times \mathbb{R}^n, \mathbb{R}^n \times \mathbb{R}^n).$$

A simple calculation shows that

$$(I.11.86) \qquad \begin{aligned} &\mu \text{ is an eigenvalue of } D_x f(0,0,\lambda_0) \Leftrightarrow \\ &\pm\sqrt{\mu} \text{ are eigenvalues of } A_0, \end{aligned}$$

and the spectral assumption (I.11.3) is satisfied if

(I.11.87)
$$-\kappa_0^2 < 0 \text{ is an eigenvalue of } D_x f(0,0,\lambda_0) \in L(\mathbb{R}^n, \mathbb{R}^n)$$
$$\text{and } N(\kappa_0^2 I + D_x f(0,0,\lambda_0)) = \text{span}[x_0].$$

Let $(\ ,\ )$ denote the Euclidean scalar product in $\mathbb{R}^n$. Its complexification $\langle\ ,\ \rangle$ is the bilinear duality between $Z = \mathbb{C}^n$ and its dual $Z' = \mathbb{C}^n$. Then $-\kappa_0^2$ is also an eigenvalue of the dual $(D_x f(0,0,\lambda_0))'$ with eigenvector x_0', and

(I.11.88)
$$\varphi_0 = \begin{pmatrix} -ix_0 \\ \kappa_0 x_0 \end{pmatrix}, \ \varphi_0' = \begin{pmatrix} i\kappa_0 x_0' \\ x_0' \end{pmatrix}$$
are eigenvectors of $A_0, A_0' \in L(\mathbb{C}^n \times \mathbb{C}^n, \mathbb{C}^n \times \mathbb{C}^n)$ with eigenvalue $i\kappa_0$, respectively.

For R as in $(I.11.84)_1$ we have $R\varphi_0 = \overline{\varphi}_0, R'\varphi_0' = \overline{\varphi}_0'$; for R as in $(I.11.84)_2$ we replace φ_0, φ_0' by $i\varphi_0, i\varphi_0'$, respectively; and we have again $R\varphi_0 = \overline{\varphi}_0, R'\varphi_0' = \overline{\varphi}_0'$; cf. (I.11.69).

Therefore, the **Constrained Hopf Bifurcation Theorem, Theorem I.11.2**, applies to nonlinear oscillations (I.11.81) satisfying (I.11.84). For convenience we give the two cases in separate Theorems: A **Center Theorem for Nonlinear Oscillations** reads as follows:

Theorem I.11.11 *For the nonlinear oscillation*

$$\ddot{x} = f(x, \dot{x}, \lambda) \quad \text{in } \mathbb{R}^n$$

assume (I.11.82), $f \in C^3(U \times V, \mathbb{R}^n)$, and that

$$f \text{ is odd in } x \text{ or}$$
$$f \text{ is even in } \dot{x}.$$

If

$$-\kappa_0^2 < 0 \text{ is an algebraically simple eigenvalue of } D_x f(0,0,\lambda_0)$$
$$\text{and if for all } m \in \mathbb{Z}\backslash\{1,-1\}, -m^2\kappa_0^2 \text{ is not an eigenvalue,}$$

then there exists a continuously differentiable surface $\{(x(r,\lambda),\lambda) \mid r \in (-\delta,\delta), \lambda \in (\lambda_0 - \delta, \lambda_0 + \delta)\}$ of nontrivial (real) $2\pi/\kappa(r,\lambda)$-periodic solutions of (I.11.81) through $(x(0,\lambda),\lambda) = (0,\lambda)$ and $\kappa(0,\lambda_0) = \kappa_0$ in $C^2_{2\pi/\kappa(r,\lambda)}(\mathbb{R},\mathbb{R}^n) \times \mathbb{R}$. Furthermore, $\kappa(-r,\lambda) = \kappa(r,\lambda)$, and $x(-r,\lambda)$ is obtained from $x(r,\lambda)$ by a phase shift of half the period $\pi/\kappa(r,\lambda)$. The bifurcation formulas are given in (I.11.47) and (I.11.79), where

(I.11.89)
$$\langle D^2_{(x,y)\lambda} F(0,0,\lambda_0)\varphi_0, \varphi_0' \rangle = -i \langle D^2_{x\lambda} f(0,0,\lambda_0)x_0, x_0' \rangle,$$
$$\langle \varphi_0, \varphi_0' \rangle = 1 \Leftrightarrow \langle x_0, x_0' \rangle = \frac{1}{2\kappa_0}.$$

Note that the parameter λ can be frozen in $(\lambda_0 - \delta, \lambda_0 + \delta)$, or in other words, a parameter λ is not necessary for the Center Theorem.

The **Hamiltonian Hopf Bifurcation for Nonlinear Oscillations**:

Theorem I.11.12 *If under the same hypotheses as stated in Theorem I.11.11 the eigenvalue $-\kappa_0^2$ of $D_x f(0, 0, \lambda_0)$ is geometrically but not necessarily algebraically simple and if the nondegeneracy*

$$(I.11.90) \qquad\qquad \langle D_{x\lambda}^2 f(0, 0, \lambda_0) x_0, x_0' \rangle \neq 0$$

holds, then there exists a continuously differentiable surface $\{(x(r, \kappa), \lambda(r, \kappa)) \mid r \in (-\delta, \delta), \kappa \in (\kappa_0 - \delta, \kappa_0 + \delta)\}$ of nontrivial (real) $2\pi/\kappa$-periodic solutions of (I.11.81) through $(x(0, \kappa), \lambda(0, \kappa)) = (0, \lambda(0, \kappa))$ and $\lambda(0, \kappa_0) = \lambda_0$ in $C_{2\pi/\kappa}^2(\mathbb{R}, \mathbb{R}^n) \times \mathbb{R}$. Furthermore, $\lambda(-r, \kappa) = \lambda(r, \kappa)$, and $x(-r, \kappa)$ is obtained from $x(r, \kappa)$ by a phase shift of half the period π/κ. The bifurcation formulas are given in (I.11.54) and (I.11.80), where we use (I.11.89)$_1$. For $D_{rr}^2 \tilde{\Phi}^0$ given in (I.9.11) use definition (I.11.83) for F, (I.11.85) for A_0, (I.11.88) for φ_0, φ_0', and finally,

$$
(I.11.91) \qquad
\begin{aligned}
\langle \varphi_0^1, \varphi_0' \rangle &= -\frac{1}{2} i \langle x_0^1, x_0' \rangle, \\
\text{where } (\kappa_0^2 I &+ D_x f(0, 0, \lambda_0)) x_0^1 = x_0.
\end{aligned}
$$

(The translation of (I.9.11) using only f, $D_x f(0, 0, \lambda_0)$, and x_0, x_0' is left to the reader.)

As discussed before, if $-\kappa_0^2$ is an algebraically double eigenvalue of the operator $D_x f(0, 0, \lambda_0)$, the second derivative of λ with respect to κ is nonzero, and the bifurcation curves $\{(\kappa, \lambda(0, \kappa))\}$ are sketched in Figure I.11.1. The eigenvalue perturbations of $A(\lambda)$ shown in Figure I.11.3 correspond to eigenvalue perturbations of $D_x f(0, 0, \lambda_0)$ sketched in Figure I.11.4.

Note that Theorem I.11.12 does not require that $-\kappa_0^2$ be an algebraically double eigenvalue of $D_x f(0, 0, \lambda_0)$.

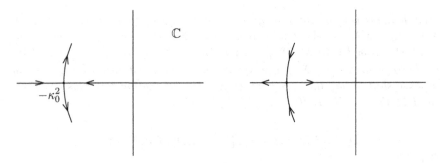

Figure I.11.4

I.11.4 Conservative Systems

There is another class of systems (I.11.1) for which the crucial assumptions (I.11.25) and (I.11.26) of the Constrained Hopf Bifurcation Theorem can be verified. Assume that a scalar product (,) is defined on the Banach space Z satisfying (I.3.1). For a function

(I.11.92)
$$H : U \times V \to \mathbb{R} \text{ in } C^2(U \times V, \mathbb{R}), \text{ where}$$
$$0 \in U \subset X, \ \lambda_0 \in V \subset \mathbb{R},$$

the gradient $\nabla_x H(x, \lambda)$ with respect to the scalar product (,) is defined as in Definition I.3.1. Then (I.11.1) is called **conservative** if

(I.11.93)
$$(\nabla_x H(x, \lambda), F(x, \lambda)) = 0$$
$$\text{for all } (x, \lambda) \in U \times V.$$

Let $x = x(t)$ be any solution of (I.11.1) such that $(x, \lambda) \in U \times V$. Then

(I.11.94)
$$\frac{d}{dt} H(x(t), \lambda) = D_x H(x(t), \lambda) \frac{dx}{dt}(t)$$
$$= (\nabla_x H(x(t), \lambda), F(x(t), \lambda)) = 0 \text{ or}$$
$$H(x(t), \lambda) = \text{const.}$$

Therefore, the function H gives a first conservation law or a first integral for solutions of (I.11.1). As seen in Section I.10, (I.10.8), Hamiltonian systems are special cases of conservative systems.

The spaces X and Z are complexified as described in Section I.8, and $\langle \ , \ \rangle$ denotes the bilinear pairing of Z and Z'. The real scalar product on Z is also complexified to a bilinear product on $Z \times Z$, and in order to distinguish it from $\langle \ , \ \rangle$ we keep the notation (,) (in contrast to Section I.10 and the paragraph about Hamiltonian Systems in this section). For the Lyapunov–Schmidt reduction of (I.11.4) to (I.11.20) and (I.11.21) we use the projections (I.11.13); i.e., we choose $\varphi_0, \varphi_0', \varphi_1, \varphi_1'$ according to (I.11.8), (I.11.9). In order to prove the assumptions (I.11.25) and (I.11.26), we follow the lines of the proof for Hamiltonian systems. For the function H defined in (I.11.92) we obtain (I.11.38), where $\langle \ , \ \rangle$ is replaced by (,). In view of assumption (I.11.93), the second term in (I.11.39) vanishes, and (I.11.40) holds accordingly: In the first integrals the pairing $\langle \ , \ \rangle$ is replaced by the scalar product (,). Defining $h(r, \kappa, \lambda)$ as (I.11.41), i.e.,

(I.11.95)
$$h(r, \kappa, \lambda) \equiv \frac{1}{2\pi} \int_0^{2\pi} (\nabla_x H(Px + \psi(Px, \kappa, \lambda), \lambda), \psi_1) dt$$
$$\text{for } Px = r(\psi_0 + \overline{\psi}_0), r \in (-\delta, \delta), (\kappa, \lambda) \in \tilde{V}_2 \subset \mathbb{R}^2,$$

then (I.11.42) holds, and we obtain as in (I.11.43),

$$\text{(I.11.96)} \qquad \begin{aligned} &D_r h(0, \kappa_0, \lambda_0) = (B_0 \overline{\varphi}_0, \varphi_1), \text{ where} \\ &B_0 = D_x \nabla_x H(0, \lambda_0) \in L(X, Z) \text{ is the Hessian of } H. \end{aligned}$$

If $i\kappa_0$ is an algebraically simple eigenvalue of $A_0 = D_x F(0, \lambda_0)$, then $\varphi_1 = \varphi_0$ and $\langle \varphi_0, \varphi_0' \rangle = 1$; cf. (I.11.5)–(I.11.9). Since the Hessian B_0 is symmetric with respect to the scalar product $(\ ,\ )$ (see Proposition I.3.3), the derivative (I.11.96) is real, and if it is nonzero, (I.11.42) implies (I.11.25). Therefore, the first case of the **Constrained Hopf Bifurcation Theorem, Theorem I.11.2**, is applicable to conservative systems if

$$\text{(I.11.97)} \qquad\qquad (D_x \nabla_x H(0, \lambda_0) \overline{\varphi}_0, \varphi_1) \neq 0.$$

It gives a **Center Theorem for Conservative Systems**, providing vertical bifurcations of $2\pi/\kappa(r, \lambda)$-periodic solutions of (I.11.1) for each λ near λ_0.

If $i\kappa_0$ is a geometrically but not necessarily an algebraically simple eigenvalue of A_0, then (I.11.42) and (I.11.96) imply (I.11.26), provided that

$$\text{(I.11.98)} \qquad \begin{aligned} &\operatorname{Re}(D_x \nabla_x H(0, \lambda_0) \overline{\varphi}_0, \varphi_1) \neq 0 \text{ and} \\ &\operatorname{Im} \langle D_{x\lambda}^2 F(0, \lambda_0) \varphi_0, \varphi_0' \rangle \neq 0, \text{ or} \\[6pt] &\operatorname{Im}(D_x \nabla_x H(0, \lambda_0) \overline{\varphi}_0, \varphi_1) \neq 0 \text{ and} \\ &\operatorname{Re} \langle D_{x\lambda}^2 F(0, \lambda_0) \varphi_0, \varphi_0' \rangle \neq 0. \end{aligned}$$

The second case of the Constrained Hopf Bifurcation Theorem then implies a **Hamiltonian Hopf Bifurcation Theorem for Conservative Systems**.

The Bifurcation Formulas (I.11.47), (I.11.79) in the first case and (I.11.54), (I.11.80) in the second case hold accordingly. If $i\kappa_0$ is an algebraically double eigenvalue of $A_0 = D_x F(0, \lambda_0)$, then formula (I.11.80)$_2$ proves a bifurcation curve as sketched in Figure I.11.1 with the typical eigenvalue perturbation shown in Figure I.11.3.

Remark I.11.13 *Hopf Bifurcation for parameter-dependent evolution equations (I.11.1) takes place in a generalized sense under much more general spectral assumptions than (I.11.3). We give a result of [102], and we refer to the literature mentioned below.*

Assume that for $n_j \in \mathbb{N}$, $j = 1, \ldots, k$, $n_1 = 1 < n_2 < \cdots < n_k$,

$$\text{(I.11.99)} \qquad \begin{aligned} &\pm i n_j \kappa_0 \text{ are all the eigenvalues of } A_0 = D_x F(0, \lambda_0) \\ &\text{that are integer multiples of } i\kappa_0. \end{aligned}$$

Let m_j denote the algebraic multiplicity of $in_j\kappa_0$. Then the eigenvalue $in_j\kappa_0$ of A_0 perturbs to an m_j-fold family of eigenvalues of $D_x F(0, \lambda) = A(\lambda)$ near $in_j\kappa_0$ when the parameter λ varies near λ_0 (the so-called $in_j\kappa_0$-group).

Assume that for all $\lambda \in (\lambda_0 - \delta, \lambda_0) \cup (\lambda_0, \lambda_0 + \delta)$ there is no element of the $in_j\kappa_0$-group on the imaginary axis. Then define for those λ,

(I.11.100)
$$n_j^>(\lambda) = \text{sum of the algebraic multiplicities}$$
$$\text{of all perturbed eigenvalues of } A(\lambda) \text{ near}$$
$$in_j\kappa_0 \text{ with positive real parts,}$$

which is constant for $\lambda \in (\lambda_0 - \delta, \lambda_0)$ and for $\lambda \in (\lambda_0, \lambda_0 + \delta)$, and define

(I.11.101)
$$n_j^>(\lambda_0 - \varepsilon) - n_j^>(\lambda_0 + \varepsilon) \equiv \chi_j(A(\lambda), \lambda_0),$$
the crossing number of the family $A(\lambda)$
at $\lambda = \lambda_0$ through $in_j\kappa_0$;

cf. Definition II.7.1. Assume that

(I.11.102)
$$\sum_{j=1}^k \chi_j(A(\lambda), \lambda_0) \neq 0.$$

Then $(0, \lambda_0)$ is a bifurcation point of a continuum $\{(x, \lambda)\}$ of nontrivial (real) $2\pi/\kappa$-periodic solutions x of (I.11.1) with periods $2\pi/\kappa$ emanating from $2\pi/\kappa_0$. (The periods are not necessarily minimal as in all cases treated in Sections I.8–I.11.)

This result is in the spirit of Theorems II.3.2, II.4.4, and II.7.3. In Theorem I.17.3 we assume $k = 1, m_1 = 1, \chi_1(A(\lambda), 0) = \pm 1$, but in contrast to Theorem I.8.2, the crossing of the simple eigenvalue through $i\kappa_0$ is degenerate; i.e., $\text{Re}\mu'(0) = \cdots = \text{Re}\mu^{(m-1)}(0) = 0$ and $\text{Re}\mu^{(m)}(0) \neq 0$ for some odd m. We obtain at least one and at most m bifurcating curves of $2\pi/\kappa$-periodic solutions with κ near κ_0. The proof of the general result given in [102] uses only analytical tools and is very involved. It admits also an interaction with stationary bifurcation, which means that 0 can be an eigenvalue of $A(\lambda_0) = A_0$, too: If the crossing number of $A(\lambda)$ through 0 is odd, then according to Theorems II.3.2, II.4.4, stationary solutions of (I.11.1) bifurcate; if the crossing number through 0 is even and (I.11.102) holds, then stationary or periodic solutions bifurcate from the trivial solution at $\lambda = \lambda_0$. Simple examples show that under the same spectral assumption the latter alternative actually occurs; see [99].

In [124] Hopf Bifurcation from a nontrivial curve of stationary solutions is proved. To be precise, the bifurcation point is a turning point of the curve of stationary solutions. This means that 0 is necessarily an eigenvalue of the linearization at the turning point.

In [105] a global version of the above general Hopf Bifurcation Theorem is given. In contrast to Theorems II.3.3, II.5.8, the global alternatives are richer: The continuum of nontrivial periodic solutions emanating at $(0, \kappa_0, \lambda_0)$ is unbounded in (x, λ)-space; it meets the trivial solution at $(0, \kappa_1, \lambda_1)$ where $(\kappa_1, \lambda_1) \neq (n_j\kappa_0, \lambda_0), j = 1, \ldots, k$, and $i\kappa_1$ is an eigenvalue of $A(\lambda_1)$; it meets some nontrivial stationary solution; or its "virtual period" is unbounded.

It is worthwhile to mention that the method of proving global Hopf Bifurcation is different from the method of proving global stationary bifurcation (although the sources are the same, and global stationary bifurcation could be proved in the same way). The idea is to perturb the operator G as defined in

(I.11.4) on a space of periodic functions so that 0 becomes a regular value; cf. Remark I.13.6. Then the solution set $G^{-1}\{0\}$ is a two-dimensional manifold without boundary, and being locally proper (due to its nonlinear Fredholm property according to Definition I.2.1), $G^{-1}\{0\}$ indeed exists globally. Without constraint this perturbation of G exists due to the Sard–Smale Theorem. The main difficulty, however, arises in keeping the S^1-equivariance (I.8.31) of G under perturbations. In this case, $G^{-1}\{0\}$ consists of so-called snakes fibered by group orbits, cf.(I.13.51), and it turns out that for S^1-equivariant mappings, Hopf bifurcations and period-doubling bifurcations are "generic" in the sense that no perturbation can avoid them; see [102], [105], for example, and our naive explanation in Remark I.14.5. Nonetheless, $G^{-1}\{0\}$ exists globally, but the snakes might split at period-doubling bifurcations or end at Hopf bifurcations in steady states.

We emphasize that the general local and global Hopf Bifurcation Theorems are not exclusively proved in [102], [105]. In [3], [21], [134], [83], [48], [85], for example, one finds equivalent results, all of which appeared before [102], [105]. We quote from [102], [105], since the terminology and the language are closest to those of this book.

Note finally that Hamiltonian Hopf Bifurcation is not included in the general Hopf Bifurcation Theorems cited above, since the crossing number of the family $A(\lambda)$ at $\lambda = \lambda_0$ through $i\kappa_0$ is not defined: At least one eigenvalue of the $i\kappa_0$-group remains on the imaginary axis for all λ near λ_0, and those that leave the imaginary axis do so in pairs to the right and left half-planes; cf. Figure I.11.3. Therefore, the Constrained Hopf Bifurcation is a special track in bifurcation theory.

I.12 The Principle of Exchange of Stability for Hopf Bifurcation

Let $x = x(t)$ be a p-periodic solution of the evolution equation

$$(I.12.1) \qquad\qquad \frac{dx}{dt} = F(x, \lambda),$$

where we assume the general setting of Section I.8. The **Principle of Linearized Stability** for a periodic solution is usually proved within "Floquet Theory" if (I.12.1) is an ODE, and it is generalized to semilinear parabolic PDEs in [76], for example. It is briefly described as follows.

The (nonlinear) Poincaré map around a p-periodic solution yields by linearization in its fixed point $x(0) = x(p)$ the period map of the linear nonautonomous evolution equation $\frac{dy}{dt} - D_x F(x(t), \lambda)y = 0$, called the "variational equation." In general, any solution of the variational equation is given by $y(t) = U(t)y_0$, where $U(0) = I$, and $U(p)$ is, by definition, its linear period map. Since (formally) $y = \frac{dx}{dt}$ is a p-periodic solution of the variational equation, we have $U(p)y(0) = y(p) = y(0)$, so that 1 is an eigenvalue of $U(p)$. If all other eigenvalues of $U(p)$ are of modulus less than 1, then the fixed point $x(0)$ of the Poincaré map is (linearly) stable, which implies the orbital stability of the p-periodic solution $x = x(t)$ of the autonomous evolution equation (I.12.1).

The eigenvalues of the period map $U(p)$ are called the *Floquet multipliers* of the p-periodic solution, and they are related to the so-called *Floquet exponents* as follows.

The eigenvalues μ of the variational operator $\frac{d}{dt} - D_x F(x(t), \lambda)$ in the space of p-periodic functions are, by definition, Floquet exponents, and they give by $e^{-p\mu}$ the Floquet multipliers of the p-periodic solution $x = x(t)$ of (I.12.1). The eigenvalue $\mu = 0$ with eigenfunction $y = \frac{dx}{dt}$ gives the multiplier $e^0 = 1$, and the Principle of Linearized Stability is equivalently stated as follows:

(I.12.2) The p-periodic solution $x = x(t)$ of (I.12.1) is (linearly) stable if the Floquet exponent $\mu = 0$ is simple and if all its Floquet exponents $\mu \neq 0$ have positive real parts.

We use (I.12.2) as a definition, since we do not know a proof for our general setting that linear stability indeed implies nonlinear orbital stability. We apply the Principle of Linearized Stability to the bifurcating curve $\{(x(r), \lambda(r))\}$ of $2\pi/\kappa(r)$-periodic solutions of (I.12.1) given by the Hopf Bifurcation Theorem, Theorem I.8.2. Again, the substitution $t/\kappa(r)$ for t fixes the period to 2π, and the stability problem amounts to the study of the eigenvalues μ of

(I.12.3)

$$\left(\kappa(r)\frac{d}{dt} - D_x F(x(r)(t), \lambda(r))\right)\psi = \mu\psi \quad \text{or of}$$

$$D_x G(x(r), \kappa(r), \lambda(r))\psi = \mu\psi, \text{ where}$$

G is defined by (I.8.10), (I.8.12) and

$$D_x G(./.) : C_{2\pi}^{1+\alpha}(\mathbb{R}, Z) \cap C_{2\pi}^\alpha(\mathbb{R}, X) \to C_{2\pi}^\alpha(\mathbb{R}, Z).$$

We introduce again the notation $W = C_{2\pi}^\alpha(\mathbb{R}, Z)$, $E = C_{2\pi}^\alpha(\mathbb{R}, X)$, and $Y = C_{2\pi}^{1+\alpha}(\mathbb{R}, Z)$.

For $r = 0$, the operator $D_x G(0, \kappa_0, \lambda_0) = \kappa_0 \frac{d}{dt} - A_0 = J_0$ has a geometrically double eigenvalue 0 with eigenvectors $\psi_0, \bar{\psi}_0$ (cf. (I.8.17), (I.8.22)). Therefore, the Principle of Linearized Stability does not apply for $r = 0$, but we show that it applies for $r \neq 0$ under the nondegeneracy condition of Section I.8 and the smoothness assumption of Section I.9: $F \in C^4(U \times V, Z)$.

Assume now that apart from the two algebraically simple eigenvalues $\pm i\kappa_0$ of $A_0 = D_x F(0, \lambda_0)$ (cf. (I.8.5)), the entire spectrum of A_0 is in the

stable left complex half-plane (cf. (I.7.2); our assumptions (I.8.8) imply that
the spectrum of A_0 consists only of isolated eigenvalues). Then the same
arguments as in (I.8.17) prove that

$$(\text{I.12.4}) \qquad \left(\kappa_0 \frac{d}{dt} - A_0\right)\psi = \mu\psi, \quad \psi(0) = \psi(2\pi) \Leftrightarrow$$
$$in\kappa_0 - \mu \text{ is an eigenvalue of } A_0 \text{ for some } n \in \mathbb{Z}.$$

This implies by our assumption that

$$(\text{I.12.5}) \qquad \begin{array}{l} \text{Re}\,\mu > 0 \text{ for all Floquet exponents } \mu \neq 0 \text{ of} \\ D_x G(0, \kappa_0, \lambda_0) = J_0. \end{array}$$

Therefore, the (linear) stability of the bifurcating curve $\{(x(r), \lambda(r))\}$ of
$2\pi/\kappa(r)$-periodic solutions of (I.12.1) is determined by the sign of the real
part of the perturbed critical eigenvalues $\mu(r)$ of $D_x G(x(r), \kappa(r), \lambda(r))$ near
$\mu(0) = 0$, at least for small $r \in (-\delta, \delta)$. The difference between this and the
situation of Section I.7 is that $\mu = 0$ is not a simple but a double eigenvalue
of $D_x G(0, \kappa_0, \lambda_0) = J_0$.

Here "double" means geometrically and algebraically double, which, in
turn, is defined by a two-dimensional eigenspace with a two-dimensional
eigenprojection. In the case in question, the eigenprojection is given by Q
(cf. (I.8.21)), which commutes with J_0 (by (I.8.22), (I.8.28)). Recall that
$Q|_{Y \cap E} = P$. An alternative way of characterizing the doubleness of the eigenvalue $\mu = 0$ of J_0 is the fact that both geometric eigenvectors $\psi_0, \overline{\psi}_0$ are not
in the range $R(J_0)$ (cf. also (I.7.4)). In this case, the eigenvalue $\mu = 0$ is
also called double but semisimple. As shown in [86], the two-dimensionality
of the eigenspace is preserved under a perturbation: There are two perturbed
eigenvalues $\mu_1(r), \mu_2(r)$ such that $\mu_1(0) = \mu_2(0) = 0$ (the so-called 0-group;
cf. [86]).

We start with the trivial Floquet exponent $\mu_1(r) \equiv 0$ of (I.12.3), which
is certainly among the perturbed eigenvalues near $\mu = 0$. By construction,
$x(r) = r(\psi_0 + \overline{\psi}_0) + \psi(r(\psi_0 + \overline{\psi}_0), \kappa(r), \lambda(r))$ (cf. (I.8.47), where t is replaced by
$t/\kappa(r)$). Then $\frac{d}{dt}x(r) = ir(\psi_0 - \overline{\psi}_0) + \frac{d}{dt}\psi(./.)$, and $\frac{d}{dt}x(r)|_{r=0} = \psi_0 + \overline{\psi}_0$ (cf.
(I.8.45)) in the topology of $C_{2\pi}^{1+\alpha}(\mathbb{R}, Z) \cap C_{2\pi}^{\alpha}(\mathbb{R}, X)$ implies $\frac{d}{dr}\frac{d}{dt}x(r)|_{r=0} =$
$i(\psi_0 - \overline{\psi}_0)$ in the topology of $C_{2\pi}^{\alpha}(\mathbb{R}, Z)$, so that

$$(\text{I.12.6}) \qquad \begin{array}{l} \psi_1(r) \equiv \dfrac{1}{r}\dfrac{d}{dt}x(r) \quad \text{for } r \neq 0 \text{ and} \\[2mm] \psi_1(0) = i(\psi_0 - \overline{\psi}_0) \text{ is in } C((-\delta, \delta), C_{2\pi}^{\alpha}(\mathbb{R}, Z)). \end{array}$$

Differentiating $\kappa\frac{dx}{dt} - F(\lambda, x) = 0$ with respect to t is a (formal) proof that $\frac{dx}{dt}$
is an eigenfunction of (I.12.3) with trivial Floquet exponent. This proof is cor-
rect in finite dimensions, i.e., for an ODE, but it might cause problems in our
general setting. We have to assume that the periodic solution $x(r)$ possesses
the trivial Floquet exponent $\mu_1(r) \equiv 0$ with a curve of eigenfunctions

(I.12.7)
$\{\psi_1(r) | r \in (-\delta, \delta)\} \subset E \cap Y$ that is twice continuously differentiable such that $P\psi_1(r) = i(\psi_0 - \overline{\psi}_0)$ and $w_1(r) \equiv (I - P)\psi_1(r)$ satisfies $w_1(0) = 0$.

Here $P = Q|_{E \cap Y}$, and Q is defined in (I.8.21). Assumption (I.12.7) is proved in Proposition I.18.3. In the context of parabolic differential equations the eigenfunction $\psi_1(r)$ is indeed given by (I.12.6); i.e., it can be shown that the periodic solution $x(r)$ is regular enough that $\psi_1(r)$ as given by (I.12.6) is in $E \cap Y$; cf. [76], Chapter 8. In any case,

(I.12.8)
$$D_x G(x(r), \kappa(r), \lambda(r))\psi_1(r) = 0$$
$$\text{for } r \in (-\delta, \delta) \text{ and } \mu_1(r) \equiv 0.$$

Next, we are interested in a linearly independent (possibly generalized) eigenfunction $\psi_2(r)$ with eigenvalue $\mu_2(r)$ such that $\mu_2(0) = 0$.

For reasons that will be clear in the sequel, we introduce the following real vectors and projections:

(I.12.9)
$$\hat{v}_1 = i(\psi_0 - \overline{\psi}_0), \quad \hat{v}_2 = \psi_0 + \overline{\psi}_0,$$

$$\hat{v}_1' = -\frac{i}{2}(\psi_0' - \overline{\psi}_0'), \quad \hat{v}_2' = \frac{1}{2}(\psi_0' + \overline{\psi}_0'),$$

$$Q_j z = \frac{1}{2\pi} \int_0^{2\pi} \langle z, \hat{v}_j' \rangle dt \hat{v}_j, \quad j = 1, 2, \quad z \in W,$$

$$Q_j|_{Y \cap E} = P_j.$$

Then $Q = Q_1 + Q_2$ (cf. (I.8.21)), $Q_1 Q_2 = Q_2 Q_1 = 0$, both Q_1 and Q_2 are real for real z, and $\psi_1(0) = \hat{v}_1, Q_2 w_1(r) = 0$ for all $r \in (-\delta, \delta)$.

Proposition I.12.1 *There is a unique twice continuously differentiable curve* $\{\mu_2(r) | r \in (-\delta, \delta), \mu_2(0) = 0\}$ *in* $\mathbb{R}$ *such that*

(I.12.10) $D_x G(x(r), \kappa(r), \lambda(r))(\hat{v}_2 + w_2(r)) = \mu_2(r)(\hat{v}_2 + w_2(r)) + \nu(r)\psi_1(r),$

where $\{w_2(r) | r \in (-\delta, \delta), w_2(0) = 0\} \subset (I - P)(E \cap Y)$ *and* $\{\nu(r) | r \in (-\delta, \delta), \nu(0) = 0\} \subset \mathbb{R}$ *are twice continuously differentiable, too.*

Before proving Proposition I.12.1, we show that $\mu_2(r)$ *is the second perturbed eigenvalue.* If $\mu_2(r) = 0$ and $\nu(r) = 0$, then $\psi_2(r) = \hat{v}_2 + w_2(r)$ is a second eigenvector with eigenvalue 0, which is geometrically double in this case. If $\mu_2(r) = 0$ and $\nu(r) \neq 0$, then $\psi_2(r) = \hat{v}_2 + w_2(r)$ is a generalized eigenvector with eigenvalue 0, which is algebraically double in this case. Finally, if $\mu_2(r) \neq 0$, then we set $\psi_2(r) = \hat{v}_2 + w_2(r) + \frac{\nu(r)}{\mu_2(r)}\psi_1(r)$, and $\psi_2(r)$ is by (I.12.8) an eigenvector with eigenvalue $\mu_2(r)$.

Proof. We define a mapping

$$H : (I - P)(E \cap Y)) \times \mathbb{R} \times \mathbb{R} \times (-\delta, \delta) \to W \text{ by}$$

(I.12.11)
$$H(w_2, \mu, \nu, r)$$
$$\equiv D_x G(x(r), \kappa(r), \lambda(r))(\hat{v}_2 + w_2) - \mu(\hat{v}_2 + w_2) - \nu\psi_1(r).$$

Then $H(0, 0, 0, 0) = 0$ and

(I.12.12)
$$\begin{aligned}
D_{w_2} H(0, 0, 0, 0) &= D_x G(0, \kappa_0, \lambda_0) = J_0, \\
D_\mu H(0, 0, 0, 0) &= -\hat{v}_2, \quad D_\nu H(0, 0, 0, 0) = -\hat{v}_1,
\end{aligned}$$

so that (I.8.28) implies that

(I.12.13) $D_{(w_2, \mu, \nu)} H(0, 0, 0, 0) : (I - P)(E \cap Y) \times \mathbb{R} \times \mathbb{R} \to W$

is an isomorphism. The statement of Proposition I.12.1 then follows by the Implicit Function Theorem.

Under the regularity assumptions of Section I.9 we know that the curve $(x(r), \kappa(r), \lambda(r))$ as well as $\mu_2(r), w_2(r)$, and $\psi_1(r)$ is twice continuously differentiable with respect to r.   □

Since $P_1(\hat{v}_2 + w_2(r)) = 0$ and $\psi_1(r) = \hat{v}_1 + w_1(r)$ (cf. (I.12.7)), equation (I.12.10) implies

(I.12.14)
$$(I - Q_1) D_x G(x(r), \kappa(r), \lambda(r))(\hat{v}_2 + w_2(r))$$
$$= \mu_2(r)(\hat{v}_2 + w_2(r)) + \nu(r) w_1(r).$$

Since $N((I - Q_1) D_x G(0, \kappa_0, \lambda_0)) = N((I - Q_1) J_0) = \text{span}[\hat{v}_2]$, equation (I.12.14) is similar to equation (I.7.10) of a simple eigenvalue perturbation: It differs only in the additive term $\nu(r) w_1(r)$. We show that this term is of higher order, which does not have any influence on $\dot{\mu}_2(0)$ or on $\ddot{\mu}_2(0)(\dot{} = \frac{d}{dr})$. Therefore, we can apply Proposition I.7.3 in order to determine $\ddot{\mu}_2(0)$, which will give us the sign of $\mu_2(r)$ for r near 0. First, we claim that

(I.12.15)
$$\left. \frac{d\mu_2}{dr}(r) \right|_{r=0} = 0, \quad \left. \frac{d\nu}{dr}(r) \right|_{r=0} = 0.$$

To prove (I.12.15) we make use of the equivariance (I.8.31). By the chain rule we obtain from (I.8.31),

(I.12.16) $D_x G(S_\theta x, \kappa, \lambda) S_\theta = S_\theta D_x G(x, \kappa, \lambda),$

which shows that (I.12.10) is equivalent to

(I.12.17)
$$D_x G(S_\theta x(r), \kappa(r), \lambda(r)) S_\theta(\hat{v}_2 + w_2(r))$$
$$= \mu_2(r) S_\theta(\hat{v}_2 + w_2(r)) + \nu(r) S_\theta \psi_1(r).$$

We choose $\theta = \pi$. Since $S_\pi x(r) = x(-r)$, $\kappa(-r) = \kappa(r)$, $\lambda(-r) = \lambda(r)$ (see Theorem I.8.2 and the arguments given after it) and also $S_\pi \psi_1(r) = -\psi_1(-r)$, $S_\pi \hat{v}_2 = -\hat{v}_2$, equation (I.12.17) shows the following:

If $\mu_2(r), \nu(r)$ solve (I.12.10) with $w_2(r)$,

(I.12.18)   then $\mu_2(r), \nu(r)$ solve (I.12.10) also for $-r$

with $w_2(-r) = -S_\pi w_2(r)$.

By uniqueness, we conclude that

$$(\text{I.12.19}) \qquad \mu_2(-r) = \mu_2(r), \quad \nu(-r) = \nu(r),$$

which proves (I.12.15). Using also $w_1(0) = 0$, we see that the additive term $\nu(r)w_1(r)$ in (I.12.14) has the properties

$$(\text{I.12.20}) \quad \nu(r)w_1(r)\big|_{r=0} = \frac{d}{dr}(\nu(r)w_1(r))\big|_{r=0} = \frac{d^2}{dr^2}(\nu(r)w_1(r))\big|_{r=0} = 0.$$

As mentioned before, the third order of $\nu(r)w_1(r)$ in (I.12.14) allows us to apply Proposition I.7.3, in particular formula (I.7.19). As a matter of fact, the second derivative of $\mu_2(r)$ at $r = 0$ is obtained in Proposition I.7.3 by differentiating equation (I.12.14) twice with respect to r at $r = 0$, and in view of (I.12.20), the additive term $\nu(r)w_1(r)$ in (I.12.14) has no influence on that procedure. Therefore, we can apply formula (I.7.19) when we define the function Ψ in such a way that it is related to the "simple eigenvalue perturbation (I.12.14)" (without the additive term $\nu(r)w_1(r)$) as definition (I.7.16) is related to (I.7.10). To that purpose, we define

$$\tilde{G}(x,r) \equiv (I - Q_1)G(x(r) + x, \kappa(r), \lambda(r)),$$

(I.12.21) $$\tilde{G} : \tilde{U}_1 \times (-\delta, \delta) \to (I - Q_1)W,$$

$$0 \in \tilde{U}_1 \subset (I - P_1)(E \cap Y), (-\delta, \delta) \subset \mathbb{R}.$$

Obviously, $\tilde{G}(0,0) = 0$, and through $(0,0) \in \tilde{U}_1 \times (-\delta, \delta)$ there is the solution curve $\{(0,r)|r \in (-\delta, \delta)\}$ of $\tilde{G}(x,r) = 0$. Equation (I.12.14) is then rewritten as

$$(\text{I.12.22}) \quad D_x\tilde{G}(0,r)(\hat{v}_2 + w_2(r)) = \mu_2(r)(\hat{v}_2 + w_2(r)) + \nu(r)w_1(r).$$

The method of Lyapunov–Schmidt for $\tilde{G}(x,r) = 0$ near $(x,r) = (0,0)$ is described as follows. By

(I.12.23)

$$D_x\tilde{G}(0,0) = (I - Q_1)J_0 \text{ we obtain}$$

$$N(D_x\tilde{G}(0,0)) = R(P_2) = \text{span}[\hat{v}_2].$$

Since $Q_1 J_0 = J_0 P_1 = 0$, we have also

$$(\text{I.12.24}) \qquad\qquad R(D_x\tilde{G}(0,0)) = R(J_0),$$

which gives us the decomposition

$$(I - Q_1)W = R(J_0) \oplus N((I - Q_1)J_0)$$

$$(I.12.25) \qquad = R(D_x\tilde{G}(0,0)) \oplus N(D_x\tilde{G}(0,0))$$

with projection Q_2 onto $N(D_x\tilde{G}(0,0))$ along $R(D_x\tilde{G}(0,0))$.

The reduced equation (cf. (I.2.9)) is then

$$(I.12.26) \qquad \Phi(v,r) = Q_2\tilde{G}(v + \tilde{\psi}(v,r),r) = 0,$$
$$\text{where } v = P_2x \in N(D_x\tilde{G}(0,0)).$$

The function $\tilde{\psi}$ is defined by (cf. (I.2.8))

$$(I.12.27) \qquad (I - Q_2)\tilde{G}(v + \tilde{\psi}(v,r),r) = 0,$$
$$\tilde{\psi}(v,r) \in R(D_x\tilde{G}(0,0)) \cap (E \cap Y) = (I - P_2)(E \cap Y).$$

Inserting definition (I.12.21) into (I.12.27), we obtain by $Q_1 + Q_2 = Q$ and $Q_2Q_1 = 0$,

$$(I.12.28) \qquad (I - Q)G(x(r) + v + \tilde{\psi}(v,r), \kappa(r), \lambda(r)) = 0.$$

Now $v = P_2x = y\hat{v}_2$ for $y \in \mathbb{R}$, $x(r) = r\hat{v}_2 + \psi(r\hat{v}_2, \kappa(r), \lambda(r))$, which implies by uniqueness (cf. (I.8.30)),

$$(I.12.29) \quad \tilde{\psi}(v,r) = \psi((r + y)\hat{v}_2, \kappa(r), \lambda(r)) - \psi(r\hat{v}_2, \kappa(r), \lambda(r)),$$

and finally,

$$\Phi(y\hat{v}_2, r)$$
$$= Q_2G((r + y)\hat{v}_2 + \psi((r + y)\hat{v}_2, \kappa(r), \lambda(r)), \kappa(r), \lambda(r)),$$

$$(I.12.30) \quad \Psi(y,r) \equiv \frac{1}{2\pi}\int_0^{2\pi} \langle \Phi(y\hat{v}_2, r), \hat{v}_2' \rangle dt$$
$$= \frac{1}{2\pi}\int_0^{2\pi} \langle \hat{\Phi}(r + y, \kappa(r), \lambda(r)), \hat{v}_2' \rangle dt$$
$$= \mathrm{Re}\hat{\Phi}(r + y, \kappa(r), \lambda(r)),$$

where we use definition (I.8.35) of $\hat{\Phi}$ and $\hat{v}_2' = \mathrm{Re}\hat{\psi}_0'$. Since the function Ψ in (I.12.30) is related to (I.12.22) in the same way as the function Ψ in (I.7.16) is related to (I.7.10) (up to terms of order two), formula (I.7.19) of Proposition I.7.3 for $y = y(r) \equiv 0$ now reads as follows:

$$(I.12.31) \qquad \mathrm{Re}\frac{d^2}{dr^2}D_r\hat{\Phi}(r, \kappa(r), \lambda(r))\big|_{r=0} = \frac{d^2}{dr^2}\mu_2(r)\big|_{r=0}.$$

Here obviously $D_y\hat{\Phi}(r + y, \kappa(r), \lambda(r)) = D_r\hat{\Phi}(r + y, \kappa(r), \lambda(r))$. In Theorem I.17.4, formula (I.12.31) is considerably generalized. The arguments of its proof are accordingly refined; cf. (I.17.31)–(I.17.49).

For the evaluation of $\frac{d^2}{dr^2} D_r\hat{\Phi}(r, \kappa(r), \lambda(r))$ at $r = 0$ we make use of the computations in Sections I.8 and I.9. By $r\tilde{\Phi}(r, \kappa, \lambda) = \hat{\Phi}(r, \kappa, \lambda)$ (cf. (I.8.39)) we obtain

(I.12.32)
$$D_r\hat{\Phi}(0, \kappa_0, \lambda_0) = \tilde{\Phi}(0, \kappa_0, \lambda_0),$$
$$D^3_{rrr}\hat{\Phi}(0, \kappa_0, \lambda_0) = 3D^2_{rr}\tilde{\Phi}(0, \kappa_0, \lambda_0) \text{ (cf. (I.9.3))}$$
$$D^2_{r\kappa}\hat{\Phi}(0, \kappa_0, \lambda_0) = D_\kappa\tilde{\Phi}(0, \kappa_0, \lambda_0),$$
$$D^2_{r\lambda}\hat{\Phi}(0, \kappa_0, \lambda_0) = D_\lambda\tilde{\Phi}(0, \kappa_0, \lambda_0).$$

Next we use $\tilde{\Phi}(0, \kappa_0, \lambda_0) = 0$ (cf. (I.8.40)), $\dot{\kappa}(0) = 0$, $\dot{\lambda}(0) = 0$ $\left(\dot{} = \frac{d}{dr}, \text{ cf.}\right.$ (I.8.46)), and also (I.8.41), (I.8.44). Thus (I.12.31) gives

(I.12.33) $\ddot{\mu}_2(0) = 3\mathrm{Re}D^2_{rr}\tilde{\Phi}(0, \kappa_0, \lambda_0) - \mathrm{Re}\mu'(\lambda_0)\ddot{\lambda}(0).$

Using the bifurcation formula $(I.9.12)_4$ for $\ddot{\lambda}(0)$ we end up with the crucial formula

(I.12.34) $\ddot{\mu}_2(0) = 2\mathrm{Re}\mu'(\lambda_0)\ddot{\lambda}(0)$ $\quad\left(' = \dfrac{d}{d\lambda}\right).$

We summarize:

Theorem I.12.2 Let $\{(x(r), \lambda(r))|r \in (-\delta, \delta)\}$ be the curve of $2\pi/\kappa(r)$-periodic solutions of (I.12.1) according to the Hopf Bifurcation Theorem, Theorem I.8.2. Let $\mu_2(r)$ be the nontrivial Floquet exponent of $x(r)$ such that $\mu_2(0) = 0$. Then $\dot{\mu}_2(0) = 0$ (and also $\dot{\lambda}(0) = 0$). The second derivatives of μ_2 and λ are linked together by formula (I.12.34). Here $\mathrm{Re}\mu'(\lambda_0) \neq 0$ is the assumed nondegeneracy (I.8.7) for the eigenvalue perturbation $\mu(\lambda)$ of $D_xF(0, \lambda)$ near $i\kappa_0$, and $\ddot{\lambda}(0) \neq 0$, provided that $\mathrm{Re}D^2_{rr}\tilde{\Phi}(0, \kappa_0, \lambda_0) \neq 0$ (cf. $(I.9.12)_4$).

By Theorem I.12.2 we easily obtain the following **Principle of Exchange of Stability**.

Corollary I.12.3 Assume that apart from the two simple eigenvalues $\pm i\kappa_0$ of $A_0 = D_xF(0, \lambda_0)$, the entire spectrum of A_0 is in the stable left complex half-plane and assume that $\mathrm{Re}\mu'(\lambda_0) > 0$; i.e., the trivial solution $\{(0, \lambda)\}$ of (I.12.1) is stable for $\lambda < \lambda_0$ and unstable for $\lambda > \lambda_0$ (locally, cf. (I.7.2)). Then

(I.12.35) $\mathrm{sign}(\lambda(r) - \lambda_0) = \mathrm{sign}\mu_2(r)$ for $r \in (-\delta, \delta)$,

which means that the bifurcating periodic solution $\{(x(r), \lambda(r))\}$ of (I.12.1) is stable, provided that the bifurcation is supercritical, and it is unstable if the *bifurcation is subcritical (cf. (I.12.2), (I.12.5)). If $\text{Re}\mu'(\lambda_0) < 0$, the stability properties of the trivial solution are reversed, and in view of*

(I.12.36) $\text{sign}(\lambda(r) - \lambda_0) = -\text{sign}\mu_2(r) \text{ for } r \in (-\delta, \delta),$

the stability of the bifurcating periodic solution is reversed, too. Thus, we have the situations sketched in Figure I.12.1.

Figure I.12.1

Remark I.12.4 *Formula (I.12.34) is valid under the hypotheses of Sections I.8 and I.9. In particular, we need only (I.8.5) and the nonresonance condition (I.8.14). However, without any knowledge of the entire spectrum of A_0, the stability property of neither the trivial nor the bifurcating periodic solution can be determined. Formula (I.12.34) describes only the relation between the critical eigenvalue $\mu(\lambda)$ near $i\kappa_0$ and the critical Floquet exponent $\mu_2(r)$ near 0 depending on the bifurcation direction $\lambda(r) - \lambda_0$.*

I.13 Continuation of Periodic Solutions and Their Stability

It is a natural question whether the local curve $\{(x(r), \lambda(r))\}$ of $p(r) = 2\pi/\kappa(r)$-periodic solutions of the evolution equation

(I.13.1) $\dfrac{dx}{dt} = F(x, \lambda),$

given by the Hopf Bifurcation Theorem, has a (global) continuation. A first step to answering that question is an Implicit Function Theorem for periodic solutions of (I.13.1). We assume that $x_0 = x_0(r)$ is a p_0-periodic solution of (I.13.1) for $\lambda = \lambda_0$. In order to apply our setting of Section I.8, we make a substitution t/κ_0 for t, where $p_0 = 2\pi/\kappa_0$, and then (without changing the notation for x)

$$\kappa \frac{dx}{dt} - F(x, \lambda) \equiv G(x, \kappa, \lambda) = 0$$

(I.13.2)   has a solution $(x_0, \kappa_0, \lambda_0)$ in

$$(C_{2\pi}^{1+\alpha}(\mathbb{R}, Z) \cap C_{2\pi}^{\alpha}(\mathbb{R}, X)) \times \mathbb{R}_+ \times \mathbb{R}$$

by assumption. A continuation of that solution for λ near λ_0 involves also a
continuation of the period $2\pi/\kappa$ near $2\pi/\kappa_0$.

The derivative $D_x G(x_0, \kappa_0, \lambda_0) = \kappa_0 \frac{d}{dt} - D_x F(x_0, \lambda_0)$ is not a bijection,
since the trivial Floquet exponent $\mu_0 = 0$ is an eigenvalue with eigenfunction
$\frac{d}{dt} x_0 \in N(D_x G(x_0, \kappa_0, \lambda_0))$. Since Proposition I.18.3 is not applicable, in
general, we assume that

$$\frac{d}{dt} x_0 \equiv \hat{\psi}_0 \in C_{2\pi}^{1+\alpha}(\mathbb{R}, Z) \cap C_{2\pi}^{\alpha}(\mathbb{R}, X)$$

(I.13.3) is an eigenfunction of $\hat{J}_0 \equiv \kappa_0 \dfrac{d}{dt} - D_x F(x_0, \lambda_0)$

with eigenvalue $\mu_0 = 0$ that is the trivial Floquet exponent of x_0.

(We do not pursue the regularity of the periodic solution x_0, but we remark
only that under reasonable assumptions, a formal differentiation of (I.13.2)
is allowed, which proves (I.13.3); cf. our comments in Section I.12. For this
reason, we change our notation from that of (I.12.8).)

The natural assumption for an Implicit Function Theorem for periodic
solutions is that

(I.13.4)
the trivial Floquet exponent $\mu_0 = 0$ of x_0 is algebraically simple;
i.e., $N(D_x G(x_0, \kappa_0, \lambda_0)) = N(\hat{J}_0) = \text{span}[\hat{\psi}_0]$ and
$\hat{\psi}_0 \notin R(D_x G(x_0, \kappa_0, \lambda_0)) = R(\hat{J}_0)$, where $\hat{\psi}_0$ is given by (I.13.3).

For our subsequent analysis we need the Fredholm property of the op-
erator $\hat{J}_0$. Recall that $x_0 = x_0(t)$, and therefore the operator $D_x F(x_0, \lambda_0)$
depends on t, which makes this operator different from the operator J_0 considered in Section I.8, in particular in Proposition I.8.1. The assumption
(I.8.8) on the operator $A_0 = D_x F(x_0, \lambda_0)$ has to be replaced by an assumption on the

(I.13.5)
2π-periodic family of operators
$A_0(t) : X \to Z$ defined by
$$A_0(t) = \frac{1}{\kappa_0} D_x F(x_0(t), \lambda_0).$$

This assumption reads as follows:

(I.13.6)
$A_0(t)$ generates a holomorphic semigroup for each fixed
$t \in [0, T]$, and this family of semigroups, in turn, generates a
"fundamental solution" $U_0(t, \tau) \in L(Z, Z)$ for $0 \le \tau \le t \le T$
such that $U_0(t, t) = I$, $U_0(t, \tau)U_0(\tau, s) = U_0(t, s)$, and any
solution of $\dfrac{dx}{dt} = A_0(t)x$ is given by $x(t) = U_0(t, 0)x(0)$.

For a construction of $U_0(t, \tau)$ (which is also called an "evolution operator") we refer to [162], [52], [7], for example. It requires some additional regularity of the family $A_0(t)$ with respect to t, which is satisfied for ODEs or parabolic PDEs under reasonable assumptions on $D_x F(x, \lambda)$. In analogy to (I.8.8) we assume also that

$$(I.13.7) \qquad U_0(t, \tau) \in L(Z, Z) \text{ is compact for } 0 \le \tau < t \le T.$$

(The time $T \ge 2\pi$ is arbitrary but finite. The compactness of (I.13.7) is given by a compact embedding $X \subset Z$ (cf. (I.8.4)).)

As mentioned before, we need the Fredholm property of $\hat{J}_0 \equiv \kappa_0(\frac{d}{dt} - A_0(t))$ as a mapping
$$(I.13.8) \qquad \hat{J}_0 : E \cap Y \to W,$$

where we again use the notation $W = C_{2\pi}^\alpha(\mathbb{R}, Z)$, $E = C_{2\pi}^\alpha(\mathbb{R}, X)$, and $Y = C_{2\pi}^{1+\alpha}(\mathbb{R}, Z)$.

The mapping (I.13.8) makes sense only if $A_0(\cdot)x \in W$ for all $x \in E$. This is satisfied by a Hölder continuity of the family $A_0(t)$ in $L(X, Z)$. (Observe an inconsistency in the notation: Whereas we write x for a function $x = x(t)$, we note explicitly the time dependence of $A_0 = A_0(t)$ in the differential equation $\frac{dx}{dt} = A_0(t)x$ or in the operator $\hat{J}_0 = \kappa_0(\frac{d}{dt} - A_0(t))$. This is in agreement with a long tradition in differential equations.) For the Fredholm property of $\hat{J}_0$ we need the following assumption:

$$(I.13.9) \qquad \begin{aligned} &\text{For } f \in C^\alpha([0, T], Z) \text{ and } \hat{\varphi} \in Z \text{ the solution of} \\ &\frac{dx}{dt} = A_0(t)x + f \text{ and } x(0) = \hat{\varphi} \text{ is given by} \\ &x(t) = U_0(t, 0)\hat{\varphi} + \int_0^t U_0(t, s)f(s)ds \\ &\text{and } x \in C^\alpha([\varepsilon, T], X) \text{ for any } \varepsilon > 0. \end{aligned}$$

For a Hölder continuous family $A_0(t)$ (with respect to the topology of $L(X, Z)$) that satisfies (I.13.6), the property (I.13.9) is proved in [7], Chapter II. As mentioned before, all conditions on $D_x F(x_0, \lambda_0)$ can be satisfied for a reasonably large class of parabolic PDEs.

By the assumption (I.13.9) we show that we can characterize the range of $\hat{J}_0$ (cf. (I.13.8)) as follows:

$$f \in R(\hat{J}_0) \subset W \Leftrightarrow \hat{J}_0 x = f \Leftrightarrow$$

$$(I - U_0(2\pi, 0))\hat{\varphi}$$

$$(I.13.10) \qquad = \frac{1}{\kappa_0} \int_0^{2\pi} U_0(2\pi, s)f(s)ds \text{ for some } x(0) = \hat{\varphi} \in Z \Leftrightarrow$$

$$\int_0^{2\pi} U_0(2\pi, s)f(s)ds \in R(I - U_0(2\pi, 0)).$$

Indeed, $(I.13.10)_2$ implies that the solution given by $(I.13.9)$ satisfies $x(0) = x(2\pi)$. By the 2π-periodicity of $A_0(t)$ as well as of f, and by the uniqueness of the solution of the initial value problem $(I.13.9)$, the equality $x(0) = x(2\pi)$ implies the 2π-periodicity of x, and the regularity assumption $(I.13.9)$ finally gives

$$(I.13.11) \qquad x \in E, \quad \frac{dx}{dt} = A_0(t)x + f \in W, \text{ or } x \in Y.$$

Proposition I.13.1 *Under the assumptions $(I.13.6)$, $(I.13.7)$, and $(I.13.9)$ the operator $\hat{J}_0$ as given in $(I.13.8)$ is a Fredholm operator of index zero.*

Proof. By $(I.13.10)$, $x \in N(\hat{J}_0)$ is equivalent to $x(0) = \hat{\varphi} \in N(I - U_0(2\pi, 0))$. The compactness of $U_0(2\pi, 0)$ (cf.$(I.13.7)$) implies that $I - U_0(2\pi, 0)$ is a Fredholm operator of index zero in $L(Z, Z)$ (Riesz–Schauder Theory). By $(I.13.3)$ dim $N(I - U_0(2\pi, 0)) \geq 1$, and assuming $\dim(I - U_0(2\pi, 0)) = n+1$, the Fredholm property implies

$$(I.13.12) \qquad \begin{aligned} N(I - U_0(2\pi, 0)) &= \operatorname{span}[\hat{\varphi}_0, \ldots, \hat{\varphi}_n], \\ R(I - U_0(2\pi, 0)) \oplus Z_0 &= Z, \text{ and} \\ Z_0 &= \operatorname{span}[\hat{\varphi}_0^*, \ldots, \hat{\varphi}_n^*]. \end{aligned}$$

By the Hahn–Banach Theorem we find vectors

$$\hat{\varphi}_0', \ldots, \hat{\varphi}_n' \in Z' \text{ (the dual space) such that}$$

$$(I.13.13) \quad \langle \hat{\varphi}_j^*, \hat{\varphi}_k' \rangle = \delta_{jk} \quad (= 1 \text{ for } j = k, = 0 \text{ for } j \neq k), \text{ and}$$

$$\langle z, \hat{\varphi}_k' \rangle = 0 \text{ for } k = 0, \ldots, n \quad \Leftrightarrow \quad z \in R(I - U_0(2\pi, 0)).$$

We set

$$\hat{\psi}_j(t) = U_0(t, 0)\hat{\varphi}_j \qquad \text{and}$$

$$(I.13.14) \quad \hat{\psi}_k'(t) = U_0'(2\pi, t)\hat{\varphi}_k', \quad j, k = 0, \ldots, n, \quad t \in [0, 2\pi],$$

$$\text{where } U_0'(2\pi, t) \in L(Z', Z') \text{ is the dual operator.}$$

Then (again by $(I.13.10)$)

$$(I.13.15) \qquad\qquad N(\hat{J}_0) = \operatorname{span}[\hat{\psi}_0, \ldots, \hat{\psi}_n]$$

and

$$\int_0^{2\pi} \langle f(t), \hat{\psi}_k'(t) \rangle dt = 0 \text{ for } k = 0, \ldots, n \Leftrightarrow$$

$$(I.13.16) \quad \left\langle \int_0^{2\pi} U_0(2\pi, t)f(t)dt, \hat{\varphi}_k' \right\rangle = 0 \text{ by definition } (I.13.14)$$

$$\Leftrightarrow \int_0^{2\pi} U_0(2\pi, t)f(t)dt \in R(I - U_0(2\pi, 0)) \text{ by } (I.13.13)$$

$$\Leftrightarrow f \in R(\hat{J}_0) \text{ by } (I.13.10).$$

We show that (I.13.16) implies $\operatorname{codim}R(\hat{J}_0) = n+1$. By the choice of $\hat{\varphi}'_k$, in particular by (I.13.13)$_3$,

$$(\text{I.13.17}) \qquad \operatorname{span}[\hat{\varphi}'_0, \ldots, \hat{\varphi}'_n] \subset N(I - U'_0(2\pi, 0)),$$

which is the easy part of the Closed Range Theorem (cf. [170]). Therefore, $\hat{\psi}'_k(0) = U'_0(2\pi, 0)\hat{\varphi}'_k = \hat{\varphi}'_k = \hat{\psi}'_k(2\pi)$ and $\hat{\psi}'_0, \ldots, \hat{\psi}'_n$ are $n+1$ linearly independent functions in $C^{\alpha}_{2\pi}([0, 2\pi], Z')$ (which we can extend to $C^{\alpha}_{2\pi}(\mathbb{R}, Z')$).

We leave it as an exercise to prove the existence of

$$\hat{\psi}^*_0, \ldots, \hat{\psi}^*_n \in C^{\alpha}_{2\pi}(\mathbb{R}, Z) \text{ such that}$$

$$(\text{I.13.18}) \qquad \frac{1}{2\pi} \int_0^{2\pi} \langle \hat{\psi}^*_j(t), \hat{\psi}'_k(t)\rangle dt = \delta_{jk}, \quad j, k = 0, \ldots, n.$$

(Hint: By $\langle \hat{\varphi}^*_j, \hat{\psi}'_k(2\pi)\rangle = \delta_{jk}$, the real-valued function $\langle \hat{\varphi}^*_j, \hat{\psi}'_k(t)\rangle$ is not in $\operatorname{span}[\{\langle \hat{\varphi}^*_j, \hat{\psi}'_k\rangle | k \neq j\}] \subset C^{\alpha}_{2\pi}(\mathbb{R}, \mathbb{R})$. Therefore, there is some $f_j \in C^{\alpha}_{2\pi}(\mathbb{R}, \mathbb{R})$ such that $\frac{1}{2\pi} \int_0^{2\pi} f_j(t)\langle \hat{\varphi}^*_j, \hat{\psi}'_k(t)\rangle dt = \delta_{jk}$ and $\hat{\psi}^*_j(t) \equiv f_j(t)\hat{\varphi}^*_j$ satisfies (I.13.18).)

Then the $(n+1)$-dimensional projection $\hat{Q} \in L(W, W)$ defined by

$$(\text{I.13.19}) \qquad (\hat{Q}z)(t) = \sum_{k=0}^{n} \frac{1}{2\pi} \int_0^{2\pi} \langle z(t), \hat{\psi}'_k(t)\rangle dt \hat{\psi}^*_k(t)$$

has the property that

$$(\text{I.13.20}) \qquad \begin{aligned} &R(\hat{J}_0) = N(\hat{Q}) \text{ by (I.13.16) and} \\ &W = R(\hat{J}_0) \oplus R(\hat{Q}), \text{ whence } \operatorname{codim}R(\hat{J}_0) = n+1. \end{aligned}$$

This completes the proof of Proposition I.13.1. $\qquad\qquad\qquad\qquad$ $\square$

As usual, we set $\hat{Q}|_{E \cap Y} = \hat{P} \in L(E \cap Y, E \cap Y)$.

Corollary I.13.2 *Let $N(\hat{J}_0) = \operatorname{span}[\hat{\psi}_0, \ldots, \hat{\psi}_n]$. Then*

$$(\text{I.13.21}) \qquad \hat{\psi}_j \notin R(\hat{J}_0) \Leftrightarrow \hat{\varphi}_j \equiv \hat{\psi}_j(0) \notin R(I - U_0(2\pi, 0)).$$

If $\hat{\psi}_j \notin R(\hat{J}_0)$ for $j = 0, \ldots, m \leq n$, then we can choose $\hat{\psi}^*_j = \hat{\psi}_j$ for $j = 0, \ldots, m$ in formula (I.13.18) and for the definition (I.13.19) of the projection $\hat{Q}$.

Proof. If $\hat{\psi}_j \in N(\hat{J}_0)$, then by (I.13.10), $\hat{\psi}_j(t) = U_0(t, 0)\hat{\varphi}_j$, so that $\hat{\varphi}_j \in N(I - U_0(2\pi, 0))$, which means that $U_0(2\pi, 0)\hat{\varphi}_j = \hat{\varphi}_j$. Therefore,

$$\int_0^{2\pi} U_0(2\pi,t)\hat{\psi}_j(t)dt = \int_0^{2\pi} U_0(2\pi,t)U_0(t,0)\hat{\varphi}_j dt$$

(I.13.22)
$$= \int_0^{2\pi} U_0(2\pi,0)\hat{\varphi}_j dt = 2\pi\hat{\varphi}_j \notin R(I - U_0(2\pi,0)) \Leftrightarrow$$

$$\hat{\psi}_j \notin R(\hat{J}_0) \text{ by (I.13.16).}$$

Thus the complement Z_0 of $R(I - U_0(2\pi,0))$ can be spanned as

(I.13.23) $$Z_0 = \text{span}[\hat{\varphi}_0,\ldots,\hat{\varphi}_m,\hat{\varphi}^*_{m+1},\ldots,\hat{\varphi}^*_n],$$

or $\hat{\varphi}^*_j = \hat{\varphi}_j$ for $j = 0,\ldots,m$. We choose again $\hat{\varphi}'_k$ according to (I.13.13), and we define $\hat{\psi}'_k$ by (I.13.14) for $k = 0,\ldots,n$. Then, for $j = 0,\ldots,m$ and $k = 0,\ldots,n$,

$$\frac{1}{2\pi}\int_0^{2\pi} \langle \hat{\psi}_j(t), \hat{\psi}'_k(t)\rangle dt$$

(I.13.24)
$$= \frac{1}{2\pi}\int_0^{2\pi} \langle U_0(t,0)\hat{\varphi}_j, U'_0(2\pi,t)\hat{\varphi}'_k\rangle dt$$

$$= \frac{1}{2\pi}\int_0^{2\pi} \langle U_0(2\pi,t)U_0(t,0)\hat{\varphi}_j, \hat{\varphi}'_k\rangle dt$$

$$= \frac{1}{2\pi}\int_0^{2\pi} \langle \hat{\varphi}_j, \hat{\varphi}'_k\rangle dt = \delta_{jk},$$

which proves that we can choose $\hat{\psi}^*_j(t) = \hat{\psi}_j(t)$ for $j = 0,\ldots,m$. (The choice $\hat{\psi}^*_j(t) = U_0(t,0)\hat{\varphi}^*_j$ for $m+1 \leq j \leq n$ does not work, since $\hat{\psi}^*_j$ is not 2π-periodic, or $\hat{\psi}^*_j \notin W$ in this case: We tacitly assume that the number m such that $\hat{\varphi}_j = \hat{\psi}_j(0) \notin R(I - U_0(2\pi,0))$ for $j = 0,\ldots,m \leq n$ is maximal or that $\dim(N(I - U_0(2\pi,0)) \cap R(I - U_0(2\pi,0))) = n - m$. Therefore, the remaining $n - m$ vectors $\hat{\varphi}^*_j$ for $m+1 \leq j \leq n$ spanning the complement Z_0 of $R(I - U_0(2\pi,0))$ cannot be in $N(I - U_0(2\pi,0))$, which means that $U_0(2\pi,0)\hat{\varphi}^*_j \neq \hat{\varphi}^*_j$ for $m+1 \leq j \leq n$. Nonetheless, there exist such functions $\hat{\psi}^*_j \in W$ for $m+1 \leq j \leq n$ such that (I.13.18) is satisfied.) $\square$

Now we are ready to prove the **Implicit Function Theorem for Periodic Solutions of (I.13.1)**.

Theorem I.13.3 *Assume that (I.13.1) has a p_0-periodic solution x_0 for $\lambda = \lambda_0$ with a simple trivial Floquet exponent $\mu_0 = 0$ in the sense of (I.13.4) (after normalization of the period p_0 to 2π). Under the assumptions (I.13.6), (I.13.7), and (I.13.9), the periodic solution x_0 has a continuation described as follows: There are continuously differentiable mappings $x(\lambda)$ and $p(\lambda)$ defined on $(\lambda_0 - \delta, \lambda_0 + \delta)$ such that $x(\lambda_0) = x_0, p(\lambda_0) = p_0$, and $x(\lambda)$ is a $p(\lambda)$-periodic solution of (I.13.1). All periodic solutions of (I.13.1) in a* *neighborhood of* (x_0, λ_0) *(in the topology defined in the proof) having a period near* p_0 *are given by a phase shift of* $(x(\lambda), \lambda)$.

Proof. Substituting t/κ_0 for t where $p_0 = 2\pi/\kappa_0$, we prove the unique continuation of the solution $(x_0, \kappa_0, \lambda_0)$ of $G(x, \kappa, \lambda) \equiv \kappa \frac{dx}{dt} - F(x, \lambda) = 0$ in $(E \cap Y) \times \mathbb{R}_+ \times \mathbb{R}$; cf. (I.13.2). We define

$$(\mathrm{I}.13.25) \qquad \hat{G}(x, \kappa, \lambda) = \left(G(x, \kappa, \lambda), \int_0^{2\pi} \langle x - x_0, \hat{\psi}_0' \rangle dt \right),$$

where G is given by (I.8.10), (I.8.12), with $x_0 \in \tilde{U}$, and $\hat{\psi}_0'$ is given by (I.13.13), (I.13.14).

By assumption (I.13.4), we have $N(\hat{J}_0) = \mathrm{span}[\hat{\psi}_0]$, $\hat{\psi}_0 = \frac{d}{dt}x_0$, and $\hat{\psi}_0 \notin R(\hat{J}_0)$. Therefore, we can apply Corollary I.13.2, and the projection

$$(\mathrm{I}.13.26) \quad (\hat{Q}z)(t) = \frac{1}{2\pi} \int_0^{2\pi} \langle z(t), \hat{\psi}_0'(t) \rangle dt \hat{\psi}_0(t) \quad \text{(cf. (I.13.19))}$$

projects W onto $N(\hat{J}_0)$ along $R(\hat{J}_0)$.

By $G(x_0, \kappa_0, \lambda_0) = 0$, we clearly have $\hat{G}(x_0, \kappa_0, \lambda_0) = (0, 0)$, and the derivative

$$D_{(x, \kappa)}\hat{G}(x_0, \kappa_0, \lambda_0) \in L((E \cap Y) \times \mathbb{R}, W \times \mathbb{R})$$
is given by the matrix

$$(\mathrm{I}.13.27)$$
$$D_{(x, \kappa)}\hat{G}(x_0, \kappa_0, \lambda_0) = \begin{pmatrix} D_x G(x_0, \kappa_0, \lambda_0) & \frac{d}{dt}x_0 \\ \int_0^{2\pi} \langle \cdot, \hat{\psi}_0' \rangle dt & 0 \end{pmatrix}.$$

We claim that $D_{(x, \kappa)}\hat{G}(x_0, \kappa_0, \lambda_0)$ is bijective. Indeed,

$$D_{(x, \kappa)}\hat{G}(x_0, \kappa_0, \lambda_0)(x, \kappa) = (0, 0) \Leftrightarrow$$
$$(\mathrm{I}.13.28)$$
$$\hat{J}_0 x + \kappa \hat{\psi}_0 = 0 \text{ and } \int_0^{2\pi} \langle x, \hat{\psi}_0' \rangle dt = 0.$$

Since $\hat{\psi}_0 \notin R(\hat{J}_0)$ by assumption (I.13.4), the first equation of (I.13.28)$_2$ implies

$$(\mathrm{I}.13.29) \qquad\qquad x \in N(\hat{J}_0), \quad \kappa = 0,$$

and by the second equation of (I.13.28)$_2$,

$$(\mathrm{I}.13.30) \qquad\qquad\qquad x = \hat{Q}x = 0.$$

Thus $N(D_{(x, \kappa)}\hat{G}(x_0, \kappa_0, \lambda_0)(x, \kappa)) = \{(0, 0)\}$.

Again by assumption (I.13.4) we have $W = R(\hat{J}_0) \oplus \text{span}[\hat{\psi}_0]$, and by $\frac{1}{2\pi} \int_0^{2\pi} \langle \hat{\psi}_0, \hat{\psi}_0' \rangle dt = 1$, we see that $R(D_{(x,\kappa)}\hat{G}(x_0, \kappa_0, \lambda_0)) = W \times \mathbb{R}$, or that $D_{(x,\kappa)}\hat{G}(x_0, \kappa_0, \lambda_0)$ is surjective.

To summarize, the assumptions of the Implicit Function Theorem are satisfied for $\hat{G}$ at $(x_0, \kappa_0, \lambda_0)$. (If $F \in C^2(U \times V, Z)$, then $G, \hat{G} \in C^1(\tilde{U} \times \tilde{V}, W)$, where U is a neighborhood of $\{x_0(t) | t \in [0, 2\pi]\} \subset X, \lambda_0 \in V \subset \mathbb{R}, x_0 \in \tilde{U} \subset E \cap Y$, and $(\kappa_0, \lambda_0) \in \tilde{V} \subset \mathbb{R}^2$.)

Therefore, there exists a continuously differentiable curve $\{(x(\lambda), \kappa(\lambda), \lambda) | \lambda \in (\lambda_0 - \delta, \lambda_0 + \delta)\}$ in $(E \cap Y) \times \mathbb{R}_+ \times \mathbb{R}$ such that $(x(\lambda_0), \kappa(\lambda_0)) = (x_0, \kappa_0)$ and

$$\hat{G}(x(\lambda), \kappa(\lambda), \lambda) = (0,0), \text{ or, by (I.13.25)},$$
$$G(x(\lambda), \kappa(\lambda), \lambda) = 0 \text{ and}$$

(I.13.31)
$$\int_0^{2\pi} \langle x(\lambda) - x_0, \hat{\psi}_0' \rangle dt = 0 \text{ for all } \lambda \in (\lambda_0 - \delta, \lambda_0 + \delta).$$

Substituting finally $\kappa(\lambda)t$ for t in $x(\lambda)(t)$, we obtain a continuously differentiable curve $(x(\lambda), \lambda)$ of $p(\lambda)$-periodic solutions of (I.13.1) through (x_0, λ_0) and the continuation of the period $p(\lambda) = 2\pi/\kappa(\lambda)$ satisfies $p(\lambda_0) = 2\pi/\kappa_0 = p_0$.

The last equation in (I.13.31) links the phase of $x(\lambda)$ to that of x_0. Due to the equivariance (I.8.41), any phase shift in $x(\lambda)$ provides another solution of $G(x, \kappa, \lambda) = 0$.   □

In Section I.4 we pursue a curve of "stationary" solutions of $F(x, \lambda) = 0$ around a **turning point** (or a **fold**) by an Implicit Function Theorem with a one-dimensional kernel. We extend this to periodic solutions of $\frac{dx}{dt} = F(x, \lambda)$. We stay in the setting of (I.13.2); i.e., we fix the period to 2π by introducing a second parameter κ. We do not assume the simplicity of the trivial Floquet exponent $\mu_0 = 0$ in the sense of (I.13.4). Therefore, we have to distinguish two cases:

(I.13.32)
$$(1) \qquad \hat{\psi}_0 = \frac{d}{dt}x_0 \notin R(\hat{J}_0),$$
$$(2) \qquad \hat{\psi}_0 \in R(\hat{J}_0).$$

We begin with case (1); i.e., we are in the situation of the proof of Theorem I.13.3. We define $\hat{G}$ as in (I.13.25), and if $D_{(x,\kappa)}\hat{G}(x_0, \kappa_0, \lambda_0)$ has a trivial kernel $\{(0,0)\}$, we are precisely back to Theorem I.13.3. Therefore, we assume now

(I.13.33) $\dim N(D_{(x,\kappa)}\hat{G}(x_0, \kappa_0, \lambda_0)) = 1.$

By (I.13.27), (I.13.28), this is equivalent to

(I.13.34)
$$N(\hat{J}_0) = \text{span}[\hat{\psi}_0, \hat{\psi}_1], \text{ i.e., } \dim N(\hat{J}_0) = 2, \text{ and}$$
$$N(D_{(x,\kappa)}\hat{G}(x_0, \kappa_0, \lambda_0)) = \text{span}[(\hat{\psi}_1, 0)].$$

Furthermore,

(I.13.35) $R(D_{(x,\kappa)}\hat{G}(x_0, \kappa_0, \lambda_0)) = (R(\hat{J}_0) \oplus \text{span}[\hat{\psi}_0]) \times \mathbb{R}$,

which proves by $\text{codim} R(\hat{J}_0) = 2$ (cf. Proposition I.13.1) that

(I.13.36) $\text{codim} R(D_{(x,\kappa)}\hat{G}(x_0, \kappa_0, \lambda_0)) = 1$.

Therefore, $D_{(x,\kappa)}\hat{G}(x_0, \kappa_0, \lambda_0)$ is a Fredholm operator of index zero. Choosing $\hat{\psi}_0^* = \hat{\psi}_0$ according to Corollary I.13.2 and $\hat{\psi}_1'$ according to (I.13.13), (I.13.14), we have $\int_0^{2\pi} \langle \hat{\psi}_0, \hat{\psi}_1' \rangle dt = 0$ and finally by (I.13.16),

$$(f, 0) \notin R(D_{(x,\kappa)}\hat{G}(x_0, \kappa_0, \lambda_0)) \Leftrightarrow$$

(I.13.37)

$$\int_0^{2\pi} \langle f, \hat{\psi}_1' \rangle dt \neq 0.$$

Theorem I.4.1 is then applicable if

$$D_\lambda \hat{G}(x_0, \kappa_0, \lambda_0) = (-D_\lambda F(x_0, \lambda_0), 0) \notin R(D_{(x,\kappa)}\hat{G}(x_0, \kappa_0, \lambda_0))$$

(I.13.38)

$$\Leftrightarrow \int_0^{2\pi} \langle D_\lambda F(x_0, \lambda_0), \hat{\psi}_1' \rangle dt \neq 0.$$

Next we consider case (2); i.e., $\hat{\psi}_0 \in R(\hat{J}_0)$. Now we cannot apply Corollary I.13.2, and in particular, $\hat{\psi}_0^* \neq \hat{\psi}_0$, $\int_0^{2\pi} \langle \hat{\psi}_0, \hat{\psi}_0^* \rangle dt = 0$ according to (I.13.16). Nonetheless, we find a function $\tilde{\psi}_0' \in C_{2\pi}^\alpha(\mathbb{R}, Z')$ such that

(I.13.39) $\dfrac{1}{2\pi} \displaystyle\int_0^{2\pi} \langle \hat{\psi}_0, \tilde{\psi}_0' \rangle dt = 1$.

(The proof is similar to that of (I.13.18): Consider the 2π-periodic real-valued function $\langle \hat{\psi}_0(t), z' \rangle$ for some $z' \in Z'$ such that $\langle \hat{\psi}_0(t), z' \rangle \not\equiv 0$.) We now define $\hat{G}$ as in (I.13.25), but we replace $\hat{\psi}_0'$ by $\tilde{\psi}_0'$ satisfying (I.13.39). By the assumption $\hat{\psi}_0 \in R(\hat{J}_0)$, the corresponding equations (I.13.28) imply $\dim N(D_{(x,\kappa)}\hat{G}(x_0, \kappa_0, \lambda_0)) \geq 1$. Again we assume (I.13.33), which is equivalent to

(I.13.40)
$$N(\hat{J}_0) = \text{span}[\hat{\psi}_0], \text{ i.e., } \dim N(\hat{J}_0) = 1, \text{ and}$$
$$N(D_{(x,\kappa)}\hat{G}(x_0, \kappa_0, \lambda_0)) = \text{span}[(\hat{\psi}_1, 1)], \ \hat{\psi}_1 \neq 0.$$

Furthermore,
(I.13.41) $R(D_{(x,\kappa)}\hat{G}(x_0, \kappa_0, \lambda_0)) = R(\hat{J}_0) \times \mathbb{R}$,

which proves, by $\text{codim} R(\hat{J}_0) = 1$, that

(I.13.42) $\text{codim} R(D_{(x,\kappa)}\hat{G}(x_0, \kappa_0, \lambda_0)) = 1$.

Therefore, $D_{(x,\kappa)}\hat{G}(x_0,\kappa_0,\lambda_0)$ is a Fredholm operator of index zero, and by (I.13.16),

$$(f,0) \notin R(D_{(x,\kappa)}\hat{G}(x_0,\kappa_0,\lambda_0)) \Leftrightarrow$$

(I.13.43)

$$\int_0^{2\pi} \langle f, \hat{\psi}_0' \rangle dt \neq 0.$$

Thus Theorem I.4.1 is applicable, provided that

(I.13.44) $$\int_0^{2\pi} \langle D_\lambda F(x_0,\lambda_0), \hat{\psi}_0' \rangle dt \neq 0.$$

Remark I.13.4 *In both cases (I.13.32), the trivial Floquet exponent $\mu_0 = 0$ is at least algebraically double: In case (1) it is geometrically double by (I.13.34); in case (2) it is algebraically double, possibly of higher multiplicity if $\hat{\psi}_1$ in (I.13.40) is in $R(\hat{J}_0)$, too.*

We summarize our results as an **Implicit Function Theorem for Periodic Solutions with Two-Dimensional Kernels: Turning Points** (and we recall that our conditions are given when the period is normalized to 2π).

Theorem I.13.5 *Assume that (I.13.1) has a p_0-periodic solution x_0 for $\lambda = \lambda_0$ with a double trivial Floquet exponent $\mu_0 = 0$ in the sense of (I.13.34) or (I.13.40). Furthermore, we assume that the 2π-periodic function $D_\lambda F(x_0,\lambda_0)$ satisfies (I.13.38) or (I.13.44). Under the remaining assumptions (I.13.6), (I.13.7), and (I.13.9) of this section there are continuously differentiable curves $\{(x(s),\lambda(s))|s \in (-\delta,\delta)\}$ and $\{p(s)|s \in (-\delta,\delta)\}$ through $(x(0),\lambda(0)) = (x_0,\lambda_0)$ and $p(0) = p_0$, respectively, such that $x(s)$ is a $p(s)$-periodic solution of (I.13.1) for $\lambda = \lambda(s)$, and any periodic solution of (I.13.1) in a neighborhood of (x_0,λ_0) having a period near p_0 is obtained by a phase shift of $(x(s), \lambda(s))$.*

We recall that we prove the existence of a curve of solutions of $G(x,\kappa,\lambda) \equiv \kappa\frac{dx}{dt} - F(x,\lambda) = 0$ through (x_0,κ_0,λ_0) in $(C_{2\pi}^{1+\alpha}(\mathbb{R},Z) \cap C_{2\pi}^\alpha(\mathbb{R},X)) \times \mathbb{R}_+ \times \mathbb{R}$. This is achieved by applying Theorem I.4.1 to $\hat{G}(x,\kappa,\lambda) = 0$, where the pair (x,κ) is combined into one variable and λ is the only parameter. Corollary I.4.2 then also gives the tangent vector of the solution curve at (x_0,κ_0,λ_0), namely, by (I.13.34) or (I.13.40),

(I.13.45)
$$\frac{dx}{ds}(s)\Big|_{s=0} = \hat{\psi}_1, \quad \frac{d\kappa}{ds}(s)\Big|_{s=0} = 0 \text{ or } 1 \text{ , respectively,}$$
$$\frac{d\lambda}{ds}(s)\Big|_{s=0} = 0.$$

Assuming more differentiability of $\hat{G}$ (which means more differentiability of F, namely, $F \in C^4(U \times V, Z)$, where $U \subset X$ is a neighborhood of

$\{x_0(t)\}$ and $\lambda_0 \in V \subset \mathbb{R}$), the solution curve $\{(x(s), \kappa(s), \lambda(s))|s \in (-\delta, \delta)\}$ of $\hat{G}(x, \kappa, \lambda) = 0$ through $(x_0, \kappa_0, \lambda_0)$ is twice continuously differentiable.

In order to determine $\ddot{\lambda}(0)$ (and also $\ddot{\kappa}(0)$ in the first case; $\dot{} = \frac{d}{ds}$), we follow the lines of (I.4.16), and we obtain in case (1) (cf. (I.4.23)),

$$
(I.13.46) \quad \ddot{\lambda}(0) = -\frac{\int_0^{2\pi} \langle D_{xx}^2 F(x_0, \lambda_0)[\hat{\psi}_1, \hat{\psi}_1], \hat{\psi}_1' \rangle dt}{\int_0^{2\pi} \langle D_\lambda F(x_0, \lambda_0), \hat{\psi}_1' \rangle dt},
$$

$$
\ddot{\kappa}(0) = \frac{1}{2\pi} \int_0^{2\pi} \langle D_{xx}^2 F(x_0, \lambda_0)[\hat{\psi}_1, \hat{\psi}_1] + D_\lambda F(x_0, \lambda_0)\ddot{\lambda}(0), \hat{\psi}_0' \rangle dt,
$$

where we used $\frac{1}{2\pi} \int_0^{2\pi} \langle \hat{\psi}_0, \hat{\psi}_0' \rangle dt = 1$, $\int_0^{2\pi} \langle \hat{\psi}_0, \hat{\psi}_1' \rangle dt = 0$.

In case (2) we obtain by $\int_0^{2\pi} \langle \hat{\psi}_0, \hat{\psi}_0' \rangle dt = 0$,

$$
(I.13.47) \quad \ddot{\lambda}(0) = -\frac{\int_0^{2\pi} \langle D_{xx}^2 F(x_0, \lambda_0)[\hat{\psi}_1, \hat{\psi}_1] - 2\frac{d}{dt}\hat{\psi}_1, \hat{\psi}_0' \rangle dt}{\int_0^{2\pi} \langle D_\lambda F(x_0, \lambda_0), \hat{\psi}_0' \rangle dt}.
$$

Observe that the denominators are nonzero by assumptions (I.13.38) and (I.13.44), respectively. If the numerators are nonzero, too, then $\ddot{\lambda}(0) \neq 0$, and in this sense the curve $\{(x(s), \lambda(s))\}$ of $p(s)$-periodic solutions has a nondegenerate **turning point** (or **fold**) at $(x(0), \lambda(0)) = (x_0, \lambda_0)$. (Here it makes no sense to call it a "saddle-node bifurcation.") For a different expression of the numerator of (I.13.47), see (I.13.59).

A closer look reveals that the turning points are "generically" of two different types: The turning point in case (1) has an extremal period at $(x_0, \kappa_0, \lambda_0)$, since $\dot{\kappa}(0) = 0$ and $\ddot{\kappa}(0) \neq 0$. At the turning point of case (2), the period $p(s) = 2\pi/\kappa(s)$ is a monotonic function, since $\dot{\kappa}(0) = 1$.

Remark I.13.6 *We can unify all cases of a continuation described in Theorems I.13.3 and I.13.5 in a single point of view. In any case, $G(x(s), \kappa(s), \lambda(s)) = 0$ for $s \in (-\delta, \delta)$ (where $s = \lambda$ in case of Theorem I.13.3). Differentiating this equation with respect to s at $s = 0$ gives*

$$
(I.13.48) \quad D_x G(x_0, \kappa_0, \lambda_0)\dot{x}(0) + \frac{d}{dt}x_0\dot{\kappa}(0) - D_\lambda F(x_0, \lambda_0)\dot{\lambda}(0) = 0, \ or
$$

$$
\hat{J}_0\dot{x}(0) + \hat{\psi}_0\dot{\kappa}(0) - D_\lambda F(x_0, \lambda_0)\dot{\lambda}(0) = 0.
$$

The total derivative $D_{(x,\kappa,\lambda)}G(x_0, \kappa_0, \lambda_0)$ as a mapping from $(E \cap Y) \times \mathbb{R}_+ \times \mathbb{R}$ into W is given by the matrix

$$
(I.13.49) \quad D_{(x,\kappa,\lambda)}G(x_0, \kappa_0, \lambda_0) = (\hat{J}_0 \quad \hat{\psi}_0 \quad -D_\lambda F(x_0, \lambda_0))
$$

having the kernel vectors

$$(\dot{x}(0), \dot{\kappa}(0), \dot{\lambda}(0))$$

$$= (\dot{x}(0), \dot{\kappa}(0), 1) \ \text{in case of Theorem I.13.3,}$$

(I.13.50) $= (\hat{\psi}_1, 0, 0) \ \text{in case (1) of Theorem I.13.5,}$

$$= (\hat{\psi}_1, 1, 0) \ \text{in case (2) of Theorem I.13.5,} \ \hat{\psi}_1 \neq 0,$$

and in all cases it has the kernel vector $(\hat{\psi}_0, 0, 0)$, $\quad \hat{\psi}_0 = \dfrac{d}{dt} x_0.$

Thus $\dim N(D_{(x,\kappa,\lambda)}G(x_0, \kappa_0, \lambda_0)) = 2$ *in all cases. On the other hand, by the assumptions on* $\hat{\psi}_0$ *or on* $D_\lambda F(x_0, \lambda_0)$, *the total derivative* $D_{(x,\kappa,\lambda)}G(x_0, \kappa_0, \lambda_0)$ *is surjective in all cases.*

In this sense, 0 *is a "regular value" of* G *at its zero* $(x_0, \kappa_0, \lambda_0)$, *and since the kernel of its total derivative is two-dimensional, the solution set of* $G(x, \kappa, \lambda) = 0$ *near* $(x_0, \kappa_0, \lambda_0)$ *is a two-dimensional manifold (=surface) through* $(x_0, \kappa_0, \lambda_0)$ *described by*

(I.13.51) $\{(x(s)(\cdot + \theta), \kappa(s), \lambda(s)) | s \in (-\delta, \delta), \theta \in [0, 2\pi]\}.$

(For more details we refer to [102], [105].) We obtain a curve by fixing the phase θ *to* $\theta = 0$. *For obvious reasons, the surface (I.13.51) is called a "snake" in [134].*

I.13.1 Exchange of Stability at a Turning Point

At the end of this section we investigate the stability of the curves of periodic solutions of (I.13.1) given by Theorems I.13.3 and I.13.5. Clearly, "stability" means as usual "linear stability," which is determined by the Floquet exponents of the p-periodic solution x. To be more precise, only the nontrivial Floquet exponents play a role, provided that $\mu_0 = 0$ is a simple Floquet exponent of x in the sense of (I.13.4) (cf. the comments at the beginning of Section I.12).

In case of the Implicit Function Theorem for periodic solutions (Theorem I.13.3), the stability of $x(\lambda)$ is the same as that of $x(\lambda_0) = x_0$ for $\lambda \in (\lambda_0 - \delta, \lambda_0 + \delta)$ (if δ is sufficiently small). Indeed, the simplicity of the trivial Floquet exponent $\mu_0(\lambda) \equiv 0$ is preserved for λ near λ_0 if we assume (I.13.3) not only for x_0 (i.e., for $\lambda = \lambda_0$) but also for $x(\lambda)$ for λ near λ_0. Then $\hat{\psi}_0(\lambda) \equiv \frac{d}{dt}x(\lambda)$ is the eigenfunction of $D_xG(x(\lambda), \kappa(\lambda), \lambda)$ with eigenvalue $\mu_0(\lambda) \equiv 0$, and by the closedness of $R(D_xG(x(\lambda), \kappa(\lambda), \lambda))$ in $C_{2\pi}^\alpha(\mathbb{R}, Z) = W$ (cf. Proposition I.13.1), the simplicity of $\mu_0(\lambda) = 0$ is inherited from the simplicity of $\mu_0(\lambda_0) = 0$, since $R(D_xG(x(\lambda), \kappa(\lambda), \lambda))$ as well as $\hat{\psi}_0(\lambda)$ depends continuously on λ in the topology of W.

Next we investigate the linear stability near a turning point; i.e., we consider the stability of the curve $\{(x(s), \lambda(s)) | s \in (-\delta, \delta)\}$ of $p(s)$-periodic solu-

tions of (I.13.1) given by Theorem I.13.5. Again we focus only on the critical
Floquet exponents emanating from the double Floquet exponent $\mu_0 = 0$. To
do so we have to be sure that $\mu_0 = 0$ is an algebraically double eigenvalue
of $\hat{J}_0 = D_x G(x_0, \kappa_0, \lambda_0)$ such that the so-called 0-group of the perturbed
Floquet exponents is twofold (cf. [86]). Therefore, assumptions (I.13.34) and
(I.13.40) have to be sharpened in the sense that

$$(I.13.52) \qquad\qquad\qquad \hat{\psi}_1 \notin R(\hat{J}_0)$$

(cf. also Remark I.13.4). Assumption (I.13.52) implies by (I.13.34) or (I.13.40)
that

$$W \times \mathbb{R} = R(D_{(x,\kappa)}\hat{G}(x_0, \kappa_0, \lambda_0)) \oplus N(D_{(x,\kappa)}\hat{G}(x_0, \kappa_0, \lambda_0)),$$

(I.13.53)   or 0 is a simple eigenvalue of $D_{(x,\kappa)}\hat{G}(x_0, \kappa_0, \lambda_0)$ in the

sense of (I.7.4) with eigenvector $(\hat{\psi}_1, 0)$ or $(\hat{\psi}_1, 1)$, respectively.

Proposition I.7.2 gives the simple eigenvalue perturbation as follows
(for $D_{(x,\kappa)}\hat{G}(x, \kappa, \lambda)$ see ((I.13.27)):

$$
(I.13.54) \quad
\begin{aligned}
D_{(x,\kappa)}\hat{G}(x(s), \kappa(s), \lambda(s)) \begin{pmatrix} \hat{\psi}_1 + w(s) \\ \nu(s) \end{pmatrix} &= \mu_1(s) \begin{pmatrix} \hat{\psi}_1 + w(s) \\ \nu(s) \end{pmatrix}, \\
(w(s), \nu(s)) &\in R(D_{(x,\kappa)}\hat{G}(x_0, \kappa_0, \lambda_0)) \cap ((E \cap Y) \times \mathbb{R}).
\end{aligned}
$$

For $s = 0$ we have $w(0) = 0$, $\nu(0) = 0$ or 1, and $\mu_1(0) = 0$.
    The first equation of (I.13.54) is

$$
(I.13.55) \quad
\begin{aligned}
D_x G(x(s), \kappa(s), \lambda(s))(\hat{\psi}_1 + w(s)) &= \mu_1(s)(\hat{\psi}_1 + w(s)) - \nu(s)\hat{\psi}_0(s), \\
\text{where } \hat{\psi}_0(s) &= \frac{d}{dt} x(s).
\end{aligned}
$$

We assume (I.13.3) not only for $x_0 = x(0)$ but also for $x(s)$ for $s \in (-\delta, \delta)$.
Therefore, $\hat{\psi}_0(s)$ is the eigenfunction with eigenvalue $\mu_0(s) \equiv 0$, which is the
trivial Floquet exponent. Equation (I.13.55) is then the analogue to equation
(I.12.10), and the arguments given after Proposition I.12.1 prove that $\mu_1(s)$
is the nontrivial perturbed Floquet exponent.
    If $\hat{G}$ and therefore $(x(s), \kappa(s), \lambda(s))$ are smooth enough (which is guaran-
teed under the assumptions on F in Section I.9), we can differentiate equation
(I.13.55) with respect to s at $s = 0$, and using $w(0) = 0, \mu_1(0) = 0$, as well
as (I.13.45), we obtain

$$
(I.13.56) \quad
\begin{aligned}
-D_{xx}^2 F(x_0, \lambda_0)[\hat{\psi}_1, \hat{\psi}_1] + D_x G(x_0, \kappa_0, \lambda_0)\dot{w}(0) + \frac{d}{dt}\hat{\psi}_1 \dot{\kappa}(0) \\
= \dot{\mu}_1(0)\hat{\psi}_1 - \dot{\nu}(0)\hat{\psi}_0 - \nu(0)\frac{d}{ds}\hat{\psi}_0(0).
\end{aligned}
$$

First we treat case (1) when $\dot\kappa(0) = 0$, $\nu(0) = 0$, and where we can choose $\hat\psi_0^* = \hat\psi_0$, $\hat\psi_1^* = \hat\psi_1$ in view of Corollary I.13.2. For the choices of $\hat\psi_0'$, $\hat\psi_1'$, formula (I.13.18) is valid. Using also (I.13.16) for $D_x G(x_0, \kappa_0, \lambda_0)\dot w(0) \in R(\hat J_0)$, equation (I.13.56) yields

$$
\begin{aligned}
\dot\mu_1(0) &= -\frac{1}{2\pi} \int_0^{2\pi} \langle D_{xx}^2 F(x_0, \lambda_0)[\hat\psi_1, \hat\psi_1], \hat\psi_1' \rangle dt \\
&= \frac{1}{2\pi} \int_0^{2\pi} \langle D_\lambda F(x_0, \lambda_0), \hat\psi_1' \rangle dt \dot\lambda(0) \quad \text{by (I.13.46)} \quad \left(\dot{} = \tfrac{d}{ds} \right).
\end{aligned}
$$
(I.13.57)

In case (2), $\dot\kappa(0) = 1$, $\nu(0) = 1$, and $\int_0^{2\pi} \langle \hat\psi_0, \hat\psi_0' \rangle dt = 0$. Furthermore,

$$
\text{(I.13.58)} \quad \frac{d}{ds}\hat\psi_0(0) = \frac{d}{ds}\frac{d}{dt}x(0) = \frac{d}{dt}\frac{d}{ds}x(0) = \frac{d}{dt}\hat\psi_1 \quad \text{by (I.13.45)},
$$

and the relation $\hat J_0\hat\psi_1 + \hat\psi_0 = \kappa_0 \frac{d}{dt}\hat\psi_1 - D_x F(x_0, \lambda_0)\hat\psi_1 + \hat\psi_0 = 0$ (cf. (I.13.40)) implies

$$
\text{(I.13.59)} \quad \int_0^{2\pi} \left\langle \frac{d}{dt}\hat\psi_1, \hat\psi_0' \right\rangle dt = \frac{1}{\kappa_0} \int_0^{2\pi} \langle D_x F(x_0, \lambda_0)\hat\psi_1, \hat\psi_0' \rangle dt.
$$

Equation (I.13.56) finally yields

$$
\begin{aligned}
\dot\mu_1(0) &= \frac{1}{2\pi} \int_0^{2\pi} \left\langle 2\frac{d}{dt}\hat\psi_1 - D_{xx}^2 F(x_0, \lambda_0)[\hat\psi_1, \hat\psi_1], \hat\psi_0' \right\rangle dt \\
&= \frac{1}{2\pi} \int_0^{2\pi} \langle D_\lambda F(x_0, \lambda_0), \hat\psi_0' \rangle dt \dot\lambda(0) \quad \text{by (I.13.47)}.
\end{aligned}
$$
(I.13.60)

Therefore, $\dot\mu_1 \neq 0$ at a nondegenerate turning point in case (1) and also in case (2). Thus the critical (nontrivial) Floquet exponent $\mu_1(s)$ changes sign at $s = 0$, i.e., at the turning point (x_0, λ_0). If the remaining Floquet exponents are in the stable right half-plane of $\mathbb{C}$ (cf. (I.12.2)), then the stability of the $p(s)$-periodic solution $x(s)$ of (I.13.1) for $\lambda = \lambda(s)$ changes at the turning point $(x(0), \lambda(0)) = (x_0, \lambda_0)$. We summarize:

Theorem I.13.7 *If a nontrivial periodic solution x_0 of (I.13.1) for $\lambda = \lambda_0$ is continued via the Implicit Function Theorem for periodic solutions (Theorem I.13.3), then the stability does not change. If it is continued around a nondegenerate turning point via the two cases described in Theorem I.13.5, then its stability changes at the turning point.*

The possibilities are sketched in Figure I.13.1, where each point on the curves represents a periodic solution of (I.13.1).

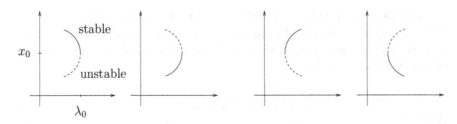

x_0

λ_0

Figure I.13.1

Remark I.13.8 *Usually, the continuation of periodic solutions is proved by the continuation of fixed points of the Poincaré map associated with (I.13.1). This fine geometric setting, however, requires the definition, existence, and smoothness of the Poincaré map generated by the nonlinear flow in a neighborhood of a periodic solution x_0 (for the parameter λ_0). Although that causes no problems for ODEs, for parabolic PDEs the existence of a nonlinear flow usually requires more assumptions on the mapping F (cf. (I.8.2)). We mention that typically, the nonlinear remainder of F (i.e., $F(x,\lambda) - D_x F(0,\lambda)x$) is defined on a domain of a fractional power of $A_0 = D_x F(0,\lambda_0)$ that is an intermediate (or interpolation) space between X and Z (see [88], [90], [76], [142], for instance). In other words, (I.13.1) represents a semilinear parabolic PDE. Then the derivative of the Poincaré map at the p_0-periodic solution x_0 is $U_0(p_0, 0)$, which is the period map of the variational equation. The continuation of a fixed point of the Poincaré map is then guaranteed if the trivial Floquet multiplier 1 of $U_0(p_0, 0)$ is simple. The occurrence of the two types of turning points is proved in an analogous way.*

We restrict our investigations a priori to periodic solutions, or in other words, we do not need the entire nonlinear flow in a neighborhood of a periodic solution. Therefore, our assumptions on F seem to be weaker. On the other hand, we need the regularity assumptions (I.13.3) on the periodic solution x_0, and (I.13.6) on the periodic family of operators $D_x F(x_0, \lambda_0)$. We do not know whether these assumptions require the same conditions on F as the construction of the Poincaré map around the periodic solution x_0. In any case, that construction requires some additional steps, and it certainly uses the linear fundamental solution (or evolution operator) $U_0(t, \tau)$ generated by $D_x F(x_0, \lambda_0)$ as well. In order to avoid these additional steps we do not prove the continuation of periodic solutions in the common way using the Poincaré map, but we stay in the setting of Fourier analysis.

The same comments refer also to the next section: Typically, the period-doubling bifurcation is proved by a continuation of a fixed point of the iterated Poincaré map (which is not a fixed point of the Poincaré map itself). For the same reasons explained before, we stay in our setting and study (I.13.1) in a space of periodic functions.

I.14 Period-Doubling Bifurcation and Exchange of Stability

We resume the situation of Section I.13, where we prove the existence of a smooth curve of periodic solutions of

$$(I.14.1) \qquad\qquad \frac{dx}{dt} = F(x, \lambda)$$

parameterized by λ or having a turning point and therefore parameterized as $\{(x(s), \lambda(s)) | s \in (-\delta, \delta)\}$. The continuation of the period $p = p(s)$ is part of the problem, and some remarks on that period are in order. Obviously, any integer multiple of a period is again a period, and so far, we did not require that the period be the minimal period. (In case of the Hopf Bifurcation, the emanating path has the **minimal** period $2\pi/\kappa(r)$, where $\kappa(0) = \kappa_0$ is determined by the imaginary eigenvalue $i\kappa_0$ of $A_0 = D_x F(0, \lambda_0)$ (cf. Theorem I.8.2).) In the situation considered in Section I.13, however, the following could happen: A p_0-periodic solution x_0 of (I.14.1) can be continued by the methods of Section I.13, but the same solution x_0 cannot be continued as a kp_0-periodic solution for some integer $k > 1$. Thus it could make a difference whether we consider some period of x_0 or the minimal period of x_0. We make this more precise now.

We define κ_0 by $p_0 = 2\pi/k\kappa_0$ for some $k \in \mathbb{N}$, and as usual, we substitute t/κ_0 for t in x_0. Without changing the notation for x, the equation $G(x, \kappa, \lambda) = \kappa \frac{dx}{dt} - F(x, \lambda) = 0$ has a solution $(x_0, \kappa_0, \lambda_0)$ in $(C^{1+\alpha}_{2\pi/k}(\mathbb{R}, Z) \cap C^{\alpha}_{2\pi/k}(\mathbb{R}, X)) \times \mathbb{R}_+ \times \mathbb{R}$ by assumption.

That solution has a unique continuation if the assumptions of Section I.13 are satisfied (when 2π is replaced by $2\pi/k$). These assumptions are summarized in Remark I.13.6: The total derivative $D_{(x,\kappa,\lambda)}G(x_0, \kappa_0, \lambda_0)$ has a two-dimensional kernel, and it is surjective.

Since x_0 is also 2π-periodic, we can consider $G(x, \kappa, \lambda)$ also as a mapping in the corresponding space of 2π-periodic functions having a zero $(x_0, \kappa_0, \lambda_0)$ in $(C^{1+\alpha}_{2\pi}(\mathbb{R}, Z) \cap C^{\alpha}_{2\pi}(\mathbb{R}, X)) \times \mathbb{R}_+ \times \mathbb{R}$. The assumptions of Section I.13 are violated if the total derivative $D_{(x,\kappa,\lambda)}G(x_0, \kappa_0, \lambda_0)$ (see (I.13.49)) is no longer surjective in the space of 2π-periodic functions. By the Fredholm property of $\hat{J}_0$ (cf. Proposition I.13.1), this is equivalent to the fact that $D_{(x,\kappa,\lambda)}G(x_0, \kappa_0, \lambda_0)$ has a kernel of dimension greater than 2. (It is easy to show that the total derivative $D_{(x,\kappa,\lambda)}G(x_0, \kappa_0, \lambda_0)$ as given by (I.13.49) is a Fredholm operator of index 2, cf. Definition I.2.1, since $\hat{J}_0$ is a Fredholm operator of index 0.)

If $D_{(x,\kappa,\lambda)}G(x_0, \kappa_0, \lambda_0)$ has a kernel of dimension greater than 2 in the space $(C^{1+\alpha}_{2\pi}(\mathbb{R}, Z) \cap C^{\alpha}_{2\pi}(\mathbb{R}, X)) \times \mathbb{R}_+ \times \mathbb{R}$, we expect a bifurcation from the curve of $2\pi/k$-periodic solutions of (I.14.1) into the space of 2π-periodic solutions, which means a multiplication of the period by k. A "simple" bi-

furcation, however, occurs only if the dimension of the kernel is 3, i.e., if its
dimension is increased by 1.

Remark I.14.1 *We give arguments for our assumption (I.14.17) below on a
simple period-doubling bifurcation; i.e., we explain why we restrict the factor
k of a period multiplication to 2. To that end we choose the setting of a Hilbert
space as in [94], [102], [105], where*

$$G : D_{2\pi} \times \mathbb{R} \times \mathbb{R} \to H_{2\pi},$$

$$H_{2\pi} = L^2[(0, 2\pi), H] \text{ with scalar product}$$

$$(\ ,\ )_0 = \frac{1}{2\pi} \int_0^{2\pi} (\ ,\ )_H dt, \quad (\ ,\ )_H = \text{scalar product in } H,$$

(I.14.2) $D \subset H$ *(densely, continuously),*

$$D_{2\pi} = \left\{ x \in H_{2\pi} | \frac{dx}{dt} \in H_{2\pi},\ x \in L^2[(0, 2\pi), D],\ x(0) = x(2\pi) \right\}$$

$$\text{with scalar product } (\ ,\ )_1 = (\frac{d}{dt}\cdot, \frac{d}{dt}\cdot)_0 + \frac{1}{2\pi} \int_0^{2\pi} (\ ,\ )_D dt,$$

$$(\ ,\ )_D = \text{scalar product in } D.$$

*Choosing a complete orthonormal system $\{f_m\}_{m \in \mathbb{N}}$ in H, we expand $x \in H_{2\pi}$
into a Fourier series (after a natural complexification) as follows:*

(I.14.3)
$$x(t) = \sum_{\substack{m \in \mathbb{N} \\ n \in \mathbb{Z}}} c_{mn} f_m e^{int}, \quad c_{mn} = (x, f_m e^{int})_0,$$
$$c_{m,-n} = \overline{c}_{mn} \text{ for real } x.$$

*(If $x \in D_{2\pi}$, the expansion (I.14.3) holds as well with a stronger convergence
of the Fourier coefficients.) It is crucial in this Hilbert space setting that
the closed subspaces $H_{2\pi/k}$ and $D_{2\pi/k}$ of $2\pi/k$-periodic functions have closed
orthogonal complements. We decompose*

$$x(t) = \sum_{\substack{m \in \mathbb{N} \\ n=k\ell, \ell \in \mathbb{Z}}} c_{mn} f_m e^{int} + \sum_{\substack{m \in \mathbb{N} \\ n \neq k\ell}} c_{mn} f_m e^{int},$$

(I.14.4)
which yields the decompositions
$$H_{2\pi} = H_{2\pi/k} \oplus \hat{H}_{2\pi/k}, \quad D_{2\pi} = D_{2\pi/k} \oplus \hat{D}_{2\pi/k}$$
with projections $Q_k, \hat{Q}_k$ such that $Q_k + \hat{Q}_k = I$,
$$P_k = Q_k|_{D_{2\pi}}, \quad \hat{P}_k = \hat{Q}_k|_{D_{2\pi}}.$$

*The total derivative $D_{(x,\kappa,\lambda)} G(x_0, \kappa_0, \lambda_0)$ given by the matrix $(\hat{J}_0 \quad \hat{\psi}_0$
$- D_\lambda F(x_0, \lambda_0))$ (cf.(I.13.49)) has the following property: Since $x_0 \in D_{2\pi/k}$,
the functions $\hat{\psi}_0 = \frac{d}{dt}x_0$, $D_\lambda F(x_0, \lambda_0)$, and $D_x F(x_0, \lambda_0)$ are in $H_{2\pi/k}$, and
since by definition, $\hat{J}_0 \equiv \kappa_0 \frac{d}{dt} - D_x F(x_0, \lambda_0)$, the subspace of $2\pi/k$-periodic
functions is invariant:*

(I.14.5) $\qquad D_{(x,\kappa,\lambda)}G(x_0,\kappa_0,\lambda_0) : D_{2\pi/k} \times \mathbb{R} \times \mathbb{R} \to H_{2\pi/k}.$

Whereas this invariance is true in the Banach space formulation as well, i.e.,
$$D_{(x,\kappa,\lambda)}G(x_0,\kappa_0,\lambda_0) : (C^{1+\alpha}_{2\pi/k}(\mathbb{R},Z) \cap C^{\alpha}_{2\pi/k}(\mathbb{R},X)) \times \mathbb{R} \times \mathbb{R} \to C^{\alpha}_{2\pi/k}(\mathbb{R},Z),$$
the following invariance of the complement can hardly be stated in that setting. We claim that

(I.14.6) $\qquad\qquad \begin{aligned} &\hat{J}_0 : \hat{D}_{2\pi/k} \to \hat{H}_{2\pi/k}, \ \textit{or equivalently,} \\ &Q_k \hat{J}_0 \hat{P}_k = 0. \end{aligned}$

Let $S_{2\pi/k}$ be the shift operator (I.8.31). By the $2\pi/k$-periodicity of x_0, we see easily that
(I.14.7) $\qquad\qquad\qquad S_{2\pi/k}\hat{J}_0 = \hat{J}_0 S_{2\pi/k}.$

The action of $S_{2\pi/k}$ on $\hat{H}_{2\pi/k}$ is described by

(I.14.8)
$$\begin{aligned} &S_{2\pi/k}x(t) = \sum_{\substack{m\in\mathbb{N} \\ n\neq k\ell}} e^{i2\pi n/k}c_{mn}f_m e^{int} \ \textit{such that} \\ &e^{i2\pi n/k} \neq 1 \ \textit{for all } n \neq k\ell \ \textit{and} \\ &I - S_{2\pi/k} \ \textit{is an isomorphism in } L(\hat{H}_{2\pi/k}, \hat{H}_{2\pi/k}). \end{aligned}$$

(Obviously, $I - S_{2\pi/k}$ is also an isomorphism in $L(\hat{D}_{2\pi/k}, \hat{D}_{2\pi/k})$.) We now prove (I.14.6). By definition of Q_k, we have

(I.14.9)
$$\begin{aligned} &S_{2\pi/k}Q_k\hat{J}_0\hat{P}_k = Q_k\hat{J}_0\hat{P}_k, \ \textit{and by (I.14.7)}, \\ &S_{2\pi/k}Q_k\hat{J}_0\hat{P}_k = Q_k\hat{J}_0\hat{P}_k S_{2\pi/k}. \end{aligned}$$

(The projections commute with any shift operator.) Therefore,

(I.14.10) $\qquad Q_k\hat{J}_0\hat{P}_k(I - S_{2\pi/k}) = 0 \ \textit{in } L(\hat{D}_{2\pi/k}, H_{2\pi/k}),$

which proves (I.14.6) by (I.14.8). Identifying "$\oplus$" and "$\times$" in an obvious way, we see that the matrix of

$$D_{(x,\kappa,\lambda)}G(x_0,\kappa_0,\lambda_0) : D_{2\pi/k} \times \hat{D}_{2\pi/k} \times \mathbb{R} \times \mathbb{R} \to H_{2\pi/k} \times \hat{H}_{2\pi/k}$$

(I.14.11)
$$\textit{is therefore given by } \begin{pmatrix} \hat{J}_0 & 0 & \hat{\psi}_0 & -D_\lambda F(x_0,\lambda_0) \\ 0 & \hat{J}_0 & 0 & 0 \end{pmatrix}.$$

This proves, in turn, the following:

(I.14.12)
$$\begin{aligned} &N(D_{(x,\kappa,\lambda)}G(x_0,\kappa_0,\lambda_0)) \\ &= N(D_{(x,\kappa,\lambda)}G(x_0,\kappa_0,\lambda_0))\big|_{D_{2\pi/k}\times\mathbb{R}\times\mathbb{R}} \times N(\hat{J}_0\big|_{\hat{D}_{2\pi/k}}). \end{aligned}$$

Returning to the theme of this section, if the first part of that kernel is two-dimensional, we expect a "simple" period-multiplying bifurcation from the curve of $2\pi/k$-periodic solutions of (I.14.1) into the complement of 2π-periodic functions (that are not $2\pi/k$-periodic), provided that the dimension of $N(\hat{J}_0|_{\hat{D}_{2\pi/k}})$ is one.

Next we investigate when this one-dimensionality is possible, or when

(I.14.13) $\qquad N(\hat{J}_0|_{\hat{D}_{2\pi/k}}) = \mathrm{span}[\hat{\psi}_2] \text{ for some } \hat{\psi}_2 \in \hat{D}_{2\pi/k}.$

By (I.14.7),

(I.14.14)
$\qquad \hat{\psi}_2 \text{ and } S_{2\pi/k}\hat{\psi}_2 \text{ are both in } N(\hat{J}_0|_{\hat{D}_{2\pi/k}}),$
$\qquad \text{and (I.14.13) is true only if}$
$\qquad \hat{\psi}_2 \text{ and } S_{2\pi/k}\hat{\psi}_2 \text{ are linearly dependent over } \mathbb{R}.$

By (I.14.8), this, in turn, is true only if

(I.14.15)
$\qquad e^{i2\pi n/k} = -1 \text{ for all } n \neq k\ell, \ \ell \in \mathbb{Z},$
$\qquad \text{which requires that } k \text{ be even and that}$
$\qquad (\hat{\psi}_2, f_m e^{int})_0 \neq 0 \text{ only for } n = \frac{k}{2}(2\ell + 1).$

Therefore, necessarily,

(I.14.16)
$\qquad S_{2\pi/k}\hat{\psi}_2 = -\hat{\psi}_2 \text{ and}$
$\qquad \hat{\psi}_2 \text{ is } 4\pi/k\text{-periodic.}$

This finishes the proof that in the Hilbert space setting the only **simple** period-multiplying bifurcation is the **period-doubling bifurcation**. For ODEs the Hilbert space setting and the Hölder space setting (which we choose) are equivalent. For parabolic PDEs, however, the Hilbert space formulation (I.14.2) restricts the class of nonlinear operators (cf. [89], [94]), which is why we prefer the Banach space formulation in Sections I.8–I.14.

We return to our usual definiton of $G(x, \kappa, \lambda) = \kappa\frac{dx}{dt} - F(x, \lambda)$ defined in $(C_{2\pi}^{1+\alpha}(\mathbb{R}, Z) \cap C_{2\pi}^{\alpha}(\mathbb{R}, X)) \times \mathbb{R}_+ \times \mathbb{R}$ and having values in $C_{2\pi}^{\alpha}(\mathbb{R}, Z)$. Again, we use the notation $W = C_{2\pi}^{\alpha}(\mathbb{R}, Z)$, $E = C_{2\pi}^{\alpha}(\mathbb{R}, X)$, and $Y = C_{2\pi}^{1+\alpha}(\mathbb{R}, Z)$.

We give the assumptions for a simple period-doubling bifurcation motivated by Remark I.14.1.

Since the normalization of the period p_0 of x_0 to $2\pi/k$ (via $p_0 = 2\pi/k\kappa_0$ and substitution t/κ_0 for t) is arbitrary, we now choose $k = 2$, so that the period of x_0 is normalized to π. We assume that

$$G(x_0, \kappa_0, \lambda_0) = 0 \text{ for some}$$
$$(x_0, \kappa_0, \lambda_0) \in (C_\pi^{1+\alpha}(\mathbb{R}, Z) \cap C_\pi^\alpha(\mathbb{R}, X)) \times \mathbb{R}_+ \times \mathbb{R},$$

(I.14.17)
$$N\left(D_{(x,\kappa,\lambda)}G(x_0, \kappa_0, \lambda_0)\big|_{(C_\pi^{1+\alpha}(\mathbb{R},Z)\cap C_\pi^\alpha(\mathbb{R},X))\times\mathbb{R}\times\mathbb{R}}\right)$$
$$= \operatorname{span}[(\hat{\psi}_0, 0, 0), (\hat{\psi}_1, \kappa_1, \lambda_1)] \text{ (cf. (I.13.50))},$$

$$N\left(D_{(x,\kappa,\lambda)}G(x_0, \kappa_0, \lambda_0)\big|_{(C_{2\pi}^{1+\alpha}(\mathbb{R},Z)\cap C_{2\pi}^\alpha(\mathbb{R},X))\times\mathbb{R}\times\mathbb{R}}\right)$$
$$= \operatorname{span}[(\hat{\psi}_0, 0, 0), (\hat{\psi}_1, \kappa_1, \lambda_1), (\hat{\psi}_2, 0, 0)],$$

and $S_\pi \hat{\psi}_2 = -\hat{\psi}_2$.

The three possible cases of $(\hat{\psi}_1, \kappa_1, \lambda_1)$ are summarized in (I.13.50), and the vector $(\hat{\psi}_1, \kappa_1, \lambda_1) = (\dot{x}(0), \dot{\kappa}(0), \dot{\lambda}(0))$ is the tangent vector of the curve of π-periodic solutions $\{(x(s), \kappa(s), \lambda(s)) | s \in (-\delta, \delta)\}$ of $G(x, \kappa, \lambda) = 0$ through $(x_0, \kappa_0, \lambda_0)$ (and an arbitrary phase shift generates a surface with the second tangent vector $(\hat{\psi}_0, 0, 0)$).

We draw some conclusions from assumption (I.14.17). As mentioned before, $D_{(x,\kappa,\lambda)}G(x_0, \kappa_0, \lambda_0)$ is a Fredholm operator of index two when considered in the space of π- or 2π-periodic functions.

Therefore, $D_{(x,\kappa,\lambda)}G(x_0, \kappa_0, \lambda_0)$ is surjective when restricted to the space $(C_\pi^{1+\alpha}(\mathbb{R}, Z) \cap C_\pi^\alpha(\mathbb{R}, X)) \times \mathbb{R} \times \mathbb{R}$, and $\operatorname{codim}R(D_{(x,\kappa,\lambda)}G(x_0, \kappa_0, \lambda_0)) = 1$ when π is replaced by 2π. Thus

$$R(D_{(x,\kappa,\lambda)}G(x_0, \kappa_0, \lambda_0))$$

(I.14.18)
$$= \left\{ z \in C_{2\pi}^\alpha(\mathbb{R}, Z) \,\Big|\, \int_0^{2\pi} \langle z, \hat{\psi}_2' \rangle dt = 0 \right\}$$
for some $\hat{\psi}_2' \in C_{2\pi}^\alpha(\mathbb{R}, Z')$ (cf. (I.13.16)), and in particular,

$$\int_0^{2\pi} \langle z, \hat{\psi}_2' \rangle dt = 0 \text{ for all } z \in C_\pi^\alpha(\mathbb{R}, Z).$$

The last relation can be restated as follows, where we use the π-periodicity of z and the shift operator S_π:

(I.14.19)
$$\int_0^\pi \langle z, \hat{\psi}_2' + S_\pi \hat{\psi}_2' \rangle dt = 0 \quad \text{for all } z \in C_\pi^\alpha(\mathbb{R}, Z),$$
which implies $S_\pi \hat{\psi}_2' = -\hat{\psi}_2'$.

Next we choose functions $\hat{\psi}_0', \hat{\psi}_1' \in C_\pi^\alpha(\mathbb{R}, Z')$ such that

(I.14.20)
$$\frac{1}{\pi} \int_0^\pi \langle \hat{\psi}_0, \hat{\psi}_0' \rangle dt = 1, \quad \frac{1}{\pi} \int_0^\pi \langle \hat{\psi}_1, \hat{\psi}_1' \rangle dt = 1,$$
whence in view of $S_\pi \hat{\psi}_2 = -\hat{\psi}_2$,
$$\int_0^{2\pi} \langle \hat{\psi}_2, \hat{\psi}_0' \rangle dt = \int_0^{2\pi} \langle \hat{\psi}_2, \hat{\psi}_1' \rangle dt = 0.$$

(The functions $\hat\psi_0'$, $\hat\psi_1'$ exist by the argument following (I.13.39).)

We now prove a simple period-doubling bifurcation from the curve of π-periodic solutions passing through $(x_0, \kappa_0, \lambda_0)$. That curve (whose existence is proved in Section I.13) is locally parameterized as $\{(x(s), \kappa(s), \lambda(s))|s \in (-\delta, \delta)\} \subset (C_\pi^{1+\alpha}(\mathbb{R}, Z) \cap C_\pi^\alpha(\mathbb{R}, X)) \times \mathbb{R} \times \mathbb{R}$, where $(x(0), \kappa(0), \lambda(0)) = (x_0, \kappa_0, \lambda_0)$. We define

$$\tilde G(u, s) = G(x(s) + x, \kappa(s) + \kappa, \lambda(s) + \lambda),$$
where $u = (x, \kappa, \lambda)$ and

(I.14.21) $\qquad \hat G(u, s) = \left(\tilde G(u, s), \int_0^{2\pi} \langle x, \hat\psi_0' \rangle dt, \int_0^{2\pi} \langle x, \hat\psi_1' \rangle dt \right),$

$$\hat G : \hat U \times (-\delta, \delta) \to W \times \mathbb{R} \times \mathbb{R},$$
$$0 \in \hat U \subset (E \cap Y) \times \mathbb{R} \times \mathbb{R}.$$

The Fredholm property of $\hat J_0 = D_x G(x_0, \kappa_0, \lambda_0)$ (cf. Proposition I.13.1) induces the Fredholm property to

$$D_u \hat G(0,0)(x, \kappa, \lambda)$$
(I.14.22)
$$= \left(D_{(x,\kappa,\lambda)} G(x_0, \kappa_0, \lambda_0)(x, \kappa, \lambda), \int_0^{2\pi} \langle x, \hat\psi_0' \rangle dt, \int_0^{2\pi} \langle x, \hat\psi_1' \rangle dt \right),$$
where $D_{(x,\kappa,\lambda)} G(x_0, \kappa_0, \lambda_0)$ is given by (I.13.49).

To be more precise, $D_u \hat G(0,0)$ is a Fredholm operator of index zero.

Obviously, $\hat G(0, s) = 0$ for $s \in (-\delta, \delta)$ (which is the "trivial" solution line), and in view of (I.14.17) and (I.14.20),

(I.14.23)
$$N(D_u \hat G(0,0)) = \operatorname{span}[(\hat\psi_2, 0, 0)] \text{ and}$$
$$R(D_u \hat G(0,0)) = R(D_{(x,\kappa,\lambda)} G(x_0, \kappa_0, \lambda_0)) \times \mathbb{R} \times \mathbb{R}.$$

We can apply Theorem I.5.1, provided that the nondegeneracy condition (I.5.3) is satisfied:

(I.14.24) $\qquad\qquad D_{us}^2 \hat G(0,0)(\hat\psi_2, 0, 0) \notin R(D_u \hat G(0,0)).$

(If $F \in C^3(U \times V, Z)$, then $\hat G \in C^2(\hat U \times (-\delta, \delta), W \times \mathbb{R} \times \mathbb{R})$, where U is a neighborhood of $\{x_0(t)|t \in [0, \pi]\} \subset X, \lambda_0 \in V \subset \mathbb{R}$.) In view of (I.14.18) and (I.14.23), this condition (I.14.24) reads as follows:

$$D_{us}^2 \tilde{G}(0,0)(\hat{\psi}_2, 0, 0)$$

$$= \dot{\kappa}(0)\frac{d}{dt}\hat{\psi}_2 - D_{xx}^2 F(x_0, \lambda_0)[\dot{x}(0), \hat{\psi}_2] - \dot{\lambda}(0)D_{x\lambda}^2 F(x_0, \lambda_0)\hat{\psi}_2$$

$$= \kappa_1 \frac{d}{dt}\hat{\psi}_2 - D_{xx}^2 F(x_0, \lambda_0)[\hat{\psi}_1, \hat{\psi}_2] - \lambda_1 D_{x\lambda}^2 F(x_0, \lambda_0)\hat{\psi}_2$$

(I.14.25)   by $(\dot{x}(0), \dot{\kappa}(0), \dot{\lambda}(0)) = (\hat{\psi}_1, \kappa_1, \lambda_1)$ $\left(\dot{} = \dfrac{d}{ds} \right)$ and

$$D_{us}^2 \hat{G}(0,0)(\hat{\psi}_2, 0, 0) \notin R(D_u \hat{G}(0,0)) \Leftrightarrow$$

$$\int_0^{2\pi} \langle \kappa_1 \frac{d}{dt}\hat{\psi}_2 - D_{xx}^2 F(x_0, \lambda_0)[\hat{\psi}_1, \hat{\psi}_2] - \lambda_1 D_{x\lambda}^2 F(x_0, \lambda_0)\hat{\psi}_2, \hat{\psi}_2' \rangle \, dt$$

$$\neq 0.$$

Theorem I.5.1 then yields the following **Period-Doubling Bifurcation Theorem**.

Theorem I.14.2 *Assume the general hypotheses of Sections I.13 and I.14 on* $G(x, \kappa, \lambda) \equiv \kappa \frac{dx}{dt} - F(x, \lambda)$, *and in particular, assume (I.14.17) and (I.14.25). Then two continuously differentiable curves of solutions of* $G(x, \kappa, \lambda) = 0$ *pass through* $(x_0, \kappa_0, \lambda_0)$:

$\{(x(s), \kappa(s), \lambda(s))|s \in (-\delta, \delta)\}$ *is the curve of* π-*periodic solutions through* $(x(0), \kappa(0), \lambda(0)) = (x_0, \kappa_0, \lambda_0)$

and $\{(x(s(\tau))+\tilde{x}(\tau), \kappa(s(\tau))+\tilde{\kappa}(\tau), \lambda(s(\tau))+\tilde{\lambda}(\tau))|\tau \in (-\tilde{\delta}, \tilde{\delta})\}$ *is the curve of* 2π-*periodic solutions through* $(x(s(0))+\tilde{x}(0), \kappa(s(0))+\tilde{\kappa}(0), \lambda(s(0))+\tilde{\lambda}(0)) = (x_0, \kappa_0, \lambda_0)$.

All 2π-*periodic solutions of* $G(x, \kappa, \lambda) = 0$ *near* $(x_0, \kappa_0, \lambda_0)$ *(in the topology of* $(E \cap Y) \times \mathbb{R} \times \mathbb{R}$) *are given by arbitrary phase shifts of solutions on these two curves.*

For a **proof** we solve by Theorem I.5.1 the equation $\hat{G}(u, s) = 0$ near $(0, 0)$ by a nontrivial curve

(I.14.26) $\{(u(\tau), s(\tau))|\tau \in (-\tilde{\delta}, \tilde{\delta}), (u(0), s(0)) = (0, 0)\},$
            where $u(\tau) = (\tilde{x}(\tau), \tilde{\kappa}(\tau), \tilde{\lambda}(\tau))$.

(We place a "$\sim$" on $(\tilde{x}(\tau), \tilde{\kappa}(\tau), \tilde{\lambda}(\tau))$ in order to distinguish this curve from the curve $(x(s), \kappa(s), \lambda(s))$.) We insert $\{(u(\tau), s(\tau))\}$ into $\hat{G}$ (cf. (I.14.21)), and by $\tilde{G}(u(\tau), s(\tau)) = 0$ we obtain the curve $\{(x(s(\tau)) + \tilde{x}(\tau), \kappa(s(\tau)) + \tilde{\kappa}(\tau), \lambda(s(\tau)) + \tilde{\lambda}(\tau))\}$ of 2π-periodic solutions of $G(x, \kappa, \lambda) = 0$. □

Next we study the tangent vector of the curve of 2π-periodic solutions at $(x_0, \kappa_0, \lambda_0)$. By Corollary I.5.2, we obtain

(I.14.27) $\dfrac{d}{d\tau}u(\tau)|_{\tau=0} = \left(\dfrac{d}{d\tau}\tilde{x}(0), \dfrac{d}{d\tau}\tilde{\kappa}(0), \dfrac{d}{d\tau}\tilde{\lambda}(0) \right) = (\hat{\psi}_2, 0, 0).$

We have to evaluate $\frac{d}{d\tau}s(0)$ in order to give the tangent vector of the period-doubling bifurcation curve.

The nontrivial solution of $\hat{G}(u,s) = 0$ is found by Lyapunov–Schmidt reduction (Theorem I.2.3). We decompose $u = \tau(\hat{\psi}_2,0,0) + \psi(\tau(\hat{\psi}_2,0,0),s)$, and the reduced equation is

$$(\text{I.14.28}) \qquad \int_0^{2\pi} \langle \tilde{G}(\tau(\hat{\psi}_2,0,0) + \psi(\tau(\hat{\psi}_2,0,0),s),s), \hat{\psi}_2' \rangle dt = 0$$

(cf. (I.2.9), (I.4.21), (I.4.22), (I.14.18), and (I.14.23)). By definition (I.14.21), equation (I.14.28) is equivalent to

$$\Psi(\tau,s) \equiv$$

$$(\text{I.14.29}) \qquad \int_0^{2\pi} \langle G(x(s) + \tau\hat{\psi}_2 + \psi_1, \kappa(s) + \psi_2, \lambda(s) + \psi_3), \hat{\psi}_2' \rangle dt = 0,$$

where $\psi(\tau(\hat{\psi}_2,0,0),s) = (\psi_1,\psi_2,\psi_3)(\tau(\hat{\psi}_2,0,0),s)$
according to the three components of u (cf. (I.14.21)).

Next we make use of the equivariance (I.8.31) of G.

Since $D_{(x,\kappa,\lambda)}G(x_0,\kappa_0,\lambda_0)(x,\kappa,\lambda) = S_\pi D_{(x,\kappa,\lambda)}G(x_0,\kappa_0,\lambda_0)(x,\kappa,\lambda)$, the range $R(D_u\hat{G}(0,0))$ is invariant under S_π (cf. (I.14.23)). Since $N(D_u\hat{G}(0,0))$ is obviously invariant under S_π, too, the Lyapunov–Schmidt complements of $N(D_u\hat{G}(0,0))$ and $R(D_u\hat{G}(0,0))$ can be chosen invariant under S_π as well. (Choose the complement $\{x + S_\pi\tilde{x}|x,\tilde{x}$ are in a complement$\}$.) This means that the Lyapunov–Schmidt projections commute with S_π. Replacing $x(s) + \tau\hat{\psi}_2 + \psi_1$ by $S_\pi(x(s) + \tau\hat{\psi}_2 + \psi_1) = x(s) - \tau\hat{\psi}_2 + S_\pi\psi_1$, we see by the equivariance of G and the uniqueness of the function $\psi = (\psi_1,\psi_2,\psi_3)$ (cf. (I.2.8)) that

$$(\text{I.14.30}) \qquad \begin{aligned} S_\pi\psi_1(\tau(\hat{\psi}_2,0,0),s) &= \psi_1(-\tau(\hat{\psi}_2,0,0),s), \\ \psi_j(\tau(\hat{\psi}_2,0,0),s) &= \psi_j(-\tau(\hat{\psi}_2,0,0),s), \quad j = 2,3. \end{aligned}$$

Inserting this into (I.14.29), we obtain by $S_\pi\hat{\psi}_2' = S_{-\pi}\hat{\psi}_2' = -\hat{\psi}_2'$ (cf. (I.14.19)),

$$(\text{I.14.31}) \qquad \Psi(-\tau,s) = -\Psi(\tau,s) \text{ for } s \in (-\delta,\delta), \tau \in (-\tilde{\delta},\tilde{\delta}).$$

In other words, the bifurcation function Ψ is odd in τ, so that $\Psi(\tau,s) = 0$ implies $\Psi(-\tau,s) = 0$. By uniqueness of the nontrivial solution curve this implies that

$$(\text{I.14.32}) \qquad s(-\tau) = s(\tau) \quad \text{and} \quad \frac{d}{d\tau}s(\tau)|_{\tau=0} = 0.$$

Therefore, the **tangent vector** of the period-doubling bifurcation curve at (x_0,κ_0,λ_0) is given by (cf. (I.14.27))

$$\text{(I.14.33)} \quad \frac{d}{d\tau}(x(s(\tau)) + \tilde{x}(\tau), \kappa(s(\tau)) + \tilde{\kappa}(\tau), \lambda(s(\tau)) + \tilde{\lambda}(\tau))|_{\tau=0}$$
$$= (\hat{\psi}_2, 0, 0).$$

Furthermore, by (I.14.30) and (I.14.32),

$$\text{(I.14.34)} \quad \begin{aligned} S_\pi \tilde{x}(\tau) &= S_\pi(\tau\hat{\psi}_2 + \psi_1(\tau(\hat{\psi}_2, 0, 0), s(\tau))) \\ &= -\tau\hat{\psi}_2 + \psi_1(-\tau(\hat{\psi}_2, 0, 0), s(-\tau)) = \tilde{x}(-\tau), \\ \tilde{\kappa}(\tau) &= \psi_2(\tau(\hat{\psi}_2, 0, 0), s(\tau)) = \tilde{\kappa}(-\tau), \\ \tilde{\lambda}(\tau) &= \psi_3(\tau(\hat{\psi}_2, 0, 0), s(\tau)) = \tilde{\lambda}(-\tau), \end{aligned}$$

so that we obtain for the period-doubling bifurcation curve

$$\text{(I.14.35)} \quad \begin{aligned} &(S_\pi(x(s(\tau)) + \tilde{x}(\tau)), \kappa(s(\tau)) + \tilde{\kappa}(\tau), \lambda(s(\tau)) + \tilde{\lambda}(\tau)) \\ &= (x(s(-\tau)) + \tilde{x}(-\tau), \kappa(s(-\tau)) + \tilde{\kappa}(-\tau), \lambda(s(-\tau)) + \tilde{\lambda}(-\tau)) \end{aligned}$$

by $S_\pi x(s(\tau)) = x(s(\tau))$.

Property (I.14.35) is restated as follows: The functions on the curve for negative τ are obtained by a phase shift of half the period from those functions for positive τ. Therefore, we sketch period-doubling bifurcations as in Figure I.14.1. The second case shows a period-doubling bifurcation at a turning point of the primary branch.

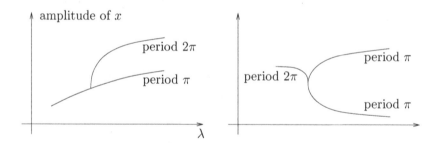

Figure I.14.1

Needless to say, the substitution $\kappa(s)t$ for t on the primary π-periodic curve and the substitution $(\kappa(s(\tau)) + \tilde{\kappa}(\tau))t$ for t on the bifurcating 2π-periodic curve gives curves

$$\text{(I.14.36)} \quad \begin{aligned} &\{(x(s), \lambda(s))|s \in (-\delta, \delta)\} \text{ of } p_1(s) = \pi/\kappa(s)\text{-periodic} \\ &\text{solutions of (I.14.1) through } (x_0, \lambda_0) \text{ and} \\ &\{(x(s(\tau)) + \tilde{x}(\tau), \lambda(s(\tau)) + \tilde{\lambda}(\tau))|\tau \in (-\tilde{\delta}, \tilde{\delta})\} \text{ of} \\ &p_2(\tau) = 2\pi/(\kappa(s(\tau)) + \tilde{\kappa}(\tau))\text{-periodic solutions} \\ &\text{of (I.14.1) through } (x_0, \lambda_0) \text{ such that} \\ &p_2(0) = 2\pi/\kappa_0 = 2p_1(0). \end{aligned}$$

If $p_1(0)$ is the minimal period of x_0, then $2p_1(0)$ is the minimal period of the bifurcating curve in the limit: By $\tilde{x}(\tau) = \tau\hat{\psi}_2 + o(\tau)$ and $S_{p_2(\tau)/2}\hat{\psi}_2 = -\hat{\psi}_2$ (cf. (I.14.17) and the substitution $(\kappa(s(\tau)) + \tilde{\kappa}(\tau))t$ for t), the solution $x(s(\tau)) + \tilde{x}(\tau)$ has indeed the minimal period $p_2(\tau)$, and $p_2(0) = 2p_1(0)$. Therefore, the nomenclature **Period-Doubling Bifurcation** is justified.

I.14.1 The Principle of Exchange of Stability for a Period-Doubling Bifurcation

Next we prove a Principle of Exchange of Stability at a "generic" period-doubling bifurcation point. Here, genericity means that the primary curve of π-periodic solutions of $G(x, \kappa, \lambda) = 0$ is given by the Implicit Function Theorem for periodic solutions (Theorem I.13.3); i.e., we exclude a turning point at $(x_0, \kappa_0, \lambda_0)$. This means for assumption (I.14.17) that

$$(I.14.37) \qquad \begin{aligned} &(\hat{\psi}_1, \kappa_1, \lambda_1) = (\hat{\psi}_1, \kappa_1, 1), \quad \text{which implicitly} \\ &\text{implies that } D_\lambda F(x_0, \lambda_0) \neq 0. \end{aligned}$$

In this special case we can prove a period-doubling bifurcation in an abbreviated way that is more convenient for our stability analysis. Define instead of (I.14.21), that

$$(I.14.38) \qquad \begin{aligned} &\tilde{G}(u, \lambda) = G(x(\lambda) + x, \kappa(\lambda) + \kappa, \lambda), \\ &\text{where } u = (x, \kappa) \text{ and} \\ &\{(x(\lambda), \kappa(\lambda), \lambda) | \lambda \in (\lambda_0 - \delta, \lambda_0 + \delta)\} \text{ is the curve} \\ &\text{of } \pi\text{-periodic solutions given by Theorem I.13.3,} \\ &\hat{G}(u, \lambda) = \left(\tilde{G}(u, \lambda), \int_0^{2\pi} \langle x, \hat{\psi}_0' \rangle dt \right), \\ &\hat{G} : \hat{U} \times (\lambda_0 - \delta, \lambda_0 + \delta) \longrightarrow W \times \mathbb{R}, \\ &0 \in \hat{U} \subset (E \cap Y) \times \mathbb{R}. \end{aligned}$$

Obviously, $N(D_u\hat{G}(0, \lambda_0)) = \text{span}[(\hat{\psi}_2, 0)]$, and all arguments (I.14.22) through (I.14.25) in their modified versions allow the application of Theorem I.5.1. (In the nondegeneracy condition (I.14.25) we replace $(u, s) = (0, 0)$ by $(u, \lambda) = (0, \lambda_0), \lambda_1 = 1$, and $\dot{} = \frac{d}{d\lambda}$.) The nontrivial solution curve of $\hat{G}(u, \lambda) = 0$,

$$(I.14.39) \qquad \begin{aligned} &\{(u(\tau), \lambda(\tau)) | \tau \in (-\tilde{\delta}, \tilde{\delta}), (u(0), \lambda(0)) = (0, \lambda_0)\} \\ &\text{where } u(\tau) = (\tilde{x}(\tau), \tilde{\kappa}(\tau)), \end{aligned}$$

gives the period-doubling bifurcation curve by

$$(I.14.40) \qquad \{(x(\lambda(\tau)) + \tilde{x}(\tau), \kappa(\lambda(\tau)) + \tilde{\kappa}(\tau), \lambda(\tau))\}.$$

Its tangent vector at the bifurcation point $(x_0, \kappa_0, \lambda_0)$ is again given by (I.14.33), and the symmetry with respect to τ is analogous to (I.14.35). In particular,

$$(I.14.41) \qquad \frac{d\lambda}{d\tau}(\tau)|_{\tau=0} = 0.$$

As usual, stability means linear stability, and for the periodic solutions this is determined by the real parts of the Floquet exponents that are the eigenvalues of $D_x G(x, \kappa, \lambda)$ (see our comments in Section I.12). For the primary curve of π-periodic solutions we have to insert $(x, \kappa, \lambda) = (x(\lambda), \kappa(\lambda), \lambda)$, which corresponds to the trivial solution line $(u, \lambda) = (0, \lambda)$ of $\hat{G}(u, \lambda) = 0$. Accordingly, $(x, \kappa, \lambda) = (x(\lambda(\tau)) + \tilde{x}(\tau), \kappa(\lambda(\tau)) + \tilde{\kappa}(\tau), \lambda(\tau))$ represents the bifurcating pitchfork $(u, \lambda) = ((\tilde{x}(\tau), (\tilde{\kappa}(\tau)), \lambda(\tau))$ of $\hat{G}(u, \lambda) = 0$.

In order to apply the results of Section I.7, we assume that

$$
\begin{aligned}
&0 \text{ is a simple eigenvalue of } D_u \hat{G}(0, \lambda_0), \\
&\text{i.e., } (\hat{\psi}_2, 0) \notin R(D_u \hat{G}(0, \lambda_0)) \Leftrightarrow
\end{aligned}
$$

$$(I.14.42) \qquad
\begin{aligned}
&\hat{\psi}_2 \notin R(D_{(x,\kappa,\lambda)} G(x_0, \kappa_0, \lambda_0)), \text{ since} \\
&D_\lambda F(x_0, \lambda_0) \in C_\pi^\alpha(\mathbb{R}, Z) \subset R(D_u \tilde{G}(0, \lambda_0)) \Leftrightarrow
\end{aligned}
$$

$$\int_0^{2\pi} \langle \hat{\psi}_2, \hat{\psi}_2' \rangle dt \neq 0 \quad (\text{cf. } (I.14.18)) \ (= 1 \text{ w.l.o.g.}).$$

We focus on the critical Floquet exponent 0, which is an eigenvalue of $\hat{J}_0 = D_x G(x_0, \kappa_0, \lambda_0)$ with eigenfunctions $\hat{\psi}_0$ and $\hat{\psi}_2$. A perturbation analysis, however, requires that this Floquet exponent be not only geometrically but also algebraically double, which means (by assumption) that

$$(I.14.43) \qquad
\begin{aligned}
&\hat{\psi}_0, \hat{\psi}_2 \notin R(\hat{J}_0), \text{ where} \\
&\hat{J}_0 = D_x G(x_0, \kappa_0, \lambda_0) : E \cap Y \to W.
\end{aligned}
$$

Whereas $\hat{\psi}_2 \notin R(\hat{J}_0)$ follows from (I.14.42), the assumption (I.13.4) for Theorem I.13.3 means that $\hat{\psi}_0 \notin R(\hat{J}_0)$ when $\hat{J}_0 : C_\pi^\alpha(\mathbb{R}, X) \cap C_\pi^{1+\alpha}(\mathbb{R}, Z) \to C_\pi^\alpha(\mathbb{R}, Z)$. However, in the Hilbert space setting as expounded in Remark I.14.1, the subspace of π-periodic functions as well as its complement are invariant for $\hat{J}_0$ (cf. (I.14.5), (I.14.6)), which means that $\hat{\psi}_0 \notin R(\hat{J}_0)$ in (I.14.43) follows from our assumption (I.13.4) on the primary curve of π-periodic solutions. Thus it is reasonable to assume (I.14.43) for our Banach space formulation as well.

We relate the simple eigenvalue perturbations of $D_u \hat{G}(u, \lambda)$ along the pitchfork to the perturbations of the nontrivial Floquet exponent along the period-doubling bifurcation. Equation (I.7.10) of Proposition I.7.2 now reads as follows:

$$D_u\hat{G}(u(\tau), \lambda(\tau))((\hat{\psi}_2, 0) + (\hat{w}(\tau), \hat{\kappa}(\tau)))$$

(I.14.44)
$$= \mu(\tau)((\hat{\psi}_2, 0) + (\hat{w}(\tau), \hat{\kappa}(\tau))),$$
$$\text{where } \mu(0) = 0, \ (\hat{w}(0), \hat{\kappa}(0)) = (0, 0),$$

and $(\hat{w}(\tau), \hat{\kappa}(\tau))$ is in the complement of $N(D_u\hat{G}(0, \lambda_0))$. Under the assumption (I.14.42), this complement can be chosen as $R(D_u\hat{G}(0, \lambda_0)) \cap ((E \cap Y) \times \mathbb{R})$, which, by (I.14.18), allows us to rewrite (I.14.44) as follows:

$$D_x G(\tau)(\hat{\psi}_2 + \hat{w}(\tau)) + \hat{\kappa}(\tau)\hat{\psi}_0(\tau) = \mu(\tau)(\hat{\psi}_2 + \hat{w}(\tau)),$$

$$\int_0^{2\pi} \langle \hat{w}(\tau), \hat{\psi}_0' \rangle dt = \mu(\tau)\hat{\kappa}(\tau) \text{ (cf. (I.14.20))},$$

(I.14.45)
$$\int_0^{2\pi} \langle \hat{w}(\tau), \hat{\psi}_2' \rangle dt = 0, \text{ where}$$

$$D_x G(\tau) \equiv D_x G(x(\lambda(\tau)) + \tilde{x}(\tau), \kappa(\lambda(\tau)) + \tilde{\kappa}(\tau), \lambda(\tau)) \text{ and}$$

$$\hat{\psi}_0(\tau) = \frac{d}{dt}(x(\lambda(\tau)) + \tilde{x}(\tau)), \ \hat{\psi}_0(0) = \frac{d}{dt}x_0 = \hat{\psi}_0.$$

The function $\hat{\psi}_0(\tau)$ is the eigenvector of $D_x G(\tau)$ with trivial Floquet exponent $\mu_0(\tau) \equiv 0$, and according to Proposition I.12.1, the perturbation of the nontrivial Floquet exponent $\mu_1(\tau)$ is given by

$$D_x G(\tau)(\hat{\psi}_2 + w(\tau)) = \mu_1(\tau)(\hat{\psi}_2 + w(\tau)) + \nu(\tau)\hat{\psi}_0(\tau),$$

(I.14.46)
$$\int_0^{2\pi} \langle w(\tau), \hat{\psi}_0' \rangle dt = 0, \quad \int_0^{2\pi} \langle w(\tau), \hat{\psi}_2' \rangle dt = 0$$
(cf. (I.12.10)),
$$\text{where } \mu_1(0) = 0, \quad w(0) = 0, \quad \nu(0) = 0.$$

Thus the perturbations (I.14.45) for $\mu(\tau)$ and (I.14.46) for $\mu_1(\tau)$ are nearly identical when we set $\nu(\tau) = -\hat{\kappa}(\tau)$. The difference is only between (I.14.45)$_2$ and (I.14.46)$_2$. Nonetheless, we show that $\mu = \mu_1$ up to order 2; i.e., $\mu(0) = \mu_1(0) = 0$, $\dot{\mu}(0) = \dot{\mu}_1(0) = 0$, $\ddot{\mu}(0) = \ddot{\mu}_1(0)$, $\dot{} = \frac{d}{d\tau}$.

To this end, we use the equivariance (I.8.31) of G. Taking the derivative of (I.8.31) with respect to $u = (x, \kappa)$, we obtain

(I.14.47) $D_{(x,\kappa)}G(S_\theta x, \kappa, \lambda)(S_\theta \hat{x}, \hat{\kappa}) = S_\theta D_{(x,\kappa)}G(x, \kappa, \lambda)(\hat{x}, \hat{\kappa}).$

In view of (I.14.35) this implies, for $\theta = \pi$,

(I.14.48)
$$D_{(x,\kappa)}G(-\tau)(S_\pi \hat{x}, \hat{\kappa}) = S_\pi D_{(x,\kappa)}G(\tau)(\hat{x}, \hat{\kappa})$$
$$\text{for every } (\hat{x}, \hat{\kappa}) \in (E \cap Y) \times \mathbb{R}.$$

In particular, $\hat{\psi}_0(-\tau) = S_\pi \hat{\psi}_0(\tau)$, which also follows directly from (I.14.35). By $S_{-\pi}\hat{\psi}_0' = \hat{\psi}_0'$ and $S_{-\pi}\hat{\psi}_2' = -\hat{\psi}_2'$ we obtain

$$\int_0^{2\pi} \langle S_\pi \hat{x}, \hat{\psi}_0' \rangle dt = \int_0^{2\pi} \langle \hat{x}, \hat{\psi}_0' \rangle dt \text{ and}$$

(I.14.49)
$$\int_0^{2\pi} \langle S_\pi \hat{x}, \hat{\psi}_2' \rangle dt = -\int_0^{2\pi} \langle \hat{x}, \hat{\psi}_2' \rangle dt$$

for every $\hat{x} \in W$.

The observations (I.14.48), (I.14.49), together with the uniqueness of the eigenvalue perturbations (I.14.45) and (I.14.46), yield the following symmetries:

(I.14.50)
$$\begin{aligned}
&\hat{w}(-\tau) = -S_\pi \hat{w}(\tau), \quad &\hat{\kappa}(-\tau) = -\hat{\kappa}(\tau), \quad &\mu(-\tau) = \mu(\tau), \\
&w(-\tau) = -S_\pi w(\tau), \quad &\nu(-\tau) = -\nu(\tau), \quad &\mu_1(-\tau) = \mu_1(\tau).
\end{aligned}$$

These properties of the eigenvalues $\mu(\tau)$ and $\mu_1(\tau)$ are in natural accord with the fact that a transition from positive to negative τ means a phase shift by π, which does not change the stability. Properties (I.14.50) imply in particular,

(I.14.51)
$$\frac{d\mu}{d\tau}(\tau)|_{\tau=0} = 0, \quad \frac{d\mu_1}{d\tau}(\tau)|_{\tau=0} = 0.$$

We claim that

(I.14.52)
$$\frac{d^2\mu}{d\tau^2}(\tau)|_{\tau=0} = \frac{d^2\mu_1}{d\tau^2}(\tau)|_{\tau=0}.$$

Indeed, taking the derivatives of (I.14.45)$_2$ with respect to τ at $\tau = 0$ gives

(I.14.53)
$$\int_0^{2\pi} \langle \dot{\hat{w}}(0), \hat{\psi}_0' \rangle dt = \int_0^{2\pi} \langle \dot{\hat{w}}(0), \hat{\psi}_0' \rangle dt = 0 \quad \left(\cdot = \frac{d}{d\tau} \right),$$

so that up to derivatives (with respect to τ) of order 2 at $\tau = 0$ the eigenvalue perturbations (I.14.45) and (I.14.46) are identical. This observation proves (I.14.52).

We now do the same analysis along the primary curve of π-periodic solutions $\{(x(\lambda), \kappa(\lambda), \lambda)\}$ that corresponds to the trivial solution line $\{(0, \lambda)\}$ of $\hat{G}(u, \lambda) = 0$. We obtain the same eigenvalue perturbations (I.14.45) and (I.14.46) with the only differences that $G(\tau)$ is replaced by the family $G(\lambda) \equiv G(x(\lambda), \kappa(\lambda), \lambda)$, and $\hat{\psi}_0(\tau)$ is replaced by $\hat{\psi}_0(\lambda) = \frac{d}{dt}x(\lambda)$ for λ near λ_0. Accordingly, we denote the corresponding eigenvalues by $\mu(\lambda)$ and $\mu_1(\lambda)$, and $\mu(\lambda_0) = \mu_1(\lambda_0) = 0$.

However, there is obviously no symmetry, analogous to (I.14.50), of the functions depending on λ. Nonetheless, $\frac{d}{d\lambda}(\mu(\lambda)\hat{\kappa}(\lambda))|_{\lambda=\lambda_0} = \mu'(\lambda_0)\hat{\kappa}(\lambda_0) + \mu(\lambda_0)\hat{\kappa}'(\lambda_0) = 0$, since $\hat{\kappa}(\lambda_0) = \mu(\lambda_0) = 0$ ($' = \frac{d}{d\lambda}$), and we obtain from (I.14.45)$_2$,

(I.14.54)
$$\int_0^{2\pi} \langle \hat{w}'(\lambda_0), \hat{\psi}_0' \rangle dt = 0,$$

so that up to the first derivative (with respect to λ) at $\lambda = \lambda_0$ the two eigenvalue perturbations (I.14.45) and (I.14.46) are identical. This proves that

$$(\text{I.14.55}) \qquad \frac{d\mu}{d\lambda}(\lambda)|_{\lambda=\lambda_0} = \frac{d\mu_1}{d\lambda}(\lambda)|_{\lambda=\lambda_0}.$$

As shown in Section I.7, the nondegeneracy condition (I.14.24) or (I.14.25) is equivalent to

$$(\text{I.14.56}) \qquad \frac{d\mu}{d\lambda}(\lambda)|_{\lambda=\lambda_0} \neq 0$$

(see (I.7.32)–(I.7.36)). For the pitchfork bifurcation of $\hat{G}(u, \lambda) = 0$ at $(u, \lambda) = (0, \lambda_0)$ given by $\{(u(\tau), \lambda(\tau))\}$ with $\frac{d\lambda}{d\tau}(0) = 0$ (cf. (I.14.41)), formula (I.7.45) gives

$$(\text{I.14.57}) \quad 2\mu'(\lambda_0)\ddot{\lambda}(0) = -\ddot{\mu}(0), \quad \text{where} \quad ' = \frac{d}{d\lambda} \text{ and } \cdot = \frac{d}{d\tau}.$$

The condition for $\ddot{\lambda}(0) \neq 0$ (i.e., for a nondegenerate pitchfork) is given by (I.6.11), and in (I.6.9) the mapping F has to be replaced by $\hat{G}$, x by $u = (x, \kappa)$, and $\hat{v}_0$ by $(\hat{\psi}_2, 0)$. We abstain from a translation of condition (I.6.11) to our situation, but we simply assume that $\ddot{\lambda}(0) \neq 0$.

By (I.14.52) and (I.14.55), relation (I.14.57) gives the crucial formula

$$(\text{I.14.58}) \qquad 2\mu_1'(\lambda_0)\ddot{\lambda}(0) = -\ddot{\mu}_1(0),$$

which links the two nontrivial critical Floquet exponents $\mu_1(\lambda)$ and $\mu_1(\tau)$ along the primary and bifurcating curves, respectively, via the bifurcation direction $\ddot{\lambda}(0)$ (recall that $\dot{\lambda}(0) = 0$).

Assume that $\mu'(\lambda_0) = \mu_1'(\lambda_0) < 0$, which means a loss of stability of the π-periodic solutions at $\lambda = \lambda_0$ in the space of 2π-periodic functions (cf. (I.12.2)). (Here we tacitly assume that all noncritical nontrivial Floquet exponents are in the stable right half-plane of $\mathbb{C}$.) Then, by (I.14.58),

$$(\text{I.14.59}) \qquad \text{sign}(\lambda(\tau) - \lambda_0) = \text{sign}\mu_1(\tau) \text{ for } 0 < \tau < \tilde{\delta},$$

which proves an exchange of stability as sketched in Figure I.14.2. If $\mu_1'(\lambda_0) > 0$, the stability properties of all solution curves are reversed.

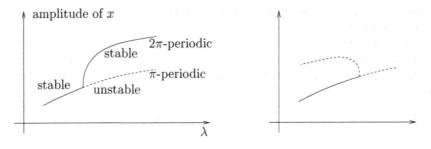

Figure I.14.2

Theorem I.14.3 *At a Generic Period-Doubling Bifurcation, the Principle of Exchange of Stability is valid. Formula (I.14.58) links the critical nontrivial Floquet exponents to the bifurcation direction.*

Remark I.14.4 *As pointed out in Remark I.13.8, the period-doubling bifurcation is commonly proved by studying the iterated Poincaré map near a periodic solution of (I.14.1). Fixed points of the Poincaré map itself stay fixed points of its iteration (but not vice versa), and in this setting a period-doubling bifurcation is a pitchfork bifurcation for the iterated Poincaré map combined with a continuation of fixed points of the Poincaré map itself.*

That means for the derivative of the Poincaré map at the critical periodic solution that 1 and −1 are simple eigenvalues. The existence of the Poincaré map causes no problems for ODEs of type (I.14.1). However, since it requires a careful analysis of the complete nonlinear flow, its existence is not so obvious if (I.14.1) represents a parabolic PDE (cf. our comments in Remark I.13.8). That is why we stay in the setting of a Fourier analysis that restricts the nonlinear flow to a periodic one.

Remark I.14.5 *In Remark I.14.1 we give the arguments why we confine ourselves to a period-doubling bifurcation and why we do not study any other period-multiplying bifurcation. In a Hilbert space setting we show that it is the only simple period-multiplying bifurcation in the sense that it is created by a one-dimensional kernel.*

Usuallys one refers to period-doubling bifurcations and Hopf bifurcations as "generic" bifurcations. What does that mean? We make some comments on it that are without mathematical rigor but that explain our point of view.

There are two properties that make a bifurcation generic:

- *an unavoidable spectral property of a family of linear operators along the primary solution branch and*
- *a robustness of the primary solution branch.*

Consider a path of algebraically simple eigenvalues of a one-parameter family of real linear operators having endpoints in different half-planes of $\mathbb{C}$. Then that path necessarily crosses the imaginary axis at some $i\kappa_0 \neq 0$ (and then there is also a complex conjugate path) or at 0. In the first case, the eigenvalue 0 is avoided, whereas in the second case, the entire path is real, and the (simple) eigenvalue 0 is unavoidable.

We know that both crossings of the imaginary axis "with nonvanishing speed" (or in a nondegenerate or transversal way) entail a Hopf bifurcation or a steady-state bifurcation from the primary solution branch, respectively (see our comments at the beginning of Sections I.8, I.16, I.17). Are these bifurcations under those "generic" linear assumptions unavoidable? In the theory of "imperfections" or "unfoldings" the diagrams of Figure I.14.3 are well known: They show that bifurcations are easily avoided by an arbitrarily small perturbation of the operator even under generic spectral assumptions on its linearization.

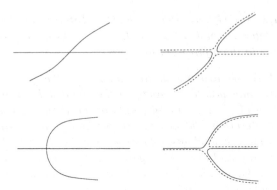

Figure I.14.3

Looking at Figure I.14.3 reveals that an abolition of bifurcations by small perturbations breaks up the primary (or trivial) solution line (and also breaks the symmetry of the diagram). In this sense the primary branch is not robust. Note that due to the Lyapunov–Schmidt reduction, that scenario indeed takes place in a plane. In the case of the Hopf bifurcation as well as of the period-doubling bifurcation, however, the total derivative along the stationary or π-periodic solutions is surjective in the space of stationary or π-periodic functions, respectively. Therefore, the primary solutions form a robust curve or surface that cannot be broken up by a small perturbation. (That robustness is also reflected by the oddness of the respective bifurcation function, which vanishes necessarily at zero for all parameters.) Accordingly the unfolding of a pitchfork cannot be realized by a bifurcation function. If a primary branch is robust, then a bifurcation can take place only in a larger space. The bifurcating branch is of a different type from that of the primary branch.

These naive arguments for the genericity of the two bifurcations help us to understand the following result: Generically, the global "net" of a continuation of any periodic solution of an evolution equation (I.14.1) consists of smooth surfaces ("snakes") having turning points, splitting at period-doubling bifurcations, ending at Hopf bifurcations in steady states, or being trapped in loops with unbounded "virtual periods" (see [134], [48], [102], [105]). We refer also to Remark I.11.13 about global Hopf Bifurcation.

I.15 The Newton Polygon

Bifurcation with a one-dimensional kernel is discovered via the method of Lyapunov–Schmidt by studying a one-dimensional bifurcation equation depending on two variables. If we introduce coordinates by choosing suitable basis vectors, the bifurcation equation can be written as

(I.15.1)
$$\Phi(s, \lambda) = 0, \text{ where}$$
$$\Phi : U \times V \to \mathbb{R} \text{ and}$$
$$(0, \lambda_0) \in U \times V \subset \mathbb{R} \times \mathbb{R} \text{ (cf. Section I.5).}$$

In Section I.5, this bifurcation equation is nontrivially solved by the Implicit Function Theorem, whose validity is guaranteed by the nondegeneracy (or transversality) condition (I.5.3). In the next section we give up that condition, and we study "degenerate bifurcation." The Implicit Function Theorem has to be replaced by a more general tool, which is called the Newton polygon method.

Replacing λ by $\lambda - \lambda_0$, we can assume w.l.o.g. that $\lambda_0 = 0$, and we study (I.15.1) in a neighborhood of $(s, \lambda) = (0, 0)$. The main difference in comparison to the Implicit Function Theorem is that we assume that

(I.15.2)
$$\Phi : U \times V \to \mathbb{R} \text{ is analytic near } (0, 0); \text{ i.e.,}$$
$$\Phi(s, \lambda) = \sum_{j,k=0}^{\infty} c_{jk} s^k \lambda^j, \text{ where } c_{jk} \in \mathbb{R},$$
and convergence holds in $U \times V$.

We assume that $\Phi(0, 0) = 0$ (i.e., $c_{00} = 0$), and we look for curves of solutions of $\Phi(s, \lambda) = 0$ passing through $(0, 0)$.

To this end we mark the powers j of λ on the ordinate and the powers k of s on the abscissa, and we mark a point (k, j) whenever $c_{jk} \neq 0$.

If $c_{jk} = 0$ for all $j \in \mathbb{N}_0, k = 0, \ldots, k_0 - 1$, i.e., if the first nonvanishing coefficients exist only for $k = k_0$, then we can divide by s^{k_0}, and for the new equation we find a smallest point $(0, j_0)$ on the ordinate.

If $k_0 = 0$, then $j_0 \geq 1$ by $c_{00} = 0$.

If $k_0 \geq 1$, we have the solution curve $\{(0, \lambda)\}$, and if $j_0 = 0$, it is the only solution curve through $(0, 0)$.

Thus we assume in the following that $j_0 \geq 1$. Next, only coefficients c_{jk} with $0 \leq j \leq j_0 - 1, k \geq 1$, are of interest. If all such coefficients vanish, we can divide by λ^{j_0}, and $\{(s, 0)\}$ is the only solution curve through $(0, 0)$.

Otherwise, we find coefficients $c_{jk} \neq 0$ with $0 \leq j \leq j_0 - 1$, and we may assume that $c_{0k_n} \neq 0$ for some smallest $k_n \geq 1$. If $c_{jk} = 0$ for all $0 \leq j \leq j_n \leq j_0 - 1, k \geq 1$, we can divide by λ^{j_n}, and for the new equation there is some $c_{0k_n} \neq 0$ for a smallest $k_n \geq 1$. Note that $j_n > 0$ implies the existence of the solution curve $\{(s, 0)\}$.

Now we are in the situation that there are a smallest point $(0, j_0)$, $j_0 \geq 1$, on the ordinate and a smallest point $(k_n, 0)$, $k_n \geq 1$, on the abscissa. Then the polygon forming the convex hull of all points (k, j) such that $c_{jk} \neq 0$ and $jk_n + j_0 k \leq j_0 k_n$ is called the Newton polygon.

(I.15.3) It joins (k_ν, j_ν), $\nu = 0, \ldots, n$, $j_0 > j_1 > \cdots > j_n = 0$, $0 = k_0 < k_1 < \cdots < k_n$, and $c_{jk} = 0$ if (k, j) is below all lines passing through two consecutive points (k_ν, j_ν).

We sketch a Newton Polygon in Figure I.15.1.

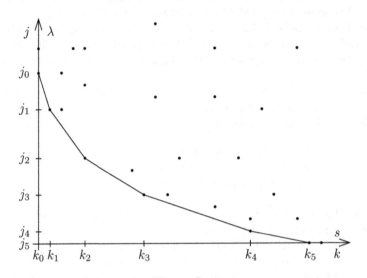

Figure I.15.1

In the following, we confine ourselves to $s \geq 0$. For $s \leq 0$ set $-s = \tilde{s} \geq 0$ and repeat the procedure for the new equation $\Phi^-(-s, \lambda) \equiv \Phi(s, \lambda) = 0$. Furthermore, we set $c_{j_\nu k_\nu} = c_\nu \neq 0$ for $\nu = 0, \ldots, n$.

Then (I.15.2) is rewritten as

(I.15.4)
$$\Phi(s, \lambda) = \sum_{\nu=0}^{n} c_\nu s^{k_\nu} \lambda^{j_\nu} + R(s, \lambda),$$
where R contains all remaining terms.

Let $-1/\gamma \in \mathbb{Q}$ be one of the negative slopes of the Newton polygon. Then for all, say $r + 1$, points $(k_\ell, j_\ell), \ldots (k_{\ell+r}, j_{\ell+r})$ on the line with slope $-1/\gamma$, we have

(I.15.5)
$$\gamma j_\ell + k_\ell = \cdots = \gamma j_{\ell+r} + k_{\ell+r} = \sigma,$$

and the ansatz

(I.15.6)
$$\lambda = s^\gamma \tilde{\lambda} \qquad \text{in (I.15.4)}$$

yields for $(s, \tilde{\lambda})$ the equation

(I.15.7)
$$s^\sigma \left(\sum_{\rho=0}^{r} c_{\ell+\rho} \tilde{\lambda}^{j_{\ell+\rho}} + R_1(t, \tilde{\lambda}) \right) = 0.$$

Here $t = s^{1/\gamma_2} \geq 0$ when $\gamma = \gamma_1/\gamma_2$ with $\gamma_1/\gamma_2 \in \mathbb{N}$, and R_1 is analytic in $(t, \tilde{\lambda})$ near $(0,0)$. By the substitution (I.15.6), the analyticity of R_1 is true in $(-\delta_1, \delta_1) \times (-L, L)$ for every $L > 0$, provided that $\delta_1 > 0$ is small enough. Therefore, we need no restriction on $\tilde{\lambda}$ in the subsequent analysis. Furthermore, by the definition (I.15.3), the exponents $\gamma j + k$ of s in R (cf. (I.15.4)) are all positive, so that $R_1(0, \tilde{\lambda}) = 0$ for all $\tilde{\lambda}$ where it is defined.

Let $\tilde{\lambda}_0$ be a real zero of the polynomial

(I.15.8) $$P_\gamma(\tilde{\lambda}) = \sum_{\rho=0}^{r} c_{\ell+\rho} \tilde{\lambda}^{j_\ell + \rho},$$

and let $q \geq 1$ denote its multiplicity.

If $q = 1$ the Implicit Function Theorem for (I.15.7) (after division by s^σ) gives a solution

(I.15.9) $$\{(t, \tilde{\lambda}(t)) | 0 \leq t < \delta_1, \tilde{\lambda}(0) = \tilde{\lambda}_0\},$$ which is analytic in t.

After inserting $\tilde{\lambda}(t) = \tilde{\lambda}(s^{1/\gamma_2})$ into (I.15.6) we obtain a solution curve of $\Phi(s, \lambda) = 0$ through $(0,0)$ in the following form:

(I.15.10) $$\{(s, s^\gamma(\tilde{\lambda}_0 + \sum_{k=1}^{\infty} a_k s^{k/\gamma_2})) | 0 \leq s < \delta_2\}.$$

Remark I.15.1 *If $\tilde{\lambda}_0 = 0$, there is a first nonvanishing coefficient in the expansion (I.15.10) such that $\lambda(s) = \tilde{\lambda}_1 s^{\tilde{\gamma}} + o(s^{\tilde{\gamma}})$ with $\tilde{\lambda}_1 \neq 0$ and $\tilde{\gamma} > \gamma$. Recall that there is no solution curve $\{(s,0)\}$ if the Newton polygon ends at $(k_n, 0)$. An easy proof shows that $-1/\tilde{\gamma}$ is then necessarily a slope in the Newton polygon, and $\tilde{\lambda}_1$ is a zero of the corresponding defining polynomial $P_{\tilde{\gamma}}$. Thus only nonzero roots $\tilde{\lambda}_0$ of the polynomial (I.15.8) are of interest, which give by (I.15.10) the lowest-order term of the solution.*

Next let $\tilde{\lambda}_0 \neq 0$ be a real zero of (I.15.8) of multiplicity $q > 1$. Then

(I.15.11) $$P_\gamma(\tilde{\lambda}) = (\tilde{\lambda} - \tilde{\lambda}_0)^q \tilde{P}_\gamma(\tilde{\lambda}), \quad \text{where} \quad \tilde{P}_\gamma(\tilde{\lambda}_0) \neq 0.$$

After division by s^σ, equation (I.15.7) becomes

(I.15.12) $$\begin{gathered}(\tilde{\lambda} - \tilde{\lambda}_0)^q = -R_1(t, \tilde{\lambda})/\tilde{P}_\gamma(\tilde{\lambda}) \\ \text{for } (t, \tilde{\lambda}) \text{ near } (0, \tilde{\lambda}_0).\end{gathered}$$

If $R_1 \equiv 0$, then $\{(t, \tilde{\lambda}_0) | 0 \leq t < \delta_1\}$ is the unique solution curve of (I.15.12) through $(0, \tilde{\lambda}_0)$. Otherwise, by $R_1(0, \tilde{\lambda}) \equiv 0$, we have $R_1(t, \tilde{\lambda}) = t^{k_0}(\tilde{\lambda} - \tilde{\lambda}_0)^{j_0} \tilde{R}_1(t, \tilde{\lambda})$ for some $k_0 > 0, j_0 \geq 0$, so that $\tilde{R}_1(0, \tilde{\lambda}_0) \neq 0$. If $j_0 > 0$, we have again the solutions $\{(t, \tilde{\lambda}_0)\}$, and if $j_0 \geq q$, it is the only solution curve of (I.15.12) through $(0, \tilde{\lambda}_0)$. Assuming $0 \leq j_0 < q$, we can divide (I.15.12) by $(\tilde{\lambda} - \tilde{\lambda}_0)^{j_0}$, and we obtain

$$(\tilde{\lambda} - \tilde{\lambda}_0)^p = t^{k_0} f(t, \tilde{\lambda}), \quad \text{where} \quad p = q - j_0,$$

(I.15.13) $f(t, \tilde{\lambda}) \equiv -\tilde{R}_1(t, \tilde{\lambda})/\tilde{P}_\gamma(\tilde{\lambda}),$ so that

$$f(0, \tilde{\lambda}_0) \neq 0.$$

If $p > 0$ is *even*, then for nontrivial solutions for $t > 0$, necessarily

(I.15.14) $f(0, \tilde{\lambda}_0) > 0,$ and (I.15.13) becomes

$$(\tilde{\lambda} - \tilde{\lambda}_0) = \pm t^{k_0/p}(f(t, \tilde{\lambda}))^{1/p}.$$

If $p > 0$ is *odd*, then in any case,

(I.15.15) $$(\tilde{\lambda} - \tilde{\lambda}_0) = t^{k_0/p}(f(t, \tilde{\lambda}))^{1/p}.$$

Since $f(t, \tilde{\lambda})$ is analytic by its definition $(\text{I.15.13})_2$ and $f(0, \tilde{\lambda}_0) \neq 0$, the function $(f(t, \tilde{\lambda}))^{1/p}$ is analytic in $(t, \tilde{\lambda})$ near $(0, \tilde{\lambda}_0)$. Setting $\tau = t^{1/p}$, we see that the right-hand sides of $(\text{I.15.14})_2$ and (I.15.15), respectively, are analytic functions of the variables $(\tau, \tilde{\lambda})$ near $(0, \tilde{\lambda}_0)$.

Finally, since $D_{\tilde{\lambda}}(\tau^{k_0}(f(\tau^p, \tilde{\lambda}))^{1/p})|_{(\tau, \tilde{\lambda}) = (0, \tilde{\lambda}_0)} = 0$, the Implicit Function Theorem gives solution curves of $(\text{I.15.14})_2$ and (I.15.15),

(I.15.16) $\{(t, \tilde{\lambda}(t)) | 0 \leq t < \delta_1, \tilde{\lambda}(0) = \tilde{\lambda}_0\},$
which are analytic in $t^{1/p}$.

After inserting $\tilde{\lambda}(t) = \tilde{\lambda}(s^{1/\gamma_2})$ into (I.15.6), we obtain solution curves of $\Phi(s, \lambda) = 0$ through $(0, 0)$ in the following form:

(I.15.17) $\{(s, s^\gamma(\tilde{\lambda}_0 + \sum_{k=1}^\infty a_k s^{k/p\gamma_2})) | 0 \leq s < \delta_2\}, p \in \mathbb{N}.$

The locally convergent series of $\lambda(s)$ in terms of rational powers of $s \geq 0$ with a common denominator are called *Puiseux series*.

For every real root $\tilde{\lambda}_0 \neq 0$ of (I.15.18) of *odd* multiplicity $q \geq 1$ we obtain at least one solution of the form (I.15.17).

Finally, if the number j_0 found for (I.15.3) is *odd*, at least one of the polynomials (I.15.8) for all slopes in the Newton polygon has a nontrivial real root $\tilde{\lambda}_0$ of *odd* multiplicity.

Taking into account that all real solution curves of $\Phi(s, \lambda) = 0$ through $(0, 0)$ can be found in this way (see [37]), we may also state that at most j_0 such solution curves (I.15.17) can exist.

So far, we have done this for $s \geq 0$. The same arguments hold also for $\Phi^-(-s, \lambda) \equiv \Phi(s, \lambda) = 0$ if $s \leq 0$; in particular, the number j_0 is the same for Φ and Φ^-. Thus an odd j_0 yields at least one solution curve through $(0, 0)$ for positive and negative s. In the next section we apply these observations to degenerate bifurcation at a simple eigenvalue.

Remark I.15.2 *If the Newton polygon (I.15.3) is established, then it is not necessary that the remainders R in (I.15.4) or R_1 in (I.15.7) be analytic. If*

they vanish to the same order as in the analytic case, we obtain continuously differentiable solution curves of the form

(I.15.18) $$\{(s, a_0 s^\gamma + o(s^\gamma))\}$$

that are not necessarily Puiseux series in that general case. However, if infinitely many coefficients c_{jk} of (I.15.2) are used in order to establish solution curves of the form $\{(0, \lambda)\}$ or $\{(s, 0)\}$, then even in the C^∞ case one has to be careful. Nonetheless, a complete picture of all local solutions of (I.15.1) (where $\lambda_0 = 0$) can be obtained by the Newton polygon method in nonanalytic cases, too (cf. [37]).

I.16 Degenerate Bifurcation at a Simple Eigenvalue and Stability of Bifurcating Solutions

We resume the situation of Section I.5, where we prove bifurcation from the trivial solution line $\{(0, \lambda) | \lambda \in \mathbb{R}\}$ for

(I.16.1) $$F(x, \lambda) = 0.$$

In addition to assumption (I.5.1), we assume a continuous embedding $X \subset Z$ and (I.7.4); i.e., 0 is an algebraically simple eigenvalue of $D_x F(0, \lambda_0)$ for some $\lambda_0 \in \mathbb{R}$. Replacing λ by $\lambda - \lambda_0$, we can assume w.l.o.g. that $\lambda_0 = 0$. The regularity of F near $(x, \lambda) = (0, 0)$, however, is much stronger than assumed before:

$$F : U \times V \to Z, \quad 0 \in U \subset X \subset Z, \quad 0 \in V \subset \mathbb{R} \text{ is analytic,}$$

which means that $F(x, \lambda) = \sum_{j,k=0}^\infty \lambda^j F_{jk}(x)$, where

(I.16.2)

$F_{jk} : X^k \to Z$ are k-linear, symmetric, and continuous, and convergence holds in Z for $x \in U = \{x \in U | \|x\| < d\}$ and $\lambda \in V = \{\lambda \in \mathbb{R} | \ |\lambda| < d\}$ for some $d > 0$.

Since $F(0, \lambda) = 0$, we assume that $F_{j0} = 0$ for all $j \in \mathbb{N}_0$. Furthermore, the linear family $D_x F(0, \lambda)$ is given by

(I.16.3)

$$D_x F(0, \lambda) = \sum_{j=0}^\infty \lambda^j F_{j1} \equiv \sum_{j=0}^\infty \lambda^j A_j \equiv A(\lambda),$$

where $A_j \in L(X, Z)$ for all $j \in \mathbb{N}_0$.

The assumption of a simple eigenvalue of $D_x F(0, 0) = A_0$ is that

(I.16.4)   0 is an algebraically simple eigenvalue of A_0;
             i.e., $\dim N(A_0) = 1$ and $Z = R(A_0) \oplus N(A_0)$.

We recall the meaning of vectors $\hat{v}_0 \in X$ and $\hat{v}_0' \in Z'$ from (I.7.8):

$$N(A_0) = \text{span}[\hat{v}_0], \quad \langle \hat{v}_0, \hat{v}_0' \rangle = 1,$$

(I.16.5)

$$R(A_0) = \{z \in Z | \langle z, \hat{v}_0' \rangle = 0\}.$$

The Lyapunov–Schmidt projections

(I.16.6)
$P : X \to N(A_0)$ along $R(A_0) \cap X$ and
$Q : Z \to N(A_0)$ along $R(A_0)$ are then
given by $Qz = \langle z, \hat{v}_0' \rangle \hat{v}_0$ for $z \in Z$
and $Px = \langle x, \hat{v}_0' \rangle \hat{v}_0$ for $x \in X$, i.e.,
$P = Q|_X$.

The simple eigenvalue perturbation

$$D_x F(0, \lambda)(\hat{v}_0 + w(\lambda)) = \mu(\lambda)(\hat{v}_0 + w(\lambda)),$$

(I.16.7)   where $\{w(\lambda)|\lambda \in (-\delta, \delta), w(0) = 0\} \subset R(A_0) \cap X$

and $\{\mu(\lambda)|\lambda \in (-\delta, \delta), \mu(0) = 0\} \subset \mathbb{R}$,

given by Proposition I.7.2 will be crucial in this section. Since $w(\lambda)$ and $\mu(\lambda)$ are obtained by the Implicit Function Theorem, the property (I.1.7) guarantees that both functions are analytic in λ:

(I.16.8)
$$w(\lambda) = \sum_{j=1}^{\infty} w_j \lambda^j, \quad w_j \in R(A_0) \cap X,$$

$$\mu(\lambda) = \sum_{j=1}^{\infty} \mu_j \lambda^j, \quad \mu_j \in \mathbb{R},$$

and convergence holds for $|\lambda| < \delta \leq d$ (w.l.o.g.).

In Section I.7 we showed that under assumption (I.16.4), the nondegeneracy (I.5.3) is equivalent to $\mu'(0) \neq 0$ (here $\lambda_0 = 0$ and $' = \frac{d}{d\lambda}$; cf. (I.7.34)–(I.7.36)). Thus a crossing of $\mu(\lambda)$ of the imaginary axis at $\mu(0) = 0$ "with nonvanishing speed" causes bifurcation of a unique nontrivial curve of solutions of (I.16.1).

The crossing of $\mu(\lambda)$ of the imaginary axis implies a loss of stability of the trivial solution at $\lambda = 0$, and by the Principle of Exchange of Stability (cf. Section I.7), that stability is taken over by the bifurcating solution. What happens to that plausible physical scenario if the loss of stability of the trivial solution is "slow" or degenerate? Mathematically, this means that $\mu(0) = \mu'(0) = \cdots = \mu^{(m-1)}(0) = 0$, but $\mu^{(m)}(0) \neq 0$ for some odd m. One expects that there should be some bifurcating solution to take over the stability. In this section we prove this expectation; i.e., we show that any possibly degenerate loss of stability gives rise to a bifurcation of nontrivial curves of solutions, which, however, are not necessarily unique. A first step is the computation of the coefficients μ_j in the perturbation series (I.16.8).

Proposition I.16.1 *The coefficients of the eigenvalue perturbation (I.16.8) are recursively given by*

$$\mu_1 = \langle A_1 \hat{v}_0, \hat{v}_0' \rangle,$$

$$\text{(I.16.9)} \quad \mu_\ell = \langle A_\ell \hat{v}_0, \hat{v}_0' \rangle + \sum_{\substack{j+k=\ell-1 \\ j\geq 1, k\geq 0}} \langle A_j A_{k0} \hat{v}_0, \hat{v}_0' \rangle$$

$$+ \sum_{n=0}^{\ell-3} \sum_{m=1}^{\ell-2-n} \sum_{\substack{j+k=n+1 \\ j\geq 1, k\geq 0}} \langle A_j A_{km} \hat{v}_0, \hat{v}_0' \rangle \sum_{\nu_1+\cdots+\nu_m=\ell-2-n} \mu_{\nu_1} \cdots \mu_{\nu_m}$$

for $\ell \geq 2$. The vectors $A_{nm}\hat{v}_0$ for $n, m \geq 0$ are defined by

$$A_{00}\hat{v}_0 = -A_0^{-1}(I-Q)A_1\hat{v}_0, \quad A_{0m}\hat{v}_0 = A_0^{-1}A_{0,m-1}\hat{v}_0,$$

$$\text{(I.16.10)} \quad A_{n0}\hat{v}_0 = -A_0^{-1} \sum_{\substack{j+k=n+1 \\ j\geq 1, k\geq 0}} (I-Q)A_j A_{k-1,0}\hat{v}_0 \text{ where } A_{-1,0}\hat{v}_0 = \hat{v}_0,$$

$$A_{nm}\hat{v}_0 = -A_0^{-1} \sum_{\substack{j+k=n \\ j\geq 1, k\geq 0}} (I-Q)A_j A_{km}\hat{v}_0 + A_0^{-1}A_{n,m-1}\hat{v}_0,$$

for $n, m \geq 1$. Here we agree upon $A_0^{-1} : R(A_0) \to R(A_0) \cap X$; i.e., $PA_{nm} = 0$ for all $n, m \geq 0$.

Proof. Let the eigenvector expansion be $\hat{v}_0 + w(\lambda)$, where $w(\lambda)$ is given by (I.16.8). Then, by (I.16.5),

$$\mu_\ell = \sum_{\substack{j+k=\ell \\ j\geq 1, k\geq 0}} \langle A_j w_k, \hat{v}_0' \rangle, \text{ where } w_0 = \hat{v}_0,$$

$$A_0 w_\ell = (I-Q)A_0 w_\ell = -\sum_{\substack{j+k=\ell \\ j\geq 1, k\geq 1}} (I-Q)A_j w_k + \sum_{\substack{j+k=\ell \\ j\geq 1, k\geq 1}} \mu_j w_k,$$

$$\text{(I.16.11)}$$

$$w_\ell = -A_0^{-1}\left(\sum_{\substack{j+k=\ell \\ j\geq 1, k\geq 0}} (I-Q)A_j w_k - \sum_{\substack{j+k=\ell \\ j\geq 1, k\geq 1}} \mu_j w_k \right) \text{ for } \ell \geq 1.$$

Using this recurrence formula for the vectors w_ℓ and the defining formulas for A_{nm}, we can prove by induction that

$$w_1 = A_{00}\hat{v}_0,$$

$$\text{(I.16.12)} \quad w_\ell = A_{\ell-1,0}\hat{v}_0 + \sum_{n=0}^{\ell-2} \sum_{m=1}^{\ell-1-n} A_{nm}\hat{v}_0 \sum_{\nu_1+\cdots+\nu_m=\ell-1-n} \mu_{\nu_1} \cdots \mu_{\nu_m},$$

for $\ell \geq 2$. We omit the lengthy but elementary proof. $\qquad\square$

These expressions show that the perturbation series of $\mu(\lambda)$ in terms of the operators A_j and the vectors $\hat{v}_0, \hat{v}_0'$ is rather complicated. For matrices, for instance, evaluation of the characteristic equation $\det(A(\lambda) - \mu(\lambda)I) =$

0 $(A(\lambda) = D_x F(0, \lambda))$ might be simpler, especially in low dimensions. For differential operators there might be a simpler method for computing $\mu(\lambda)$, too (see our example (III.2.15)).

Next, we relate the perturbation series $\mu(\lambda)$ to the bifurcation equation obtained by the method of Lyapunov–Schmidt. Using the projections (I.16.6), we see that problem (I.16.1) is equivalent to

$$(I.16.13) \quad \begin{aligned} QF(v + w, \lambda) &= 0, \quad x = Px + (I - P)x = v + w, \\ (I - Q)F(v + w, \lambda) &= 0 \quad (\text{cf. (I.2.6)}). \end{aligned}$$

The second equation solved by the Implicit Function Theorem yields $w = \psi(v, \lambda)$, which is analytic in (v, λ) near $(0, 0)$. We compute its expansion by $(I.16.13)_2$, which is, in view of (I.16.2) and (I.16.3),

$$(I.16.14) \quad A_0 w + (I - Q) \sum_{j=1}^{\infty} \lambda^j A_j(v + w) + (I - Q) \sum_{\substack{j=0 \\ k=2}}^{\infty} \lambda^j F_{jk}(v + w) = 0.$$

With the ansatz

$$(I.16.15) \quad \begin{aligned} w &= \sum_{j,k=0}^{\infty} \lambda^j G_{jk}(v), \quad \text{where} \\ G_{jk} &: N(A_0)^k \to R(A_0) \cap X \text{ are } k\text{-linear,} \\ &\quad \text{symmetric, and continuous,} \end{aligned}$$

we obtain, by insertion into (I.16.14) and comparing like powers,

$$(I.16.16) \quad \begin{aligned} G_{j0} &= 0 \quad \text{for all} \quad j \in \mathbb{N}_0, \quad G_{01} = 0, \\ G_{11}v &= -A_0^{-1}(I - Q)A_1 v, \\ G_{n1}v &= -A_0^{-1}(I - Q)A_n v - \sum_{\substack{j+k=n \\ j \geq 1, k \geq 1}} A_0^{-1}(I - Q)A_j G_{k1}v, \quad n \geq 2. \end{aligned}$$

For later use we give also the recurrence formula for G_{0n}:

$$(I.16.17) \quad \begin{aligned} G_{02}(v) &= -A_0^{-1}(I - Q)F_{02}(v), \quad \text{and for} \quad n \geq 2, \\ G_{0n}(v) &= -A_0^{-1}(I - Q)\sum_{k=2}^{n}\sum_{m=0}^{k}\sum_{\nu_1+\cdots+\nu_m=n-k+m} \binom{k}{m} F_{0k}^{(k)}[./.], \\ &\quad \text{where } [./.] = [G_{0\nu_1}(v), \ldots, G_{0\nu_m}(v), v, \ldots, v] \end{aligned}$$

and $F_{0k}^{(k)}[v, \ldots, v] = F_{0k}(v)$ is symmetric.

For $m = 0$ we agree upon $\nu_1 + \cdots + \nu_m = 0$ (i.e., $k = n$) and that $G_{0\nu_1}(v), \ldots, G_{0\nu_m}(v)$ are not present in $F_{0k}^{(k)}$.

We leave the general recurrence formula for G_{jn} to the reader; in principle, it is clear how to construct it, but it is tedious to write it down.

When $w = \psi(v, \lambda)$ is inserted into $(I.16.13)_1$, we obtain the bifurcation equation (cf. (I.5.6)). Setting $v = s\hat{v}_0 \in N(A_0)$ and using the explicit representation (I.16.6) of the projection Q, we see that this equation reads as follows:

$$(I.16.18) \qquad \langle F(s\hat{v}_0 + \sum_{j,k=0}^{\infty} \lambda^j s^k G_{jk}(\hat{v}_0), \lambda), \hat{v}_0' \rangle = 0.$$

Using (I.16.2) and $F_{j0} = 0, G_{j0} = 0$ for all $j \in \mathbb{N}_0$, we write it as

$$(I.16.19) \qquad \begin{aligned} &\Phi(s, \lambda) \\ &= \sum_{j=1}^{\infty} \langle A_j \hat{v}_0, \hat{v}_0' \rangle \lambda^j s + \sum_{n=2}^{\infty} \sum_{\substack{j+k=n \\ j\geq 1, k\geq 1}} \langle A_j G_{k1} \hat{v}_0, \hat{v}_0' \rangle \lambda^n s \\ &+ \sum_{\substack{j=0 \\ k=2}}^{\infty} H_{jk} \lambda^j s^k = 0 \quad \text{with coefficients} \quad H_{jk} \in \mathbb{R}. \end{aligned}$$

(The mapping Φ is not identical to the bifurcation function $\Phi(v, \lambda)$ of Section I.5. Here $\Phi(s, \lambda)$ is real-valued for real s, since we simply identify v and Φ with their coordinates, $s = \langle v, \hat{v}_0' \rangle$ and $\langle \Phi, \hat{v}_0' \rangle$, respectively. We call Φ in (I.16.19) the "scalar bifurcation function.")

In particular,

$$H_{02} = \langle F_{02}^{(2)}[\hat{v}_0, \hat{v}_0], \hat{v}_0' \rangle, \quad \text{and for} \quad n \geq 2,$$

$$(I.16.20) \qquad \begin{aligned} &H_{0n} = \sum_{k=2}^{n} \sum_{m=0}^{k} \sum_{\nu_1+\cdots+\nu_m=n-k+m} \binom{k}{m} \langle F_{0k}^{(k)}[./.], \hat{v}_0' \rangle, \\ &\text{where } [./.] = [G_{0\nu_1}(\hat{v}_0), \ldots, G_{0\nu_m}(\hat{v}_0), \hat{v}_0, \ldots, \hat{v}_0], \end{aligned}$$

with the same agreement as for (I.16.17).

Remark I.16.2 *For $n = 3$, formula (I.16.20) gives in view of (I.16.17),*

$$(I.16.21) \qquad \begin{aligned} H_{03} &= \langle F_{03}^{(3)}[\hat{v}_0, \hat{v}_0, \hat{v}_0], \hat{v}_0' \rangle + 2\langle F_{02}^{(2)}[G_{02}(\hat{v}_0), \hat{v}_0], \hat{v}_0' \rangle \\ &= \langle F_{03}^{(3)}[\hat{v}_0, \hat{v}_0, \hat{v}_0], \hat{v}_0' \rangle - 2\langle F_{02}^{(2)}[A_0^{-1}(I - Q)F_{02}^{(2)}[\hat{v}_0, \hat{v}_0], \hat{v}_0], \hat{v}_0' \rangle. \end{aligned}$$

Here $A_0 = D_x F(0,0) : (I - P)X \to (I - Q)Z$, cf. Proposition I.16.1, and formula (I.16.21) seems to be different from (I.6.9) for $\lambda_0 = 0$: The structure of the formula is the same, but instead of a factor 2, formula (I.6.9) has a factor 3. Since this has led to unnecessary confusion, we reveal the mystery: Insert

$$(I.16.22) \qquad \begin{aligned} H_{03} &= \frac{1}{3!} D_{sss}^3 \Phi(0,0) = \frac{1}{3!} D_{vvv}^3 \Phi(0,0)[\hat{v}_0, \hat{v}_0, \hat{v}_0], \\ F_{02}^{(2)} &= \frac{1}{2!} D_{xx}^2 F(0,0), \quad F_{03}^{(3)} = \frac{1}{3!} D_{xxx}^3 F(0,0), \end{aligned}$$

into (I.16.21), and we regain formula (I.6.9).

Equation (I.16.19) has the solution $\{(0, \lambda)\}$ for all $|\lambda| < d$, which recovers the trivial solution. After dividing (I.16.19) by s, we obtain

$$\tilde{\Phi}(s,\lambda) = \sum_{j=1}^{\infty} c_{j0}\lambda^j + \sum_{\substack{j=0 \\ k=1}}^{\infty} c_{jk}\lambda^j s^k = 0 \text{ (cf. (I.5.10))}, \text{ where}$$

(I.16.23)

$$c_{10} = \langle A_1 \hat{v}_0, \hat{v}_0' \rangle,$$

$$c_{n0} = \Big\langle A_n \hat{v}_0, \hat{v}_0' \Big\rangle + \sum_{\substack{j+k=n \\ j\geq 1, k\geq 1}} \langle A_j G_{k1}\hat{v}_0, \hat{v}_0' \rangle \quad \text{for } n \geq 2,$$

$$c_{jk} = H_{j,k+1} \quad \text{for} \quad j \geq 0, k \geq 1.$$

Comparing the recurrence formulas for A_{n0} (I.16.10) and for G_{n1} (I.16.16), we see that

(I.16.24) $$G_{n1}\hat{v}_0 = A_{n-1,0}\hat{v}_0 \quad \text{for all} \quad n \geq 2.$$

Finally, we can use the coefficients $c_{\ell 0}$ in (I.16.23) for the expansion of the eigenvalue perturbations $\mu(\lambda)$; cf. (I.16.9):

$$\mu_1 = c_{10}, \quad \text{and for} \quad \ell \geq 2,$$

$$\mu_\ell = c_{\ell 0} + \sum_{n=0}^{\ell-3} \sum_{m=1}^{\ell-2-n} \sum_{\substack{j+k=n+1 \\ j\geq 1, k\geq 0}} \langle A_j A_{km}\hat{v}_0, \hat{v}_0' \rangle \sum_{\nu_1+\cdots+\nu_m=\ell-2-n} \mu_{\nu_1}\cdots\mu_{\nu_m}.$$

(I.16.25)

We summarize the result:

Theorem I.16.3 Let $\mu(\lambda) = \sum_{j=1}^{\infty}\mu_j\lambda^j$ be the perturbation series of the critical eigenvalue $\mu = 0$ of $A(\lambda) = \sum_{j=0}^{\infty}\lambda^j A_j = D_x F(0,\lambda)$, which is the linearization of $F(x,\lambda)$ along the trivial solution $\{(0,\lambda)\}$. Let $\tilde{\Phi}(s,\lambda)$ be the bifurcation function $\Phi(s,\lambda)$ divided by s. Then $\tilde{\Phi}(0,\lambda) = \sum_{j=1}^{\infty} c_{j0}\lambda^j$ along the trivial solution. If $\mu_1 = \cdots = \mu_{m-1} = 0$, then $\mu_j = c_{j0}$ for $j = 1,\ldots,m+1$. Conversely, if $c_{10} = \cdots = c_{m-1,0} = 0$, then $\mu_j = c_{j0}$ for $j = 1,\ldots,m+1$. If all μ_j are zero, then all c_{j0} are zero and vice versa. In other words, the first two nonvanishing coefficients c_{m0} and $c_{m+1,0}$ of $\tilde{\Phi}(0,\lambda)$ are precisely the two first nonvanishing coefficients μ_m and μ_{m+1} of $\mu(\lambda)$.

If m is odd, then $\mu(\lambda)$ changes sign at $\lambda_0 = 0$, which means that the trivial solution line loses or gains stability (provided that all other eigenvalues in the spectrum of $D_x F(0,0)$ are in the stable left complex half-plane). Now we prove that any degenerate crossing of $\mu(\lambda)$ of the imaginary axis at $\mu(0) = 0$ causes bifurcation. By Theorem I.16.3, the bifurcation equation to determine nontrivial solutions is of the form

(I.16.26)

$$\tilde{\Phi}(s,\lambda) = \sum_{j=m}^{\infty} c_{j0}\lambda^j + \sum_{\substack{j=0 \\ k=1}}^{\infty} c_{jk}\lambda^j s^k = 0$$

for some odd $m \geq 1$ and $c_{m0} = \mu_m$.

We apply the Newton polygon method to solve (I.16.26) locally near $(s,\lambda) = (0,0)$ (cf. Section I.15). We have a smallest point $(0,m)$ on the ordinate (which corresponds to $(0,j_0)$ of the previous section). After possibly dividing by λ^{j_n} (if $j_n > 0$, a solution curve $\{(s,0)|\ |s| < d\}$ exists), we also find a smallest point $(k_n, 0)$ on the abscissa. The two points $(0, j_0)$ and $(k_n, 0)$ are connected by a convex polygon, which is the so-called Newton polygon. As expounded in the previous section, an odd m yields at least one and at most m solution curves of the following form:

$$\{(s, s^\gamma \textstyle\sum_{k=0}^{\infty} a_k s^{k/p\gamma_2})|0 \le s < \delta\}\ \text{and}$$

$$\{(s, (-s)^{\tilde\gamma} \textstyle\sum_{k=0}^{\infty} \tilde a_k (-s)^{k/\tilde p\tilde\gamma_2}|-\delta < s \le 0\}$$

(I.16.27)
for some $\gamma, \tilde\gamma \in \mathbb{Q},\ p, \tilde p \in \mathbb{N}, a_k, \tilde a_k \in \mathbb{R}$.

The numbers $\gamma = \gamma_1/\gamma_2,\ \tilde\gamma = \tilde\gamma_1/\tilde\gamma_2$ are given by slopes $-1/\gamma$, $-1/\tilde\gamma$ of the Newton polygon for $\check\Phi(s,\lambda),\ \check\Phi^-(-s,\lambda) \equiv \check\Phi(s,\lambda)$, respectively. The first coefficients $a_0 \ne 0, \tilde a_0 \ne 0$ are zeros of the defining polynomials $P_\gamma, \hat P_{\tilde\gamma}$, respectively.

This proves the following theorem:

Theorem I.16.4 *Let* $\mu(\lambda) = \sum_{j=m}^{\infty} \mu_j \lambda^j$ *be the perturbation series of the critical eigenvalue* $D_x F(0, \lambda)$. *If m is odd and $\mu_m \ne 0$, then at least one and at most m nontrivial solution curves*

$$\{(x(s), \lambda(s))|-\delta < s < \delta\}\ \ of\ F(x, \lambda) = 0$$
$$bifurcate\ at\ (x, \lambda) = (0, 0).\ In\ particular,$$

(I.16.28)
$$x(s) = s\hat v_0 + \psi(s\hat v_0, \lambda(s)) = s\hat v_0 + o(|s|),$$

$$\lambda(s) = s^\gamma \textstyle\sum_{k=0}^{\infty} a_k s^{k/p\gamma_2}\ \ \ for\ 0 \le s < \delta,\ and$$

$$\lambda(s) = (-s)^{\tilde\gamma} \textstyle\sum_{k=0}^{\infty} \tilde a_k (-s)^{k/\tilde p\tilde\gamma_2}\ \ \ for\ -\delta < s \le 0.$$

The meaning of the parameters is explained in (I.16.27). The function ψ is analytic near $(0,0)$. If $m \ge 1$ is arbitrary, then at most m nontrivial solution curves of the form (I.16.28) bifurcate. (We include here possible vertical bifurcation; i.e., $\lambda(s) \equiv 0$.)

We give some special cases.
Case 1. $m = 1$.
Here we resume the nondegeneracy $\mu'(0) = \mu_1 = c_{10} \ne 0$ (cf. (I.7.36)). If all c_{0k} in (I.16.26) vanish, then we have vertical bifurcation $\lambda(s) \equiv 0$. Otherwise, we find some first $c_{0k_1} \ne 0$, and the Newton polygon is the line connecting $(0, 1)$ and $(k_1, 0)$. Here $\gamma = k_1$, and the defining polynomials (I.15.8) are

(I.16.29)
$$P_\gamma(\tilde\lambda) = \begin{cases} c_{10}\tilde\lambda + c_{0k_1} & \text{for } s \ge 0, \\ c_{10}\tilde\lambda + c_{0k_1} & \text{for } s \le 0,\ \text{if } k_1 \text{ is even}, \\ c_{10}\tilde\lambda - c_{0k_1} & \text{for } s \le 0,\ \text{if } k_1 \text{ is odd}. \end{cases}$$

For the nontrivial solution curve (I.16.28) we obtain in all cases

$$\lambda(s) = -(s^{k_1} c_{0k_1}/c_{10}) + o(|s|^{k_1}),$$

(I.16.30)   where $c_{10} = \mu_1 = \langle A_1 \hat{v}_0, \hat{v}_0' \rangle$, $A_1 = D^2_{x\lambda} F(0,0)$ (cf. (I.16.3)),
and $c_{0k_1} = H_{0,k_1+1}$ is given by (I.16.20).

For $k_1 = 1$ we regain formula (I.6.3), and for $k_1 = 2$ we recover formula
(I.6.11) (see Remark I.16.2). In general, we have a **transcritical bifurcation**
whenever k_1 is **odd** and a **pitchfork bifurcation** whenever k_1 is **even**.

Case 2. $m > 1, (k_1, j_1) = (1,0)$.

This case occurs for any degeneracy of the eigenvalue perturbation $\mu(\lambda) = \sum_{j=m}^{\infty} \mu_j \lambda^j$ and if $c_{01} = H_{02} = \langle F_{02}^{(2)}[\hat{v}_0, \hat{v}_0], \hat{v}_0' \rangle \neq 0$ (cf. (I.16.20), (I.16.23)).
The defining polynomial to be solved is

(I.16.31)
$$\begin{aligned}
&P_\gamma(\tilde{\lambda}) = \mu_m \tilde{\lambda}^m + H_{02} \text{ for } s \geq 0, \\
&P_\gamma(\tilde{\lambda}) = \mu_m \tilde{\lambda}^m - H_{02} \text{ for } s \leq 0, \text{ and} \\
&\gamma = 1/m.
\end{aligned}$$

If m is **odd**, we obtain without any restriction

(I.16.32) $$\lambda(s) = -(s^{1/m}(H_{02}/\mu_m)^{1/m}) + o(|s|^{1/m}).$$

If m is **even**, then

(I.16.33)
$$\begin{aligned}
&\lambda(s) = \pm(s^{1/m}(-H_{02}/\mu_m)^{1/m}) + o(|s|^{1/m}) \text{ for } s \geq 0 \\
&\text{if } -H_{02}/\mu_m > 0, \text{ and} \\
&\lambda(s) = \pm((-s)^{1/m}(H_{02}/\mu_m)^{1/m}) + o(|s|^{1/m}) \text{ for } s \leq 0 \\
&\text{if } H_{02}/\mu_m > 0.
\end{aligned}$$

The bifurcation diagrams in the (s, λ)-plane are depicted in Figure I.16.1. The
bifurcating curve and the trivial solution line are no longer "transversal."

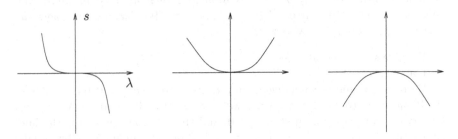

Figure I.16.1

Remark I.16.5 *We recall Remark I.15.2: For these two special cases, analyticity of F (and thus of Φ) is certainly not needed. It suffices that $\tilde{\Phi}$ be of class C^ℓ, where $\ell > k_1$ or m, respectively.*

I.16.1 The Principle of Exchange of Stability for Degenerate Bifurcation

Next, we consider solutions of (I.16.1) as stationary solutions of

(I.16.34)
$$\frac{dx}{dt} = F(x, \lambda),$$

and we investigate their stability as we did in Section I.7. As a matter of fact, only the perturbation of the critical eigenvalue $\mu = 0$ of $D_x F(0,0)$ is studied for $D_x F(x(s), \lambda(s))$, where $\{(x(s), \lambda(s))\}$ is any curve of solutions of (I.16.1) through $(x, \lambda) = (0, 0)$.

For the trivial solution line $\{(0, \lambda)\}$, Theorem I.16.3 relates its stability to the bifurcation function $\Phi(s, \lambda)$. Obviously,

(I.16.35)
$$\tilde{\Phi}(0, \lambda) = D_s \Phi(0, \lambda),$$

and Theorem I.16.3 can be restated as follows: If $\mu'(0) = \cdots = \mu^{(m-1)}(0) = 0$, then $D_\lambda \tilde{\Phi}(0,0) = \cdots = D_\lambda^{m-1} \tilde{\Phi}(0,0) = 0$, and vice versa. In this case,

(I.16.36)
$$\frac{d^m}{d\lambda^m} D_s \Phi(0, \lambda)|_{\lambda=0} = \frac{d^m}{d\lambda^m} \mu(\lambda)|_{\lambda=0}.$$

Formula (I.16.36) generalizes formulas (I.7.18) and (I.7.19) for the trivial solution line. We show that the generalization is also valid for all bifurcating solution curves given by Theorem I.16.4.

Introducing $t = s^{1/p_{\gamma_2}}$ (for $s \geq 0$), we see that the curve $\{(x(s), \lambda(s))\}$ is analytic in t (cf. (I.16.28)). Writing that curve as $\{(\hat{x}(t), \hat{\lambda}(t))\}$, we see that the expansions in t converge not only for $0 \leq t < \tilde{\delta}$ but also for $|t| < \tilde{\delta}$. By analyticity of $F(\hat{x}(t), \hat{\lambda}(t))$ in t, the equation $F(\hat{x}(t), \hat{\lambda}(t)) = 0$ holds for $|t| < \tilde{\delta}$. Therefore, $\{(\hat{x}(t), \hat{\lambda}(t))| \, |t| < \tilde{\delta}\}$ represents a solution curve through $(x, \lambda) = (0, 0)$ that is among all curves (I.16.28) given by Theorem I.16.4. Later, we confine ourselves again to $s \geq 0$. Next we define

(I.16.37)
$$\hat{F}(x, t) \equiv F(\hat{x}(t) + x, \hat{\lambda}(t)),$$
which is analytic in the sense of (I.16.2) in a neighborhood
$$x \in \hat{U} = \{\hat{x} \in X| \, \|\hat{x}\| < \tilde{d}\} \text{ and } t \in \hat{V} = \{t \in \mathbb{R}| \, |t| < \tilde{\delta}\}.$$

For the trivial solution $\{(\hat{x}(t), \hat{\lambda}(t))\} = \{(0, \lambda)\}$ we regain for $\hat{\lambda}(t) \equiv \lambda$ and $t = \lambda$ the original function F. For any solution curve $\{(\hat{x}(t), \hat{\lambda}(t))\}$, we have $\hat{F}(0, t) = 0$ for all $|t| < \tilde{\delta}$, which is the trivial solution line for $\hat{F}(x, t) = 0$. We obtain also $D_x \hat{F}(0, 0) = D_x F(0, 0) = A_0$. Accordingly, we have the simple eigenvalue perturbation

(I.16.38)
$$D_x \hat{F}(0, t)(\hat{v}_0 + \hat{w}(t)) = \hat{\mu}(t)(\hat{v}_0 + \hat{w}(t)) \text{ (cf. (I.16.7))}$$
$$\text{for } D_x \hat{F}(0, t) = D_x F(\hat{x}(t), \hat{\lambda}(t)).$$

Thus $\hat{\mu}(t) = \sum_{j=1}^{\infty} \hat{\mu}_j t^j$ is the perturbation of the critical eigenvalue $\mu = 0$ of $D_x F(0,0)$ in which we are interested for the stability analysis of $\{(\hat{x}(t), \hat{\lambda}(t))\}$. As before, we can relate the perturbation series $\hat{\mu}(t)$ to the bifurcation function $\hat{\Phi}$ obtained by the method of Lyapunov–Schmidt. Since $D_x \hat{F}(0,0) = A_0$, the projections P and Q are the same as before, and an easy computation shows that for

(I.16.39)
$$
\begin{aligned}
&x = s\hat{v}_0 + \hat{w}, \quad \hat{w} \in R(A_0) \cap X, \\
&\hat{\Phi}(s,t) = \Phi(\hat{s}(t) + s, \hat{\lambda}(t)) \text{ with } \hat{s}(t) = \langle \hat{x}(t), \hat{v}_0' \rangle \\
&\text{is the bifurcation function for } \hat{F}(x,t) = 0 \text{ when} \\
&\Phi(s,\lambda) \text{ is the bifurcation function for } F(x,\lambda) = 0.
\end{aligned}
$$

We apply Theorem I.16.3 to the trivial solution $\{(0,t)\}$, which relates $\hat{\mu}(t)$ to $D_s \hat{\Phi}(0,t) = D_s \Phi(\hat{s}(t), \hat{\lambda}(t))$, as follows (cf. (I.16.35), (I.16.36)):

Theorem I.16.6 *Let $\{(\hat{x}(t), \hat{\lambda}(t)) \mid |t| < \tilde{\delta}\}$ be any analytic solution curve of $F(x,\lambda) = 0$ through $(x,\lambda) = (0,0)$, and let $\hat{\mu}(t) = \sum_{j=1}^{\infty} \hat{\mu}_j t^j$ be the simple eigenvalue perturbation of the critical eigenvalue $\mu = 0$ of $D_x F(\hat{x}(t), \hat{\lambda}(t))$. Let $\Phi(s,\lambda)$ be the bifurcation function obtained by method of Lyapunov–Schmidt (cf. (I.16.19)) and let $\hat{s}(t) = \langle \hat{x}(t), \hat{v}_0' \rangle$ (cf. (I.16.5), (I.16.6)). Then the following holds:*

(I.16.40)
$$
\hat{\mu}'(0) = \cdots = \hat{\mu}^{(\hat{m}-1)}(0) = 0 \quad \left(' = \frac{d}{dt}, \ \hat{m} \geq 2 \right) \Leftrightarrow
$$
$$
\frac{d}{dt} D_s \Phi(\hat{s}(t), \hat{\lambda}(t))\big|_{t=0} = \cdots
$$
$$
= \frac{d^{\hat{m}-1}}{dt^{\hat{m}-1}} D_s \Phi(\hat{s}(t), \hat{\lambda}(t))\big|_{t=0} = 0, \text{ and in this case,}
$$
$$
\frac{d^{\hat{m}}}{dt^{\hat{m}}} D_s \Phi(\hat{s}(t), \hat{\lambda}(t))\big|_{t=0} = \frac{d^{\hat{m}}}{dt^{\hat{m}}} \hat{\mu}(t)\big|_{t=0}.
$$

(The last equation holds also for $\hat{m} + 1$.)

This is the generalization of Proposition I.7.3 announced in its proof. (For a different proof see also Remark I.18.2.) Obviously, we can draw the same conclusion as (I.7.48): Under the assumptions of Theorem I.16.6, we obtain from $\frac{d^{\hat{m}}}{dt^{\hat{m}}} \hat{\mu}(t)\big|_{t=0}/\hat{m}! = \hat{\mu}_{\hat{m}} \neq 0$,

(I.16.41)
$$
\begin{aligned}
&\text{sign}\hat{\mu}(t) = \text{sign} D_s \Phi(\hat{s}(t), \hat{\lambda}(t)) \neq 0 \\
&\text{for } t \in (-\tilde{\delta}, \tilde{\delta}) \backslash \{0\}.
\end{aligned}
$$

Then Theorem I.7.4 holds for our setting as well. Before stating the general Principle of Exchange of Stability, we describe under what condition we have $\hat{\mu}_{\hat{m}} \neq 0$ for some $\hat{m} \geq 1$.

We recall that solutions $(x(s), \lambda(s))$ of $F(x,\lambda) = 0$ are obtained by solving $\tilde{\Phi}(s,\lambda) = 0$ given in (I.16.26). When the Newton polygon for $\tilde{\Phi}(s,\lambda)$ is established, we write it as

$$\text{(I.16.42)} \qquad \tilde{\Phi}(s, \lambda) = \sum_{\nu=0}^{n} c_\nu s^{k_\nu} \lambda^{j_\nu} + R(s, \lambda) \quad \text{(cf. (I.15.4))},$$

where $j_0 = m$, $k_0 = 0$ (but possibly $j_n > 0$ if there is vertical bifurcation). In the following, we confine ourselves to $s \geq 0$.

Let $-1/\gamma$ be a slope in the Newton polygon with a defining polynomial P_γ (cf. (I.15.5)–(I.15.8)). We assume that

$$\text{(I.16.43)} \qquad \tilde{\lambda}_0 \neq 0 \text{ is a simple zero of } P_\gamma.$$

In the nomenclature of Section I.15, we assume that for $\tilde{\lambda}_0$ we have a multiplicity $q = 1$, and that the solution curve $\{(x(s), \lambda(s))\}$ is of the form

$$\text{(I.16.44)} \qquad \begin{aligned} x(s) &= s\hat{v}_0 + \psi(s\hat{v}_0, \lambda(s)), \\ \lambda(s) &= s^\gamma(\tilde{\lambda}_0 + \sum_{k=1}^{\infty} a_k s^{k/\gamma_2}) \text{ for } 0 \leq s < \delta, \end{aligned}$$

where $\gamma = \gamma_1/\gamma_2$ (cf. (I.15.10)). Introducing $t = s^{1/\gamma_2}$, we obtain an analytic solution curve $\{(\hat{x}(t), \hat{\lambda}(t))\}$.

Since $\tilde{\Phi}(s, \lambda(s)) = 0$ for $0 \leq s < \delta$ and $\Phi(s, \lambda) = s\tilde{\Phi}(s, \lambda)$, we have

$$\text{(I.16.45)} \qquad D_s\Phi(s, \lambda(s)) = sD_s\tilde{\Phi}(s, \lambda(s)),$$

and by $\lambda(s) = s^\gamma\tilde{\lambda}(s)$, we get, in view of (I.15.5)–(I.15.7),

$$\text{(I.16.46)} \qquad \begin{aligned} &sD_s\tilde{\Phi}(s, \lambda(s)) \\ &= s^\sigma \left(\sum_{\rho=0}^{r} c_{\ell+\rho} k_{\ell+\rho} \tilde{\lambda}(s)^{j_\ell+\rho} + R_1(t, \tilde{\lambda}(s)) \right), \\ &\text{where } \sigma \text{ is defined by (I.15.5)}, t = s^{1/\gamma_2}, \\ &\text{and } R_1 \text{ is analytic in } (t, \tilde{\lambda}) \text{ with } R_1(0, \tilde{\lambda}) = 0. \end{aligned}$$

Switching to the variable t, we see that

$$\text{(I.16.47)} \quad D_s\Phi(\hat{s}(t), \hat{\lambda}(t)) = t^{\sigma\gamma_2} \sum_{\rho=0}^{r} c_{\ell+\rho} k_{\ell+\rho} \tilde{\lambda}_0^{j_\ell+\rho} + \text{higher-order terms}.$$

Now, by (I.15.5), (I.15.8), and assumption (I.16.43),

$$\text{(I.16.48)} \qquad \begin{aligned} &\sum_{\rho=0}^{r} c_{\ell+\rho} k_{\ell+\rho} \tilde{\lambda}_0^{j_\ell+\rho} \\ &= \sigma P_\gamma(\tilde{\lambda}_0) - \gamma\tilde{\lambda}_0 P_\gamma'(\tilde{\lambda}_0) = -\gamma\tilde{\lambda}_0 P_\gamma'(\tilde{\lambda}_0) \neq 0 \quad \left(' = \frac{d}{d\tilde{\lambda}} \right). \end{aligned}$$

Thus we have proved the following theorem:

Theorem I.16.7 *Let a solution curve $\{(x(s), \lambda(s))\}$ of $F(x, \lambda) = 0$ through $(0, 0)$ be obtained by a simple zero $\tilde{\lambda}_0 \neq 0$ of a defining polynomial P_γ for some slope $-1/\gamma$ of the Newton polygon for $\tilde{\Phi}(s, \lambda)$ (cf. Theorem I.16.4). If $\gamma = \gamma_1/\gamma_2$ and $t = s^{1/\gamma_2}$, then $\{(x(s), \lambda(s))\} = \{(\hat{x}(t), \hat{\lambda}(t))\}$ is an analytic*

solution curve of $F(x, \lambda) = 0$ (in t), and for the simple eigenvalue perturba-
tion $\hat{\mu}(t) = \sum_{j=1}^{\infty} \hat{\mu}_j t^j$ of $D_x F(\hat{x}(t), \hat{\lambda}(t))$, we obtain the following:

$$\hat{\mu}_1 = \cdots = \hat{\mu}_{\hat{m}-1} = 0 \ for \ \hat{m} = \gamma_1 j_\ell + \gamma_2 k_\ell,$$
(I.16.49) *where $\gamma j_\ell + k_\ell = \cdots = \gamma j_{\ell+r} + j_{\ell+r} = \sigma$ (cf (I.15.5)), and*
$$\hat{\mu}_{\hat{m}} = -\gamma \tilde{\lambda}_0 P_\gamma'(\tilde{\lambda}_0) \neq 0.$$

In particular, $\hat{\mu}(t) \neq 0$ for $t \neq 0$, so that (I.16.41) holds.

We apply Theorem I.16.6 to the special cases studied before.

Case 1. $m = 1$.
We exclude vertical bifurcation by assuming $c_{0k_1} = H_{0,k_1+1} \neq 0$ for some minimal $k_1 \geq 1$. Then $\gamma = k_1$ and $j_0 = m = 1, k_0 = 0$, so that by (I.16.49), $\hat{m} = k_1$. In the first case of (I.16.29) we have $t = s$; in the two last cases of (I.16.29) we set $t = -s$. Therefore, we can unify all cases to $\hat{\mu}(t) = \hat{\mu}(s)$, and formula (I.16.49) gives, for the first nonvanishing coefficient,

$$(I.16.50) \qquad\qquad \hat{\mu}_{k_1} = k_1 c_{0k_1}.$$

Since by Theorem I.16.3, $\mu_1 = c_{10} \neq 0$, we obtain, in view of (I.16.30),

$$(I.16.51) \qquad k_1 \frac{d\mu}{d\lambda}(0) \frac{d^{k_1}\lambda}{ds^{k_1}}(0) = -\frac{d^{k_1}\hat{\mu}}{ds^{k_1}}(0) \neq 0,$$

which generalizes formulas (I.7.40) and (I.7.45). As shown in Section I.7, the relation (I.16.51) proves an **exchange of stability** for a **transcritical bifurcation** (when k_1 is odd) as well as for a **pitchfork bifurcation** (when k_1 is even). The situation is sketched in Figure I.7.2 and Figure I.7.3.

Case 2. $m > 1, (k_1, j_1) = (1, 0)$.
Here $\gamma = 1/m, j_0 = m, k_0 = 0$, so that $\hat{m} = m$. In the first case of (I.16.31) we have $t = s^{1/m}$, and in the second case of (I.16.31) we set $t = (-s)^{1/m}$. For **odd** $m = \hat{m}$ we can unify both cases, and we obtain from (I.16.49), for $\hat{\mu}(t) = \hat{\mu}(s^{1/m})$,

$$(I.16.52) \qquad \begin{array}{l} \hat{\mu}_m = H_{02}, \ or \\ \hat{\mu}(s^{1/m}) = sH_{02} + o(|s|). \end{array}$$

By (I.16.32), $\hat{\lambda}(t) = -t(H_{02}/\mu_m)^{1/m} + o(|t|)$, which proves that

$$(I.16.53) \qquad \frac{d^m\mu}{d\lambda^m}(0)\left(\frac{d\lambda}{dt}(0)\right)^m = -\frac{d^m\hat{\mu}}{dt^m}(0) \neq 0.$$

For the cases (I.16.33) for **even** m, we obtain

$$(I.16.54) \qquad \begin{array}{l} \hat{\mu}_m = H_{02} \ for \ \hat{\mu}(t) = \hat{\mu}(s^{1/m}) \ for \ s \geq 0, \\ \hat{\mu}_m = -H_{02} \ for \ \hat{\mu}(t) = \hat{\mu}((-s)^{1/m}) \ for \ s \leq 0. \end{array}$$

The bifurcation diagrams are sketched in Figure I.16.1. Possible stabilities are marked in Figure I.16.2.

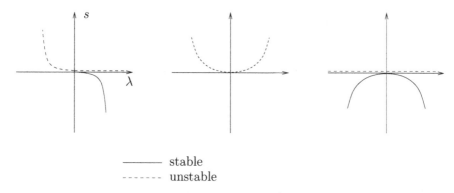

————— stable

------- unstable

Figure I.16.2

If m is even, the stability of the trivial solution does not change at $\lambda = 0$. In view of (I.16.33), (I.16.54), the stability of the bifurcating curve does not change either. In all cases (I.16.32) and (I.16.33), the stability properties of the trivial solution line and of the bifurcating curve for some fixed value of $\lambda \neq 0$ are opposite. This **Principle of Exchange of Stability** is formulated in the next Theorem.

Theorem I.16.8 *Consider all local solution curves of $F(x, \lambda) = 0$ through $(x, \lambda) = (0, 0)$ except a vertical curve $\{(x(s), 0)\}$ but including the trivial line $\{(0, \lambda)\}$. The nontrivial curves are given by Theorem I.16.4, and for fixed $\lambda \neq 0$ these curves can be ordered in the (s, λ)-plane. Then consecutive curves have opposite stability properties in a weakened sense: Let $\mu_i(s)$ be the perturbed eigenvalues of $D_x F(x_i(s), \lambda_i(s))$, $i = 1, 2$, for consecutive curves in the above sense. Then $\mu_1(s)\mu_2(s) \leq 0$, with equality only if $D_s \Phi(s, \lambda_i(s)) \equiv 0$ for $i = 1$ or 2 and for all $0 \leq s < \delta$ or $-\delta < s \leq 0$. Here Φ is the bifurcation function (I.16.19).*

Proof. We refer to the proof of Theorem I.7.4: If $D_s \Phi(s, \lambda_i(s)) \not\equiv 0$, then by (I.16.40), $\mu_i(s) \not\equiv 0$, and derivatives of $\Phi(\cdot, \lambda)$ at consecutive zeros cannot be both positive or both negative. The claim then follows from (I.16.41). □

Thus, if the total bifurcation diagram and the stability properties of the trivial solution line are known, then the stability of each bifurcating curve can be derived by Theorem I.16.8. On the other hand, if a single curve is computed by the Newton polygon method, Theorem I.16.7 gives its stability without any knowledge of other solution curves.

We give an **example**. Note that in view of (I.16.16), (I.16.17),

(I.16.55)
if $\langle F_{j,k+1}(\hat{v}_0), \hat{v}_0' \rangle \neq 0$ and $F_{\ell m} = 0$ for $0 \leq \ell \leq j$,
$1 \leq m \leq k+1$, $(\ell, m) \neq (j, k+1)$,
then the lowest-order terms c_{jk} of (I.16.23) for $k \geq 1$ are
$c_{jk} = H_{j,k+1} = \langle F_{j,k+1}(\hat{v}_0), \hat{v}_0' \rangle$.

Assume (I.16.55) and that the coefficients c_ν, $\nu = 0, \ldots, n$, on the Newton polygon are $c_0 = c_{m0} = \mu_m$, $c_\nu = \langle F_{j_\nu, k_\nu+1}(\hat{v}_0), \hat{v}_0 \rangle$ for $\nu = 1, \ldots, n$ (cf. (I.15.3), (I.15.4)). Then the solution curves can be computed directly from F, and their stability follows from Theorem I.16.8 or I.16.7. Let, for instance,

$$F(x, \lambda) = A(\lambda)x + \lambda^4 F_{42}(x) + \lambda^2 F_{24}(x) + F_{07}(x) + \text{h.o.t.},$$

(I.16.56) $A(\lambda)(\hat{v}_0 + w(\lambda)) = (\mu_7 \lambda^7 + \text{h.o.t.})(\hat{v}_0 + w(\lambda))$, where $\mu_7 > 0$,

$$\langle F_{42}(\hat{v}_0), \hat{v}_0' \rangle < 0, \quad \langle F_{24}(\hat{v}_0), \hat{v}_0' \rangle > 0, \quad \langle F_{07}(\hat{v}_0), \hat{v}_0' \rangle < 0.$$

(Here h.o.t. means higher-order terms.)

Then the Newton polygon for $\tilde{\Phi}(s, \lambda) = 0$ (cf. ((I.16.23))) is given by the points $(0, 7), (1, 4), (3, 2), (6, 0)$ with coefficients $c_0 > 0, c_1 < 0, c_2 > 0, c_3 < 0$. The values γ of the three slopes are $\gamma = \frac{1}{3}, \gamma = 1$, and $\gamma = \frac{3}{2}$. The defining polynomials have simple nonzero roots, so that Theorem I.16.6 is applicable. The bifurcation diagram is sketched in Figure I.16.3. A concrete example is given in Section III.2 in (III.2.20).

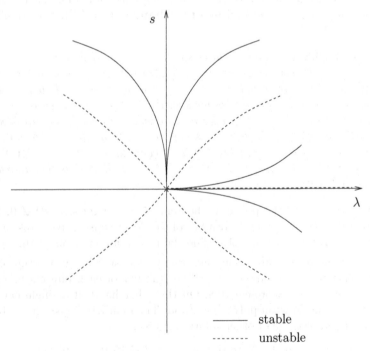

Figure I.16.3

I.17 Degenerate Hopf Bifurcation and Floquet Exponents of Bifurcating Periodic Orbits

Here we assume the situation of Section I.8, where we proved Hopf bifurcation for the evolution equation

$$(I.17.1) \qquad \frac{dx}{dt} = F(x, \lambda)$$

from the trivial solution line $\{(0, \lambda) | \lambda \in \mathbb{R}\}$ at some $(0, \lambda_0)$. For simplicity, we assume now that $\lambda_0 = 0$ and that the mapping F satisfies the hypotheses of Section I.8, in particular, (I.8.2)–(I.8.6), (I.8.8), and (I.8.14) with $\lambda_0 = 0$. Instead of the regularity (I.8.13), we assume now the analyticity (I.16.2) of F.

Apart from these technical assumptions, the nondegeneracy (I.8.7) of the simple eigenvalue perturbation $\mu(\lambda)$ (I.8.6) is crucial for the solution of the bifurcation equation (I.8.34) or (I.8.35) (cf. (I.8.43), (I.8.44)). A crossing of the imaginary axis of $\mu(\lambda)$ at $\lambda = 0$ occurs also if $\mu(0) = i\kappa_0$ and $\text{Re}\mu'(0) = \cdots = \text{Re}\mu^{(m-1)}(0) = 0$ but $\text{Re}\mu^{(m)}(0) \neq 0$ for some odd m. As proved in the previous section, a possibly degenerate loss of stability of the trivial solution line at $\lambda = 0$ through $\mu(0) = 0$ causes bifurcation of stationary solutions of (I.17.1). Therefore, one expects a bifurcation of periodic solutions of (I.17.1) if the stability of the trivial solution is lost at $\lambda = 0$ through $\mu(0) = i\kappa_0(\neq 0)$ also in a degenerate way. In order to prove this expectation, we have to study the eigenvalue perturbation (I.8.6) near $\lambda_0 = 0$. Choosing the vector φ_0' as in (I.8.18), we write (I.8.6) as

$$A(\lambda)(\varphi_0 + \textstyle\sum_{j=1}^{\infty} \varphi_j \lambda^j) = \mu(\lambda)(\varphi_0 + \textstyle\sum_{j=1}^{\infty} \varphi_j \lambda^j) \text{ (cf. (I.16.3))},$$

$$(I.17.2) \qquad \mu(\lambda) = i\kappa_0 + \textstyle\sum_{j=1}^{\infty} \mu_j \lambda^j, \quad \mu_j \in \mathbb{C},$$

$$\langle \varphi_j, \varphi_0' \rangle = 0 \text{ for all } j \geq 1; \text{ i.e.,}$$

$$\varphi_j \in R(i\kappa_0 I - A_0) \cap X \quad \text{(complexified)}.$$

For the subsequent analysis we need the projections

$$(I.17.3) \quad \begin{array}{l} Q_{01} : Z \to N(i\kappa_0 I - A_0) \text{ along } R(i\kappa_0 I - A_0) \\ \text{given by } Q_{01} z = \langle z, \varphi_0' \rangle \varphi_0 \text{ for } z \in Z, \text{ cf. (I.8.18)}, \\ Q_{01}|_X = P_{01} : X \to N(i\kappa_0 I - A_0) \text{ along } R(i\kappa_0 I - A_0) \cap X. \end{array}$$

Then in analogy to Proposition I.16.1, we have the following result:

Proposition I.17.1 *The coefficients μ_ℓ for $\ell \geq 1$ of the eigenvalue perturbation I.17.2 are recursively given by (I.16.9), where we replace $\hat{v}_0$ by φ_0, and $\hat{v}_0'$ by φ_0', and where we substitute A_0 by $A_0 - i\kappa_0 I$ in the definition of the vectors $A_{nm}\varphi_0$ for $n, m \geq 0$.*

Proof. Writing (I.17.2) as

$$(\text{I.17.4}) \quad \begin{aligned} \Big((A_0 - i\kappa_0 I) + \sum_{j=1}^{\infty} \lambda^j A_j \Big) \Big(\varphi_0 + \sum_{j=1}^{\infty} \varphi_j \lambda^j \Big) \\ = \Big(\sum_{j=1}^{\infty} \mu_j \lambda^j \Big) \Big(\varphi_0 + \sum_{j=1}^{\infty} \varphi_j \lambda^j \Big), \end{aligned}$$

then we are precisely in the same situation as (I.16.7), (I.16.8) when the substitutions are made according to Proposition I.17.1.   □

The Lyapunov–Schmidt reduction for

$$(\text{I.17.5}) \qquad\qquad G(x, \kappa, \lambda) \equiv \kappa \frac{dx}{dt} - F(x, \lambda) = 0$$

near $(0, \kappa_0, 0)$ follows Section I.8 completely. Here we use the same notation, in particular,

$$Q : W \to N(J_0) \text{ along } R(J_0),$$

$$(\text{I.17.6}) \qquad \text{where } J_0 \equiv \kappa_0 \frac{d}{dt} - A_0 \text{ and } Q \text{ is given by (I.8.21);}$$

$$Q|_{Y \cap E} = P : Y \cap E \to N(J_0) \text{ along } R(J_0) \cap (Y \cap E).$$

Decomposing $x \in E \cap Y$ as $x = Px + (I - P)x = v + w$, we see then that (I.17.5) is equivalent to

$$(\text{I.17.7}) \qquad \begin{aligned} QG(v + w, \kappa, \lambda) &= 0, \\ (I - Q)G(v + w, \kappa, \lambda) &= 0, \end{aligned}$$

and the second equation is solved for $w = \psi(v, \kappa, \lambda)$ by the Implicit Function Theorem. By the assumed analyticity of F (which induces analyticity for G), the function $w = \psi(v, \kappa, \lambda)$ is analytic, too, and we give some coefficients in its expansion. Inserting

$$(\text{I.17.8}) \qquad \begin{aligned} w = \sum_{j,\ell,k=0}^{\infty} \lambda^j (\kappa - \kappa_0)^\ell G_{j\ell k}(v), \text{ where} \\ G_{j\ell k} : N(J_0)^k \to R(J_0) \cap (E \cap Y) \\ \text{are } k\text{-linear, symmetric, and continuous,} \end{aligned}$$

into $(\text{I.17.7})_2$, which is

$$(\text{I.17.9}) \qquad \begin{aligned} J_0 w + (\kappa - \kappa_0) \frac{d}{dt} w - (I - Q) \sum_{j=1}^{\infty} \lambda^j A_j (v + w) \\ - (I - Q) \sum_{\substack{j=0 \\ k=2}}^{\infty} \lambda^j F_{jk}(v + w) = 0, \end{aligned}$$

we obtain, by comparing like powers,

$$G_{j\ell 0} = 0 \text{ for all } j, \ell \in \mathbb{N}_0, \quad G_{001} = 0,$$
$$G_{101}v = J_0^{-1}(I - Q)A_1v, \quad G_{0\ell 1} = 0 \text{ for } \ell \geq 1,$$
$$G_{111}v = -J_0^{-1}\frac{d}{dt}J_0^{-1}(I - Q)A_1v,$$

(I.17.10)
$$G_{1\ell 1}v = -J_0^{-1}\frac{d}{dt}G_{1,\ell-1,1} \text{ for } \ell \geq 1,$$

$$G_{n01}v = J_0^{-1}(I - Q)A_nv + \sum_{\substack{j+k=n\\j\geq 1, k\geq 1}} J_0^{-1}(I - Q)A_jG_{k01}v,$$

$$G_{nm1}v = -J_0^{-1}\frac{d}{dt}G_{n,m-1,1}v + \sum_{\substack{j+k=n\\j\geq 1, k\geq 1}} J_0^{-1}(I - Q)A_jG_{km1}v,$$

for $n \geq 2, m \geq 1$. The recurrence formulas for G_{00n} for $n \geq 2$ are

$$G_{002}(v) = J_0^{-1}(I - Q)F_{02}(v),$$

(I.17.11) $\quad G_{00n}(v) = J_0^{-1}(I - Q)\sum_{k=2}^{n}\sum_{m=0}^{k}\sum_{\nu_1+\cdots+\nu_m=n-k+m}\binom{k}{m}F_{0k}^{(k)}[./.],$

where $[./.] = [G_{00\nu_1}(v), \ldots, G_{00\nu_m}(v), v, \ldots, v]$; cf.(I.16.17).

We insert $w = \psi(v, \kappa, \lambda)$ into (I.17.7)$_1$, and we obtain the bifurcation equation (cf. (I.8.30)). For real v, this two-dimensional system is equivalent to the one-dimensional complex equation (I.8.34), which by $Q\frac{d}{dt}w = \frac{d}{dt}Pw = 0$, $QJ_0w = 0$ reduces to

(I.17.12)
$$\frac{1}{2\pi}\int_0^{2\pi}\Big\langle(\kappa - \kappa_0)\frac{d}{dt}v - \sum_{j=1}^{\infty}\lambda^j A_j(v + w)$$
$$- \sum_{\substack{j=0\\k=2}}^{\infty}\lambda^j F_{jk}(v + w), \psi_0'\Big\rangle dt = 0.$$

Clearly, by (I.17.8), the function (I.17.12) is analytic in its variables $(v, \kappa - \kappa_0, \lambda)$, but before studying its coefficients we investigate its structure when we insert $v = c\psi_0 + \overline{c}\overline{\psi}_0$, $c \in \mathbb{C}$. According to (I.8.35), we denote this complex-valued funtion by $\hat{\Phi}(c, \kappa, \lambda)$, and due to equivariance of G, it has the properties (I.8.36) and (I.8.37). In particular, it is odd in $c \in \mathbb{C}$. We rediscover this oddness as follows: Let $G^{(k)}[v, \ldots, v]$ be a k-linear, symmetric, and continuous mapping in the expansion of $G(v + \psi(v, \kappa, \lambda), \kappa, \lambda)$. When we insert $v = c\psi_0 + \overline{c}\overline{\psi}_0$, we obtain, for $k = 2\ell + 1$,

(I.17.13)
$$\frac{1}{2\pi}\int_0^{2\pi}\langle G^{(k)}[v, \ldots, v], \psi_0'\rangle dt$$
$$= c|c|^{2\ell}\binom{2\ell+1}{\ell}\langle G^{(k)}[\underbrace{\varphi_0, \ldots, \varphi_0}_{\ell+1}, \underbrace{\overline{\varphi}_0, \ldots, \overline{\varphi}_0}_{\ell}], \varphi_0'\rangle,$$

and 0 for $k = 2\ell$ (cf. (I.8.18), (I.8.21)). Therefore, the **Bifurcation Equation for Hopf Bifurcation** (I.17.12) is of the form

$$\text{(I.17.14)} \quad \begin{aligned} \hat{\Phi}(c,\kappa,\lambda) &= c \sum_{j,\ell,k=0}^{\infty} H_{j\ell k} \lambda^j (\kappa - \kappa_0)^\ell |c|^{2k} \\ &= c\tilde{\Phi}(c,\kappa,\lambda) \quad \text{(cf. (I.8.39))} \end{aligned}$$
with coefficients $H_{j\ell k} \in \mathbb{C}$.

(We know that there is convergence for $|c|, |\kappa - \kappa_0|, |\lambda| < \tilde{\delta}$.) From (I.17.12) we deduct for the coefficients of the linear terms (in c):

$$H_{000} = 0, \quad H_{010} = i, \quad H_{0\ell 0} = 0 \text{ for } \ell \geq 2,$$

$$H_{100} = -\langle A_1 \varphi_0, \varphi_0' \rangle, \quad H_{1\ell 0} = 0 \quad \text{for } \ell \geq 1,$$

$$\text{(I.17.15)} \quad H_{n00} = -\langle A_n \varphi_0, \varphi_0' \rangle - \sum_{\substack{j+k=n \\ j\geq 1, k\geq 1}} \langle A_j G_{k01} \varphi_0, \varphi_0' \rangle,$$

$$H_{nm0} = - \sum_{\substack{j+k=n \\ j\geq 1, k\geq 1}} \langle A_j G_{km1} \varphi_0, \varphi_0' \rangle, \text{ for } n \geq 2, m \geq 1.$$

The operators G_{nm1} are given by (I.17.10). In order to relate them to the operators A_{nm} of (I.16.10), we make use of the following observations:

$$\text{(I.17.16)} \quad \begin{aligned} &\text{For } v = c\psi_0 + \bar{c}\bar{\psi}_0, \quad c \in \mathbb{C}, \\ &\frac{1}{2\pi} \int_0^{2\pi} \langle S J_0^{-1}(I - Q)Tv, \psi_0' \rangle dt \\ &= c\langle S(i\kappa_0 I - A_0)^{-1}(I - Q_{01})T\varphi_0, \varphi_0' \rangle, \\ &\frac{1}{2\pi} \int_0^{2\pi} \left\langle S J_0^{-1} \frac{d}{dt}(I - Q)Tv, \psi_0' \right\rangle dt \\ &= ic\langle S(i\kappa_0 I - A_0)^{-1}(I - Q_{01})T\varphi_0, \varphi_0' \rangle \end{aligned}$$

for every $S, T \in L(X, Z)$ and for Q_{01} given by (I.17.3).

Considering $i\kappa_0 I - A_0 : (I - P_{01})X \to (I - Q_{01})Z$, we see that the preimages are in $(I - P_{01})X = R(i\kappa_0 I - A_0) \cap X$. We claim that

$$\text{(I.17.17)} \quad \begin{aligned} &\frac{1}{2\pi} \int_0^{2\pi} \langle A_j G_{km1} v, \psi_0' \rangle dt = \langle A_j G_{km1} \varphi_0, \varphi_0' \rangle \\ &= -i^m \langle A_j A_{k-1,m} \varphi_0, \varphi_0' \rangle \text{ for } j \geq 1, k \geq 1, m \geq 0, \end{aligned}$$

where A_{nm} are given by (I.16.10) with the substitutions listed in Proposition I.17.1.

The proof is by induction, using (I.17.16) and the recurrence formulas (I.16.10) and (I.17.10).

Next, we choose $c = r \in \mathbb{R}$ and decompose $\tilde{\Phi}(r, \kappa, \lambda) = \hat{\Phi}(r, \kappa, \lambda)/r = 0$ into real and imaginary parts:

$$\sum_{\substack{j=1\\\ell=0}}^{\infty} \mathrm{Re} H_{j\ell 0} \lambda^j (\kappa - \kappa_0)^\ell + \sum_{\substack{j,\ell=0\\k=1}}^{\infty} \mathrm{Re} H_{j\ell k} \lambda^j (\kappa - \kappa_0)^\ell r^{2k} = 0,$$

(I.17.18)

$$\kappa - \kappa_0 + \sum_{\substack{j=1\\\ell=0}}^{\infty} \mathrm{Im} H_{j\ell 0} \lambda^j (\kappa - \kappa_0)^\ell + \sum_{\substack{j,\ell=0\\k=1}}^{\infty} \mathrm{Im} H_{j\ell k} \lambda^j (\kappa - \kappa_0)^\ell r^{2k} = 0.$$

This is system (I.8.39), and it is solved in Section I.8 by the Implicit Function Theorem when $\mathrm{Re} H_{100} \neq 0$. By (I.8.44), this assumption is equivalent to $\mathrm{Re}\mu'(0) = \mathrm{Re}\mu_1 \neq 0$ (cf. (I.17.2)). Here we allow a degeneracy of the eigenvalue perturbation, and we have to relate this to the bifurcation equations (I.17.18).

The second equation (I.17.18) allows us to express $\kappa - \kappa_0$ in terms of (λ, r^2) (by an application of the Implicit Function Theorem). By analyticity,

$$\kappa = \kappa_0 + \sum_{j,k=0}^{\infty} d_{jk} \lambda^j r^{2k} \equiv \tilde{\kappa}(r, \lambda), \quad \text{where} \quad d_{00} = 0,$$

(I.17.19)
$$d_{10} = -\mathrm{Im} H_{100}, \quad d_{20} = -\mathrm{Im} H_{200},$$

$$d_{\ell 0} = -\mathrm{Im} H_{\ell 00} - \sum_{n=2}^{\ell-1} \sum_{m=1}^{\ell-n} \mathrm{Im} H_{nm0} \sum_{\nu_1 + \cdots + \nu_m = \ell - n} d_{\nu_1 0} \cdots d_{\nu_m 0},$$

for $\ell \geq 3$ (in view of $H_{1\ell 0} = 0$).

When we insert $\kappa - \kappa_0$ as given by (I.17.19) into (I.17.18)$_1$, we obtain the **Reduced Bifurcation Equation**

$$\tilde{\Phi}_{re}(r, \lambda)$$
$$\equiv \sum_{j,k=0}^{\infty} c_{jk} \lambda^j r^{2k} = 0 \quad \text{with} \quad c_{00} = 0,$$

(I.17.20) $\quad c_{10} = \mathrm{Re} H_{100}, \quad c_{20} = \mathrm{Re} H_{200},$

$$c_{\ell 0} = \mathrm{Re} H_{\ell 00} + \sum_{n=2}^{\ell-1} \sum_{m=1}^{\ell-n} \mathrm{Re} H_{nm0} \sum_{\nu_1 + \cdots + \nu_m = \ell - n} d_{\nu_1 0} \cdots d_{\nu_m 0},$$

for $\ell \geq 3$.

Finally, we express μ_ℓ as given by (I.16.9) (with the substitutions of Proposition I.17.1 by the coefficients H_{nm0}, where we use (I.17.15) and (I.17.17)):

$$\mu_1 = -H_{100}, \quad \mu_2 = -H_{200},$$

(I.17.21) $\mu_\ell = -H_{\ell 00} - \displaystyle\sum_{n=2}^{\ell-1} \sum_{m=1}^{\ell-n} H_{nm0} \sum_{\nu_1 + \cdots + \nu_m = \ell - n} (-i\mu_{\nu_1}) \cdots (-i\mu_{\nu_n}),$

for $\ell \geq 3$.

Assume that $\mathrm{Re}\mu_1 = \cdots = \mathrm{Re}\mu_{m-1} = 0$. Then, in view of (I.17.19) and (I.17.21),

(I.17.22) $\qquad\qquad \mathrm{Im}\mu_\ell = d_{\ell 0}$ for $\ell = 1, \ldots, m+1,$

and by (I.17.20) and (I.17.21),

(I.17.23) $\operatorname{Re}\mu_\ell = -c_{\ell 0}$ for $\ell = 1,\ldots,m+1$.

We summarize the result:

Theorem I.17.2 *Let* $\mu(\lambda) = i\kappa_0 + \sum_{j=1}^\infty \mu_j \lambda^j$ *be the perturbation series of the critical eigenvalue* $i\kappa_0$ *of* $A(\lambda) = \sum_{j=0}^\infty \lambda^j A_j = D_x F(0,\lambda)$, *which is the linearization of* $F(x,\lambda)$, *along the trivial solution* $\{(0,\lambda)\}$. *Let* $\tilde{\Phi}_{re}(r,\lambda)$ *be the Reduced Bifurcation Function (I.17.20). Then* $\tilde{\Phi}_{re}(0,\lambda) = \sum_{j=1}^\infty c_{j0}\lambda^j$ *along the trivial solution. If* $\operatorname{Re}\mu_1 = \cdots = \operatorname{Re}\mu_{m-1} = 0$, *then* $\operatorname{Re}\mu_j = -c_{j0}$ *for* $j = 1,\ldots,m+1$. *Conversely, if* $c_{10} = \cdots = c_{m-1,0} = 0$, *then* $\operatorname{Re}\mu_j = -c_{j0}$ *for* $j = 1,\ldots,m+1$. *If all* $\operatorname{Re}\mu_j$ *are zero, then all* c_{j0} *are zero and vice versa. In other words, the first two nonvanishing coefficients* c_{m0} *and* $c_{m+1,0}$ *of* $\tilde{\Phi}_{re}(0,\lambda)$ *are precisely the negatives of the two first nonvanishing coefficients* $\operatorname{Re}\mu_m$ *and* $\operatorname{Re}\mu_{m+1}$ *of* $\operatorname{Re}\mu(\lambda)$.

If m is odd, then $\operatorname{Re}\mu(\lambda)$ changes sign at $\lambda_0 = 0$, which means that the trivial solution line loses or gains stability (provided that all other eigenvalues in the spectrum of $D_x F(0,0)$ are in the stable left complex half-plane). The proof that a degenerate crossing of $\mu(\lambda)$ of the imaginary axis at $\mu(0) = i\kappa_0$ causes bifurcation of periodic solutions of (I.17.1) is the same as that in Section I.16: $\tilde{\Phi}_{re}(r,\lambda)$ corresponds, in view of Theorem I.17.2, precisely to $\tilde{\Phi}(s,\lambda)$ given in (I.16.26) when s is replaced by r^2, and the arguments for the following theorem are the same as for Theorem I.16.4.

Theorem I.17.3 *Let* $\mu(\lambda) = i\kappa_0 + \sum_{j=1}^\infty \mu_j \lambda^j$ *be the perturbation series of the critical eigenvalue* $i\kappa_0$ *of* $D_x F(0,\lambda)$, *and let* $\operatorname{Re}\mu(\lambda) = \sum_{j=m}^\infty \operatorname{Re}\mu_j \lambda^j$ *with* $\operatorname{Re}\mu_m \neq 0$. *If* m *is odd, then at least one and at most* m *nontrivial curves* $\{(x(r),\lambda(r))\}$ *of real* $2\pi/\kappa(r)$-*periodic solutions of (I.17.1) through* $(x(0),\lambda(0)) = (0,0)$ *and* $2\pi/\kappa(0) = 2\pi/\kappa_0$ *in* $(C_{2\pi/\kappa(r)}^{1+\alpha}(\mathbb{R},Z) \cap C_{2\pi/\kappa(r)}^\alpha(\mathbb{R},X))$
$\times \mathbb{R}$ *bifurcate. In particular,*

        $x(r)$ *is given by (I.8.47), where* ψ *is analytic near* $(0,\kappa_0,0)$;

        $\kappa(r) = \tilde{\kappa}(r,\lambda(r))$, *where* $\tilde{\kappa}$ *is analytic near* $(0,0)$,

(I.17.24) $\lambda(r) = r^{2\gamma}\sum_{k=0}^\infty a_k r^{2k/p\gamma_2}$ *for* $0 \leq r < \delta$

        with exponents explained in (I.16.27) and
        $a_0 \neq 0$ *a zero of the defining polynomial* P_γ.
        Finally, $x(-r) = S_{\pi/\kappa(r)}x(r)$, $\kappa(-r) = \kappa(r)$,
        and $\lambda(-r) = \lambda(r)$ *(cf. Theorem I.8.2).*

If $m \geq 1$ *is arbitrary, then at most* m *nontrivial periodic solution curves of the form (I.17.23) bifurcate (we include here possible vertical bifurcation, i.e.,* $\lambda(r) \equiv 0$).

We have the analogous special cases here as expounded in Section I.16 when the Newton polygon for $\tilde{\Phi}_{re}(r, \lambda) = 0$ consists of a single line connecting $(0, 1)$ and $(2k_1, 0)$ or $(0, m)$ and $(2, 0)$. Whereas the coefficient of the defining polynomial at $(0, 1)$ or $(0, m)$ is given by Theorem I.17.2 (it is $-\operatorname{Re}\mu_1$ or $-\operatorname{Re}\mu_m$, respectively), the coefficient at $(2k_1, 0)$ or $(2, 0)$ is $\operatorname{Re}H_{00k_1}$ or $\operatorname{Re}H_{001}$. The computation of these coefficients is not trivial; it follows the lines of (I.16.20), where the expressions for G_{00n} correspond to (I.16.17). Here A_0 is replaced by J_0. For $v = r(\psi_0 + \overline{\psi}_0)$, the computation of $J_0^{-1}(I - Q)F_{0k}^{(k)}[G_{00\nu_1}(v), \ldots, G_{00\nu_m}(v), v, \ldots, v]$ is accomplished as in Section I.9, in particular (I.9.9). We give the formula for H_{001}, which follows from (I.9.11) with the observations of Remark I.16.2:

$$
\begin{aligned}
H_{001} = &-3\langle F_{03}^{(3)}[\varphi_0, \varphi_0, \overline{\varphi}_0], \varphi_0' \rangle \\
&-2\langle F_{02}^{(2)}[\overline{\varphi}_0, (2i\kappa_0 - A_0)^{-1}F_{02}^{(2)}[\varphi_0, \varphi_0]], \varphi_0' \rangle \\
&+4\langle F_{02}^{(2)}[\varphi_0, A_0^{-1}F_{02}^{(2)}[\varphi_0, \overline{\varphi}_0]], \varphi_0' \rangle.
\end{aligned}
$$
(I.17.25)

Case 1

For $m = 1$ and any $k_1 \geq 1$ we always have a **pitchfork bifurcation** by formula (I.16.30), where s^{k_1} is replaced by r^{2k_1}. If $k_1 = 1$, we have, in view of (I.17.23) and (I.17.25),

$$
\begin{aligned}
\lambda(r) &= \frac{\operatorname{Re}H_{001}}{\operatorname{Re}\mu_1}r^2 + O(r^4), \\
\kappa(r) &= \kappa_0 + d_{10}\lambda(r) + d_{01}r^2 + O(r^4) \\
&= \kappa_0 + \left(\operatorname{Im}\mu_1\frac{\operatorname{Re}H_{001}}{\operatorname{Re}\mu_1} - \operatorname{Im}H_{001}\right)r^2 + O(r^4)
\end{aligned}
$$
(I.17.26)

by (I.17.19), (I.17.22), and $d_{01} = -\operatorname{Im}H_{001}$, which follows from (I.17.18)$_2$. Thus, in the nondegenerate case in which $\operatorname{Re}\mu_1 = \operatorname{Re}\mu'(0) \neq 0$, we regain the bifurcation formulas (I.9.12). The general case is given by

$$
\lambda(r) = \frac{\operatorname{Re}H_{00k_1}}{\operatorname{Re}\mu_1}r^{2k_1} + O\left(r^{2k_1+2}\right),
$$
(I.17.27)

and $\kappa(r)$ is given by (I.17.19); the lowest-order term of $\kappa(r)$ depends on (I.17.27) and the first nonvanishing coefficient in the expansion (I.17.19); cf. also (I.17.22).

Case 2

For the case $m > 1, k_1 = 1$, we have $c_{01} = \operatorname{Re}H_{001}$, and in view of $s = r^2 \geq 0$, the cases (I.16.31)–(I.16.33) reduce to

$$\lambda(r) = r^{2/m} \left(\frac{\text{Re}H_{001}}{\text{Re}\mu_m} \right)^{1/m} + o(r^{2/m})$$

if m is odd, and

(I.17.28)

$$\lambda(r) = \pm r^{2/m} \left(\frac{\text{Re}H_{001}}{\text{Re}\mu_m} \right)^{1/m} + o(r^{2/m})$$

if m is even and $\text{Re}H_{001}/\text{Re}\mu_m > 0$.

If $\text{Re}H_{001}/\text{Re}\mu_m < 0$, then for **even** m, there is **no** bifurcation of real periodic solutions of (I.17.1). Recall that in this case there is no exchange of stability of the trivial solution line $\{(0,\lambda)\}$, since the critical eigenvalue $\mu(\lambda)$ of $D_xF(0,\lambda)$ does not cross the imaginary axis.

However, if m is **odd**, the stability of $\{(0,\lambda)\}$ changes (provided that all other eigenvalues of $A_0 = D_xF(0,0)$ are in the stable left complex half-plane of $\mathbb{C}$; in this case, the stability of the bifurcating periodic solutions of (I.17.1) is determined by the perturbation of the critical Floquet exponents near 0; cf. (I.12.4), (I.12.5)). Since

$$\tilde{\Phi}_{re}(0,\lambda) = D_r(r\tilde{\Phi}_{re}(r,\lambda))|_{r=0} \text{ and}$$

(I.17.29) $r\tilde{\Phi}_{re}(r,\lambda) = \text{Re}\hat{\Phi}(r,\tilde{\kappa}(r,\lambda),\lambda)$ (cf. (I.17.14), (I.17.18)

where $\tilde{\kappa}(r,\lambda)$ is given by (I.17.19) (cf. (I.17.20)),

Theorem I.17.2 can be restated as follows:

If $\text{Re}\mu'(0) = \cdots = \text{Re}\mu^{(m-1)}(0) = 0$, then $D_\lambda D_r\text{Re}\hat{\Phi}(0,\tilde{\kappa}(0,0),0) = \cdots$
$= D_\lambda^{m-1}D_r\text{Re}\hat{\Phi}(0,\tilde{\kappa}(0,0),0) = 0$, and vice versa. In this case,

(I.17.30) $$\frac{d^m}{d\lambda^m}\text{Re}D_r\hat{\Phi}(0,\tilde{\kappa}(0,\lambda),\lambda)|_{\lambda=0} = -\frac{d^m}{d\lambda^m}\text{Re}\mu(\lambda)|_{\lambda=0}.$$

We show now how formula (I.17.30) generalizes to the bifurcating curves of periodic solutions given by Theorem I.17.3 and their critical Floquet exponents.

I.17.1 The Principle of Exchange of Stability for Degenerate Hopf Bifurcation

If we substitute $\tau = r^{2/p_{\gamma_2}}$ into (I.17.24), the curve $\{(x(r),\kappa(r),\lambda(r))|\ |r| < \delta\}$ is a solution curve of $G(x,\kappa,\lambda) = 0$ (cf. (I.17.5)), where we normalized the period again to 2π), which is analytic in τ. We write this curve as $\{(\hat{x}(\tau),\hat{\kappa}(\tau),\hat{\lambda}(\tau))\}$, and its stability is determined by the critical Floquet exponent μ_2 that is the nontrivial perturbed eigenvalue of

(I.17.31) $D_x G(\hat{x}(\tau), \hat{\kappa}(\tau), \hat{\lambda}(\tau))\psi = \mu\psi$ (cf. (I.12.3)).

We adopt the terminology of Section I.12. The existence of the second critical Floquet exponent $\mu_2 = \hat{\mu}_2(\tau)$ is given by Proposition I.12.1, and by its construction via the Implicit Function Theorem, all functions that are involved depend analytically on τ for $|\tau| < \tilde{\delta}$. For simplicity, we stay with the curve $\{(x(r), \kappa(r), \lambda(r))\}$ as long as possible, and we have in mind that analyticity (or even differentiability) is given only when it is parameterized by τ (i.e., $r = r(\tau) = \tau^{p\gamma_2/2}$). In the sequel we use the following notation: $D_x G(r) \equiv D_x G(x(r), \kappa(r), \lambda(r))$, $\psi_1(r) = \hat{v}_1 + w_1(r)$, and $\psi_2(r) = \hat{v}_2 + w_2(r)$, where $Pw_i(r) = 0, i = 1, 2$. Then, by (I.12.6), (I.12.9), and (I.12.10),

$$D_x G(r)\psi_2(r) = \nu(r)\psi_1(r) + \mu_2(r)\psi_2(r), \text{ or, by (I.12.8)},$$

$$D_x G(r)[\nu(r)\psi_1(r) + \mu_2(r)\psi_2(r)] = \mu_2(r)[\nu(r)\psi_1(r) + \mu_2(r)\psi_2(r)].$$
(I.17.32)
By the equivariance (I.8.31) or (I.12.16), equation (I.17.32)$_2$ is equivalent to

(I.17.33) $$\begin{aligned} D_x G(e^{i\theta}r)[\nu(r)S_\theta\psi_1(r) + \mu_2(r)S_\theta\psi_2(r)] \\ = \mu_2(r)[\nu(r)S_\theta\psi_1(r) + \mu_2(r)S_\theta\psi_2(r)]. \end{aligned}$$

(Cf. (I.8.32), (I.8.33), and recall that $\kappa(e^{i\theta}r) = \kappa(r)$, $\lambda(e^{i\theta}r) = \lambda(r)$, which is also seen from (I.17.14).)

By the definitions of $\psi_1(r), \psi_2(r)$ and by (I.12.9), we obtain

(I.17.34) $$\begin{aligned} &\nu(r)S_\theta\psi_1(r) + \mu_2(r)S_\theta\psi_2(r) \\ &= (\nu(r)\cos\theta + \mu_2(r)\sin\theta)\hat{v}_1 + (\mu_2(r)\cos\theta - \nu(r)\sin\theta)\hat{v}_2 \\ &\quad + w(\theta, r), \\ &\text{so that } Pw(\theta, r) = 0 \text{ for } \theta \in [0, 2\pi], |r| < \delta. \end{aligned}$$

In the sequel we distinguish the cases $(\nu(r), \mu_2(r)) \equiv (0, 0)$ and $(\nu(r), \mu_2(r)) \neq (0, 0)$ for $0 < |r| < \delta$. In the first case, we choose $\theta = \theta(r) \equiv 0$, and we have equation (I.17.32)$_1$ with zero right-hand side. In the second case, we choose the phase $\theta = \theta(r)$ by

(I.17.35) $\nu(r)\cos\theta + \mu_2(r)\sin\theta = 0$ for $0 < |r| < \delta$,

and substituting $r = r(\tau) = \tau^{p\gamma_2/2}$, we see that the phase

(I.17.36) $\theta = \theta(r) = \hat{\theta}(\tau) \in \left[-\dfrac{\pi}{2}, \dfrac{\pi}{2}\right]$ is analytic in τ for $|\tau| < \tilde{\delta}$.

Since the eigenfunction (I.17.34) with the phase (I.17.35) has no $\hat{v}_1$-component, equation (I.17.33) implies

(I.17.37) $$\begin{aligned} &(I - Q_1)D_x G(e^{i\theta}r)\psi(r) = \mu_2(r)\psi(r), \text{ where} \\ &\psi(r) = (\mu_2(r)\cos\theta - \nu(r)\sin\theta)\hat{v}_2 + w(\theta, r) \\ &\text{is in } (I - P_1)(E \cap Y) \text{ for } |r| < \delta. \end{aligned}$$

The real coefficient $\mu_2(r)\cos\theta - \nu(r)\sin\theta \equiv \beta(r)$ satisfies the equation $|\beta(r)| = \sqrt{\nu(r)^2 + \mu_2(r)^2}$ if $\theta = \theta(r)$ is chosen according to (I.17.35). Therefore, $\beta(r) \neq 0$ for $0 < |r| < \delta$ in the second case, and $\hat{\beta}(\tau) = \beta(r(\tau))$ (with $r = r(\tau) = \tau^{p\gamma_2/2}$) is analytic and does not vanish for $0 < |\tau| < \tilde{\delta}$.

By definition (I.17.34), the remainder $w(\hat{\theta}(\tau), r(\tau))$ is given by

(I.17.38)
$$w(\hat{\theta}(\tau), r(\tau)) = \hat{\nu}(\tau)S_{\hat{\theta}(\tau)}w_1(r(\tau)) + \hat{\mu}_2(\tau)S_{\hat{\theta}(\tau)}w_2(r(\tau)),$$
$$\text{where } \hat{\nu}(\tau) = \nu(r(\tau)), \ \hat{\mu}_2(\tau) = \mu_2(r(\tau)).$$

Since

(I.17.39)
$$|\nu(r)|/|\beta(r)| \leq 1 \text{ and } |\mu_2(r)|/|\beta(r)| \leq 1$$
$$\text{for } 0 < |r| < \delta,$$

we obtain

(I.17.40)
$$w(\hat{\theta}(\tau), r(\tau))/\hat{\beta}(\tau) \to 0 \text{ as } \tau \to 0$$

by $w_1(0) = w_2(0) = 0$.

To summarize, when dividing equation (I.17.37) by the coefficient $\hat{\beta}(\tau) \neq 0$, we obtain an analytic simple eigenvalue perturbation

(I.17.41)
$$(I - Q_1)D_xG(e^{i\hat{\theta}(\tau)}r(\tau))\hat{\psi}(\tau) = \hat{\mu}_2(\tau)\hat{\psi}(\tau),$$
$$\hat{\psi}(\tau) = \hat{v}_2 + \hat{w}(\tau), \ \hat{w}(0) = 0, \ \hat{\mu}_2(0) = 0,$$
$$\text{with some } \hat{w}(\tau) \in (I - P_1)(E \cap Y).$$

Recall that in the case $\hat{\beta}(\tau) \equiv 0$ (i.e., when $(\nu(r), \mu_2(r)) \equiv (0,0)$), we obtain (I.17.41) with $\hat{\theta}(\tau) \equiv 0$ by (I.17.32)$_1$ as well.

Thus for the analytic family $A(\tau) \equiv (I - Q_1)D_xG(e^{i\hat{\theta}(\tau)}r(\tau))$, the results of Section I.16 are applicable, in particular Theorem I.16.3. The calculation of the bifurcation function related to (I.17.41) is already accomplished in Section I.12, (I.12.21)–(I.12.30). We repeat the main steps (where we stay for simplicity with the dependence on r rather than on τ). We define

(I.17.42)
$$\mathcal{G}(x, \theta, r) \equiv (I - Q_1)G(x(e^{i\theta}r) + x, \kappa(r), \lambda(r)) \text{ and}$$
$$\mathcal{G}(0, \theta, r) = 0, \text{ which is the trivial solution line.}$$

Then

(I.17.43)
$$D_x\mathcal{G}(0, \theta, r) = (I - Q_1)D_xG(e^{i\theta}r) \text{ and}$$
$$D_x\mathcal{G}(0, \theta(0), 0) = (I - Q_1)J_0 \quad (\theta = \theta(r))$$

and

(I.17.44)
$$N(D_x\mathcal{G}(0, \theta(0), 0)) = R(P_2) = \text{span}[\hat{v}_2],$$
$$R(D_x\mathcal{G}(0, \theta(0), 0)) = R(J_0), \text{ and}$$
$$(I - Q_1)W = R(D_x\mathcal{G}(0, \theta(0), 0)) \oplus N(D_x\mathcal{G}(0, \theta(0), 0))$$
$$\text{with projection } Q_2 \text{ onto } N \text{ along } R.$$

The Lyapunov–Schmidt reduction yields the bifurcation function

$$\Phi(v,\theta,r) = Q_2 G(v + \tilde{\psi}(v,\theta,r),\theta,r),$$

and for $v = s\hat{v}_2$ the computations (I.12.27)–(I.12.29) give

(I.17.45)
$$\Phi(s\hat{v}_2,\theta,r)$$
$$= Q_2 G((e^{i\theta}r + s)\psi_0 + (e^{-i\theta}r + s)\overline{\psi}_0$$
$$+ \psi((e^{i\theta}r + s)\psi_0 + (e^{-i\theta}r + s)\overline{\psi}_0, \kappa(r), \lambda(r)), \kappa(r), \lambda(r)).$$

Using the definition (I.12.9) of the projection Q_2, and using the definitions (I.8.34), (I.8.35) of the bifurcation function $\hat{\Phi}$, we see that the definition (I.16.18) of the bifurcation function for (I.17.42) and (I.17.43) finally gives

(I.17.46)
$$\Psi(s,\tau) \equiv \frac{1}{2\pi}\int_0^{2\pi} \langle \Phi(s\hat{v}_2, \hat{\theta}(\tau), r(\tau)), \hat{v}_2' \rangle dt$$
$$= \mathrm{Re}\hat{\Phi}(e^{i\hat{\theta}(\tau)}r(\tau) + s, \kappa(r(\tau)), \lambda(r(\tau))).$$

Theorem I.16.3 then relates the first two nonvanishing coefficients in the expansion of $\hat{\mu}_2(\tau)$ to the first two nonvanishing coefficients of $D_s\Psi(0,\tau)$. In order to compute $D_s\Psi(0,\tau)$, we make use of the equivariance (I.8.36) of $\hat{\Phi}$, which implies

(I.17.47)
$$\hat{\Phi}(e^{i\theta}r + s, \kappa, \lambda) = e^{i\theta}\hat{\Phi}(r + e^{-i\theta}s, \kappa, \lambda).$$

Therefore, differentiation of (I.17.47) with respect to s at $s = 0$ and taking the real part gives

(I.17.48)
$$D_s\Psi(0,\tau) = \mathrm{Re}D_r\hat{\Phi}(r(\tau), \kappa(r(\tau)), \lambda(r(\tau))).$$

By definition (I.17.19) of $\kappa = \tilde{\kappa}(r,\lambda)$ (cf. (I.17.29)) and by $\hat{\kappa}(\tau) = \kappa(r(\tau))$, $\hat{\lambda}(\tau) = \lambda(r(\tau))$, we can rewrite (I.17.48) as

(I.17.49)
$$D_s\Psi(0,\tau) = \mathrm{Re}D_r\hat{\Phi}(r(\tau), \hat{\kappa}(\tau), \hat{\lambda}(\tau))$$
$$= \mathrm{Re}D_r\hat{\Phi}(r(\tau), \tilde{\kappa}(r(\tau), \hat{\lambda}(\tau)), \hat{\lambda}(\tau))$$
$$= D_r(r(\tau)\tilde{\Phi}_{re}(r(\tau), \hat{\lambda}(\tau))) \text{ for } |\tau| < \tilde{\delta}.$$

Now application of Theorem I.16.3 gives the following theorem, which extends formula (I.17.30) for the trivial solution line to any bifurcating curve of periodic solutions.

Theorem I.17.4 *Let* $\{(\hat{x}(\tau), \hat{\kappa}(\tau), \hat{\lambda}(\tau))\}$ *be any analytic solution curve of* $G(x, \kappa, \lambda) = 0$ *through* $(0, \kappa_0, 0)$ *(cf. (I.17.5)) and let* $\hat{\mu}_2(\tau) = \sum_{j=1}^\infty \hat{\mu}_j \tau^j$ *be the nontrivial critical Floquet exponent of* $D_x G(\hat{x}(\tau), \hat{\kappa}(\tau), \hat{\lambda}(\tau))$. *Let* $\hat{\Phi}(r, \kappa, \lambda)$ *be the complex bifurcation function obtained by the method of Lyapunov–Schmidt (cf. (I.17.14)) and let*

$$r(\tau) = \frac{1}{2\pi}\int_0^{2\pi} \langle \hat{x}(\tau), \hat{v}_2' \rangle dt; \quad cf. \ (I.12.9).$$

Then the following holds:

$$\hat{\mu}_2'(0) = \cdots = \hat{\mu}_2^{(\hat{m}-1)}(0) = 0 \quad \left(' = \frac{d}{d\tau}, \hat{m} \geq 2\right) \Leftrightarrow$$

$$\frac{d}{d\tau}\mathrm{Re}D_r\hat{\Phi}(r(\tau), \hat{\kappa}(\tau), \hat{\lambda}(\tau))|_{\tau=0} = \cdots$$

(I.17.50)
$$= \frac{d^{\hat{m}-1}}{d\tau^{\hat{m}-1}}\mathrm{Re}D_r\hat{\Phi}(r(\tau), \hat{\kappa}(\tau), \hat{\lambda}(\tau))|_{\tau=0} = 0,$$

and in this case,

$$\frac{d^{\hat{m}}}{d\tau^{\hat{m}}}\mathrm{Re}D_r\hat{\Phi}(r(\tau), \hat{\kappa}(\tau), \hat{\lambda}(\tau))|_{\tau=0} = \frac{d^{\hat{m}}}{dt^{\hat{m}}}\hat{\mu}_2(\tau))|_{\tau=0}.$$

(The last equation also holds for $\hat{m} + 1$.)

This generalizes formula (I.12.31) (here $\tau = r$). (For a different proof see also Remark I.18.5.)

If $\frac{d^{\hat{m}}}{dt^{\hat{m}}}\hat{\mu}_2(\tau)|_{\tau=0}/\hat{m}! = \hat{\mu}_{\hat{m}} \neq 0$, the sign of the nontrivial critical Floquet exponent is given by

(I.17.51)
$$\mathrm{sign}\hat{\mu}_2(\tau) = \mathrm{sign}\mathrm{Re}D_r\hat{\Phi}(r(\tau), \hat{\kappa}(\tau), \hat{\lambda}(\tau)) \neq 0$$
$$\text{for } \tau \in (-\tilde{\delta}, \tilde{\delta})\backslash\{0\}.$$

As in Section I.16, we can give a criterion for when $\hat{\mu}_{\hat{m}} \neq 0$ for some $\hat{m} \geq 1$. The advantage of that criterion is that it is part of the construction of the analytic curve $\{(\hat{x}(\tau), \hat{\kappa}(\tau), \hat{\lambda}(\tau))\}$. The proof is the same as in Section I.16, (I.16.42)–(I.16.48), when we replace $s\tilde{\Phi}(s, \lambda)$ by $r\tilde{\Phi}_{re}(r, \lambda)$. The peculiarity here is that $\tilde{\Phi}_{re}(r, \lambda)$ depends only on r^2 (cf. (I.17.20)). Observe also that $r\tilde{\Phi}_{re}(r, \lambda) = \mathrm{Re}\hat{\Phi}(r, \tilde{\kappa}(r, \lambda), \lambda)$ (cf. (I.17.29)). We give the result:

Theorem I.17.5 *Let a solution curve $\{(x(r), \kappa(r), \lambda(r))\}$ of $G(x, \kappa, \lambda) = 0$ (cf. (I.17.5)) be obtained by a simple zero $\tilde{\lambda}_0 \neq 0$ of a defining polynomial P_γ for some slope $-1/\gamma$ of the Newton polygon for $\tilde{\Phi}_{re}(r, \lambda)$ (cf. Theorem I.17.3). If $\gamma = \gamma_1/\gamma_2$ and $\tau = r^{2/\gamma_2}$, then $\{(x(r), \kappa(r), \lambda(r))\} = \{(\hat{x}(\tau), \hat{\kappa}(\tau), \hat{\lambda}(\tau))\}$ is an analytic solution curve through $(0, \kappa_0, 0)$ (in τ), and for the nontrivial critical Floquet exponent $\hat{\mu}_2(\tau) = \sum_{j=1}^{\infty} \hat{\mu}_j\tau^j$ of $D_xG(\hat{x}(\tau), \hat{\kappa}(\tau), \hat{\lambda}(\tau))$, we obtain the following:*

(I.17.52)
$$\hat{\mu}_1 = \cdots = \hat{\mu}_{\hat{m}-1} = 0, \text{ for } \hat{m} = \gamma_1 j_\ell + \gamma_2 k_\ell \text{ and where}$$
$$\gamma j_\ell + k_\ell = \cdots = \gamma j_{\ell+r} + k_{\ell+r} = \sigma \text{ (cf. (I.15.5)), and}$$
$$\hat{\mu}_{\hat{m}} = -2\gamma\tilde{\lambda}_0 P_\gamma'(\tilde{\lambda}_0) \neq 0.$$

In particular, $\hat{\mu}_2(\tau) \neq 0$ for $\tau \neq 0$, so that (I.17.51) holds.

We apply Theorem I.17.5 to the special cases (I.17.26), (I.17.27), studied before.

Case 1

Let $m = 1$ and $k_1 = 1$. Then (according to (I.17.20)) $\gamma = 1, j_0 = m = 1, k_0 = 0$, so that $\hat{m} = 1$. Furthermore, $\tau = r^2$, and from (I.17.52) we obtain, by $P_1(\tilde{\lambda}) = c_{10}\tilde{\lambda} + \text{Re}H_{100} = -\text{Re}\mu_1\tilde{\lambda} + \text{Re}H_{001}$,

$$(I.17.53) \qquad \mu_2(r) = 2\text{Re}H_{001}r^2 + O(r^4).$$

In view of $(I.17.26)_1$ this gives, with (I.17.2),

$$(I.17.54) \qquad \frac{d^2\mu_2}{dr^2}(0) = 2\text{Re}\frac{d\mu}{d\lambda}(0)\frac{d^2\lambda}{dr^2}(0),$$

which recovers formula (I.12.34).

For $k_1 \geq 1$ we have $\gamma = k_1$ and $\hat{m} = k_1$. Again $\tau = r^2$, and formulas (I.17.52) and (I.17.27) give in this case

$$\mu_2(r) = 2k_1\text{Re}H_{00k_1}r^{2k_1} + O(r^{2k_1+2}),$$

$$(I.17.55)$$
$$\frac{d^{2k_1}\mu_2}{dr^{2k_1}}(0) = 2k_1\text{Re}\frac{d\mu}{d\lambda}(0)\frac{d^{2k_1}\lambda}{dr^{2k_1}}(0).$$

Formula $(I.17.55)_2$ links the nontrivial Floquet exponent $\mu_2(r)$ along the bifurcating pitchfork $(x(r), \kappa(r), \lambda(r))$ to $\lambda(r)$ via the nondegeneracy $\text{Re}\mu'(0) \neq 0$ (cf. (I.8.7)).

Finally, we have the analogue to Theorem I.16.8: There is a general **Principle of Exchange of Stability** for degenerate Hopf bifurcation formulated in the next theorem.

Theorem I.17.6 *Consider all local periodic solution curves of (I.17.1) through* $(x, \lambda) = (0, 0)$ *except a vertical curve* $\{(x(r), 0)\}$ *but including the trivial line* $\{(0, \lambda)\}$. *The nontrivial curves are given by Theorem I.17.3, and for fixed* $\lambda \neq 0$, *these are ordered in the* (r, λ)-plane for positive r. *Then consecutive curves have opposite stability properties in a possibly weakened sense (if the Floquet exponent* $\mu_2(r)$ *vanishes along that curve).*

Thus the total diagram shows also the stability properties of the curves if the stability of the trivial solution line is known. On the other hand, Theorem I.17.5 gives the stability of a curve by its construction.

We give an **example**:

$$(I.17.56) \qquad \ddot{x} - \lambda^7\dot{x} + x = -\lambda^3\dot{x}^3 + \lambda\dot{x}^5 - \dot{x}^9,$$

which is a "nonlinear oscillation" for the scalar function $x = x(t)$. Transformed to a first-order system in $\mathbb{R}^2$, it takes the form (I.17.1). Here $\kappa_0 = 1$ and

$$(I.17.57) \qquad A(\lambda) = \begin{pmatrix} 0 & 1 \\ -1 & \lambda^7 \end{pmatrix}, \text{ so that } \text{Re}\mu(\lambda) = \frac{1}{2}\lambda^7.$$

Thus, in view of Theorem I.17.2, the Reduced Bifurcation Function satisfies $\tilde{\Phi}_{re}(0,\lambda) = -\frac{1}{2}\lambda^7 +$ h.o.t.. As in (I.16.55), the lowest-order coefficients c_{jk} of $\tilde{\Phi}_{re}(r,\lambda)$ for $k \geq 1$ are of the form

$$\text{Re}H_{j0k} = -\binom{2k+1}{k}\text{Re}\Big\langle F^{(2k+1)}_{j,2k+1}[\underbrace{\varphi_0,\ldots,\varphi_0}_{k+1},\underbrace{\overline{\varphi}_0,\ldots,\overline{\varphi}_0}_{k}],\varphi_0'\Big\rangle$$

(I.17.58)

if $F_{\ell m} = 0$ for $0 \leq \ell \leq j$, $1 \leq m \leq 2k+1$, $(\ell,m) \neq (j,2k+1)$.

Here $\varphi_0 = \frac{1}{\sqrt{2}}\binom{i}{1}, \varphi_0' = \frac{1}{\sqrt{2}}\binom{-i}{1}$, and $\langle \ , \ \rangle$ denotes the real bilinear scalar product on $\mathbb{R}^2$ extended to $\mathbb{C}^2$.

For our example, the Newton polygon for $\tilde{\Phi}_{re}(r,\lambda)$ consists of the points $(0,7),(1,3),(2,1),(4,0)$ with coefficients $-\frac{1}{2}, \frac{3}{4}, -\frac{5}{4}, \frac{63}{16}$. The values of the three slopes are $\gamma = \frac{1}{4}, \gamma = \frac{1}{2}$, and $\gamma = 2$. The defining polynomials have simple nonzero roots, so that Theorem I.17.5 is applicable. The bifurcation diagram is sketched in Figure I.17.1.

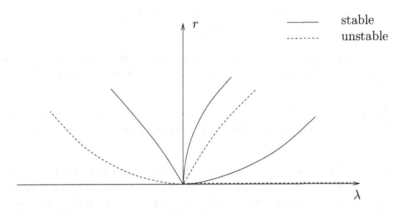

Figure I.17.1

Remark I.17.7 *For the planar system (I.17.56) the alternating stability of consecutive periodic solutions (which are closed curves containing the origin and each of which is contained in the next one) is not surprising: a simple geometric intuition "proves" the Principle of Exchange of Stability, since there is "apparently" only one possibility for how the trajectories spiral in the rings between the closed orbits. For higher-dimensional systems (I.17.1), however, there is obviously not such a geometric intuition for such a principle, and our assumption of nonresonance (cf. (I.8.14)) does not exclude that the (parameterized) center manifold for (I.17.1) is of high dimension (provided that it exists).*

Remark I.17.8 *A degenerate bifurcation as discussed in the last two sections is not "generic." A small perturbation of the linearization $A(\lambda)$ reduces* the problem to a generic bifurcation where the critical eigenvalue crosses the imaginary axis transversally ("with nonvanishing speed"). For the concrete example (I.17.56), a perturbed bifurcation diagram of Figure I.17.1 is sketched in Figure I.17.2.

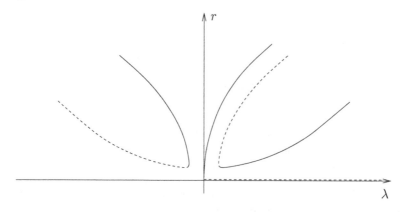

Figure I.17.2

We observe a generic Hopf bifurcation (cf. Remark I.14.5) with "imperfections" that contain stable branches of periodic solutions, too. It seems to be natural to consider a solution set as sketched in Figure I.17.2 as a perturbation of a degenerate bifurcation; cf. also Remark III.2.2.

I.18 The Principle of Reduced Stability for Stationary and Periodic Solutions

As in Section I.7 we consider formally

$$(I.18.1) \qquad \frac{dx}{dt} = F(x, \lambda),$$

where $F : U \times V \to Z$ with open sets $U \subset X$, $V \subset \mathbb{R}$, and we assume that the Banach space X is continuously embedded in the Banach space Z. Let $F(x_0, \lambda_0) = 0$; i.e., $x_0 \in X$ is an equilibrium of (I.18.1) for the parameter $\lambda_0 \in \mathbb{R}$. We normalize w.l.o.g. (x_0, λ_0) to $(0, 0)$. According to the Principle of Linearized Stability (cf. (I.7.2)), the stability of this equilibrium is determined by the spectrum of $D_x F(0, 0)$.

In this section we assume a degeneration or bifurcation at $(0, 0)$ in the sense of Sections I.2, I.4, I.5, and I.8: There is an eigenvalue of $D_x F(0, 0)$

on the imaginary axis that might cause a loss of stability. (If there is no
eigenvalue of $D_xF(0,0)$ on the imaginary axis, the Implicit Function Theorem
provides a unique continuation of the solution $F(0,0) = 0$ preserving its
stability.) Here we do not prove the existence of new stationary or periodic
solutions of (I.18.1), but we are only interested in their stability, provided
they exist near $(0,0)$. In the first part of this section an eigenvalue 0 might
generate stationary solutions, whereas in the second part a pair of complex
conjugate purely imaginary eigenvalues might create periodic solutions of
(I.18.1) near $(0,0)$.

We generalize the assumption of a one-dimensional kernel of $D_xF(0,0)$ to
the following:

(I.18.2)
$$0 \text{ is a semisimple eigenvalue of } D_xF(0,0)$$
$$\text{of multiplicity } n \geq 1; \text{ i.e., } \dim N(D_xF(0,0)) = n$$
$$\text{and } N(D_xF(0,0)) \cap R(D_xF(0,0)) = \{0\}.$$

Assuming that $F(\cdot,0)$ is a Fredholm operator of index zero (cf. Definition
I.2.1), this implies the decompositions

(I.18.3)
$$X = N(D_xF(0,0)) \oplus (R(D_xF(0,0)) \cap X),$$
$$Z = R(D_xF(0,0)) \oplus (N(D_xF(0,0)) \quad \text{(cf. (I.2.2))},$$

which, in turn, define projections

(I.18.4)
$$P : X \to N, \quad P = Q|_X, \quad \text{where}$$
$$Q : Z \to N \quad \text{along} \quad R.$$

(Here $N = N(D_xF(0,0))$, $R = R(D_xF(0,0))$; cf. (I.2.3).)

The reduction method of Lyapunov–Schmidt described in Theorem I.2.3
yields that

(I.18.5)
all solutions of $F(x,\lambda) = 0$ near $(0,0) \in X \times \mathbb{R}$ are obtained
by solving $\Phi(v,\lambda) = 0$ near $(0,0) \in N \times \mathbb{R}$.
The bifurcation function Φ is derived from F
as in (I.2.9) and $v = Px$.

The **Principle of Reduced Stability for Stationary Solutions** is
now formulated as follows: Are the stability of x as an equilibrium of (I.18.1)
and the stability of $v = Px$ as an equilibrium of the reduced system

(I.18.6)
$$\frac{dv}{dt} = \Phi(v,\lambda)$$

for (x,λ) near $(0,0)$ the same?

We give an answer to this question by the Principle of Linearized Stability, cf. (I.7.2), provided that the solutions (x,λ) form a smooth curve
$\{(x(s),\lambda(s))|s \in (-\delta,\delta)\}$ through $(0,0)$. For the regularity of F near $(0,0)$ we

assume analyticity as described in (I.16.2). We also assume that the solution curve

(I.18.7) $F(x(s), \lambda(s)) = 0$ through $(x(0), \lambda(0)) = (0, 0)$

depends analytically on $s \in (-\delta, \delta)$.

Thus the problem amounts to the study of the eigenvalues of the analytic families of operators

(I.18.8) $\begin{aligned} T(s) &\equiv D_x F(x(s), \lambda(s)) \quad \text{and} \\ R(s) &\equiv D_v \Phi(v(s), \lambda(s)). \end{aligned}$

To be more precise, as in Section I.7, we confine ourselves to the eigenvalue perturbation of the critical eigenvalue $\mu = 0$ of $T(0) = D_x F(0, 0)$ and of $D_v \Phi(0, 0) = 0$ (cf. $(I.2.10)_2$), assuming that the rest of the spectrum of $D_x F(0, 0)$ is in the left complex half-plane. In contrast to the investigations of Sections I.7 and I.16, the multiplicity n of the eigenvalue $\mu = 0$ (cf. (I.18.2)) makes the eigenvalue perturbations more involved. However, the assumed analyticity of $T(s) \in L(X, Z)$ and of $R(s) \in L(N, N)$ allows us to make use of Newton polygons in studying the characteristic equations for the perturbed eigenvalues. In order to simplify the comparison of the characteristic equations we introduce the following notation:

(I.18.9) $\begin{aligned} T_{11}(s) &\equiv QT(s)|_N \in L(N, N), \\ T_{12}(s) &\equiv QT(s)|_{R \cap X} \in L(R \cap X, N), \\ T_{21}(s) &\equiv (I - Q)T(s)|_N \in L(N, R), \\ T_{22}(s) &\equiv (I - Q)T(s)|_{R \cap X} \in L(R \cap X, R). \end{aligned}$

Then, in view of (I.18.3), $T(s) \in L(X, Z)$ is given by the matrix operator

(I.18.10) $T(s) = \begin{pmatrix} T_{11}(s) & T_{12}(s) \\ T_{21}(s) & T_{22}(s) \end{pmatrix}$,

where the entries have the properties
$T_{11}(0) = 0$, $T_{12}(0) = 0$, $T_{21}(0) = 0$, and $T_{22}(0)$ is an isomorphism.

That last property holds also for $T_{22}(s)$ when s is near zero.

The eigenvalue problem for $T(s)$ leads to the study of

(I.18.11) $T(s) - \mu I = \begin{pmatrix} T_{11}(s) - \mu I_1 & T_{12}(s) \\ T_{21}(s) & T_{22}(s) - \mu I_2 \end{pmatrix}$, $\mu \in \mathbb{C}$,

where I denotes the embedding $X \subset Z$,
I_1 is the identity in N, and $I_2 = I|_{R \cap X}$.

Multiplying (I.18.11) on the left by the isomorphism (for small $|s|$ and $|\mu|$)

$$(\text{I.18.12}) \qquad \begin{pmatrix} I_1 & -T_{12}(s)(T_{22}(s) - \mu I_2)^{-1} \\ 0 & \tilde{I}_2 \end{pmatrix} \in L(Z, Z),$$

where $\tilde{I}_2$ is the identity in R, we obtain the operator

$$(\text{I.18.13}) \quad \begin{pmatrix} T_{11}(s) - \mu I_1 - T_{12}(s)(T_{22}(s) - \mu I_2)^{-1}T_{21}(s) & 0 \\ T_{21}(s) & T_{22}(s) - \mu I_2 \end{pmatrix},$$

which is in $L(X, Z)$. Thus the critical eigenvalues μ of $T(s)$ near 0 satisfy the equation

$$(\text{I.18.14}) \quad \det(T_{11}(s) - \mu I_1 - T_{12}(s)(T_{22}(s) - \mu I_2)^{-1}T_{21}(s)) = 0.$$

Observe that the operator in (I.18.14) is in $L(N, N)$, and dim $N = n$. Finally, since

$$(\text{I.18.15}) \qquad (T_{22}(s) - \mu I_2)^{-1} = \sum_{\nu=0}^{\infty} \mu^{\nu} T_{22}(s)^{-\nu-1},$$

equation (I.18.14) is transformed into

$$h(s, \mu) \equiv \det(T_{11}(s) - \mu I_1 - T_{12}(s) \sum_{\nu=0}^{\infty} \mu^{\nu} T_{22}(s)^{-\nu-1} T_{21}(s))$$

$$(\text{I.18.16}) \qquad = \det(T_{11}(s) - \mu I_1 - \sum_{\nu=0}^{\infty} \mu^{\nu} B_{\nu}(s)) = 0,$$

$$\text{where } B_{\nu}(s) = T_{12}(s) T_{22}(s)^{-\nu-1} T_{21}(s) = O(s^2).$$

The function h is analytic for (s, μ) near $(0, 0)$ in the sense of (I.15.2), with the difference, however, that s is real and μ is complex. The Newton polygon method expounded in Section I.15 for the real case gives also all solution curves of $h(s, \mu) = 0$ emanating at $(0, 0)$ if μ is complex: Simply take all nonzero complex roots $\tilde{\mu}_0$ of the defining polynomials (I.15.8) and proceed in the same way as in the real case.

We call $h(s, \mu) = 0$ the characteristic equation for the eigenvalue perturbation of the critical eigenvalue $\mu = 0$ of $T(0) = D_x F(0, 0)$.

Next, we derive the characteristic equation for the eigenvalues of $R(s)$. From (I.2.9) we obtain

$$(\text{I.18.17}) \quad D_v \Phi(v, \lambda) = Q D_x F(v + \psi(v, \lambda), \lambda)(I_1 + D_v \psi(v, \lambda)),$$

and since $(I - Q)F(v + \psi(v, \lambda), \lambda) \equiv 0$ (cf. (I.2.8)), differentiation with respect to v yields

$$(\text{I.18.18}) \qquad (I - Q)D_x F(v + \psi(v, \lambda), \lambda)(I_1 + D_v \psi(v, \lambda)) \equiv 0.$$

For the assumed solution curve $\{(x(s), \lambda(s))\}$ we have $v(s) = Px(s)$ and $v(s) + \psi(v(s), \lambda(s)) = x(s)$, so that (I.18.18) implies, in view of definitions (I.18.9),

(I.18.19)
$$D_v\psi(v(s), \lambda(s)) = -T_{22}(s)^{-1}T_{21}(s) \text{ , whence from (I.18.17),}$$
$$R(s) = T_{11}(s) - T_{12}(s)T_{22}(s)^{-1}T_{21}(s).$$

The characteristic equation for the eigenvalues of $R(s)$ is therefore given by

(I.18.20)
$$g(s, \mu) \equiv \det(T_{11}(s) - \mu I_1 - B_0(s)) = 0.$$

We have to compare the solutions of the equations $h(s, \mu) = 0$ and $g(s, \mu) = 0$ emanating at $(0, 0)$. We give our main result:

Theorem I.18.1 *Let $R(s) = s^{k_0} R_{k_0} + O(s^{k_0+1})$ for some $k_0 \geq 1$. If 0 is an eigenvalue of $R_{k_0} \in L(N, N)$ of at most algebraic multiplicity one, then the series expansions of all critical eigenvalues of $T(s) = D_x F(x(s), \lambda(s))$ and of all eigenvalues of $R(s) = D_v\Phi(v(s), \lambda(s))$ have the same first nonvanishing terms. For all but possibly one, these first terms are given by*

(I.18.21)
$$\mu_\rho s^{k_0}, \quad \rho = 1, \dots, r, \quad r \leq n,$$

where $\mu_\rho \in \mathbb{C}$ are the nonvanishing eigenvalues of R_{k_0}. If $\det R(s) \equiv 0$, then $\mu \equiv 0$ is an eigenvalue for $R(s)$ and $T(s)$. If $\det R(s) = r_\ell s^\ell + O(s^{\ell+1})$, $\ell \geq nk_0$, the last eigenvalue has the first term

(I.18.22)
$$\mu_0 s^{\ell-(n-1)k_0}, \text{ where } \mu_0 = r_\ell \mu_1^{-m_1} \cdots \mu_r^{-m_r}.$$

Here m_ρ denotes the algebraic multiplicity of μ_ρ, $\rho = 1, \dots, r \leq n - 1$.

Thus, if $\mathrm{Re}\,\mu_\rho \neq 0$ for $\rho = 0, \dots, r$, where $\mu_1, \dots, \mu_r$ are the nonvanishing eigenvalues of R_{k_0} and where μ_0 is given by (I.18.22), then *the Principle of Reduced Stability for stationary solutions is true.*

Proof. We consider the case in which $\mu = 0$ is an eigenvalue of R_{k_0}. The modification in the other case is obvious. The Newton polygon for

(I.18.23) $\det(s^{k_0} R_{k_0} - \mu I_1) = -\mu(\mu_1 s^{k_0} - \mu)^{m_1} \cdots (\mu_r s^{k_0} - \mu)^{m_r}$

contains the line connecting $(0, n)$ and $((n-1)k_0, 1)$, since the coefficients at these endpoints do not vanish: They are

(I.18.24)
at $(0, n)$, $(-1)^n$;
at $((n-1)k_0, 1)$, $-\mu_1^{m_1} \cdots \mu_r^{m_r}$ with $m_1 + \cdots + m_r = n - 1$.

For the polygon of $g(s, \mu) = \det(R(s) - \mu I_1)$ we have to distinguish the two cases $\det R(s) \equiv 0$ and $\det R(s) = r_\ell s^\ell + O(s^{\ell+1})$. In the first case, $\mu = 0$ is an eigenvalue for $R(s)$, too, and the Newton polygons for $g(s, \mu)$ and $\det(s^{k_0} R_{k_0} - \mu I_1)$ are identical. In the second case, there is an additional line connecting $((n-1)k_0, 1)$ and $(\ell, 0)$ on the s-axis. Observe that $\ell > nk_0$, so that the polygon connecting $(0, n), ((n-1)k_0, 1)$, and $(\ell, 0)$ is convex. The only slopes are $-1/k_0$ and possibly $-1/(\ell - (n-1)k_0)$, and for the slope $-1/k_0$ the defining polynomial is

$$(I.18.25) \qquad P_{k_0}(\tilde{\mu}) = -\tilde{\mu}(\mu_1 - \tilde{\mu})^{m_1} \cdots (\mu_r - \tilde{\mu})^{m_r} \quad \text{(cf. (I.15.8))}.$$

According to Remark I.15.1, only the nonzero roots of $P_{k_0}(\tilde{\mu})$ are of interest, and they give the nonzero eigenvalues of $R(s)$ in lowest order by $\mu_\rho s^{k_0}$, $\rho = 1, \ldots, r$; cf. (I.15.17). Thus in the first case, the eigenvalues of $R(s)$ are

$$(I.18.26) \qquad \mu \equiv 0 \quad \text{and} \quad \mu_\rho s^{k_0} + \text{h.o.t.}, \quad \rho = 1, \ldots, r.$$

(Since the roots μ_ρ of $P_{k_0}(\tilde{\mu})$ have multiplicity m_ρ, they give m_ρ complex solution curves $\{(s, \mu(s))\}$ of $g(s, \mu) = 0$. This follows from the procedure described in Section I.15. Recall that all complex roots of defining polynomials are of interest now. The m_ρ (complex) eigenvalues of $R(s)$ with the same lowest-order terms $\mu_\rho s^{k_0}$ are not necessarily identical.)

In the second case, the additional slope $-1/(\ell - (n-1)k_0)$ gives the defining polynomial

$$(I.18.27) \qquad P_{\ell-(n-1)k_0}(\tilde{\mu}) = -\mu_1^{m_1} \cdots \mu_r^{m_r} \tilde{\mu} + r_\ell,$$

since the coefficient at $(\ell, 0)$ is r_ℓ. This proves that in this case, the eigenvalues of $R(s)$ are

$$(I.18.28) \qquad \mu_0 s^{\ell-(n-1)k_0} + \text{h.o.t.}, \quad \mu_\rho s^{k_0} + \text{h.o.t.}, \quad \rho = 1, \ldots, r,$$

where μ_0 is given by (I.18.22).

In order to determine the solutions of $h(s, \mu) = 0$, we write it as $h(s, \mu) = \det(R(s) - \mu I_1 - \sum_{\nu=1}^{\infty} \mu^\nu B_\nu(s))$, and we recall that $B_\nu(s) = O(s^2)$; cf. (I.18.16). The additional term $\sum_{\nu=1}^{\infty} \mu^\nu B_\nu(s)$ cannot influence the Newton polygon for $\det(R(s) - \mu I_1)$: It leads only to terms of order $O(s^k \mu^j)$ with $k_0 j + k \geq n k_0 + 2$ and $j \geq 1$. The corresponding points (k, j) are all above or to the right of the line connecting $(0, n)$ and $((n-1)k_0, 1)$. Since $h(s, 0) = g(s, 0) = \det R(s)$, the operators $R(s)$ and $T(s)$ both have the eigenvalue $\mu \equiv 0$ if and only if $\det R(s) \equiv 0$. If $\det R(s) = r_\ell s^\ell + O(s^\ell)$ for some $\ell > n k_0$, then the Newton polygons for $h(s, \mu)$ and $g(s, \mu)$ end at $(\ell, 0)$ on the s-axis. $\qquad \square$

Remark I.18.2 *For $n = 1$ the eigenvalue 0 of $D_x F(0,0)$ is algebraically simple, and Theorem I.18.1 gives the lowest term of the critical eigenvalue perturbation $\mu(s)$ of $D_x F(x(s), \lambda(s))$ by the eigenvalue of $R(s) = D_v \Phi(v(s), \lambda(s)) \in L(N, N)$. Using $N(D_x F(0,0)) = \text{span}[\hat{v}_0]$ and the projection Q as given by (I.16.6), we see that the eigenvalue of $R(s)$ is $\langle D_v \Phi(v(s), \lambda(s))\hat{v}_0, \hat{v}_0' \rangle \in \mathbb{R}$. If $Px(s) = v(s) = s\hat{v}_0$, then in the terminology of Section I.16, we can rewrite $\langle D_v \Phi(v(s), \lambda(s))\hat{v}_0, \hat{v}_0' \rangle = D_s \Phi(s, \lambda(s))$. (Here we identify $\Phi(\cdot, \lambda) : N \to N$ with its representation via the basis vector $\hat{v}_0 : \Phi(s, \lambda) = \langle \Phi(s\hat{v}_0, \lambda), \hat{v}_0' \rangle$.) Let $\{(x(s), \lambda(s))\}$ be a solution curve of $F(x, \lambda) = 0$ through $(0,0)$. If*

$Px(s) = s\hat{v}_0$, then Theorem I.16.4 tells that in general, $\lambda(s)$ does not depend analytically on s but on $t = s^{1/p\gamma_2}$ (for $s \geq 0$; for $s \leq 0$ see the modification in (I.16.27)).

According to Section I.16, we write the solution curve as $\{(\hat{x}(t), \hat{\lambda}(t))\}$, and $P\hat{x}(t) = t^{p\gamma_2}\hat{v}_0 = \hat{s}(t)\hat{v}_0 = s\hat{v}_0$. By Theorem I.16.6, the first two nonvanishing coefficients of the analytic simple eigenvalue perturbation $\hat{\mu}(t)$ of the eigenvalue 0 of $D_x F(\hat{x}(t), \hat{\lambda}(t))$ and of $D_s \Phi(\hat{s}(t), \hat{\lambda}(t))$ are the same or both vanish identically.

In view of $h(s, \mu) = g(s, \mu) - b(s, \mu)$, where $b(s, \mu) = \sum_{\nu=1}^{\infty} \mu^\nu \langle B_\nu(s)\hat{v}_0, \hat{v}'_0 \rangle = O(s^2\mu)$, we obtain $h(t, \mu) = D_s\Phi(\hat{s}(t), \hat{\lambda}(t)) - \mu(1 + \hat{b}(t, \mu))$, where $\hat{b}(t, \mu) = O(t^2)$ in any case. Thus $\hat{\mu}(t)$ solving $h(t, \hat{\mu}(t)) = 0$ and $D_s\Phi(\hat{s}(t), \hat{\lambda}(t))$ have indeed the same first two nonvanishing terms. Therefore, Theorem I.18.1 provides a new proof of Theorem I.16.6.

Next, we consider $n = 2$ and the case that $\mu \equiv 0$ is an eigenvalue for both $R(s)$ and $T(s)$; i.e., $\det R(s) \equiv 0$. Then the remaining (real) eigenvalue of $R(s)$ is $\operatorname{tr} R(s)$, and the real perturbed eigenvalue $\mu(s)$ of $D_x F(x(s), \lambda(s))$ near 0 is given by $\mu(s) = \operatorname{tr} R(s) + O(s^{k_0+2})$. This means that they have the same first two nonvanishing terms or both vanish identically. We shall make use of this observation below in studying the Principle of Reduced Stability for periodic solutions.

The following **counterexample** shows that Theorem I.18.1 cannot be improved by allowing zero to be an eigenvalue of R_{k_0} of algebraic multiplicity two. Let

$$F : \mathbb{R}^4 \times \mathbb{R} \to \mathbb{R}^4 \text{ be given by}$$

$$(I.18.29) \qquad F(x, \lambda) = \begin{pmatrix} \lambda^3 & -\lambda & 0 & 0 \\ \frac{1}{5}\lambda^5 & \frac{1}{5}\lambda^5 & 0 & 0 \\ \lambda & 0 & 2 & \frac{1}{5} \\ 0 & 0 & -1 & 0 \end{pmatrix} x - \begin{pmatrix} x_1^4 - x_2^2 \\ \frac{1}{5}x_1^6 + \frac{1}{5}x_2^6 + x_1 x_3 \\ x_1^2 \\ 0 \end{pmatrix}.$$

Here assumption (I.18.2) is satisfied for $n = 2$. We consider the solution curve $\{(x(s), \lambda(s))\} = \{((s, s, 0, 0), s)\}$. Then

$$T(s) = D_x F(x(s), \lambda(s)) = \begin{pmatrix} -3s^3 & s & 0 & 0 \\ -s^5 & -s^5 & -s & 0 \\ -s & 0 & 2 & \frac{1}{5} \\ 0 & 0 & -1 & 0 \end{pmatrix},$$

(I.18.30)   and the four matrices $T_{ij}(s)$ of (I.18.10) are precisely the four 2×2 blocks of $T(s)$.

Moreover , $T_{22}(s)^{-1} = \begin{pmatrix} 0 & -1 \\ 5 & 10 \end{pmatrix}$.

Therefore, $B_0(s) \equiv 0$, and in view of (I.18.16) and (I.18.20),

$$g(s, \mu) = \det \begin{pmatrix} -3s^3 - \mu & s \\ -s^5 & -s^5 - \mu \end{pmatrix},$$

$$h(s, \mu) = \det \left(\begin{pmatrix} -3s^3 - \mu & s \\ -s^5 & -s^5 - \mu \end{pmatrix} - \mu \begin{pmatrix} 0 & 0 \\ -5s^2 & 0 \end{pmatrix} + O(s^2 \mu^2) \right).$$

(I.18.31)

The zeros of $g(s, \mu) = 0$ are $\mu(s) = \left(-\frac{3}{2} \pm \frac{1}{2}\sqrt{5} \right) s^3 +$ h.o.t., whereas the zeros of $h(s, \mu) = 0$ are $\tilde{\mu}(s) = s^3 +$ h.o.t. The difference in the lowest-order terms is due to the degeneration of R_1; namely, by (I.18.19) and $B_0(s) \equiv 0$ we have

(I.18.32)
$$R(s) = \begin{pmatrix} 0 & 1 \\ 0 & 0 \end{pmatrix} s + O(s^3), \text{ and zero is an eigenvalue}$$

of R_1 of algebraic multiplicity two.

In the next section we show that the operator R_{k_0} is known when the solution curve $\{(v(s), \lambda(s))\}$ of $\Phi(v, \lambda) = 0$ is constructed by the Implicit Function Theorem.

I.18.1 The Principle of Reduced Stability for Periodic Solutions

Next we consider again (I.18.1) with $F(0, 0) = 0$, and we assume that the stability of a curve of equilibria through $(0, 0)$ (whose existence we do not prove) is possibly lost by the assumption that

(I.18.33)
$\pm i\kappa_0(\neq 0)$ are semisimple eigenvalues of $D_x F(0, 0)$
of multiplicity $n \geq 1$; i.e., $\dim N(\pm i\kappa_0 I - D_x F(0, 0)) = n$
and $N(\pm i\kappa_0 I - D_x F(0, 0)) \cap R(\pm i\kappa_0 I - D_x F(0, 0)) = \{0\}$.
Furthermore,
$\pm i\kappa_0 I - D_x F(0, 0)$ are Fredholm operators of index zero.

As shown in Section I.8, assumption (I.18.33) for $n = 1$ can give rise to periodic solutions of (I.18.1) with periods near $2\pi/\kappa_0$. As before in the stationary case, we do not prove the existence of such periodic solutions, but we are merely interested in their stability. We assume in addition to (I.18.33) that

(I.18.34)
$A_0 = D_x F(0, 0) \in L(X, Z)$ generates an analytic (holomorphic) semigroup $e^{A_0 t}$, $t \geq 0$, on Z that is compact for $t > 0$.

Clearly, a necessary condition for (I.18.34) is that X be densely embedded into Z. We introduce again the spaces

(I.18.35) $E = C_{2\pi}^\alpha(\mathbb{R}, X), \quad W = C_{2\pi}^\alpha(\mathbb{R}, Z), \quad Y = C_{2\pi}^{1+\alpha}(\mathbb{R}, Z),$

with norms as in (I.8.11). The obvious generalization of the proof of Proposition I.8.1 gives the following result:

(I.18.36)
If for all $m \in \mathbb{Z}\backslash\{1, -1\}$, $im\kappa_0$ is not an eigenvalue of $A_0 = D_x F(0, 0)$, then the linear operator $J_0 \equiv \kappa_0 \dfrac{d}{dt} - A_0 : E \cap Y \to W$ is continuous and is a Fredholm operator of index zero, with $\dim N(J_0) = 2n$. Furthermore, $W = N(J_0) \oplus R(J_0)$.

This defines the projection $Q : W \to N$ $(N = N(J_0))$ along R $(R = R(J_0))$ and also $P = Q|_{E \cap Y} : E \cap Y \to N$ along $R \cap (E \cap Y)$. We remark that all real spaces have to be complexified in order to make a complex spectral analysis possible and that the dimension of $N(J_0)$ is $2n$ as a complex space; i.e., $N(J_0) \cong \mathbb{C}^{2n}$.

As in Section I.8, property (I.18.36) allows us to apply the method of Lyapunov–Schmidt in order to obtain all periodic solutions of (I.18.1) of small amplitude for λ near 0 and with period near $2\pi/\kappa_0$. Using the simple substitution $\tilde{x}(t) = x(\kappa t)$ this is reduced to solving

(I.18.37)
$$G(x, \kappa, \lambda) \equiv \kappa \frac{dx}{dt} - F(x, \lambda) = 0, \quad \text{where}$$
$$G : \tilde{U} \times \tilde{V} \to W \quad \text{and}$$
$$0 \in \tilde{U} \subset E \cap Y, \ (\kappa_0, 0) \in \tilde{V} \subset \mathbb{R}^2;$$

cf. (I.18.9)–(I.18.13). The method of Lyapunov–Schmidt described in Theorem I.2.3 yields that

(I.18.38)
all solutions of $G(x, \kappa, \lambda) = 0$ near $(0, \kappa_0, 0)$ in $(E \cap Y) \times \mathbb{R}^2$ are obtained by solving $\Phi(v, \kappa, \lambda) = 0$ near $(0, \kappa_0, 0)$ in $N \times \mathbb{R}^2$. The function Φ is derived from G as in (I.2.9) and $v = Px$.

The **Principle of Reduced Stability for Periodic Solutions** is now the analogue of the principle for stationary solutions, namely, the determination of the stability of x by the stability of v as a solution of the reduced system. To that end we make use of the Principle of Linearized Stability of periodic solutions as explained in Section I.12, in particular, in (I.12.2).

As before, we assume the existence of an analytic curve $\{(x(r), \kappa(r), \lambda(r))| \ r \in (-\delta, \delta)\} \subset (E \cap Y) \times \mathbb{R}^2$ of solutions of $G(x(r), \kappa(r), \lambda(r)) = 0$ through $(0, \kappa_0, 0)$. Then the stability of $x(r)$ is determined by the Floquet exponents that are the eigenvalues of

(I.18.39)
$$T(r) \equiv D_x G(x(r), \kappa(r), \lambda(r)), \quad \text{where}$$
$$T(0) = D_x G(0, \kappa_0, 0) = J_0.$$

As before, we confine ourselves to the eigenvalue perturbations of the critical eigenvalue $\mu = 0$ of $T(0) = J_0$, assuming that the rest of the Floquet exponents are in the stable right complex half-plane. We relate these eigenvalue perturbations of $T(r)$ to the eigenvalues of

(I.18.40)
$$R(r) \equiv D_v\Phi(v(r), \kappa(r), \lambda(r)) \in L(N, N),$$
$$\text{where } Px(r) = v(r).$$

Before we apply Theorem I.18.1 to this particular situation, we make the following observations: Due to the free phase of any solution of the autonomous differential equation (I.18.1), the mapping G is S^1-equivariant (cf. (I.8.31)). This means that

(I.18.41)
$$G(S_\theta x, \kappa, \lambda) = S_\theta G(x, \kappa, \lambda) \text{ for}$$
$$(S_\theta x)(t) = x(t + \theta), \quad \theta \in \mathbb{R}(\text{mod } 2\pi).$$

Choosing a basis $\{\varphi_1, \ldots, \varphi_n\}$ in $N(i\kappa_0 I - A_0)$, we see that the complex conjugates $\{\overline{\varphi}_1, \ldots, \overline{\varphi}_n\}$ form a basis in $N(-i\kappa_0 I - A_0)$ (recall that A_0 is a real operator).

Then as shown in the proof of (I.18.36), the functions $\{\varphi_1 e^{it}, \ldots, \varphi_n e^{it},$ $\overline{\varphi}_1 e^{-it}, \ldots, \overline{\varphi}_n e^{-it}\}$ are a basis in $N(J_0)$. Next we identify $\Phi(\cdot, \kappa, \lambda) : N(J_0) \to N(J_0)$ with its representation by that basis; i.e., we identify $\Phi(\cdot, \kappa, \lambda) : N(J_0) \to N(J_0)$ with a mapping in $\mathbb{C}^{2n}$. (As shown in (I.8.33)–(I.8.34), a real function x has a real projection $Px = v$, and for any real function v the mapping $\Phi(\cdot, \kappa, \lambda)$ decomposes into n complex components $\hat{\Phi}(\cdot, \kappa, \lambda)$ together with their complex conjugates; cf. (I.8.35). Thus for real v, the equation $\Phi(v, \kappa, \lambda) = 0$ is equivalent to the n-dimensional complex system $\hat{\Phi}(v, \kappa, \lambda) = 0$, which, in turn, is usually transformed into a $2n$-dimensional real system by decomposition into real and imaginary parts.)

The equivariance (I.18.41) of the mapping G is inherited by the reduced function Φ as follows:

(I.18.42)
$$\Phi(M_\theta v, \kappa, \lambda) = M_\theta \Phi(v, \kappa, \lambda), \text{ where}$$
$$M_\theta = \text{diag}(e^{i\theta}, \ldots, e^{i\theta}, e^{-i\theta}, \ldots, e^{-i\theta}) \in L(\mathbb{C}^{2n}, \mathbb{C}^{2n}).$$

Differentiation of (I.18.42) with respect to v gives

(I.18.43)
$$D_v\Phi(M_\theta v, \kappa, \lambda)M_\theta = M_\theta D_v\Phi(v, \kappa, \lambda),$$

and for some solution of $\Phi(v, \kappa, \lambda) = 0$, differentiation of (I.18.42) with respect to θ at $\theta = 0$ yields

(I.18.44)
$$D_v\Phi(v, \kappa, \lambda)Dv = D\Phi(v, \kappa, \lambda) = 0, \text{ where}$$
$$D = \frac{d}{d\theta}M_\theta|_{\theta=0} = i(E, -E) \in L(\mathbb{C}^{2n}, \mathbb{C}^{2n}).$$

The conclusion of (I.18.44) is that $D_v\Phi(v, \kappa, \lambda)$ has an eigenvalue 0 for every nontrivial solution $v \neq 0$ of $\Phi(v, \kappa, \lambda) = 0$. In view of the basis in $N(J_0)$ we have the eigenfunction $Dv = \frac{dv}{dt}$.

The preceding analysis, in particular the proof of Theorem I.18.1, shows that 0 is also an eigenvalue of $D_xG(x, \kappa, \lambda)$ if $G(x, \kappa, \lambda) = 0$. The argument does not need any analytic dependence on some parameter r, and we repeat it for convenience.

We set $T = D_xG(x, \kappa, \lambda)$, and we decompose it as in (I.18.9), (I.18.10), using the projections defined by $W = N(J_0) \oplus R(J_0)$. Then, as in (I.18.19), $D_v\Phi(v, \kappa, \lambda) = T_{11} - T_{12}T_{22}^{-1}T_{21}$, and following (I.18.11)–(I.18.13), we obtain

$$(I.18.45) \qquad \begin{pmatrix} I_1 & -T_{12}T_{22}^{-1} \\ 0 & \tilde{I}_2 \end{pmatrix} T = \begin{pmatrix} T_{11} - T_{12}T_{22}^{-1}T_{21} & 0 \\ T_{21} & T_{22} \end{pmatrix}.$$

This proves that 0 is an eigenvalue of T with eigenfunction $\psi = Dv - T_{22}^{-1}T_{21}Dv \in E \cap Y$. (Here we identify $N \times R$ with $N \oplus R = W$ and $N \times (R \cap (E \cap Y))$ with $N \oplus R \cap (E \cap Y) = E \cap Y$.)

Obviously, the eigenvalue 0 is the trivial Floquet exponent of $D_xG(x, \kappa, \lambda)$ $= \kappa\frac{d}{dt} - D_xF(x, \lambda)$ if $G(x, \kappa, \lambda) = \kappa\frac{dx}{dt} - F(x, \lambda) = 0$. However, it is not clear that the eigenfunction $\psi = (I_1 - T_{22}^{-1}T_{21})\frac{dv}{dt}$ is equal to $\frac{dx}{dt}$, which follows from the formal proof for the trivial Floquet exponent. What is needed in Sections I.12 and I.17 is $P\psi = \frac{dv}{dt}$ if $Px = v$. For $(I - P)\psi = -T_{22}^{-1}T_{21}\frac{dv}{dt}$, we have the following estimate:

$$(I.18.46) \qquad \begin{aligned} &\|(I - P)\psi\|_{E \cap Y} \leq \|T_{22}^{-1}\|\|T_{21}\|\|v\|, \quad \text{where} \\ &\|T_{22}^{-1}\| = \|((I - Q)D_xG(x, \kappa, \lambda))^{-1}\|_{L(R, R \cap (E \cap Y))} \leq C, \\ &\|T_{21}\| = \|(I - Q)D_xG(x, \kappa, \lambda)\|_{L(N, R)} \to 0 \\ &\text{as } \|v\| \to 0 \text{ and } (\kappa, \lambda) \to (\kappa_0, 0) \\ &\text{(recall that } D_xG(0, \kappa_0, 0) = J_0). \end{aligned}$$

Since $N = N(J_0)$ is finite-dimensional, all norms on N are equivalent. In view of its construction by the method of Lyapunov–Schmidt, cf. (I.18.38), $\|v\| \to 0$ is equivalent to $\|x\|_{E \cap Y} \to 0$. We summarize:

Proposition I.18.3 *Assume for the mapping F in (I.18.1) that $F(0, 0) = 0$, that it has the regularity (I.8.13), and that it satisfies (I.18.33) and (I.18.34). Let $x \in E \cap Y$ be a 2π-periodic solution of $\kappa\frac{dx}{dt} - F(x, \lambda) = 0$ obtained by the method of Lyapunov–Schmidt; cf. (I.18.38). Then x has the trivial Floquet exponent $\mu = 0$ with an eigenfunction ψ in $E \cap Y$ such that $P\psi = \frac{d}{dt}Px$ and $(I - P)\psi = o(\|Px\|)$ in $E \cap Y$ as $(x, \kappa, \lambda) \to (0, \kappa_0, 0)$ in $(E \cap Y) \times \mathbb{R}^2$. Finally, by its construction, ψ depends continuously (smoothly) on (x, κ, λ) in the topology of $E \cap Y$ if D_xG depends continuously (smoothly) on (x, κ, λ).*

In the special case of Section I.12, when $G(x(r), \kappa(r), \lambda(r)) = 0$ is a bifurcation curve through $(0, \kappa_0, \lambda_0)$ given by the Hopf Bifurcation Theorem, Theorem I.8.2, we have $Dv = \frac{dv}{dt} = ir(\psi_0 - \overline{\psi}_0) = r\hat{v}_1$ (see (I.12.9)), and an eigenfunction $\psi = r(\hat{v}_1 - T_{22}^{-1}T_{21}\hat{v}_1)$, so that (I.12.7) is true with $\psi_1(r) = \hat{v}_1 + w_1(r)$, where $w_1(r) = -T_{22}^{-1}T_{21}\hat{v}_1$. The operators $T_{jk} = T_{jk}(r)$ are defined as in (I.18.9) for $T(r) = D_x G(x(r), \kappa(r), \lambda(r))$; cf. (I.18.39). If $T(r)$ is twice continuously differentiable with respect to $r \in (-\delta, \delta)$, then $w_1(r)$ is twice continuously differentiable, too.

Now we come back to the general case of an analytic solution curve $\{(x(r), \kappa(r), \lambda(r))\}$ through $(0, \kappa_0, 0)$ defining the analytic families of operators $T(r)$ (I.18.39) and $R(r)$ (I.18.40), provided that F is analytic near $(0, 0)$ in the sense of (I.16.2). By (I.18.44) and Proposition I.18.3, $\mu \equiv 0$ is an eigenvalue for $R(r)$ and $T(r)$ for all $r \in (-\delta, \delta)$. This means that $\det R(r) \equiv 0$ and that R_{k_0} has an eigenvalue 0 if $R(r) = r^{k_0}R_{k_0} + O(r^{k_0+1})$. Application of Theorem I.18.1 gives the following result:

Theorem I.18.4 *Let $R(r) = r^{k_0}R_{k_0} + O(r^{k_0+1})$ for some $k_0 \geq 1$. If 0 is an algebraically simple eigenvalue of R_{k_0}, then the series expanding the critical nontrivial Floquet exponents of $T(r) = D_x G(x(r), \kappa(r), \lambda(r)) = \kappa(r)\frac{d}{dt} - D_x F(x(r), \lambda(r))$ near 0 and of all nontrivial eigenvalues of $R(r)$ have the same first nonvanishing terms. They are given by*

$$(I.18.47) \qquad\qquad \mu_\rho r^{k_0}, \quad 2 \leq \rho \leq 2n,$$

where $\mu_\rho \neq 0$ are the nonvanishing eigenvalues of R_{k_0}.

Thus, if $\mathrm{Re}\,\mu_\rho \neq 0$ for all nonvanishing eigenvalues of R_{k_0}, then *the Principle of Reduced Stability for periodic solutions is true.*

Remark I.18.5 *If assumption (I.18.33) is satisfied with $n = 1$, then according to (I.18.36), $\dim N(J_0) = 2$ (which means its dimension over the complex field $\mathbb{C}$). Since $\mu \equiv 0$ is an eigenvalue both for $R(r)$ and $T(r)$, the observation made in Remark I.18.2 is true here as well: The remaining nontrivial eigenvalue $\mu_2(r)$ of $R(r)$ and of $T(r)$ is given in lowest terms by $\mu_2(r) = r^{k_0}\mathrm{tr}R_{k_0} + r^{k_0+1}\mathrm{tr}R_{k_0+1} + O(r^{k_0+2})$. Here $\Phi(v, \kappa, \lambda) = 0$ is the 2-dimensional reduced system $QG(v + \psi(v, \kappa, \lambda), \kappa, \lambda) = 0$; cf. (I.8.30). We tacitly assume that x and $v = Px$ are real functions, which means that the two complex components of Φ are complex conjugates; cf. (I.8.34). Its first component is denoted by $\hat{\Phi}(c, \kappa, \lambda)$ if $v = c\varphi_0 e^{it} + \bar{c}\,\overline{\varphi_0}e^{-it}$; cf. (I.8.35). In order to compute the four complex components of the matrix $D_v\Phi$, we have to consider c and $\bar{c}$ as independent complex variables. By the computations in Section I.17, in particular (I.17.7)–(I.17.18), we obtain as the first complex component of $\Phi(v, \kappa, \lambda)$ for $v = c\varphi_0 e^{it} + \bar{c}\,\overline{\varphi_0}e^{-it}$,*

$$(I.18.48) \quad \begin{aligned} &\hat{\Phi}(c, \bar{c}, \kappa, \lambda) \\ &= i(\kappa - \kappa_0)c + \sum_{j,\ell,k=0}^{\infty} H_{j\ell k}\lambda^j(\kappa - \kappa_0)^\ell c^{k+1}\bar{c}^k; \text{ cf. (I.17.18).} \end{aligned}$$

If $c = r \in \mathbb{R}$ and if $\hat{\Phi}(r, \kappa, \lambda) = r\tilde{\Phi}(r, \kappa, \lambda) = 0$, then

$$
\begin{aligned}
D_c\hat{\Phi}(r,\kappa,\lambda) &= D_{\overline{c}}\hat{\Phi}(r,\kappa,\lambda) \\
\text{(I.18.49)} \qquad &= \sum_{\substack{j,\ell=0 \\ k=1}}^{\infty} k H_{j\ell k}\lambda^j(\kappa-\kappa_0)^\ell r^{2k} = \frac{1}{2}D_r\hat{\Phi}(r,\kappa,\lambda).
\end{aligned}
$$

Since the second complex component of $\Phi(v,\kappa,\lambda)$ is the complex conjugate of the first component, we obtain

$$
\begin{aligned}
\operatorname{tr}D_v\Phi(v,\kappa,\lambda) &= 2\operatorname{Re}D_c\hat{\Phi}(r,\kappa,\lambda) \\
\text{(I.18.50)} \qquad &= \operatorname{Re}D_r\hat{\Phi}(r,\kappa,\lambda) \\
&\text{if } v = r(\varphi_0 e^{it} + \overline{\varphi}_0 e^{-it}) \\
&\text{and if } \hat{\Phi}(r,\kappa,\lambda) = 0.
\end{aligned}
$$

Let $\{(x(r),\kappa(r),\lambda(r))\}$ be a solution curve of $G(x,\kappa,\lambda) = 0$ through $(0,\kappa_0,0)$. If $Px(r) = r(\varphi_0 e^{it} + \overline{\varphi}_0 e^{-it})$, then Theorem I.17.3 tells that in general, $\kappa(r)$ and $\lambda(r)$ do not depend analytically on r but on $t = r^{2/p\gamma_2}$. According to Section I.17 we write the solution curve as $\{(\hat{x}(\tau),\hat{\kappa}(\tau),\hat{\lambda}(\tau))\}$ and $P\hat{x}(\tau) = \tau^{p\gamma_2/2}\hat{v}_0 = r(\tau)\hat{v}_0$ with $\hat{v}_0 = \varphi_0 e^{it} + \overline{\varphi}_0 e^{-it}$. Thus the first two nonvanishing coefficients of the nontrivial eigenvalue (= Floquet exponent) $\hat{\mu}_2(\tau)$ of $D_x G(\hat{x}(\tau),\hat{\kappa}(\tau),\lambda(\tau))$ and of $\operatorname{tr}D_v\Phi(P\hat{x}(\tau),\hat{\kappa}(\tau),\hat{\lambda}(\tau)) = \operatorname{Re}D_r\hat{\Phi}(r(\tau),\hat{\kappa}(\tau),\hat{\lambda}(\tau))$ are the same or both vanish identically. This provides a new proof of Theorem I.17.4.

Remark I.18.6 *The results of Proposition I.18.3 and Theorem I.18.4 are extended to the more general case in which (I.18.33) is replaced by the following assumption:*

$$
\text{(I.18.51)} \qquad
\begin{aligned}
&\pm i\kappa_0, \pm im_2\kappa_0, \ldots, \pm im_\ell\kappa_0 \text{ are semisimple eigenvalues of} \\
&D_x F(0,0) \text{ of multiplicities } n_j \geq 1, \ j=1,\ldots,\ell, \text{ where} \\
&1 = m_1 < m_2 < \cdots < m_\ell \text{ are integers and no other} \\
&\text{eigenvalue of } D_x F(0,0) \text{ has the form } \pm im\kappa_0 \text{ with an integer } m.
\end{aligned}
$$

Then $J_0 = \kappa_0\frac{d}{dt} - A_0$, $A_0 = D_x F(0,0)$, is a Fredholm operator of index zero, and $\dim N(J_0) = 2n$, where $n = n_1 + \cdots + n_\ell$, which is the sum of the multiplicities of $im_j\kappa_0$, $j=1,\ldots,\ell$. In particular,

$$
\text{(I.18.52)} \qquad
\begin{aligned}
&N(J_0) = N_1 \oplus \cdots \oplus N_\ell, \text{ where } N_j = \\
&\operatorname{span}\{\varphi_{m_j,1}e^{im_jt}, \ldots, \varphi_{m_j,n_j}e^{im_jt}, \overline{\varphi}_{m_j,1}e^{-im_jt}, \ldots, \overline{\varphi}_{m_j,n_j}e^{-im_jt}\},
\end{aligned}
$$

and the equivariance of the reduced function Φ is now described by

$$
\begin{aligned}
&\Phi(M_\theta v,\kappa,\lambda) = M_\theta\Phi(v,\kappa,\lambda), \text{ with} \\
\text{(I.18.53)} \quad &M_\theta = \operatorname{diag}(M_{m_1\theta}, \ldots, M_{m_\ell\theta}), \text{ where} \\
&M_{m_j\theta} = \operatorname{diag}(e^{im_j\theta}, \ldots, e^{im_j\theta}, e^{-im_j\theta}, \ldots, e^{-im_j\theta}) \in L(\mathbb{C}^{2n_j},\mathbb{C}^{2n_j}).
\end{aligned}
$$

The arguments for Proposition I.18.3 and Theorem I.18.4 hold clearly in this more general situation as well. A way to construct curves of solutions of

$G(x, \kappa, \lambda)$ through $(0, \kappa_0, 0)$ by the method of Lyapunov–Schmidt is given in [94], [102], for example.

I.19 Bifurcation with High-Dimensional Kernels, Multiparameter Bifurcation, and Application of the Principle of Reduced Stability

Analytical tools to solve high-dimensional bifurcation equations (obtained by the method of Lyapunov–Schmidt) exist, but they are rarely applied, since the verification of their assumptions is hard in high dimensions. For the existence of bifurcating solutions, topological tools as expounded in the next chapter are much more adequate.

High-dimensional kernels often occur if symmetries cause a degeneration. In this case, a reduction of the problem to fixed-point spaces of symmetry subgroups is appropriate, since it reduces the dimension of the kernel considerably. The basic notions of these so-called equivariant problems are given in the monographs [164], [58], [59], [18], and examples with symmetries can be found in Chapter III.

Nonetheless, we present some analytical methods for treating high-dimensional bifurcation equations since they are constructive compared to the topological methods. These equations are obtained by the method of Lyapunov–Schmidt expounded in Section I.2. We confine ourselves to stationary solutions of (I.18.1); the solution of high-dimensional bifurcation equations for periodic solutions can be found in [94]. We consider $F : U \times V \to Z$, where $0 \in U \subset X$ and $\lambda_0 \in V \subset \mathbb{R}$ and

$$
\begin{aligned}
&F(0, \lambda) = 0 \text{ for all } \lambda \in V, \\
\text{(I.19.1)} \quad &\dim N(D_x F(0, \lambda_0)) = \operatorname{codim} R(D_x F(0, \lambda_0)) = n \geq 1; \\
&\text{i.e., } D_x F(0, \lambda_0) \text{ is a Fredholm operator of index zero.}
\end{aligned}
$$

The assumed regularity of F is

$$
\text{(I.19.2)} \qquad F \in C^k(U \times V, Z), \text{ where } k \geq 1 \text{ is large enough.}
$$

As a matter of fact, the method itself defines the value of k in (I.19.2). By Theorem I.2.3, the problem

$$
\text{(I.19.3)} \quad
\begin{aligned}
&F(x, \lambda) = 0 \quad \text{near } (0, \lambda_0) \in X \times \mathbb{R} \text{ is reduced to} \\
&\Phi(v, \lambda) = 0 \quad \text{near } (0, \lambda_0) \in N \times \mathbb{R},
\end{aligned}
$$

where as usual, $N = N(D_x F(0, \lambda_0))$. The function Φ maps a neighborhood of $(0, \lambda_0)$ into $Z_0 \subset Z$, which is a complement of $R = R(D_x F(0, \lambda_0))$ and

$\dim N = \dim Z_0 = n$. By $(I.19.1)_1$ we have the trivial solution $\Phi(0, \lambda) = 0$ for all λ near λ_0.

Choosing a basis $\{\hat{v}_1, \ldots, \hat{v}_n\}$ in N, a basis $\{\hat{v}_1^*, \ldots, \hat{v}_n^*\}$ in Z_0, and vectors $\{\hat{v}_1', \ldots, \hat{v}_n'\}$ in Z' such that

(I.19.4)
$$\begin{aligned} \langle \hat{v}_j^*, \hat{v}_k' \rangle &= \delta_{jk} \text{ (Kronecker's symbol)}, \\ \langle z, \hat{v}_k' \rangle &= 0 \text{ for all } z \in R, \ j, k = 1, \ldots, n, \end{aligned}$$

we see that the projection Q in (I.2.3) is given by $Qz = \sum_{k=1}^n \langle z, \hat{v}_k' \rangle \hat{v}_k^*$ and the Bifurcation Equation (I.2.9) is written in coordinates as

$$\Phi(y, \lambda) = (\Phi_1(y, \lambda), \ldots, \Phi_n(y, \lambda)),$$

(I.19.5) $\Phi_k(y, \lambda) \equiv \left\langle F(\sum_{j=1}^n y_j \hat{v}_j + \psi(\sum_{j=1}^n y_j \hat{v}_j, \lambda), \lambda), \hat{v}_k' \right\rangle = 0$

for $k = 1, \ldots, n, \ y = (y_1, \ldots, y_n)$,

where ψ is the function to solve $(I - Q)F(v + \psi(v, \lambda), \lambda) \equiv 0$; cf. (I.2.6)–(I.2.9). We identify the function Φ in $(I.19.3)_2$ with its representation in coordinates (I.19.5), and we consider it as a mapping from a neighborhood of $(0, \lambda_0)$ in $\mathbb{R}^n \times \mathbb{R}$ into $\mathbb{R}^n$. We call Φ in (I.19.5) the "scalar bifurcation function." In order to simplify the notation, however, we stay with $\Phi(v, \lambda)$ rather than with $\Phi(y, \lambda)$ for $y = (y_1, \ldots, y_n) \in \mathbb{R}^n$. Finally, we normalize λ_0 to 0 and we write the Bifurcation Equation in the following form:

(I.19.6)
$$\Phi(v, \lambda) = \sum_{\substack{k \geq 1 \\ j \geq 0}} \lambda^j \Phi_{jk}(v) + R(v, \lambda) = 0, \text{ where}$$

$\Phi_{jk} : N^k \to Z_0$ are k-linear, symmetric, the sum is finite, and $R(v, \lambda)$ is a continuous remainder of higher order.

By Corollary I.2.4 we have $D_v\Phi(0,0) = \Phi_{01} = 0$.

The computation of the mappings Φ_{jk} in terms of the original function F is complicated in general. We refer to Section I.16, where we do it for a one-dimensional kernel N; cf. (I.16.19). Fortunately we need to know only special terms; see (I.19.9) and (I.19.26) below. As in Section I.15, we mark points (j, k) in a lattice whenever $\Phi_{jk}(v) \not\equiv 0$, with the difference, however, that the powers j are on the abscissa and k is on the ordinate. The *Newton polygon method* to solve $\Phi(v, \lambda) = 0$ near $(0, 0)$ nontrivially is more restricted than in the case $n = 1$, since we cannot divide by v or v^{k_0} if $\Phi_{jk}(v) \equiv 0$ for all $k < k_0$. On the other hand, we can divide by λ^{j_0} if $\Phi_{jk}(v) \equiv 0$ for all $j < j_0$ and if the remainder $R(v, \lambda)$ contains only terms of order $|\lambda|^{j_0}$. If $j_0 > 0$, then we have solutions $\{(v, 0)\}$; i.e., vertical bifurcation occurs.

Next, we can assume that there is a first point $(0, k_0)$ on the ordinate such that $\Phi_{0k_0}(v) \not\equiv 0$.

Case 1

Assume that the Bifurcation Equation is of the following form:

$$(I.19.7) \qquad \Phi(v, \lambda) = \Phi_{0k_0}(v) + \lambda \Phi_{1k_0}(v) + R(v, \lambda) = 0,$$

where $R(v, \lambda)$ contains only terms of order $|\lambda|^j \|v\|^k$ with $k_0 < k$ for $j = 0$ or 1 and $k_0 \leq k$ for $j \geq 2$. A sufficient condition for (I.19.7) is

$$(I.19.8) \quad D_x^k D_\lambda^j F(0,0)(v) = 0 \quad \text{for all } j \geq 0, \ 1 \leq k < k_0, \text{ and } v \in N,$$

whence (with $[D_x F(0,0)]^{-1} : (I - Q)Z \to (I - P)X$)

$$\Phi_{0k_0}(v) = \frac{1}{k_0!} Q D_x^{k_0} F(0,0)(v),$$

$$(I.19.9) \qquad \Phi_{1k_0}(v) = \frac{1}{k_0!} Q D_x^{k_0} D_\lambda F(0,0)(v)$$

$$- \frac{1}{k_0!} Q D_{x\lambda}^2 F(0,0)[D_x F(0,0)]^{-1}(I - Q) D_x^{k_0} F(0,0)(v).$$

Note that we have to insert only elements $v \in N$ into (I.19.8) and (I.19.9). In this case, we make the substitutions

$$(I.19.10) \qquad \|v\| = |s|, \quad v = s\tilde{v} \quad \text{with } \|\tilde{v}\| = 1,$$

yielding the equation

$$(I.19.11) \qquad s^{k_0} \left(\Phi_{0k_0}(\tilde{v}) + \lambda \Phi_{1k_0}(\tilde{v}) + R_1(\tilde{v}, \lambda, s) \right) = 0,$$

with some remainder $R_1(\tilde{v}, 0, 0) = 0$.

Since we are interested only in nontrivial solutions, we divide by s^{k_0}, and we obtain the system

$$(I.19.12) \quad \tilde{\Phi}(\tilde{v}, \lambda, s) \equiv \begin{pmatrix} \Phi_{0k_0}(\tilde{v}) + \lambda \Phi_{1k_0}(\tilde{v}) + R_1(\tilde{v}, \lambda, s) \\ \|\tilde{v}\|^2 - 1 \end{pmatrix} = \begin{pmatrix} 0 \\ 0 \end{pmatrix} = 0.$$

Here we can take any norm $\| \ \|$ in N, in particular, $\|v\|^2 = \sum_{j=1}^n y_j^2$; cf. (I.19.5).

Let $\Phi_{0k_0}(\tilde{v}_0) = 0$, where $\|\tilde{v}_0\| = 1$. (By the homogeneity of Φ_{0k_0}, any nontrivial zero can be normalized to $\|\tilde{v}_0\| = 1$.) Then $\tilde{\Phi}(\tilde{v}_0, 0, 0) = 0$, and if

$$(I.19.13) \qquad D_{(\tilde{v}, \lambda)} \tilde{\Phi}(\tilde{v}_0, 0, 0) = \begin{pmatrix} D_{\tilde{v}} \Phi_{0k_0}(\tilde{v}_0) & \Phi_{1k_0}(\tilde{v}_0) \\ 2\tilde{v}_0 & 0 \end{pmatrix}$$

is regular in $L(N \times \mathbb{R}, Z_0 \times \mathbb{R})$,

then the Implicit Function Theorem gives a solution curve $\tilde{\Phi}(\tilde{v}(s), \lambda(s), s) = 0$ through $(\tilde{v}_0, 0, 0)$ for s near 0. (We note that $D_{\tilde{v}} \Phi_{0k_0}(\tilde{v}_0) \in L(N, Z_0)$, $\Phi_{1k_0}(\tilde{v}_0)$

$\in Z_0$, and the vector $\tilde{v}_0$ acts on $h \in N$ via the scalar product $\sum_{j=1}^{n} y_j^0 h_j$ if $\hat{v}_0 = \sum_{j=1}^{n} y_j^0 \hat{v}_j$ and $h = \sum_{j=1}^{n} h_j \hat{v}_j$; cf. (I.19.5).)

This solution finally defines a nontrivial solution curve

(I.19.14) $\{(s\tilde{v}(s), \lambda(s))\}$ through $(0,0)$ of $\Phi(v, \lambda) = 0$.

We remark that due to the homogeneity of Φ_{0k_0}, the derivative $D_{\tilde{v}}\Phi_{0k_0}(\tilde{v}_0)$ has at most rank $n-1$, since 0 is necessarily an eigenvalue with eigenvector $\tilde{v}_0$.

If F and Φ are analytic near $(0,0)$, then the solution curve (I.19.14) is analytic, too, and the method of Lyapunov–Schmidt provides an analytic (nontrivial) solution curve $F(x(s), \lambda(s)) = 0$ (cf. (I.19.3)). Thus it is natural to ask whether the **Principle of Reduced Stability** is applicable. According to the hypotheses of Section I.18, we have to require that $X \subset Z$ and that 0 be a semisimple eigenvalue of $D_x F(0,0)$ of multiplicity n (cf. (I.18.2); this assumption is stronger than (I.19.1)). Since

(I.19.15) $D_v \Phi(v(s), \lambda(s)) = s^{k_0-1} D_v \Phi_{0k_0}(\tilde{v}_0) + O(s^{k_0})$,

the assumption of Theorem I.18.1 is that

(I.19.16) $\begin{aligned} &R_{k_0-1} = D_v \Phi_{0k_0}(\tilde{v}_0) \text{ has an eigenvalue } 0 \\ &\text{of at most algebraic multiplicity one.} \end{aligned}$

In view of the homogeneity of Φ_{0k_0}, this means that 0 is an eigenvalue of $D_v \Phi_{0k_0}(\tilde{v}_0)$ of algebraic multiplicity one with eigenvector $\tilde{v}_0$. In this case, the Principle of Reduced Stability is valid for the solution curve (I.19.14).

Case 2

Next, we come back to the general situation that $\Phi_{0k_0}(v) \not\equiv 0$ but that there are points (j, k) in the Newton diagram for $k < k_0$. As in (I.15.3) the Newton polygon forms by definition the convex hull of all such points (but we do not require that the polygon end on the abscissa; as a matter of fact, it can end at most in some $(j_0, 1)$; cf. (I.19.6); horizontal lines do not belong to Newton's polygon by definition). The remainder contains only terms of higher order than given by the points on the Newton polygon.

Let $-1/\gamma \in \mathbb{Q}$ be one of the slopes of the Newton polygon of $\Phi(v, \lambda)$. Then, as in Section I.15, a substitution

(I.19.17) $v = \lambda^\gamma \tilde{v}$ for $\lambda \geq 0$

leads to

(I.19.18) $\Phi(v, \lambda) = \lambda^\sigma \left(\sum_{j+k\gamma=\sigma} \Phi_{jk}(\tilde{v}) + R_1(\tilde{v}, \lambda) \right) = 0$

with some remainder $R_1(\tilde{v}, 0) = 0$. Let $\tilde{v}_0 \neq 0$ be a root of the n-dimensional polynomial

(I.19.19)
$$P_\gamma(\tilde{v}) = \sum_{j+k\gamma=\sigma} \Phi_{jk}(\tilde{v}) \quad \text{such that}$$

$$D_{\tilde{v}} P_\gamma(\tilde{v}_0) \text{ is regular in } L(N, Z_0).$$

Then the Implicit Function Theorem for (I.19.18) (after division by λ^σ) gives a solution

(I.19.20) $\{(\tilde{v}(\lambda), \lambda)|0 \leq \lambda < \delta_1\} \quad \text{such that} \quad \tilde{v}(0) = \tilde{v}_0.$

Then

(I.19.21)
$$\{(\lambda^\gamma \tilde{v}(\lambda), \lambda)|0 \leq \lambda < \delta_1\} \quad \text{is a nontrivial}$$

solution curve of $\Phi(v, \lambda) = 0$ emanating at $(0, 0)$.

Defining $\Phi^-(v, -\lambda) \equiv \Phi(v, \lambda)$, we carry out the same procedure for $\Phi^-(v, \tilde{\lambda}) = 0$ for $\tilde{\lambda} \geq 0$, and we obtain possibly a solution curve of $\Phi(v, \lambda) = 0$ for $\lambda \leq 0$ emanating at $(0, 0)$.

If F and Φ are analytic near $(0, 0)$, then $\tilde{v}(\lambda)$ is analytic in λ, and setting $s = \lambda^{1/\gamma_2}$ if $\gamma = \gamma_1/\gamma_2$ for $\gamma_i \in \mathbb{N}$, $i = 1, 2$, we see that

(I.19.22) $v(s) = s^{\gamma_1} \tilde{v}(s^{\gamma_2}) \quad \text{and} \quad \lambda(s) = s^{\gamma_2}$

provide an analytic solution curve $\Phi(v(s), \lambda(s)) = 0$ (for $s \geq 0$). By the method of Lyapunov–Schmidt, this yields a solution curve $F(x(s), \lambda(s)) = 0$ (cf. (I.19.3)), and it is natural to ask whether the **Principle of Reduced Stability** holds for the curves $\{(v(s), \lambda(s))\}$ and $\{(x(s), \lambda(s))\}$.

Again, we assume that $X \subset Z$ and that 0 is a semisimple eigenvalue of $D_x F(0, 0)$ of multiplicity n (cf. (I.18.2)). In view of (I.19.18), (I.19.19), we obtain

(I.19.23) $D_v \Phi(v(s), \lambda(s)) = s^{\sigma\gamma_2 - \gamma_1} D_{\tilde{v}} P_\gamma(\tilde{v}_0) + O(s^{\sigma\gamma_2 - \gamma_1 + 1}),$

so that $R_{k_0} = D_{\tilde{v}} P_\gamma(\tilde{v}_0)$ with $k_0 = \sigma\gamma_2 - \gamma_1 > 0$. By assumption (I.19.19), R_{k_0} is regular, and therefore, the Principle of Reduced Stability is valid for all bifurcating solution curves constructed by the Newton polygon method described earlier.

If (I.19.19) cannot be satisfied, there is another method for constructing a bifurcating solution curve of $\Phi(v, \lambda) = 0$ in special cases. Assume that one line in the Newton polygon joins $(0, k_0)$, $k_0 \geq 2$, and $(1, k_1)$ with $1 \leq k_1 < k_0$. In this case, we have to solve

(I.19.24) $\Phi(v, \lambda) = \Phi_{0k_0}(v) + \lambda \Phi_{1k_1}(v) + R(v, \lambda) = 0,$

where $R(v, \lambda)$ contains all remaining terms of higher order, i.e., $R(v, \lambda)$ contains only terms of order $|\lambda|^j \|v\|^k$ with $k_0 + j(k_1 - k_0) < k$ for $0 \leq j \leq (k_0 - 1)/(k_0 - k_1)$ and $1 \leq k$ for $j > (k_0 - 1)/(k_0 - k_1)$.

A sufficient condition for (I.19.24) is

$$\text{(I.19.25)} \quad \begin{aligned} D_x^k F(0,0)(v) &= 0 \quad \text{for } k = 1,\ldots,k_0-1, \\ D_x^k D_\lambda F(0,0)(v) &= 0 \quad \text{for } k = 1,\ldots,k_1-1, \text{ and all } v \in N \end{aligned}$$

(where $(\text{I.19.25})_2$ is redundant for $k_1 = 1$), and in this case

$$\text{(I.19.26)} \quad \begin{aligned} \Phi_{0k_0}(v) &= \frac{1}{k_0!} Q D_x^{k_0} F(0,0)(v), \\ \Phi_{1k_1}(v) &= \frac{1}{k_1!} Q D_x^{k_1} D_\lambda F(0,0)(v). \end{aligned}$$

Here the slope yields $\gamma = 1/(k_0 - k_1)$ and $P_\gamma(\tilde{v}) = \Phi_{0k_0}(\tilde{v}) + \Phi_{1k_1}(\tilde{v})$. As mentioned before, if (I.19.19) does not hold for all roots $\tilde{v}_0 \neq 0$ of $P_\gamma(\tilde{v}) = 0$, we proceed as follows: We set

$$\text{(I.19.27)} \qquad \|v\| = |s|, \quad v = s\tilde{v}, \text{ and } \lambda = s^{k_0 - k_1} \tilde{\lambda},$$

and we obtain from (I.19.24),

$$\text{(I.19.28)} \qquad s^{k_0}(\Phi_{0k_0}(\tilde{v}) + \tilde{\lambda}\Phi_{1k_1}(\tilde{v}) + R_1(\tilde{v}, \tilde{\lambda}, s)) = 0$$

such that $R_1(\tilde{v}, \tilde{\lambda}, 0) = 0$. This leads to the system

$$\text{(I.19.29)} \quad \tilde{\Phi}(\tilde{v}, \tilde{\lambda}, s) \equiv \begin{pmatrix} \Phi_{0k_0}(\tilde{v}) + \tilde{\lambda}\Phi_{1k_1}(\tilde{v}) + R_1(\tilde{v}, \tilde{\lambda}, s) \\ \|\tilde{v}\|^2 - 1 \end{pmatrix} = \begin{pmatrix} 0 \\ 0 \end{pmatrix} = 0,$$

which we solve in the same way as we do for (I.19.12). The corresponding condition (I.19.13), namely, the existence of a regular zero $(\tilde{v}_0, \tilde{\lambda}_0)$ of $\tilde{\Phi}(\tilde{v}, \tilde{\lambda}, 0) = 0$, differs from the existence of a nontrivial regular zero of $P_\gamma(\tilde{v}) = 0$. If it is satisfied we obtain a nontrivial solution curve

$$\text{(I.19.30)} \qquad \{(s\tilde{v}(s), s^{k_0 - k_1}\tilde{\lambda}(s))\} \quad \text{through } (0,0) \text{ of } \Phi(v, \lambda) = 0$$

(observe that $\tilde{v}(0) = \tilde{v}_0 \neq 0$). Along that solution curve $\{(v(s), \lambda(s))\}$,

$$\text{(I.19.31)} \quad D_v\Phi(v(s), \lambda(s)) = s^{k_0-1}(D_v\Phi_{0k_0}(\tilde{v}_0) + \tilde{\lambda}_0 D_v\Phi_{1k_1}(\tilde{v}_0)) + O(s^{k_0}),$$

and according to Theorem I.18.1, the **Principle of Reduced Stability** is valid, provided that 0 is an eigenvalue of $R_{k_0-1} = D_v\Phi_{0k_0}(\tilde{v}_0) + \tilde{\lambda}_0 D_v\Phi_{1k_1}(\tilde{v}_0)$ of algebraic multiplicity at most one.

Remark I.19.1 *We demonstrate the two methods for a* **Generic Bifurcation Equation** *for which the Newton polygon consists of the line joining* $(0,2)$ *and* $(1,1)$:

$$\text{(I.19.32)} \qquad \Phi(v, \lambda) = \Phi_{02}^{(2)}[v, v] + \lambda\Phi_{11}v + R(v, \lambda)$$

(recall that $\Phi_{01} = 0$ by Corollary I.2.4). Our notation indicates that $\Phi_{02}^{(2)}$ is bilinear and that Φ_{11} is linear. Here the slope is -1 (i.e., $\gamma = 1$), and the

method (I.19.17)–(I.19.19) requires the conditions

(I.19.33)
$$\Phi_{02}^{(2)}[\tilde{v}_0, \tilde{v}_0] + \Phi_{11}\tilde{v}_0 = 0 \quad \text{for some } \tilde{v}_0 \neq 0,$$

$$2\Phi_{02}^{(2)}[\tilde{v}_0, \cdot] + \Phi_{11} \quad \text{is regular in } L(N, Z_0),$$

yielding a nontrivial solution curve (by $\tilde{v}(0) = \tilde{v}_0$)

(I.19.34) $\{(\lambda\tilde{v}(\lambda), \lambda)|\lambda \in (-\delta_1, \delta_1)\} \quad \text{of} \quad \Phi(v, \lambda) = 0.$

The method (I.19.27)–(I.19.29) requires the conditions

(I.19.35)
$$\Phi_{02}^{(2)}[\tilde{v}_0, \tilde{v}_0] + \tilde{\lambda}_0\Phi_{11}\tilde{v}_0 = 0, \; \|\tilde{v}_0\| = 1,$$

$$2\Phi_{02}^{(2)}[\tilde{v}_0, h] + \tilde{\lambda}_0\Phi_{11}h + \mu\Phi_{11}\tilde{v}_0 = 0, \; (\tilde{v}_0, h) = 0$$

$$\Leftrightarrow h = 0, \; \mu = 0 \text{ for } h \in N, \; \mu \in \mathbb{R}.$$

(Here (,) denotes the scalar product in N introduced after (I.19.13).) If (I.19.35) is satisfied, it yields a nontrivial solution curve (by $\tilde{v}(0) = \tilde{v}_0 \neq 0$, $\tilde{\lambda}(0) = \tilde{\lambda}_0$)

(I.19.36) $\{(s\tilde{v}(s), s\tilde{\lambda}(s))|s \in (-\delta_2, \delta_2)\} \quad \text{of} \quad \Phi(v, \lambda) = 0.$

Note that the second method for $n = 1$ recovers Theorem I.5.1 (whose assumption is simply $\Phi_{11} \neq 0$).

According to (I.19.23) and the statement after it, the Principle of Reduced Stability is valid for the solution curve (I.19.34). If $2\Phi_{02}^{(2)}[\tilde{v}_0, \cdot] + \tilde{\lambda}_0\Phi_{11}$ is regular or has an algebraically simple eigenvalue 0, then that principle is valid for the solution curve (I.19.36), too; cf. (I.19.31). (Observe that $Z_0 = N$ under the assumption that 0 is a semisimple eigenvalue of $D_xF(0,0)$.)

Finally, we remark that a necessary condition for bifurcation at $(0,0)$ is the existence of some $\tilde{v}_0 \neq 0$ in N such that $\Phi_{02}^{(2)}[\tilde{v}_0, \tilde{v}_0] + \tilde{\lambda}_0\Phi_{11}\tilde{v}_0 = 0$ for some $\tilde{\lambda}_0 \in \mathbb{R}$ or $\Phi_{11}\tilde{v}_0 = 0$.

A last case (formally $\tilde{\lambda}_0 = \infty$) is treated as follows: Setting $\|v\| = |s\lambda|$, $v = s\lambda\tilde{v}$, we see that (I.19.32) leads to a system yielding a solution curve

(I.19.37) $\{(s(\lambda)\lambda\tilde{v}(\lambda), \lambda)|\lambda \in (-\delta_3, \delta_3)\} \quad \text{of} \quad \Phi(v, \lambda) = 0,$

where $s(0) = 0$ and $v(0) = \tilde{v}_0$, provided that

(I.19.38)
$$\Phi_{11}\tilde{v}_0 = 0, \; \|\tilde{v}_0\| = 1,$$

$$\mu\Phi_{02}^{(2)}[\tilde{v}_0, \tilde{v}_0] + \Phi_{11}h = 0, \; (\tilde{v}_0, h) = 0$$

$$\Leftrightarrow h = 0, \; \mu = 0 \text{ for } h \in N, \; \mu \in \mathbb{R}.$$

However, the curve (I.19.37) could be trivial. A sufficient condition for $s'(0) \neq 0$ $('= \frac{d}{d\lambda})$ is that

$$(I.19.39) \qquad \Phi_{21}\tilde{v}_0 \notin R(\Phi_{11}) \subset Z_0,$$

where the range of Φ_{11} has codimension one in Z_0. In this case, $v(\lambda) \equiv s(\lambda)\lambda\tilde{v}(\lambda) = s'(0)\lambda^2\tilde{v}_0 + \text{h.o.t.}$, and $D_v\Phi(v(\lambda), \lambda) = 2s'(0)\lambda^2\Phi_{02}^{(2)}[\tilde{v}_0, \cdot] + O(\lambda^3)$ allows the application of the Principle of Reduced Stability, provided that $\Phi_{02}^{(2)}[\tilde{v}_0, \cdot]$ has an eigenvalue 0 of algebraic multiplicity at most one.

Our selection of methods making use of the Implicit Function Theorem provides curves bifurcating from the trivial solution line. For equivariant problems whose solution sets consist of connected group orbits, these methods are not adequate without further reductions; cf. the Equivariant Bifurcation Theory. For bifurcation problems with a discrete symmetry obtained by a forced symmetry-breaking, for example, they are directly useful.

I.19.1 Bifurcation with a Two-Dimensional Kernel

For the special case of a two-dimensional kernel we can show the existence of bifurcating branches under conditions that are easier to verify than the regularity required by the Implicit Function Theorem. We assume that the Bifurcation Equation $\Phi(v, \lambda) = 0$ is of the following form, cf. (I.19.24) with $k_1 = 1$:

$$(I.19.40) \qquad \Phi(v, \lambda) = \Phi_{0k_0}(v) + \lambda\Phi_{11}v + R(v, \lambda), \ k_0 \geq 2,$$

where $\Phi_{0k_0}(v)$ is k_0-linear and symmetric and $\Phi_{11}v = QD_{x\lambda}^2F(0,0)v$ is linear in $v \in N$. A sufficient condition for (I.19.40) is given in $(I.19.25)_1$ (where $(I.19.25)_2$ is redundant), whence $\Phi_{0k_0}(v)$ is of the form $(I.19.26)_1$. The remainder $R(v, \lambda)$ contains only terms of order $|\lambda|^j\|v\|^k$ with $k_0 - j(k_0 - 1) < k$ for $j = 0$ or 1 and $1 \leq k$ for $j > 1$.

According to (I.19.27) we make the substitutions

$$(I.19.41) \qquad v = s\tilde{v} \text{ with } \|\tilde{v}\| = 1, \ \lambda = s^{k_0 - 1}\tilde{\lambda},$$

and we obtain from (I.19.40)

$$(I.19.42) \qquad \tilde{\Phi}(\tilde{v}, \tilde{\lambda}, s) \equiv s^{k_0}(\Phi_{0k_0}(\tilde{v}) + \tilde{\lambda}\Phi_{11}\tilde{v} + R_1(\tilde{v}, \tilde{\lambda}, s)),$$

where $R_1(\tilde{v}, \tilde{\lambda}, 0) = 0$. Instead of requiring the regularity of a zero $(\tilde{v}_0, \tilde{\lambda}_0)$ of $\Phi_{0k_0}(\tilde{v}) + \tilde{\lambda}\Phi_{11}\tilde{v}$, $\|\tilde{v}\| = 1$, and applying the Implicit Function Theorem as for (I.19.29) or (I.19.35), we proceed as follows. Choosing a basis $\{\hat{v}_1, \hat{v}_2\}$ in N, a basis $\{\hat{v}_1^*, \hat{v}_2^*\}$ in Z_0, and vectors $\{\hat{v}_1', \hat{v}_2'\}$ in Z' such that $\langle\hat{v}_j^*, \hat{v}_k'\rangle = \delta_{jk}$, $\langle z, \hat{v}_k'\rangle = 0$ for all $z \in R$, $j, k = 1, 2$, we see that the projection Q is given by

$Qz = \langle z, \hat{v}_1' \rangle \hat{v}_1^* + \langle z, \hat{v}_2' \rangle \hat{v}_2^*$. When $v \in N$ and $\Phi(v, \lambda) \in Z_0$ are represented in coordinates (with respect to the chosen bases), then Φ maps a neighborhood of $(0, 0)$ in $\mathbb{R}^2 \times \mathbb{R}$ into $\mathbb{R}^2$; cf. (I.19.4), (I.19.5). Furthermore, the Euclidean scalar product in $\mathbb{R}^2$ defines via coordinates a scalar product (,) and an equivalent norm $\| \ \|$ in N and Z_0, respectively. Finally, the geometric rotation by the angle $\pi/2$ in $\mathbb{R}^2$ defines the following rotation in Z_0:

(I.19.43)
$$R_{\pi/2}z = -z_2\hat{v}_1^* + z_1\hat{v}_2^* \quad \text{for}$$
$$z = z_1\hat{v}_1^* + z_2\hat{v}_2^*.$$

Observe that $(z, R_{\pi/2}z) = 0$ for all $z \in Z_0$. Our main result is the following:

Theorem I.19.2 *Assume that*

(I.19.44)
$$\Phi_{11} = QD^2_{x\lambda}F(0, 0) : N \to Z_0$$
is an isomorphism,

and that there exist two vectors $\tilde{v}_1, \tilde{v}_2 \in N$ with $\|\tilde{v}_1\| = \|\tilde{v}_2\| = 1$ such that

(I.19.45)
$$(\Phi_{0k_0}(\tilde{v}_1), R_{\pi/2}\Phi_{11}\tilde{v}_1) < 0,$$
$$(\Phi_{0k_0}(\tilde{v}_2), R_{\pi/2}\Phi_{11}\tilde{v}_2) > 0.$$

Then there exists a local continuum $\mathcal{C} \subset X \times \mathbb{R}$ of nontrivial solutions of $F(x, \lambda) = 0$ through $(0, 0)$, and $\mathcal{C} \backslash \{(0, 0)\}$ consists of at least two components.

Proof. We solve $\Phi(v, \lambda) = 0$ near $(0, 0)$ in $N \times \mathbb{R}$. With the substitutions (I.19.41) we obtain (I.19.42) for $s \in (-\delta, \delta)$. Defining

(I.19.46)
$$\Psi(\tilde{v}, \tilde{\lambda}) = \Phi_{0k_0}(\tilde{v}) + \tilde{\lambda}\Phi_{11}\tilde{v}, \quad \text{then}$$
$$\tilde{\Phi}(\tilde{v}, \tilde{\lambda}, s) = 0 \quad \text{for } s \neq 0 \Leftrightarrow$$
$$f_1(\tilde{v}, \tilde{\lambda}, s) \equiv (\Psi(\tilde{v}, \tilde{\lambda}) + R_1(\tilde{v}, \tilde{\lambda}, s), \Phi_{11}\tilde{v}) = 0,$$
$$f_2(\tilde{v}, \tilde{\lambda}, s) \equiv (\Psi(\tilde{v}, \tilde{\lambda}) + R_1(\tilde{v}, \tilde{\lambda}, s), R_{\pi/2}\Phi_{11}\tilde{v}) = 0.$$

Here we use (I.19.44), i.e., the vectors $\Phi_{11}\tilde{v}$ and $R_{\pi/2}\Phi_{11}\tilde{v}$ form a basis in Z_0 for $\tilde{v} \neq 0$.

For every $\tilde{v}_0 \in N$ with $\|\tilde{v}_0\| = 1$ and for

(I.19.47)
$$\tilde{\lambda}_0 = -\frac{(\Phi_{0k_0}(\tilde{v}_0), \Phi_{11}\tilde{v}_0)}{\|\Phi_{11}\tilde{v}_0\|^2}$$

we obtain

(I.19.48)
$$f_1(\tilde{v}_0, \tilde{\lambda}_0, 0) = (\Psi(\tilde{v}_0, \tilde{\lambda}_0), \Phi_{11}\tilde{v}_0) = 0 \quad \text{and}$$
$$D_{\tilde{\lambda}}f_1(\tilde{v}_0, \tilde{\lambda}_0, 0) = \|\Phi_{11}\tilde{v}_0)\|^2 \neq 0.$$

The Implicit Function Theorem gives the existence of a continuous scalar function

(I.19.49) $\quad \tilde{\lambda}(\tilde{v}, s)$, where $\tilde{\lambda}(\tilde{v}_0, 0) = \tilde{\lambda}_0$, such that
$$f_1(\tilde{v}, \tilde{\lambda}(\tilde{v}, s), s) = 0 \text{ for } \|\tilde{v} - \tilde{v}_0\| < \delta_1, \quad s \in (-\delta_1, \delta_1).$$

This can be done for all $\tilde{v}_0 \in S_1 = \{\tilde{v} \in N | \|\tilde{v}\| = 1\} \subset N$. Clearly $\delta_1 > 0$ depends on $\tilde{v}_0$, but by the compactness of S_1 we can find a uniform $\delta_2 > 0$ such that

(I.19.50) $\quad f_1(\tilde{v}, \tilde{\lambda}(\tilde{v}, s), s) = 0 \text{ for all } \tilde{v} \in S_1, \; s \in (-\delta_2, \delta_2).$

In order to solve

(I.19.51) $\quad g(\tilde{v}, s) \equiv f_2(\tilde{v}, \tilde{\lambda}(\tilde{v}, s), s) = 0 \text{ for } \tilde{v} \in S_1, \; s \in (-\delta_3, \delta_3),$

for some $0 < \delta_3 \leq \delta_2$, we observe that g is continuous and that

(I.19.52) $\quad \begin{aligned} &g(\tilde{v}, 0) = (\Phi_{0k_0}(\tilde{v}), R_{\pi/2}\Phi_{11}\tilde{v}), \text{ and by (I.19.45)}, \\ &g(\tilde{v}_1, 0) < 0 \text{ and } g(\tilde{v}_2, 0) > 0. \end{aligned}$

By $(\text{I}.19.52)_1$ the function $g(\cdot, 0)$ is a $(k_0 + 1)$-linear mapping from N into $\mathbb{R}$, i.e., it is a homogeneous polynomial of two variables of order $k_0 + 1$. By $(\text{I}.19.52)_2$ and the mean value theorem it has a zero on S_1, but its nondegeneracy for the application of the Implicit Function Theorem to solve (I.19.51) is not easy to verify in general. It turns out that its nondegeneracy is not necessary for the persistence of a zero for $s \in (-\delta_3, \delta_3)$.

We denote by $[\tilde{v}_1, \tilde{v}_2]$ the segment on S_1 with a counterclockwise orientation. By continuity of g, there is some $0 < \delta_3 \leq \delta_2$ such that

(I.19.53) $\quad \begin{aligned} &g(\tilde{v}_1, s) \leq d_1 < 0 \text{ and } g(\tilde{v}_2, s) \geq d_2 > 0 \\ &\text{for all } s \in [-\delta_3, \delta_3]. \end{aligned}$

If we connect the sides $\{\tilde{v}_1\} \times [-\delta_3, \delta_3]$ and $\{\tilde{v}_2\} \times [-\delta_3, \delta_3]$ of the "rectangle" $[\tilde{v}_1, \tilde{v}_2] \times [-\delta_3, \delta_3]$ by a continuous curve or any connected set in that rectangle, then in view of the mean value theorem the function g has a zero on that curve or in that set. It seems therefore "evident" not only that the set

(I.19.54) $\quad \tilde{S} = \{(\tilde{v}, s) \in [\tilde{v}_1, \tilde{v}_2] \times [-\delta_3, \delta_3] | g(\tilde{v}, s) = 0\}$

is not empty but also that $\tilde{S}$ contains a continuum $\tilde{C}$ that connects the "bottom" $B = [\tilde{v}_1, \tilde{v}_2] \times \{-\delta_3\}$ and the "top" $T = [\tilde{v}_1, \tilde{v}_2] \times \{\delta_3\}$ and that separates domains where g is positive and negative, see Figure I.19.1. To see this, let $\tilde{C}^-$ denote the component of B in $\tilde{S} \cup B$. Assume that $\tilde{C}^- \cap T = \emptyset$.

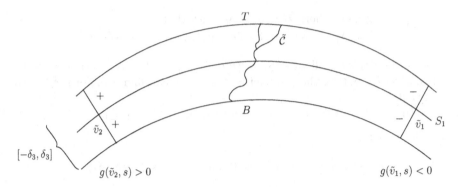

Figure I.19.1

Let W be an open neighborhood of $\tilde{C}^-$ in $N \times \mathbb{R}$ such that $W \cap T = \emptyset$. Define $K = \overline{W} \cap (\tilde{S} \cup B)$. Then K is compact, $\tilde{C}^- \subset K$, and $\partial W \cap \tilde{C}^- = \emptyset$. By the so-called Whyburn Lemma [160] there exist compact subsets $K_1, K_2 \subset K$ satisfying

$$(\text{I.19.55}) \qquad \begin{aligned} K_1 \cap K_2 &= \emptyset, \quad K_1 \cup K_2 = K, \\ \tilde{C}^- &\subset K_1, \quad \partial W \cap \tilde{S} \subset K_2. \end{aligned}$$

Choose an open neighborhood W_1 of K_1 such that

$$(\text{I.19.56}) \qquad \begin{aligned} \tilde{C}^- &\subset K_1 \subset W_1 \subset W, \quad \overline{W}_1 \cap K_2 = \emptyset, \\ \overline{W}_1 \cap T &= \emptyset, \quad \text{and} \quad \partial W_1 \cap \tilde{S} = \emptyset. \end{aligned}$$

The connected set $L = \partial W_1 \cap ([\tilde{v}_1, \tilde{v}_2] \times [-\delta_3, \delta_3])$ (which can be taken to be a continuous curve, w.l.o.g.) connects two points $(\tilde{v}_1, s_1)$ and $(\tilde{v}_2, s_2)$ with $s_1, s_2 \in (-\delta_3, \delta_3)$ and L is contained in the rectangle $[\tilde{v}_1, \tilde{v}_2] \times [-\delta_3, \delta_3]$. As mentioned before, in view of (I.19.53) and the mean value theorem, g has a zero in L, or in other words, $L \cap \tilde{S} \neq \emptyset$. But this contradicts $L \cap \tilde{S} = \emptyset$, which follows from (I.19.56)$_2$.

Therefore the assumption $\tilde{C}^- \cap T = \emptyset$ is false, which proves that $\tilde{S}$ contains a continuum $\tilde{C}$ that connects the bottom B and the top T. By definition, $\tilde{C} \subset \tilde{S}$, i.e., $g(\tilde{v}, s) = 0$ for all $(\tilde{v}, s) \in \tilde{C}$.

The set

$$(\text{I.19.57}) \qquad \mathcal{C} = \{(v, \lambda) \,|\, v = s\tilde{v}, \; \lambda = s^{k_0 - 1}\tilde{\lambda}(\tilde{v}, s), (\tilde{v}, s) \in \tilde{C}\}$$

is then a continuum of solutions of $\Phi(v, \lambda) = 0$ that contains $(0, 0)$. Furthermore, $\mathcal{C} \backslash \{(0, 0)\}$ consists of nontrivial solutions and is not connected. Since the Lyapunov–Schmidt reduction preserves connectedness of (local) solutions sets, Theorem I.19.2 is proved. $\qquad\square$

Corollary I.19.3 *If k_0 is odd, then under the hypotheses of Theorem I.19.2 there exist at least two local continua $\mathcal{C} \subset X \times \mathbb{R}$ of nontrivial solutions of $F(x, \lambda) = 0$ through $(0, 0)$ and each $\mathcal{C} \backslash \{(0, 0)\}$ consists of at least two components.*

If k_0 is even, then assumption (I.19.45) can be reduced to
$(\Phi_{0k_0}(\tilde{v}_1), R_{\pi/2}\Phi_{11}\tilde{v}_1) \neq 0$ for some $\tilde{v}_1 \in S_1$, or equivalently to the assumption
that $\Phi_{0k_0}(\tilde{v}_1)$ and $\Phi_{11}\tilde{v}_1$ are linearly independent in N, and the conclusion of
Theorem I.19.2 holds.

Proof. Assume that k_0 is odd. Then the $(k_0 + 1)$-linear form
$(\Phi_{0k_0}(\tilde{v}), R_{\pi/2}\Phi_{11}\tilde{v})$ is even, so that assumption (I.19.45) is true also for antipodal vectors $-\tilde{v}_1, -\tilde{v}_2$. Therefore the proof of Theorem I.19.2 applies to
the four segments $[\tilde{v}_1, \tilde{v}_2]$, $[\tilde{v}_2, -\tilde{v}_1]$, $[-\tilde{v}_1, -\tilde{v}_2]$, and $[-\tilde{v}_2, \tilde{v}_1]$ on S_1, yielding
four continua of the form (I.19.57) arising in each of the four segments. How-
ever, two antipodal segments provide the same continuum by the following
symmetries: By (I.19.41), (I.19.42), (I.19.46)$_3$ we obtain for odd k_0

(I.19.58)
$$\begin{aligned}
\tilde{\Phi}(-\tilde{v}, \tilde{\lambda}, -s) &= \tilde{\Phi}(\tilde{v}, \tilde{\lambda}, s), \\
f_1(-\tilde{v}, \tilde{\lambda}, -s) &= f_1(\tilde{v}, \tilde{\lambda}, s), \quad \text{whence} \\
\tilde{\lambda}(-\tilde{v}, -s) &= \tilde{\lambda}(\tilde{v}, s) \text{ by (I.19.50), and} \\
g(-\tilde{v}, -s) &= g(\tilde{v}, s); \quad \text{cf. (I.19.51).}
\end{aligned}$$

(We remark that (I.19.58)$_3$ follows from uniqueness, which is part of the
Implicit Function Theorem.) The zeros of g exist in antipodal pairs $\{(\tilde{v}, s),$
$(-\tilde{v}, -s)\}$, which, by (I.19.58)$_3$, yield only one continuum $\mathcal{C}$ (I.19.57).

Assume now that k_0 is even. Then the $(k_0 + 1)$-linear form
$(\Phi_{0k_0}(\tilde{v}), R_{\pi/2}\Phi_{11}\tilde{v})$ is odd, so that assumption (I.19.45) is satisfied by an
antipodal pair $\tilde{v}_1$ and $\tilde{v}_2 = -\tilde{v}_1$. The two antipodal segements $[\tilde{v}_1, -\tilde{v}_1]$ and
$[-\tilde{v}_1, \tilde{v}_1]$ on S_1 provide the same continuum by the following symmetries: By
(I.19.41), (I.19.42). (I.19.46)$_3$ we obtain for even k_0,

(I.19.59)
$$\begin{aligned}
\tilde{\Phi}(-\tilde{v}, -\tilde{\lambda}, -s) &= \tilde{\Phi}(\tilde{v}, \tilde{\lambda}, s), \\
f_1(-\tilde{v}, -\tilde{\lambda}, -s) &= -f_1(\tilde{v}, \tilde{\lambda}, s), \quad \text{whence} \\
\tilde{\lambda}(-\tilde{v}, -s) &= -\tilde{\lambda}(\tilde{v}, s) \text{ by (I.19.50), and} \\
g(-\tilde{v}, -s) &= -g(\tilde{v}, s); \quad \text{cf. (I.19.51).}
\end{aligned}$$

The antipodal pairs $\{(\tilde{v}, s), (-\tilde{v}, -s)\}$ of zeros of g give by (I.19.59)$_3$ only one
continuum $\mathcal{C}$ of the form (I.19.57). □

The following corollary gives an alternative condition that is more convenient for applications; cf. Section III.2.2.

Corollary I.19.4 *Assume (I.19.44) and that there exists a $\tilde{v}_1 \in S_1 \subset N$
such that*
(I.19.60) $\Phi_{0k_0}(\tilde{v}_1)$ and $\Phi_{11}\tilde{v}_1$ are linearly independent.

Let $w(t)$ for $t \in [0, 2\pi]$ be a continuous parameterization of $S_1 \subset N$. If

(I.19.61)
$$\int_0^{2\pi} (\Phi_{0k_0}(w(t)), R_{\pi/2}\Phi_{11}w(t))dt = 0,$$

then the statements of Theorem I.19.2 and of Corollary I.19.3 hold.

Proof. It is clear that (I.19.60) and (I.19.61) imply (I.19.45): By (I.19.60) the integrand of (I.19.61) does not vanish identically, whence it has to change sign on S_1.   □

Next we show that a modification of Corollary I.19.4 is applicable to a potential operator $F(\cdot, \lambda) : U \to Z$, where $U \subset X$, X is continuously embedded into Z, and Z is endowed with a continuous scalar product $(\ ,\ )$; cf. Definition I.3.1. Then there exists a function $f \in C^1(U \times V, \mathbb{R})$ such that

$$(I.19.62) \quad D_x f(x, \lambda)h = (F(x, \lambda), h) \quad \text{for all } (x, \lambda) \in U \times V,\ h \in X.$$

The function $f(\cdot, \lambda)$ is the potential of $F(\cdot, \lambda)$ with respect to the scalar product $(\ ,\ )$. If $F \in C^1(U \times V, Z)$ then $D_x F(x, \lambda)$ is symmetric with respect to $(\ ,\ )$ for all $(x, \lambda) \in U \times V$; cf. Proposition I.3.2. Therefore we may assume

(I.19.63)
$$\begin{aligned} &Z = R(D_x F(0,0)) \oplus N(D_x F(0,0)), \text{ where} \\ &R \text{ and } N \text{ are orthogonal with respect to } (\ ,\ ), \\ &Q : Z \to N \quad \text{along } R, \\ &P : X \to N \quad \text{along } R \cap X, \text{ i.e., } P = Q|_X; \end{aligned}$$

cf. (I.3.8)–(I.3.11). By Theorem I.3.4 the bifurcation function $\Phi(\cdot, \lambda)$ is also a potential operator and

(I.19.64)
$$\begin{aligned} &\varphi(v, \lambda) = f(v + \psi(v, \lambda), \lambda) \\ &\text{is the potential of } \Phi(\cdot, \lambda) \\ &\text{with respect to the scalar product } (\ ,\ ) \text{ on } N, \text{ i.e.,} \\ &D_v \varphi(v, \lambda)h = (\Phi(v, \lambda), h) \quad \text{for all } h \in N \\ &\text{and for all } (v, \lambda) \text{ near } (0,0) \in N \times \mathbb{R}. \end{aligned}$$

Here ψ is the function to solve $(I - Q)F(v + \psi(v, \lambda), \lambda) = 0$; cf. (I.2.6)–(I.2.9). By the symmetry of $D_x F(x, \lambda)$ and of the projection Q with respect to $(\ ,\ )$, the operator $\Phi_{11} = Q D_{x\lambda}^2 F(0,0) : N \to N$ is symmetric, too.

Theorem I.19.5 *Assume that $F(\cdot, \lambda)$ is a potential operator in the sense of (I.19.62) and that*

(I.19.65)
$$\begin{aligned} &\Phi_{11} = Q D_{x\lambda}^2 F(0,0) : N \to N \\ &\text{is positive (or negative) definite.} \end{aligned}$$

If there exists a $\tilde{v}_1 \in S_1 \subset N$ such that

$$(I.19.66) \qquad \Phi_{0k_0}(\tilde{v}_1) \quad and \quad \Phi_{11}\tilde{v}_1 \text{ are linearly independent,}$$

then the statements of Theorem I.19.2 and of Corollary I.19.3 hold.

Proof. We use the scalar product $(\ ,\ )$ of Z restricted to $N = Z_0$. If the chosen basis $\{\hat{v}_1, \hat{v}_2\}$ is orthonormal with respect to $(\ ,\ )$, then $(\ ,\ )$ is the same as the scalar product defined before Theorem I.19.2.

Let $E_1 = \{\tilde{v} \in N | (\tilde{v}, \Phi_{11}\tilde{v}) = 1 \text{ (or } -1)\}$ and let $w(t)$ for $t \in [0, 2\pi]$ be a smooth parameterization of the ellipse E_1. Then, by the symmetry of Φ_{11},

(I.19.67)
$$\frac{d}{dt}(w(t), \Phi_{11}w(t)) = 2(\dot{w}(t), \Phi_{11}w(t)) = 0 \quad \text{or}$$
$$\dot{w}(t) = \alpha(t)R_{\pi/2}\Phi_{11}w(t) \quad \text{for } t \in [0, 2\pi] \text{ and}$$

for some 2π-periodic smooth function $\alpha(t) \neq 0$.

Then

(I.19.68)
$$\int_0^{2\pi} \alpha(t)(\Phi_{0k_0}(w(t)), R_{\pi/2}\Phi_{11}w(t))dt$$
$$= \int_0^{2\pi} \frac{1}{k_0!}(D_v^{k_0}\Phi(0,0)(w(t)), \dot{w}(t))dt \qquad \left(\cdot = \frac{d}{dt}\right)$$
$$= \int_0^{2\pi} \frac{1}{k_0!}D_v^{k_0+1}\varphi(0,0)[w(t),\dots,w(t),\dot{w}(t)]dt \text{ by (I.19.64)}_4$$
$$= \int_0^{2\pi} \frac{1}{(k_0+1)!}\frac{d}{dt}D_v^{k_0+1}\varphi(0,0)[w(t),\dots,w(t)]dt = 0$$

by periodicity of $w(t)$.

Since $\alpha(t) \neq 0$ for all $t \in [0, 2\pi]$, $\alpha(t)$ has one sign, and therefore the function $(\Phi_{0k_0}(w(t)), R_{\pi/2}\Phi_{11}w(t))$, which does not vanish identically by (I.19.66), has to change sign on E_1. This provides (I.19.45).   □

At the end we remark that under assumption (I.19.25)$_1$ we have

(I.19.69) $\quad D_v^{k_0+1}\varphi(0,0)[v,\dots,v] = D_x^{k_0+1}f(0,0)[v,\dots,v] \quad \text{for } v \in N.$

Here f is the potential of F satisfying (I.19.62).

I.19.2 A Multiparameter Bifurcation Theorem with a High-Dimensional Kernel

A natural ingredient for overcoming the difficulties of a degeneration to a high-dimensional kernel is the dependence of the mapping F on more than one parameter, to be more precise, on as many parameters as the dimension of the cokernel, which is the dimension of the kernel in our case.

Looking at the proof of the Crandall–Rabinowitz Theorem (Theorem I.5.1) with a one-dimensional kernel, we see that a generalization is readily found, and we obtain a Multiparameter Bifurcation Theorem with a High-Dimensional Kernel.

We consider $F : U \times V \to Z$, where $0 \in U \subset X$ and $\lambda_0 \in V \subset \mathbb{R}^n$, and we assume (I.19.1). For the subsequent analysis we need only the regularity

(I.5.2), where now

(I.19.70) $$D^2_{x\lambda}F \in C(U \times V, L(\mathbb{R}^n, L(X, Z))).$$

Clearly, $L(\mathbb{R}^n, L(X, Z)) \cong L(\mathbb{R}^n \times X, Z)$ in a natural way, and $D^2_{x\lambda}F(0, \lambda_0)x \in L(\mathbb{R}^n, Z)$ is represented by the n vectors $D^2_{x\lambda_i}F(0, \lambda_0)x \in Z$, $i = 1, \ldots, n$, for $x \in X$ (cf. (I.4.3)).

The generalization of the assumption (I.5.3) is the following:

(I.19.71) There is a $\hat{v}_0 \in N(D_x F(0, \lambda_0))$, $\|\hat{v}_0\| = 1$, such that a complement Z_0 of $R(D_x F(0, \lambda_0))$ (cf. (I.2.2)) is spanned by the vectors $D^2_{x\lambda_i}F(0, \lambda_0)\hat{v}_0$, $i = 1, \ldots, n$.

Let Z_0 be a complement of $R(D_x F(0, \lambda_0))$ in Z. When the projection $Q : Z \to Z_0$ along $R(D_x F(0, \lambda_0))$ is given by

(I.19.72) $$Qz = \sum_{k=1}^n \langle z, \hat{v}'_k \rangle \hat{v}^*_k, \quad \text{cf. (I.19.4),}$$

then assumption (I.19.71) is equivalent to the following:

(I.19.73)   The matrix $(\langle D^2_{x\lambda_i}F(0, \lambda_0)\hat{v}_0, \hat{v}'_k \rangle)_{i,k=1,\ldots,n}$ is regular.

Next, we follow the lines of the proof of Theorem I.5.1. The Lyapunov–Schmidt reduction yields the n-dimensional problem $\Phi(v, \lambda) = 0$ near $(0, \lambda_0) \in N \times \mathbb{R}^n$. Setting $v = s\hat{v}_0$, we get nontrivial solutions by solving

(I.19.74) $$\tilde{\Phi}(s, \lambda) = \int_0^1 D_v \Phi(st\hat{v}_0, \lambda)\hat{v}_0 dt = 0$$
$$\text{for nontrivial } s \in (-\delta, \delta) \text{ (cf. (I.5.9), (I.5.10)).}$$

(We suppress the dependence of $\tilde{\Phi}$ on the choice of $\hat{v}_0$.) As in (I.5.11), we have $\tilde{\Phi}(0, \lambda_0) = 0$, and by (I.5.13),

(I.19.75) $$D_\lambda \tilde{\Phi}(0, \lambda_0) = Q D^2_{x\lambda}F(0, \lambda_0)\hat{v}_0$$
is regular in $L(\mathbb{R}^n, Z_0)$ by assumption (I.19.71).

The Implicit Function Theorem for (I.19.74) gives a continuously differentiable solution curve

(I.19.76) $$\tilde{\Phi}(s, \lambda(s)) = 0 \quad \text{for} \quad s \in (-\delta, \delta), \ \lambda(0) = \lambda_0,$$
$$\Phi(s\hat{v}_0, \lambda(s)) = s\tilde{\Phi}(s, \lambda(s)) = 0.$$

Clearly, $x(s) = s\hat{v}_0 + \psi(s\hat{v}_0, \lambda(s))$ (cf. (I.5.16)) provides a nontrivial solution curve $F(x(s), \lambda(s)) = 0$ through $(x(0), \lambda(0)) = (0, \lambda_0)$. We obtain an n-**Parameter Bifurcation Theorem with n-Dimensional Kernel**:

Theorem I.19.6 *Under the assumptions (I.5.2), (I.19.1), and (I.19.71), the equation $F(x, \lambda) = 0$ possesses a continuously differentiable solution curve*

$\{(x(s), \lambda(s)) | s \in (-\delta, \delta)\} \subset X \times \mathbb{R}^n$ through $(x(0), \lambda(0)) = (0, \lambda_0)$ and $\dot{x}(0) = \hat{v}_0$; cf. (I.5.18).

Remark I.19.7 *If assumption (I.19.71) is satisfied for some* $\hat{v}_0 \in N, \|\hat{v}_0\| = 1$, *then it is also satisfied for all* $\hat{v} \in N$ *in a neighborhood of* $\hat{v}_0$ *on the unit sphere in* N. *Therefore we obtain an "n-dimensional" solution set* $\{(s\hat{v}, \lambda(s, \hat{v})\}$ *for* $\Phi(v, \lambda) = 0$ *that depends in a continuously differentiable way on* $(s, \hat{v})$. *The vector* $\hat{v} \in N$ *for which (I.19.71) is valid is tangent to* $x(s) = s\hat{v} + \psi(s\hat{v}, \lambda(s, \hat{v}))$ *at* $x(0) = 0$; *cf. Corollary I.5.2.*

Restricting (I.19.1) to the assumption that 0 is a semisimple eigenvalue of $D_x F(0, \lambda_0) = 0$ of multiplicity n (which requires a continuous embedding $X \subset Z$) and assuming also analyticity of F near $(0, \lambda_0)$, we ask whether the **Principle of Reduced Stability** is applicable to the analytic solution curve $F(x(s), \lambda(s)) = 0$ through $(x(0), \lambda(0)) = (0, \lambda_0)$. A quick look at the proof of Theorem I.18.1 assures us that the condition on the validity of that principle given in Theorem I.18.1 does not depend on whether the parameter $\lambda(s)$ is a scalar or a vector. We simply have to check the condition on R_{k_0}, where $D_v \Phi(v(s), \lambda(s)) = s^{k_0} R_{k_0} + O(s^{k_0+1})$. Here $v(s) = s\hat{v}_0$.

We consider the case $k_0 = 1$.

Following the computations of Section I.6 (in particular, (I.6.1), (I.6.2)), we find that the tangent $\dot{\lambda}(0)$ to $\lambda(s)$ at $\lambda(0) = \lambda_0$ is the unique solution of

$$(I.19.77) \qquad [QD^2_{x\lambda}F(0, \lambda_0)\hat{v}_0]\dot{\lambda}(0) + \frac{1}{2}QD^2_{xx}F(0, \lambda_0)[\hat{v}_0, \hat{v}_0] = 0.$$

(Recall that upon our agreement, $QD^2_{x\lambda}F(0, \lambda_0)\hat{v}_0 \in L(\mathbb{R}^n, Z_0)$, and that it is represented as $[QD^2_{x\lambda}F(0, \lambda_0)\hat{v}_0]\lambda = \sum_{i=1}^n QD^2_{x\lambda_i}F(0, \lambda_0)\hat{v}_0\lambda_i$. Equation (I.19.77) is uniquely solvable for $\dot{\lambda}(0)$ by assumption (I.19.71).)

If $QD^2_{xx}F(0, \lambda_0)[\hat{v}_0, \hat{v}_0] \neq 0$, then $\dot{\lambda}(0) \neq 0$ and

$$(I.19.78) \qquad \begin{aligned} R_1 v &= QD^2_{xx}F(0, \lambda_0)[\hat{v}_0, v] + [QD^2_{x\lambda}F(0, \lambda_0)v]\dot{\lambda}(0) \\ &\text{for } v \in N \text{ and } R_1\hat{v}_0 = \frac{1}{2}QD^2_{xx}F(0, \lambda_0)[\hat{v}_0, \hat{v}_0]. \end{aligned}$$

Thus $R_1 \neq 0$ if $\dot{\lambda}(0) \neq 0$, and if 0 is an eigenvalue of R_1 of algebraic multiplicity at most one, then the Principle of Reduced Stability is valid for the solution (I.19.76).

We can use the computations of Section I.6 also for the case $k_0 = 2$ when $\dot{\lambda}(0) = 0$ and $R_1 = 0$. In this case, $\ddot{\lambda}(0)$ is the unique solution of

$$(I.19.79) \qquad [QD^2_{x\lambda}F(0, \lambda_0)\hat{v}_0]\ddot{\lambda}(0) + \frac{1}{3}D^3_{vvv}\Phi(0, \lambda_0)[\hat{v}_0, \hat{v}_0, \hat{v}_0] = 0$$

(cf. (I.6.4) and (I.6.10)). The formula (I.6.9) gives $D^3_{vvv}\Phi(0, \lambda_0)$ in terms of the mapping F. If the vector $D^3_{vvv}\Phi(0, \lambda_0)[\hat{v}_0, \hat{v}_0, \hat{v}_0]$ is nonzero, then $\ddot{\lambda}(0) \neq 0$ and

(I.19.80) $$R_2 v = \frac{1}{2} Q D_{vvv}^3 \Phi(0, \lambda_0)[\hat{v}_0, \hat{v}_0, v] + \frac{1}{2}[Q D_{x\lambda}^2 F(0, \lambda_0)v]\ddot{\lambda}(0)$$
            for $v \in N$ and $R_2 \hat{v}_0 = -[Q D_{x\lambda}^2 F(0, \lambda_0)\hat{v}_0]\ddot{\lambda}(0) \neq 0$.

If 0 is an eigenvalue of R_2 of algebraic multiplicity at most one, then the
Principle of Reduced Stability is valid for the solution (I.19.76).

I.20 Bifurcation from Infinity

The notion of bifurcation from infinity is misleading, since we prove the existence of a solution curve of $F(x, \lambda) = 0$ tending to infinity, and the solution
set is actually not bifurcating. We follow the general terminology in calling
it bifurcation from infinity. We consider

(I.20.1) $F : X \times V \to Z$, where
            X, Z are Banach spaces and
            $V \subset \mathbb{R}$ is an open neighborhood of λ_0.

We assume that $F \in C^2(X \times V, Z)$, and we decompose

(I.20.2) $F(x, \lambda) = A(\lambda)x + R(x, \lambda)$
            for $(x, \lambda) \in X \times V$ with a remainder $R \in C^2(X \times V, Z)$.

We assume also that

(I.20.3) $A(\lambda_0) \in L(X, Z)$ is a Fredholm operator
            of index zero (cf. Definition I.2.1) and
            $N(A(\lambda_0)) = \mathrm{span}[\hat{v}_0]$, i.e., $\dim N(A(\lambda_0)) = 1$.

The mapping $A(\lambda) \in L(X, Z)$ is assumed to be a derivative of $F(x, \lambda)$ at
"(∞, λ)" in the following sense: Let $U \subset X$ be an open neighborhood of $\hat{v}_0$.
Then, by assumption,

$$sR\left(\frac{v}{s}, \lambda\right) \to 0 \quad \text{in } Z,$$

(I.20.4) $$sD_\lambda R\left(\frac{v}{s}, \lambda\right) \to 0 \quad \text{in } Z, \text{ and}$$

$$D_x R\left(\frac{v}{s}, \lambda\right) \to 0 \quad \text{in } L(X, Z)$$
            as $s \to 0$ in $\mathbb{R}$ for all $(v, \lambda) \in U \times V$.

The **Theorem on Bifurcation from Infinity** then reads as follows:

Theorem I.20.1 *Assume in addition to (I.20.3) and (I.20.4) that*

(I.20.5) $\dfrac{d}{d\lambda} A(\lambda_0)\hat{v}_0 \notin R(A(\lambda_0))$ *(cf. (I.5.3))*.

Then there is a unique continuous curve

(I.20.6) $\{(v(s), \lambda(s)) | s \in (-\delta, \delta), (v(0), \lambda(0)) = (\hat{v}_0, \lambda_0)\} \subset X \times \mathbb{R}$

such that

(I.20.7) $F\left(\dfrac{v(s)}{s}, \lambda(s)\right) = 0$ *for* $s \in (-\delta, \delta)\backslash\{0\}$.

In other words, the solution curve $\{(x(s), \lambda(s))\} = \{(\frac{v(s)}{s}, \lambda(s))\}$ *satisfies*

(I.20.8) $\|x(s)\| \to \infty, \quad \lambda(s) \to \lambda_0 \quad as \quad s \to 0.$

Proof. Let $\hat{v}_0' \in X'$ (= the dual space) such that $\langle \hat{v}_0, \hat{v}_0' \rangle = 1$. We define for $s \neq 0$,

(I.20.9) $\tilde{F}(v, \lambda, s) = \left(A(\lambda)v + sR\left(\dfrac{v}{s}, \lambda\right), \langle v, \hat{v}_0' \rangle - 1\right),$

and in view of (I.20.4), we extend $\tilde{F}$ continuously to $s = 0$ for all $v \neq 0$ in a neighborhood U of $\hat{v}_0$ by

(I.20.10) $\tilde{F}(v, \lambda, 0) = (A(\lambda)v, \langle v, \hat{v}_0' \rangle - 1).$

Thus $\tilde{F} : U \times V \times \mathbb{R} \to Z \times \mathbb{R}$ is continuous. Next, we show that $D_{(v,\lambda)}\tilde{F} \in C(U \times V \times \mathbb{R}, L(X \times \mathbb{R}, Z \times \mathbb{R}))$. Indeed, written as a matrix in an obvious way, we obtain for $s \neq 0$,

$$D_{(v,\lambda)}\tilde{F}(v, \lambda, s) = \begin{pmatrix} A(\lambda) + D_x R\left(\dfrac{v}{s}, \lambda\right) & \dfrac{d}{d\lambda} A(\lambda)v + sD_\lambda R\left(\dfrac{v}{s}, \lambda\right) \\ \langle \cdot, \hat{v}_0' \rangle & 0 \end{pmatrix},$$

(I.20.11)

which is extended continuously to $s = 0$ for all $(v, \lambda) \in U \times V$ by

(I.20.12) $D_{(v,\lambda)}\tilde{F}(v, \lambda, 0) = \begin{pmatrix} A(\lambda) & \dfrac{d}{d\lambda} A(\lambda)v \\ \langle \cdot, \hat{v}_0' \rangle & 0 \end{pmatrix}$; cf. (I.20.4).

By (I.20.3) and the choice of $\hat{v}_0'$, we have

(I.20.13) $\tilde{F}(\hat{v}_0, \lambda_0, 0) = (0, 0) \in Z \times \mathbb{R}.$

We show that $D_{(v,\lambda)}\tilde{F}(\hat{v}_0, \lambda_0, 0)$ is bijective. Let

            $D_{(v,\lambda)}\tilde{F}(\hat{v}_0, \lambda_0, 0)(v, \lambda) = (0, 0)$ for some $(v, \lambda) \in X \times \mathbb{R}$ or

(I.20.14)

            $A(\lambda_0)v + \lambda \dfrac{d}{d\lambda} A(\lambda_0)\hat{v}_0 = 0$ and $\langle v, \hat{v}_0' \rangle = 0.$

By assumption (I.20.5), $\lambda = 0$ and $v \in N(A(\lambda_0))$, which by (I.20.3) and $\langle \hat{v}_0, \hat{v}_0' \rangle = 1$ yields $v = 0$. On the other hand, the Fredholm property of $A(\lambda_0)$ implies that assumption (I.20.5) guarantees a solution $(v, \lambda) \in X \times \mathbb{R}$ of

$$(I.20.15) \qquad A(\lambda_0)v + \lambda \frac{d}{d\lambda}A(\lambda_0)\hat{v}_0 = w \quad \text{and} \quad \langle v, \hat{v}_0' \rangle = \mu$$

$$\text{for all } (w, \mu) \in Z \times \mathbb{R}.$$

The Implicit Function Theorem provides a continuous curve (I.20.6) such that

$$(I.20.16) \qquad \tilde{F}(v(s), \lambda(s), s) = 0 \quad \text{for all } s \in (-\delta, \delta).$$

Definition (I.20.9) then gives

$$(I.20.17) \qquad F\left(\frac{v(s)}{s}, \lambda(s)\right) = A(\lambda(s))\frac{v(s)}{s} + R\left(\frac{v(s)}{s}, \lambda(s)\right) = 0$$
$$\text{for all } s \in (-\delta, \delta) \backslash \{0\}.$$

Property (I.20.8) is obvious. $\qquad \qquad \square$

We remark that the existence of $D_s\tilde{F}(v, \lambda, s)$ at $s = 0$ does not follow from our assumptions on the remainder R. Therefore, the curve $\{(v(s), \lambda(s))\}$ is not necessarily differentiable at $s = 0$. In particular, the sign of $\dot{\lambda}(0)$ is not determined as in Sections I.4 and I.6, for example. It is possible that the curve $\{(\frac{v(s)}{s}, \lambda(s))\}$ is oscillating around $\lambda(0) = \lambda_0$ when tending to (∞, λ_0) and $\dot{\lambda}(0)$ does not exist.

Since assumptions (I.20.4) are unusual, we refer to applications in Section III.7. A more general result on Bifurcation from Infinity using degree theory is mentioned in Remark III.7.3.

I.21 Bifurcation with High-Dimensional Kernels for Potential Operators: Variational Methods

In this section, we prove a new type of bifurcation at $(0, \lambda_0)$ for $F(x, \lambda) \equiv F(x) - \lambda x = 0$ with $F(0) = 0$, namely, that $(0, \lambda_0)$ is a cluster point of non-trivial solutions $(x, \lambda) \in X \times \mathbb{R}$, $x \neq 0$. In other words, there is not necessarily a curve of nontrivial solutions through $(0, \lambda_0)$, and examples show that this sort of more general bifurcation actually occurs under the hypotheses of this section, see [14].

The subsequent analysis is motivated by the following well-known facts: Let $S_r = \partial B_r(0)$ be the boundary of the ball $B_r(0) = \{x \in \mathbb{R}^n | \|x\| < r\}$ and let $f : \mathbb{R}^n \to \mathbb{R}$ be a smooth function. Since S_r is compact, there are minimizers and maximizers of $f : S_r \to \mathbb{R}$, and according to Lagrange's multiplier rule, for every such extremal $x_r \in S_r$, there is a $\lambda_r \in \mathbb{R}$ such that

$$(\text{I}.21.1) \qquad \nabla f(x_r) = \lambda_r x_r, \quad \|x_r\| = r.$$

Here we use Definition I.3.1 of the gradient of f with respect to the scalar product on $\mathbb{R}^n$ (which induces its norm). The right-hand side of (I.21.1) is λ_r times the gradient of the constraint $\frac{1}{2}(x, x) = \frac{1}{2}r^2$. If $\lambda_r = (\nabla f(x_r), x_r)/r^2$ converges to some λ_0 as $r \searrow 0$ (provided that $\nabla f(0) = 0$), then $(0, \lambda_0)$ is a bifurcation point for $F(x, \lambda) = \nabla f(x) - \lambda x = 0$ in the above sense (and λ_0 is necessarily an eigenvalue of the Hessian $D\nabla f(0)$). However, since the sphere $S_r = \partial B_r(0)$ is not compact in an infinite-dimensional Banach space, this simple argument cannot be used for an infinite-dimensional problem.

We assume the situation of Section I.3: Let $X \subset Z$ be real Banach spaces and let X be continuously embedded into Z. A scalar product $(\ , \ )$ is defined on Z such that $F : U \to Z$ with $0 \in U \subset X$ is a potential operator according to Definition I.3.1: There is an

$$
(\text{I}.21.2) \qquad
\begin{aligned}
&f \in C^2(U, \mathbb{R}) \text{ such that} \\
&Df(x)h = (F(x), h) \quad \text{for all } x \in U \subset X, h \in X, \\
&\text{and } F \in C^1(U, Z).
\end{aligned}
$$

We assume, furthermore, that

$$
(\text{I}.21.3) \qquad
\begin{aligned}
&F(0) = 0, \\
&\lambda_0 \in \mathbb{R} \text{ is an isolated eigenvalue of } A_0 = DF(0), \\
&A_0 - \lambda_0 I \in L(X, Z) \text{ is a Fredholm operator of index zero, and} \\
&Z = R(A_0 - \lambda_0 I) \oplus N(A_0 - \lambda_0 I).
\end{aligned}
$$

By Proposition I.3.2 the operator $A_0 \in L(X, Z)$ is symmetric with respect to the scalar product $(\ , \ )$.

Remark I.21.1 *Let Z be a Hilbert space with scalar product $(\ , \ )$. Then (I.21.3)$_{3,4}$ is satisfied under the following assumptions:*

$$
(\text{I}.21.4) \qquad
\begin{aligned}
&A_0 : Z \to Z \text{ with domain of definition} \\
&D(A_0) = X \subset Z \text{ is closed and} \\
&\dim N(A_0 - \lambda_0 I) < \infty.
\end{aligned}
$$

We give briefly the arguments: By the results of [86], [170] for isolated eigenvalues, the following holds in this case:

$$
(\text{I}.21.5) \qquad
\begin{aligned}
&Z = E_{\lambda_0} \oplus Z_{\lambda_0}, \\
&X = E_{\lambda_0} \oplus X_{\lambda_0}, \quad X_{\lambda_0} = Z_{\lambda_0} \cap X,
\end{aligned}
$$

where E_{λ_0} is the generalized eigenspace of A_0 with eigenvalue λ_0, and Z_{λ_0} is a closed complement that is invariant for A_0 : $A_0 \in L(X_{\lambda_0}, Z_{\lambda_0})$ and λ_0 is in *the resolvent set of the restriction of A_0 to Z_{λ_0}. Since $A_0 \in L(E_{\lambda_0}, E_{\lambda_0})$ is symmetric and therefore self-adjoint (see Proposition I.3.2), we obtain $E_{\lambda_0} = N(A_0 - \lambda_0 I)$, and being finite-dimensional, λ_0 is a pole of finite order of the resolvent of A_0. In this case, $Z_{\lambda_0} = R((A_0 - \lambda_0 I)^m)$ for some $m \in \mathbb{N}$ and $N((A_0 - \lambda_0 I)^m) = N(A_0 - \lambda_0 I)$ for all $m \in \mathbb{N}$ implies also $R((A_0 - \lambda_0 I)^m) = R(A_0 - \lambda_0 I) = Z_{\lambda_0}$. (For details see [86] or [170].)*

In this section we prove the following **Bifurcation Theorem for Potential Operators**:

Theorem I.21.2 *Under the assumptions (I.21.2) and (I.21.3), for every sufficiently small $\varepsilon > 0$ there exist at least two solutions $(x(\varepsilon), \lambda(\varepsilon)) \in X \times \mathbb{R}$ of*

(I.21.6)
$$F(x) = \lambda x,$$
$$(x(\varepsilon), x(\varepsilon)) = \varepsilon^2, \ \lambda(\varepsilon) \to \lambda_0 \ as \ \varepsilon \searrow 0.$$

Proof. We start with a Lyapunov–Schmidt reduction. Setting $R = R(A_0 - \lambda_0 I)$ and $N = N(A_0 - \lambda_0 I)$, we use the decompositions

(I.21.7)
$$X = N \oplus (R \cap X),$$
$$Z = R \oplus N,$$

with projections $Q : Z \to N$ along R and $P = Q|_X : X \to N$ along $R \cap X$. Both projections are continuous in X and Z, respectively, and orthogonal with respect to the scalar product $(\ ,\ )$; cf. Section I.3.

As in Sections I.2, I.3 we set $Px = v, (I - P)x = w$, and according to Theorem I.2.3,

(I.21.8)
$$F(x) = \lambda x \quad \text{for } (x, \lambda) \text{ near } (0, \lambda_0)$$
$$\text{is equivalent to}$$
$$QF(v + \psi(v, \lambda)) = \lambda v \quad \text{for } (v, \lambda) \text{ near } (0, \lambda_0).$$

We need some estimates for $\psi : \tilde{U}_2 \times (\lambda_0 - \delta, \lambda_0 + \delta) \to R \cap X$. To this end we set $F(x) = A_0 x + G(x)$ with $DG(0) = 0$. Then by its construction, the function $w = \psi(v, \lambda)$ satisfies

(I.21.9)
$$(A_0 - \lambda I)\psi(v, \lambda) + (I - Q)G(v + \psi(v, \lambda)) \equiv 0,$$
$$\psi(v, \lambda) = -(A_0 - \lambda I)^{-1}(I - Q)G(v + \psi(v, \lambda)),$$
$$\text{since } A_0 - \lambda I \in L(R \cap X, R) \text{ is an isomorphism}$$
$$\text{for } \lambda \in (\lambda_0 - \delta, \lambda_0 + \delta), \text{ and by differentiation,}$$
$$D_v\psi(v, \lambda)$$
$$= -(A_0 - \lambda I)^{-1}(I - Q)DG(v + \psi(v, \lambda))(I_N + D_v\psi(v, \lambda)),$$

where I_N denotes the identity in N, and $D_v\psi(v, \lambda) \in L(N, R \cap X)$. By uniqueness, $\psi(0, \lambda) = 0$ for $\lambda \in (\lambda_0 - \delta, \lambda_0 + \delta)$, and by $DG(0) = 0$, we obtain

$$(I.21.10) \quad \begin{array}{l} D_v\psi(0,\lambda) = 0 \text{ for all } \lambda \in (\lambda_0 - \delta, \lambda_0 + \delta), \text{ whence} \\ \|\psi(v,\lambda)\| \leq \varepsilon\|v\| \text{ for all } \|v\| \leq \delta(\varepsilon), \ \lambda \in (\lambda_0 - \delta, \lambda_0 + \delta). \end{array}$$

Differentiation of $(I.21.9)_1$ with respect to λ yields, after some simple calculation,

$$(I.21.11) \quad \begin{array}{l} D_\lambda\psi(v,\lambda) = \\ (A_0 - \lambda I)^{-1}\{\psi(v,\lambda) - (I - Q)DG(v + \psi(v,\lambda))D_\lambda\psi(v,\lambda)\}, \\ \|D_\lambda\psi(v,\lambda)\| \leq C_1\{\varepsilon\|v\| + \varepsilon C_2\|D_\lambda\psi(v,\lambda)\|\}, \\ \|D_\lambda\psi(v,\lambda)\| \leq C_3\varepsilon\|v\| \text{ for all } \|v\| \leq \delta(\varepsilon), \ \lambda \in (\lambda_0 - \delta, \lambda_0 + \delta). \end{array}$$

Clearly, $\varepsilon > 0$ is sufficiently small, and C_i are constants for $i = 1, 2, 3$. The Bifurcation Equation $\Phi(v,\lambda) = 0$ (cf. (I.2.9) and (I.21.8)) in this case is

$$(I.21.12) \quad QG(v + \psi(v,\lambda)) = (\lambda - \lambda_0)v.$$

All norms in the finite-dimensional space $N \subset X$ are equivalent. According to our convention, the norm for $v \in N \subset X$ in the estimates of (I.21.10), (I.21.11) is the norm of X, but in the sequel, we switch to $\|v\|^2 = (v,v)$. In view of (I.21.12), we define

$$(I.21.13) \quad \begin{array}{l} g(v,\lambda) \equiv \lambda - \lambda_0 - (G(v + \psi(v,\lambda)), v)/\|v\|^2, v \neq 0, \\ g(0,\lambda) \equiv \lambda - \lambda_0, \quad \text{which is continuous near } (0,\lambda_0), \\ D_\lambda g(v,\lambda) = 1 - (DG(v + \psi(v,\lambda))D_\lambda\psi(v,\lambda), v)/\|v\|^2, v \neq 0, \\ D_\lambda g(0,\lambda) = 1, \quad \text{which is continuous by } (I.21.11)_3 \text{ as well.} \end{array}$$

Clearly, $g(0,\lambda_0) = 0$, and by the Implicit Function Theorem for $g : \tilde{U}_2 \times (\lambda_0 - \delta, \lambda_0 + \delta) \to \mathbb{R}$ there is a continuous solution $\lambda : \tilde{U}_2 \to (\lambda_0 - \delta, \lambda_0 + \delta)$ (where $0 \in \tilde{U}_2 \subset N$ is shrunk, if necessary) such that

$$(I.21.14) \quad g(v, \lambda(v)) = 0 \text{ for all } v \in \tilde{U}_2 \subset N, \ \lambda(0) = \lambda_0.$$

Since $D_v g(v,\lambda)$ exists and is continuous for $v \in \tilde{U}_2 \backslash \{0\}$, we obtain also that the function $\lambda = \lambda(v)$ is continuously differentiable on $\tilde{U}_2 \backslash \{0\}$. Next, we define

$$(I.21.15) \quad \chi : \tilde{U}_2 \to R \cap X \text{ by } \chi(v) = \psi(v, \lambda(v)).$$

The function χ is continuous on $\tilde{U}_2, \chi(0) = 0$, and by (I.21.10), $D\chi(0) = 0$. The following estimates for $D\chi(v)$ for $v \neq 0$ prove that $\chi \in C^1(\tilde{U}_2, R \cap X)$. For $z \in N$, equation $(I.21.9)_1$ gives for $v \neq 0$,

$$(I.21.16) \quad \begin{array}{l} (A_0 - \lambda(v + tz)I)\chi(v + tz) + (I - Q)G(v + tz + \chi(v + tz)) = 0, \\ \text{and after differentiation with respect to } t \text{ at } t = 0, \\ (A_0 - \lambda(v)I)D\chi(v)z - (\nabla\lambda(v), z)\chi(v) \\ +(I - Q)DG(v + \chi(v))(z + D\chi(v)z) = 0. \end{array}$$

Equation (I.21.14) gives, for $v \neq 0$,

$$\lambda(v + tz) - \lambda_0 = (G(v + tz + \chi(v + tz)), v + tz)/\|v + tz\|^2,$$

(I.21.17) and after differentiation with respect to t at $t = 0$,

$$(\nabla\lambda(v), z) = (DG(v + \chi(v))(z + D\chi(v)z), v)/\|v\|^2$$
$$+ (G(v + \chi(v)), z)/\|v\|^2 - 2(G(v + \chi(v)), v)(v, z)/\|v\|^4.$$

Inserting this expression into $(I.21.16)_3$ and using

(I.21.18) $\|(A_0 - \lambda(v)I)^{-1}\|_{L(R, R\cap X)} \leq C_4$ for all $v \in \tilde{U}_2,$

we obtain the estimate

$$\|D\chi(v)z\| \leq C_5\{\ \|DG(v + \chi(v))\|\ (\|z\| + \|D\chi(v)z\|)$$
$$+ \|DG(v + \chi(v))\|\ (\|z\| + \|D\chi(v)z\|)\|\chi(v)\|/\|v\|$$
$$+ 3(\|G(v + \chi(v))\|/\|v\|)(\|\chi(v)\|/\|v\|)\|z\|\}.$$

(I.21.19)

Concerning the choice of norms in (I.21.19) we have to be careful: (I.21.18) is true if R is endowed with the norm of Z and $R \cap X$ is given the norm of X. Consequently, the operator norm $DG(v + \chi(v))$ is the norm of $L(X, Z)$, and the norm of $G(v + \chi(v))$ is the norm of Z. Finally, the norms of $\chi(v)$ and $D\chi(v)z$ are the norms of X. In order to obtain (I.21.19), observe that for $y \in Z, v \in N$, by continuity of the scalar product on Z, we can estimate $|(y, v)| \leq C_6\|y\|\|v\|$ with both norms in Z. For $x \in X$ the same argument leads to $|(x, v)| \leq C_7\|x\|\|v\|$ with norm of x in X. For the elements $v, z \in N$ we can switch to the equivalent norm induced by $(\ ,\ )$.

By $DG(0) = 0, \chi(0) = 0$, and $D\chi(0) = 0$, we have

(I.21.20) $\|DG(v + \chi(v))\| \to 0, \ \|G(v + \chi(v))\|/\|v\| \to 0,$
$\|\chi(v)\|/\|v\| \to 0$ as $v \to 0,$

so that (I.21.19) implies

(I.21.21) $\|D\chi(v)z\|/\|z\| \to 0$ as $v \to 0,$
for all $z \in N\backslash\{0\}$, proving that
$\chi \in C^1(\tilde{U}_2, R \cap X)$ with $D\chi(0) = 0.$

Choose $\delta_1 > 0$ such that $\{v \in N|(v, v) \leq \delta_1^2\} \subset \tilde{U}_2$ and define

(I.21.22) $M = \{v + \chi(v)|v \in N, (v, v) \leq \delta_1^2\} \subset X \subset Z,$
$S_\varepsilon = \{y \in Z|(y, y) = \varepsilon^2\} \subset Z,$
$M_\varepsilon = M \cap S_\varepsilon \subset X \subset Z.$

Since dim$N = n$, say, property (I.21.21) implies that M and M_ε are compact in X. But both sets clearly have the structure of manifolds. Being the injective image of an open set in N with injective derivative $I + D\chi(v)$, the set $\overset{\circ}{M} = \{v + \chi(v)|(v, v) < \delta_1^2\}$ is an n-dimensional manifold with tangent space

(I.21.23) $T_x \overset{\circ}{M} = (I + D\chi(v))N$ at $x = v + \chi(v) \in \overset{\circ}{M}$.

The sphere S_ε is a manifold in Z with codimension one and tangent space $T_y S_\varepsilon = \{z \in Z | (z, y) = 0\}$. Now

(I.21.24)
$$
\begin{aligned}
(v + D\chi(v)v, x) &= (v + D\chi(v)v, v + \chi(v)) \\
&= \|v\|^2 + (D\chi(v)v, \chi(v)) \\
&\geq \|v\|^2 - C_8 \|D\chi(v)\| \, \|\chi(v)\| \|v\| \geq \tfrac{1}{2} \|v\|^2 > 0
\end{aligned}
$$
by (I.21.20), (I.21.21) if $\delta_1 > 0$ is small enough.

This means that $(I + D\chi(v))v \in T_x \overset{\circ}{M} \setminus T_x S_\varepsilon$, that $\overset{\circ}{M}$ and S_ε intersect transversally (provided that $0 < \varepsilon \leq \varepsilon_0$ is small), and that

(I.21.25) $M_\varepsilon \subset U \subset X$ is a compact $(n-1)$-dimensional C^1-manifold with tangent space
$$
T_x M_\varepsilon = T_x \overset{\circ}{M} \cap T_x S_\varepsilon = \{z \in T_x \overset{\circ}{M} \, |(z, x) = 0\}.
$$

Finally, (I.19.28) shows also that

(I.21.26) $T_x \overset{\circ}{M} = \operatorname{span}[(I + D\chi(v))v, T_x M_\varepsilon]$ if $x = v + \chi(v) \in M_\varepsilon$.

By compactness, the potential $f : M_\varepsilon \to \mathbb{R}$ (cf. (I.21.2)) has a minimizer and a maximizer on M_ε. Let $x = x(\varepsilon)$ satisfying $(x, x) = \varepsilon^2$ be one of the at least two such extremals. Then a simple argument considering curves in M_ε through x proves that

(I.21.27) $(F(x), z) = 0$ for all $z \in T_x M_\varepsilon$,
              since $F(x) = \nabla f(x)$ with respect to $(\ ,\ )$.

By orthogonality of $T_x M_\varepsilon$ to x, cf. (I.21.25)$_3$,

(I.21.28) $\left(F(x) - \dfrac{1}{\varepsilon^2}(F(x), x)x, z \right) = 0$ for all $z \in \operatorname{span}[x, T_x M_\varepsilon]$.

Let $x = v + \chi(v)$, $\|v\|^2 = (v, v) < \delta_1^2$. By definition (I.21.15) and (I.21.9)$_1$,

(I.21.29) $(I - Q)F(x) = \lambda \chi(v)$ with $\lambda = \lambda(v)$, whence
              $(F(x), \chi(v)) = \lambda \|\chi(v)\|^2$, since $Q\chi(v) = 0$.

By construction of $\lambda = \lambda(v)$ via (I.21.13), (I.21.14),

(I.21.30)
$$
\begin{aligned}
(F(x), v) &= (A_0 x + G(x), v) \\
&= (x, A_0 v) + (G(v + \psi(v, \lambda)), v) \\
&= (v, A_0 v) + (\lambda - \lambda_0)\|v\|^2 \\
&= \lambda_0 \|v\|^2 + (\lambda - \lambda_0)\|v\|^2 = \lambda \|v\|^2,
\end{aligned}
$$

which implies by (I.21.29),

$$(I.21.31) \qquad \begin{aligned} (F(x), x) &= (F(x), v + \chi(v)) \\ &= \lambda(\|v\|^2 + \|\chi(v)\|^2) = \lambda\|x\|^2 = \lambda\varepsilon^2, \end{aligned}$$

since v and $\chi(v)$ are orthogonal with respect to $(\ ,\ )$. (Observe that we switched to the norm induced by the scalar product $(\ ,\ )$.) Inserting (I.21.31) into (I.21.28) gives

$$(I.21.32) \qquad (F(x) - \lambda x, z) = 0 \quad \text{for all } z \in \text{span}[x, T_x M_\varepsilon].$$

By $(I - Q)(F(x) - \lambda x) = 0$ (cf. $(I.21.9)_1$) we finally obtain

$$(I.21.33) \qquad (F(x) - \lambda x, z) = 0 \quad \text{for all } z \in \text{span}[x, T_x M_\varepsilon, R].$$

We claim that $\text{span}[x, T_x M_\varepsilon, R] = Z$. Let $y \in Z$. Then

$$(I.21.34) \qquad y = (I + D\chi(v))Qy + (I - Q)y - D\chi(v)Qy.$$

Since $Qy \in N$, the first summand belongs to $T_x \overset{\circ}{M}$, cf. (I.21.23), and the second and third summands are in R. Therefore,

$$(I.21.35) \qquad\qquad \text{span}[T_x \overset{\circ}{M}, R] = Z.$$

Furthermore, for $x = v + \chi(v)$ the vector $(I + D\chi(v))v \in \text{span}[v, R] = \text{span}[x, R]$ so that in view of (I.21.26), $T_x \overset{\circ}{M} \subset \text{span}[x, T_x M_\varepsilon, R]$. By (I.21.35) this proves that $\text{span}[x, T_x M_\varepsilon, R] = Z$.

Since $(\ ,\ )$ is a scalar product (cf. (I.3.1)), relation (I.21.33) implies (I.21.6) with $\lambda(\varepsilon) = \lambda(v(\varepsilon))$ such that $v(\varepsilon) + \chi(v(\varepsilon)) = x(\varepsilon)$ with $(x(\varepsilon), x(\varepsilon)) = \varepsilon^2$. By $\|v(\varepsilon)\|^2 = (x(\varepsilon), v(\varepsilon)) \le \varepsilon\|v(\varepsilon)\|$ we see that $v(\varepsilon) \to 0$ and $\lambda(\varepsilon) \to \lambda_0$ as $\varepsilon \searrow 0$, cf. (I.21.14). $\qquad\qquad \square$

Theorem I.21.2 is false, in general, if the operator F has no potential. This is seen by the simple counterexample

$$(I.21.36) \quad F : \mathbb{R}^2 \to \mathbb{R}^2 \quad \text{given by} \quad F(x) = F(x_1, x_2) = \begin{pmatrix} -x_2^3 \\ x_1^3 \end{pmatrix}.$$

Then for every $\lambda \in \mathbb{R}$ the equation $F(x) = \lambda x$ has only the trivial solution $x = 0$, although (I.21.3) is satisfied for $A_0 = DF(0) = 0$ and $\lambda_0 = 0$.

Equation $(I.21.6)_1$ is special in the sense that the parameter λ appears linearly. This is due to the fact that in this approach λ plays the role of a Lagrange multiplier. The linear dependence on x on the right-hand side of $(I.21.6)_1$, however, can be generalized to a more general constraint; cf. [14].

The existence of a curve or a connected set of nontrivial solutions through $(0, \lambda_0)$ is discussed in [14], [9]. In view of the counterexample in [14], one needs more assumptions on F.

In Section II.7 we alter the linear dependence on the parameter λ; i.e., we consider general equations $F(x, \lambda) = 0$, where $F(\cdot, \lambda)$ is a potential operator in the sense of Definition I.3.1 for all $\lambda \in (\lambda_0 - \delta, \lambda_0 - \delta)$.

By Theorem I.21.2, equation (I.21.6) has at least two solutions (x, λ). This is due to the fact that the potential $f : M_\varepsilon \to \mathbb{R}$ has at least two "critical points," namely, its two extremals, and not more, in general. A (relatively) critical point of f is, by definition, an $x \in M_\varepsilon$ such that $(\nabla f(x), z) = 0$ for all $z \in T_x M_\varepsilon$; cf. (I.21.27).

If we assume more symmetry of the problem, however, the number of critical points is considerably increased. It is known, in particular, that even functionals on the $(n - 1)$-dimensional sphere $S_1 \subset \mathbb{R}^n$ have at least n pairs $(x, -x)$ of critical points. Introducing the notion of "genus" permits a generalization to even functions on symmetric manifolds $M_\varepsilon = -M_\varepsilon$ that are diffeomorphic to S_1 via an odd diffeomorphism; cf. [118], [150] for more details.

Assume that $0 \in U = -U \subset X$ and that

(I.21.37) f and F in (I.21.2) satisfy
$$f(-x) = f(x), \ F(-x) = -F(x).$$

Then, by uniqueness, the function ψ in (I.21.8) is odd,

(I.21.38) $$\psi(-v, \lambda) = -\psi(v, \lambda),$$

and for the same reason, the function λ solving (I.21.14) is even,

(I.21.39) $\lambda(-v) = \lambda(v)$, so that
$$\chi(-v) = \psi(-v, \lambda(-v)) = -\psi(v, \lambda(v)) = -\chi(v).$$

Therefore, M_ε as defined in (I.21.22) satisfies $M_\varepsilon = -M_\varepsilon$, and it is diffeomorphic to $S_1 \subset N$ via

(I.21.40) $h(x) = v/\sqrt{\varepsilon^2 - \|\chi(v)\|^2}$ for
$x = v + \chi(v) \in M_\varepsilon$; i.e., $\|v\|^2 + \|\chi(v)\|^2 = \varepsilon^2$.

Consequently, if $\dim N = n$, the even potential $f : M_\varepsilon \to \mathbb{R}$ has at least n pairs $(x, -x)$ of critical points in M_ε for which (I.21.27) is valid. This proves the following corollary:

Corollary I.21.3 *If under the same assumptions as for Theorem I.21.2 the mapping F is odd, i.e.,*

$$F(-x) = -F(x) \quad \text{for all } x \in U = -U \subset X,$$

then for every sufficiently small $\varepsilon > 0$ there exist at least n pairs of solutions $(\pm x(\varepsilon), \lambda(\varepsilon)) \in X \times \mathbb{R}$ of

$$F(x) = \lambda x,$$
$$(x(\varepsilon), x(\varepsilon)) = \varepsilon^2, \ \lambda(\varepsilon) \to \lambda_0 \ as \ \varepsilon \searrow 0.$$

The number n is given by

$$n = \dim N(DF(0) - \lambda_0 I).$$

I.22 Notes and Remarks to Chapter I

The method of Lyapunov–Schmidt goes back to the beginning of the last century, when both authors used this reduction to study bifurcation for integral equations.

It was known to most experts of that reduction principle that it preserves the existence of a potential if the complementary spaces are chosen to be orthogonal; see Section 4.11 of [19], [101], [146], for example.

The Implicit Function Theorem with a one-dimensional kernel is explicitly stated in [28], and bifurcation with a one-dimensional kernel is proved in [27].

The Principle of Exchange of Stability in the nondegenerate case was known in special cases [155] and stated in general in [28].

The Hopf Bifurcation, going back to E. Hopf [79], was generalized to infinite dimensions in [29], whose proof, however, is different from that presented here.

It was observed in [62], [157], [19] that the Lyapunov Center Theorem can be proved by a Lyapunov–Schmidt reduction, yielding a "vertical" Hopf Bifurcation due to the Hamiltonian structure. A survey on Constrained Hopf Bifurcation for Hamiltonian, conservative, and reversible systems can be found in [165], [57]. An application to nonlinear oscillations is explicitly stated in [94] (Center Theorem) and in [132] (Hamiltonian Hopf Bifurcation). A different approach to bifurcation of periodic solutions of reversible systems, including a global version, is given in [117].

We give some references for the general local and global Hopf Bifurcation Theorems described in Remark I.11.13: [3], [21], [134], [83], [48], [85].

The Principle of Exchange of Stability for nondegenerate Hopf Bifurcation was first proved in [29].

As remarked in the text, the Continuation of Periodic Solutions as well as Period-Doubling Bifurcation is commonly proved in the literature via a continuation of fixed points of the Poincaré map and a bifurcation of fixed points of its first iterate. The setting to consider zeros of a map in a loop space (where the perturbed period is a parameter) is taken from [94], [102], [105]. The exchange of stability of periodic solutions at turning points and at period-doubling bifurcations then follows as in the stationary case.

The Newton polygon method for the real case is described in [37], [144].

The results on Degenerate Bifurcation of stationary solutions and the general Principle of Exchange of Stability are taken from [95]. A different proof for that principle in a more general setting is given in [167].

Degenerate Hopf Bifurcation together with a general Principle of Exchange of Stability was proved in [93], [97]. Examples of degenerate stationary as well as periodic bifurcation are found in [96].

The Principle of Reduced Stability is taken from [111].

Bifurcation with a two-dimensional kernel is taken from [119]. Theorem I.19.5 is essentially due to [9].

That one can overcome the degeneracy of a high-dimensional kernel by as many parameters as the codimension of the range is well known. We refer to [129], for example.

Bifurcation for potential operators using Lagrange's multiplier rule was first proved for completely continuous (compact) operators in [118], for non-compact operators in [14], [135]. The proof here is adapted from [150]. The refinement for odd operators goes back to Lyusternik; see [118], [150]. For more information about the structure of bifurcating solutions as the parameter varies, we refer to [151]. The preservation of bifurcating solutions under perturbations that have no potential is studied in [113].

We recommend the classical book on "branching" [163] and the actual books on bifurcation [15], [131], [47] and [65].

Chapter II
Global Theory

II.1 The Brouwer Degree

Comparable to the importance of the Implicit Function Theorem for the local analysis is the degree of a mapping for any global analysis. Although the theory was originally invented and defined in topology we present an analytical theory developed later. For finite-dimensional continuous mappings it is the Brouwer degree; its extension to infinite dimensions is the Leray–Schauder degree. Both degrees are special cases of a degree for proper Fredholm operators if the mappings are of class C^2. The proofs for this general class are given in Section II.5, and they apply in a simplified form to the Brouwer and the Leray–Schauder degree. Therefore we give in the next two sections only the results and we refer to the literature; see [34],[6] for instance.

Step 1

We assume that

(II.1.1)
$$F : \overline{U} \to \mathbb{R}^n \text{ is in } C^1(\tilde{U}, \mathbb{R}^n), \text{ where}$$
$$\text{the closure } \overline{U} \subset \tilde{U} \subset \mathbb{R}^n \text{ and}$$
$$U, \tilde{U} \text{ are open and bounded sets in } \mathbb{R}^n.$$

The vector $0 \in \mathbb{R}^n$ is assumed to be a regular value for F; i.e.,

(II.1.2)
$$DF(x) \in L(\mathbb{R}^n, \mathbb{R}^n) \text{ is regular (bijective) for all}$$
$$x \in \overline{U} \text{ with } F(x) = 0,$$

which could be possibly the empty set. Since $\overline{U}$ is compact, the solution set of $F(x) = 0$ is finite, and the following definition is adequate:

Definition II.1.1 *Assume (II.1.1), (II.1.2), and $0 \notin F(\partial U)$, where ∂U is the boundary of $U \subset \mathbb{R}^n$. Then*

(II.1.3) $$d(F, U, 0) = \sum_{\substack{x \in U \\ F(x)=0}} \operatorname{sign} \det DF(x),$$

where $\sum_{\emptyset} \cdot / \cdot = 0$. The integer $d(F, U, 0)$ is called the Brouwer degree of F with respect to U and 0. The local degree around any (isolated) solution $x_0 \in U$ of $F(x) = 0$, namely,

(II.1.4) $i(F, x_0) = \operatorname{sign} \det DF(x_0)$ when $F(x_0) = 0$,

is called the index of F at x_0. Thus the degree of F with respect to U is the sum of its indices.

The crucial property of the degree (II.1.3) is that it is continuous with respect to its entries; i.e., it is locally constant in the following sense:

Let $\tilde{F} : \overline{U} \to \mathbb{R}^n$ satisfy (II.1.1), (II.1.2), and $0 \notin \tilde{F}(\partial U)$. Then there is a $\delta > 0$ such that

(II.1.5) $d(\tilde{F}, U, 0) = d(F, U, 0)$, provided that
                   $\|\tilde{F} - F\|_{0,0} \equiv \max_{x \in \overline{U}} \|\tilde{F}(x) - F(x)\| < \delta$

(with some norm $\| \ \|$ on $\mathbb{R}^n$). This local constancy is also referred to as homotopy invariance of the Brouwer degree. A main tool for its proof is Sard's Theorem, which also plays an essential role in the next step, when the degree is extended to continuous mappings.

Step 2
Now we assume that

(II.1.6) $F : \overline{U} \to \mathbb{R}^n$ is in $C(\overline{U}, \mathbb{R}^n)$, where
                   U is a bounded and open set in $\mathbb{R}^n$, and
                   $0 \notin F(\partial U)$.

By the Stone–Weierstrass Approximation Theorem and by Sard's Theorem, there is a sequence of mappings $(F_k)_{k \in \mathbb{N}}$ satisfying (II.1.1), (II.1.2), and $\lim_{k \to \infty} \|F_k - F\|_{0,0} = 0$.

By (II.1.5), the sequence $(d(F_k, U, 0))_{k \in \mathbb{N}}$ is constant for $k \geq k_0$, so that the following definition makes sense:

Definition II.1.2 *Assume (II.1.6) and a sequence of mappings $(F_k)_{k \in \mathbb{N}}$ satisfying (II.1.1), (II.1.2), and approximating the mapping F uniformly on $\overline{U}$:*

(II.1.7) $\lim_{k \to \infty} \|F_k - F\|_{0,0} = 0.$

Then the Brouwer degree of F with respect to U and 0 is defined as

(II.1.8) $d(F, U, 0) = \lim_{k \to \infty} d(F_k, U, 0).$

Finally, for every $y \notin F(\partial U)$, define

(II.1.9) $$d(F, U, y) = d(F - y, U, 0).$$

Needless to say, due to (II.1.5), definition (II.1.8) is independent of the choice of a sequence (F_k), and the degree (II.1.9) is continuous with respect to F (using the maximum norm (II.1.5)), with respect to U (using the distance $\text{dist}(\tilde{U}, U) = \sup_{\tilde{x} \in \tilde{U}} \text{dist}(\tilde{x}, U) + \sup_{x \in U} \text{dist}(x, \tilde{U})$ where $\text{dist}(\tilde{x}, U) = \inf_{x \in U} \|\tilde{x} - x\|$), and, finally, with respect to y in any component of $\mathbb{R}^n \setminus F(\partial U)$ using any norm $\| \quad \|$ on $\mathbb{R}^n$.

We summarize the essential properties of the Brouwer degree:

(1) Let I be the identity on $\mathbb{R}^n$. Then for $y \notin \partial U$,

(II.1.10) $$d(I, U, y) = \begin{cases} 1 \text{ if } y \in U, \\ 0 \text{ if } y \notin \overline{U} \end{cases}$$

(normalization).

(2) Let U_1, U_2 be open and bounded sets in $\mathbb{R}^n$ such that $U_1 \cap U_2 = \emptyset$ and $F \in C(\overline{U}_1 \cup \overline{U}_2, \mathbb{R}^n)$ and $y \notin F(\partial U_1 \cup \partial U_2)$. Then

(II.1.11) $$d(F, U_1 \cup U_2, y) = d(F, U_1, y) + d(F, U_2, y)$$

(additivity).

(3) The degree is *homotopy invariant*, which means that it is continuous ($=$ constant) with respect to its entries (F, U, y) in the sense described above. A special case is the following homotopy invariance with respect to "non-cylindrical domains":
Let $\mathbb{R}^{n+1} = \{(\tau, x) | \tau \in \mathbb{R}, x \in \mathbb{R}^n\}$, $U \subset \mathbb{R}^{n+1}$, be open and $U_\tau = \{x \in \mathbb{R}^n | (\tau, x) \in U\}$ be uniformly bounded (possibly empty) for τ in finite intervals of $\mathbb{R}$. Assume for $F : \overline{U} \to \mathbb{R}^n$ that $F \in C(\overline{U}, \mathbb{R}^n)$ and that for a continuous curve $y : \mathbb{R} \to \mathbb{R}^n$,

(II.1.12) $$y(\tau) \neq F(\tau, x) \quad \text{for all } (\tau, x) \in \partial U.$$

Then
(II.1.13) $$d(F(\tau, \cdot), U_\tau, y(\tau)) = \text{const} \quad \text{for all } \tau \in \mathbb{R},$$

and $d(F(\tau, \cdot), U_\tau, y(\tau)) = 0$ if $U_{\tau_0} = \emptyset$ for some $\tau_0 \in \mathbb{R}$.
We sketch the proof: For $a < b$ let $U_{a,b} = U \cap ((a, b) \times \mathbb{R}^n)$, which is a "noncylindrical" domain bounded in $\mathbb{R}^{n+1}$. Solutions of $F(\tau, x) = y(\tau)$ for $(\tau, x) \in \partial U_{a,b}$ exist at most in U_a or U_b, and the proof of (II.1.13), using a regular approximation via Sard's Theorem, is then the same as for cylindrical domains (see also Section II.5 below, in particular Figure II.5.1). Thus $d(F(a, \cdot), U_a, y(a)) = d(F(b, \cdot), U_b, y(b))$, and (II.1.13) is proved. (If $U_a = \emptyset$ or $U_b = \emptyset$, this proof shows also that $d(F(\tau, \cdot), U_\tau, y(\tau)) = 0$ for $\tau \in [a, b]$.)

(4) The following property is crucial for the solution of nonlinear problems:

$$\text{(II.1.14)} \qquad \begin{array}{l} \text{If } d(F, U, y) \neq 0, \text{ then there is some} \\ x \in U \text{ such that } F(x) = y. \end{array}$$

(5) Let $F_1, F_2 \in C(\overline{U}, \mathbb{R}^n)$, $U \subset \mathbb{R}^n$, such that $F_1(x) = F_2(x)$ for all $x \in \partial U$. Then for $y \notin F_1(\partial U) = F_2(\partial U)$,

$$\text{(II.1.15)} \qquad\qquad d(F_1, U, y) = d(F_2, U, y),$$

i.e., the degree depends only on the values of F on the boundary ∂U.

(6) If $\{x \in \overline{U} | F(x) = y\} \subset \tilde{U} \subset U$, then

$$\text{(II.1.16)} \qquad\qquad d(F, U, y) = d(F, \tilde{U}, y)$$

(*excision*).

Finally, the notion of the *index* given in Definition II.1.1 is extended to every isolated solution $x_0 \in U$ of $F(x) = y$. It is simply the local degree around $x_0 \in U$:

$$\text{(II.1.17)} \quad \begin{array}{l} d(F, B_r(x_0), y) = i(F, x_0), \\ \text{where } B_r(x_0) = \{x \in \mathbb{R}^n | \|x - x_0\| < r\} \text{ for small } r > 0 \\ \text{such that } x_0 \text{ is the only solution of } F(x) = y \text{ in } \overline{B}_r(x_0). \end{array}$$

(We omit the dependence on y.) Clearly, $i(F, x_0) \in \mathbb{Z}$ but not necessarily in $\{1, -1\}$.

A degree having the properties (1)–(6) is uniquely defined [8]. For $U \subset \mathbb{R}^2$ the number $d(F, U, 0)$ coincides with the geometric notion of the *winding number* of F along ∂U.

For a practical computation of the degree or the index, the following *multiplicativity* is useful. It follows immediately from the corresponding property of the determinant in Step 1 and by approximation in Step 2. Let

$$\text{(II.1.18)} \quad \begin{array}{l} F_i : \overline{U}_i \to \mathbb{R}^{n_i} \text{ be in } C(\overline{U}_i, \mathbb{R}^{n_i}), \text{ where} \\ U_i \subset \mathbb{R}^{n_i} \text{ is bounded and open, } i = 1, 2. \\ \text{Define } F = (F_1, F_2) : \overline{U} = \overline{U}_1 \times \overline{U}_2 \to \mathbb{R}^{n_1} \times \mathbb{R}^{n_2} \text{ by} \\ F(x_1, x_2) = (F_1(x_1), F_2(x_2)). \\ \text{If } y_i \notin F_i(\partial U_i), \text{ then } y = (y_1, y_2) \notin F(\partial U), \text{ and} \\ d(F, U, y) = d(F_1, U_1, y_1) d(F_2, U_2, y_2). \end{array}$$

II.2 The Leray–Schauder Degree

The extension of the Brouwer degree to infinite dimensions gives a powerful tool for solving infinite-dimensional nonlinear problems. We summarize the steps taken by Leray and Schauder in 1934 [125] (see also [34], e.g.).

There does not exist a degree for continuous vector fields on an infinite-dimensional Banach space having all properties of the Brouwer degree. (There are several topological reasons for this: There are no two components of the group of regular endomorphisms corresponding to a negative and positive determinant; there exists a continuous mapping from the closed unit ball of an infinite-dimensional Banach space into itself having no fixed point, although that mapping is homotopic to the identity in the sense of Section II.1, and so on.)

A crucial part of the extension to infinite dimensions is the characterization of the class of vector fields for which a degree can be defined. This class should be large enough for reasonable applications, and the degree should be as powerful as the Brouwer degree.

Definition II.2.1 Let $f : \overline{U} \to X$, where $U \subset X$ is open and X is a (real) Banach space. The mapping f is called completely continuous (compact) if f is continuous and if f maps bounded subsets of $\overline{U}$ onto relatively compact sets (whose closure is compact). If

$$(II.2.1) \qquad\qquad F = I + f : \overline{U} \to X,$$

where I is the identity on X and f is completely continuous, the mapping F is called a compact perturbation of the identity.

For compact perturbations of the identity, Leray and Schauder succeeded in defining a useful degree. A crucial observation is that a completely continuous mapping $f : \overline{U} \to X$ on bounded domains $\overline{U}$ is uniformly approximated by finite-dimensional mappings $f_\varepsilon : \overline{U} \to X_{n_\varepsilon} \subset X$ such that $\dim X_{n_\varepsilon} = n_\varepsilon < \infty$ and $\|f - f_\varepsilon\|_{0,0} = \sup_{x \in U} \|f(x) - f_\varepsilon(x)\| < \varepsilon$, where $\varepsilon > 0$ is arbitrarily small. (Here $\| \ \|$ denotes the norm of the Banach space X.) We define $F_\varepsilon = I + f_\varepsilon$. If $0 \notin F(\partial U)$, then also (for small $\varepsilon > 0$) $0 \notin F_\varepsilon(\partial U_{n_\varepsilon})$ for $U_{n_\varepsilon} = U \cap X_{n_\varepsilon}$, which follows from the facts that due to the complete continuity of f, $F(\partial U)$ is closed and that 0 has a positive distance to $F(\partial U)$.

Therefore, if $\varepsilon > 0$ is small enough, the Brouwer degree $d(F_\varepsilon, U_{n_\varepsilon}, 0)$ for F_ε restricted on $\overline{U}_{n_\varepsilon}$ exists. Since this degree does not depend on the choice of $f_\varepsilon : \overline{U} \to X_{n_\varepsilon}$, it is a good definition of a degree for F.

Definition II.2.2 Let $U \subset X$ be open and bounded. Then for any compact perturbation of the identity (II.2.1) with $0 \notin F(\partial U)$ the Leray–Schauder degree

$$(II.2.2) \qquad\qquad d(F, U, 0) \in \mathbb{Z}$$

is defined as described above. For $y \notin F(\partial U)$ we define

$$(\text{II}.2.3) \qquad\qquad d(F, U, y) = d(F - y, U, 0).$$

We summarize the essential properties of that degree.

(1) The *normalization* (II.1.10) holds as for the Brouwer degree.
(2) The *additivity* (II.1.11) holds as for the Brouwer degree.
(3) We state its *homotopy invariance* as follows (adequate for our applications in bifurcation theory):
Let $U \subset \mathbb{R} \times X = \{(\tau, x) | \tau \in \mathbb{R}, x \in X\}$ be open and $U_\tau = \{x \in X | (\tau, x) \in U\}$ be uniformly bounded (possibly empty) for τ in finite intervals of $\mathbb{R}$. We assume that

$$(\text{II}.2.4) \qquad
\begin{aligned}
&f : \overline{U} \to X \text{ is completely continuous;} \\
&y : \mathbb{R} \to X \text{ is a continuous curve such that} \\
&y(\tau) \neq x + f(\tau, x) \text{ for all } (\tau, x) \in \partial U.
\end{aligned}$$

Define $\tilde{f}(\tau, x) = f(\tau, x) - y(\tau)$. Then $\tilde{f} : \overline{U} \to X$ is completely continuous, and therefore, for every $\varepsilon > 0$, there exists a finite-dimensional approximation $\tilde{f}_\varepsilon : U_{a,b} \to X_{n_\varepsilon}$ on any bounded "noncylindrical" domain $U_{a,b} = U \cap ((a, b) \times X)$ such that $\|\tilde{f} - \tilde{f}_\varepsilon\|_{0,0} = \sup_{(\tau,x) \in U_{a,b}} \|\tilde{f}(\tau, x) - \tilde{f}_\varepsilon(\tau, x)\| < \varepsilon$. We define

$$(\text{II}.2.5) \qquad\qquad F(\tau, x) = x + f(\tau, x) \quad \text{for } (\tau, x) \in U$$

and $\tilde{F}(\tau, x) = x + \tilde{f}(\tau, x)$, $\tilde{F}_\varepsilon(\tau, x) = x + \tilde{f}_\varepsilon(\tau, x)$. By the closedness of $F(\partial U)$ and by $(\text{II}.2.4)_3$, the curve $\{y(\tau) | \tau \in [a, b]\}$ has a positive distance to $F((\partial U)_{a,b})$, where $(\partial U)_{a,b} = \partial U \cap ([a, b] \times X)$. Therefore, if $\varepsilon > 0$ is small enough, the only solutions of $\tilde{F}_\varepsilon(\tau, x) = 0$ for $(\tau, x) \in \overline{U}_{a,b}$ exist at most in U_τ for $\tau \in [a, b]$ (and not in $\partial U_\tau \subset (\partial U)_\tau$). By definition,

$$(\text{II}.2.6) \qquad
\begin{aligned}
d(F(\tau, \cdot), U_\tau, y(\tau)) &= d(\tilde{F}(\tau, \cdot), U_\tau, 0) \\
&= d(\tilde{F}_\varepsilon(\tau, \cdot), U_\tau \cap X_{n_\varepsilon}, 0) = \text{ const for } \tau \in [a, b],
\end{aligned}$$

by the homotopy invariance of the Brouwer degree described in Section II.1. Therefore, under the assumption (II.2.4), we conclude for the Leray–Schauder degree that

$$(\text{II}.2.7) \qquad d(I + f(\tau, \cdot), U_\tau, y(\tau)) = \text{const} \quad \text{for all } \tau \in \mathbb{R}$$

and $d(I + f(\tau, \cdot), U_\tau, y(\tau)) = 0$ if $U_{\tau_0} = \emptyset$ for some $\tau_0 \in \mathbb{R}$.
(4) As for the Brouwer degree, we can easily conclude by approximation and complete continuity that

$$(\text{II}.2.8) \qquad
\begin{aligned}
&\text{if } d(F, U, y) \neq 0, \text{ then there is some} \\
&x \in U \text{ such that } F(x) = y.
\end{aligned}$$

Finally, we state that

(5) the Leray–Schauder degree depends only on the values of F on the boundary ∂U, and

(6) the *excision* (II.1.16) holds as for the Brouwer degree.

As for the Brouwer degree, the local Leray–Schauder degree around an isolated solution $x_0 \in U$ of $F(x) = y$ is called the *index* of F at x_0 (with respect to y):

(II.2.9) $\quad \begin{aligned} &d(F, B_r(x_0), y) = i(F, x_0), \\ &\text{where } B_r(x_0) = \{x \in X| \ \|x - x_0\| < r\} \text{ for small } r > 0. \end{aligned}$

(We omit the dependence on y.) By the additivity and the excision property, the Leray–Schauder degree is the sum of its indices if there are only finitely many solutions of $F(x) = y$ in U.

If $F(x_0) = y$ for some $x_0 \in U$, F is Fréchet differentiable at x_0, and $DF(x_0) \in L(X, X)$ is an isomorphism (is regular), then the index $i(F, x_0)$ is computed as follows:

Clearly, $DF(x_0) = I + Df(x_0)$ for a compact perturbation of the identity (II.2.1). If f is completely continuous, then $Df(x_0) \in L(X, X)$ is a compact operator (see [125], [34]). The regularity of $DF(x_0) = I + Df(x_0)$ implies that $\|DF(x_0)z\| \geq m\|z\|$ for some $m > 0$, whence $\tau(F(x) - y) + (1 - \tau)DF(x_0)(x - x_0) \neq 0$ for $\tau \in [0, 1]$ and $\|x - x_0\| = r$ for sufficiently small $r > 0$. Thus the homotopy invariance implies that

(II.2.10) $\quad \begin{aligned} i(F, x_0) &= d(F, B_r(x_0), y) = d(F - y, B_r(x_0), 0) \\ &= d(DF(x_0), B_r(0), 0) = i(DF(x_0), 0), \end{aligned}$

where we make the substitution $x - x_0 = z$.

The computation of the index of $DF(x_0) = I + Df(x_0) = I + K$ for some compact $K \in L(X, X)$ was already accomplished by Leray and Schauder [125]. We define a homotopy $I + \tau K$ for $\tau \in [0, 1]$ from the identity I to $I + K$. If for some $\tau_k \in (0, 1)$ the operator $I + \tau_k K$ is not regular, then $\mu_k = -1/\tau_k \in (-\infty, -1)$ is an eigenvalue of K. By the Riesz–Schauder Theory, there are only finitely many such eigenvalues in $(-\infty, -1)$ all having a finite algebraic multiplicity. Starting from 1 for the identity I, the index changes at each $\tau_k = -1/\mu_k$ by $(-1)^{m_k}$, where m_k is the algebraic multiplicity of μ_k. This gives finally the Leray–Schauder formula

(II.2.11) $\quad \begin{aligned} &i(I + K, 0) = (-1)^{m_1 + \cdots + m_\ell}, \text{ where} \\ &\mu_1, \ldots, \mu_\ell \text{ are all eigenvalues of } K \text{ in } (-\infty, -1), \\ &\text{and } m_k \text{ is the algebraic multiplicity of } \mu_k, k = 1, \ldots, \ell. \end{aligned}$

If there is no eigenvalue of K in $(-\infty, -1)$, then $i(I + K, 0) = 1$.

We give the proof of formula (II.2.11):

Let $A(\tau) = I + \tau K$ for τ near τ_k such that $-1/\tau_k$ is an eigenvalue of K of algebraic multiplicity m_k. Denote by $E_k \subset X$ the generalized eigenspace of dimension m_k. Then $1 - (\tau/\tau_k)$ is an eigenvalue of $A(\tau)$ of algebraic multiplicity m_k with the same generalized eigenspace E_k for all τ near τ_k. Decompose $X = E_k \oplus X_k$ so that X_k is a complement that is invariant for $A(\tau)$ for all τ near τ_k (choose X_k invariant for K). Identify $E_k \oplus X_k$ with $E_k \times X_k$ and $x = v + w$ according to this decomposition with $x = (v, w)$. The operator $A(\tau) \in L(X, X)$ then decomposes as

$$
\begin{aligned}
&A(\tau) = (A(\tau)|_{E_k}, A(\tau)|_{X_k}) \equiv (A_{1,k}(\tau), A_{2,k}(\tau)), \\
&(A_{1,k}(\tau), A_{2,k}(\tau)) : E_k \times X_k \to E_k \times X_k, \\
&A_{1,k}(\tau) \in L(E_k, E_k), A_{2,k}(\tau) \in L(X_k, X_k),
\end{aligned}
$$

(II.2.12)

which is an isomorphism for all τ near τ_k.

Therefore, the index $i(A_{2,k}(\tau), 0)$ is defined, and by (II.2.10), it is given by

$$
(\text{II.2.13}) \qquad i(A_{2,k}(\tau), 0) = d((I + \tau K)|_{X_k}, U_{2,k}, 0),
$$

where $U_{2,k}$ is a neighborhood of 0 in X_k. Clearly, $i(A_{2,k}(\tau), 0) = \text{const}$ for τ near τ_k by homotopy invariance. The degree (II.2.13) is defined by a finite-dimensional approximation of K on $\overline{U}_{2,k}$,

$$
(\text{II.2.14}) \qquad K_\varepsilon : \overline{U}_{2,k} \to X_{k,n_\varepsilon} \subset X_k, \quad \dim X_{k,n_\varepsilon} < \infty,
$$

which is not linear, in general:

$$
(\text{II.2.15}) \quad d((I + \tau K)|_{X_k}, U_{2,k}, 0) = d(I + \tau K_\varepsilon, U_{2,k} \cap X_{k,n_\varepsilon}, 0).
$$

Set $I + \tau K_\varepsilon = A_{2,\varepsilon}(\tau)$ and $U_{2,k} \cap X_{k,n_\varepsilon} = U_{2,\varepsilon}$. Then, for a small neighborhood $U_{1,k}$ of 0 in E_k, we obtain from (II.1.18) (recall that $\dim E_k = m_k$)

$$
\begin{aligned}
i(I + \tau K, 0) &= i(A(\tau), 0) \\
&= d(A(\tau), U_{1,k} \times U_{2,k}, 0) \\
&= d((A_{1,k}(\tau), A_{2,k}(\tau)), U_{1,k} \times U_{2,k}, (0, 0)) \\
&= d((A_{1,k}(\tau), A_{2,\varepsilon}(\tau)), U_{1,k} \times U_{2,\varepsilon}, (0, 0)) \\
&= d(A_{1,k}(\tau), U_{1,k}, 0) d(A_{2,\varepsilon}(\tau), U_{2,\varepsilon}, 0) \\
&= i((A_{1,k}(\tau), 0) i(A_{2,k}(\tau), 0).
\end{aligned}
$$

(II.2.16)

By (II.1.4),

$$
(\text{II.2.17}) \qquad i(A_{1,k}(\tau), 0) = \text{sign} \det((I + \tau K)|_{E_k}) = \text{sign}\left(1 - \frac{\tau}{\tau_k}\right)^{m_k},
$$

which changes by $(-1)^{m_k}$ at $\tau = \tau_k$. Since $i(A_{2,k}(\tau), 0) = \text{const}$ for τ near τ_k, this proves (II.2.11).

The eigenvalues μ of $K = Df(x_0)$ in $(-\infty, -1)$ give eigenvalues $1 + \mu$ in $(-\infty, 0)$ of $I + f(x_0) = DF(x_0)$ and vice versa. Formula (II.2.11) motivates the definition of an index for a class of admissible Fredholm operators given in Section II.5.

II.3 Application of the Degree in Bifurcation Theory

As in Chapter I, we study nonlinear parameter-dependent problems

(II.3.1)
$$F(x, \lambda) = 0, \text{ where}$$
$$F : U \times V \to X \text{ with open sets}$$
$$U \subset X, \ V \subset \mathbb{R}.$$

We assume the "trivial solution" $\{(0, \lambda) | \lambda \in V\}$ and that U is a neighborhood of 0, possibly $U = X$ and w.l.o.g. $V = \mathbb{R}$. Furthermore, we need $F \in C(U \times V, X)$ and that $D_x F(0, \lambda)$ exists such that $D_x F(0, \cdot) \in C(V, L(X, X))$. If $X = \mathbb{R}^n$, then we use the Brouwer degree; if X is an infinite-dimensional (real) Banach space, then we assume that $F(x, \lambda) = x + f(x, \lambda)$ and that $f : U \times V \to X$ is completely continuous. Thus for $F(\cdot, \lambda)$ the Leray–Schauder degree is applicable.

A necessary condition for bifurcation from the trivial solution line is (I.1.6). In our case, the Riesz–Schauder Theory implies that

(II.3.2)   0 is an isolated eigenvalue of finite algebraic
          multiplicity m of $D_x F(0, \lambda_0)$ for some $\lambda_0 \in \mathbb{R}$.

(For a spectral theory, the real space X is complexified as in Section I.8.)

It is crucial for bifurcation at $(0, \lambda_0)$ how the eigenvalue 0 perturbs for $D_x F(0, \lambda)$ when λ varies in a neighborhood of λ_0. As shown in [86], Sections II.5.1 and III.6.4, the number m is an invariant in the following sense: The generalized eigenspace E_{λ_0} of the eigenvalue 0 of $D_x F(0, \lambda_0)$ having dimension m is perturbed to an invariant space E_λ of $D_x F(0, \lambda)$ of dimension m, too, and all perturbed eigenvalues near 0 (the so-called 0-group) are eigenvalues of the finite-dimensional operator $D_x F(0, \lambda)$ restricted to the m-dimensional invariant space E_λ. The eigenvalues in that 0-group depend continuously on λ.

We shall prove that a necessary and sufficient condition for bifurcation is that the sign of $\det(D_x F(0, \lambda)|_{E_\lambda})$ change at $\lambda = \lambda_0$. (The necessity is seen by the counterexample (II.7.18) below.) In terms of the eigenvalues in the 0-group, this condition is expressed by the notion of an odd crossing number:

Definition II.3.1 *Define $\sigma^<(\lambda) = 1$ if there are no negative real eigenvalues in the 0-group of $D_x F(0, \lambda)$ and $\sigma^<(\lambda) = (-1)^{m_1 + \cdots + m_k}$ if $\mu_1, \ldots, \mu_k$ are all negative real eigenvalues in the 0-group having algebraic multiplicities $m_1, \ldots, m_k$, respectively. If*

(II.3.3)   *$D_x F(0, \lambda)$ is regular for $\lambda \in (\lambda_0 - \delta, \lambda_0) \cup (\lambda_0, \lambda_0 + \delta)$*
          and if $\sigma^<(\lambda)$ changes at $\lambda = \lambda_0$,

then $D_x F(0, \lambda)$ has an odd crossing number at $\lambda = \lambda_0$.

By $(\text{II.3.3})_1$, $\sigma^<(\lambda)$ is constant on $(\lambda_0 - \delta, \lambda_0)$ and on $(\lambda_0, \lambda_0 + \delta)$. It can change only if an odd number of negative real eigenvalues leave the half-axis $(-\infty, 0)$ through 0. Since nonreal eigenvalues exist only in complex conjugate pairs of equal multiplicity, an odd crossing number of a family $D_x F(0, \lambda)$ in the sense of Definition II.3.1 means that an odd number of eigenvalues in the 0-group (counting multiplicities) leave the left complex half-plane when λ passes through λ_0. (It does not mean that the total number of eigenvalues in the 0-group of $D_x F(0, \lambda)$ in the negative complex half-plane is constant on $(\lambda_0 - \delta, \lambda_0)$ and changes only at $\lambda = \lambda_0$. If this is the case, we say that the local Morse index of $D_x F(0, \lambda)$ changes at $\lambda = \lambda_0$; cf. Definition II.7.1 below.) If $X = \mathbb{R}^n$, then $D_x F(0, \lambda)$ has an odd crossing number at $\lambda = \lambda_0$ if and only if $\det D_x F(0, \lambda) \neq 0$ for $\lambda \in (\lambda_0 - \delta, \lambda_0) \cup (\lambda_0, \lambda_0 + \delta)$ and $\det D_x F(0, \lambda)$ changes sign at $\lambda = \lambda_0$.

By $(\text{II.3.3})_1$, the index $i(D_x F(0, \lambda), 0)$ exists locally for $\lambda \neq \lambda_0$, and formula (II.2.11) (or (II.1.4)) shows that

(II.3.4)
$$
\begin{aligned}
&\text{an odd crossing number of } D_x F(0, \lambda) \\
&= I + D_x f(0, \lambda) \equiv I + K(\lambda) \text{ at } \lambda = \lambda_0 \\
&\text{means that the index } i(D_x F(0, \lambda), 0) \text{ jumps} \\
&\text{at } \lambda = \lambda_0 \text{ from } +1 \text{ to } -1 \text{ or vice versa.}
\end{aligned}
$$

This change of the index is the key for the following **Krasnosel'skii Bifurcation Theorem**:

Theorem II.3.2 *If $D_x F(0, \lambda)$ has an odd crossing number at $\lambda = \lambda_0$, then $(0, \lambda_0)$ is a bifurcation point for $F(x, \lambda) = 0$ in the following sense: $(0, \lambda_0)$ is a cluster point of nontrivial solutions $(x, \lambda) \in X \times \mathbb{R}$, $x \neq 0$, of $F(x, \lambda) = 0$.*

Proof. Assume that there is a neighborhood of $(0, \lambda_0)$ in $X \times \mathbb{R}$ containing only the trivial solutions $(0, \lambda)$. Then there exist an interval $[\lambda_0 - \delta, \lambda_0 + \delta]$ and a ball $B_r(0)$ such that there is no solution of $F(x, \lambda) = 0$ on $\partial B_r(0)$ for all $\lambda \in [\lambda_0 - \delta, \lambda_0 + \delta]$. By the homotopy invariance of the Leray–Schauder degree,

(II.3.5) $d(F(\cdot, \lambda), B_r(0), 0) = \text{const for } \lambda \in [\lambda_0 - \delta, \lambda_0 + \delta]$,

contradicting the assumption that

(II.3.6) $i(D_x F(0, \lambda), 0) = d(F(\cdot, \lambda), B_r(0), 0)$ (cf. (II.2.10))

changes at $\lambda = \lambda_0$ (see (II.3.4)). $\square$

The classical special case of Theorem II.3.2 is the following: $F(x, \lambda) = x + \lambda f(x)$, $f : U \to X$, is completely continuous for some neighborhood U of 0, and 0 is an eigenvalue of $I + \lambda_0 D f(0) = D_x F(0, \lambda_0)$ of odd algebraic multiplicity for some $\lambda_0 \neq 0$. (Equivalently, we find in the literature that $-1/\lambda_0$ is an eigenvalue of $Df(0)$ or that $-\lambda_0$ is a characteristic value of $Df(0)$.) By the linear dependence on the parameter λ, the derivative

$D_x F(0, \lambda) = I + \lambda Df(0) = I + \lambda K$ has an odd crossing number at $\lambda = \lambda_0$ (see the Leray–Schauder formula (II.2.11)), which implies bifurcation.

A drawback of Theorem II.3.2 is that it gives no information on the structure of the set of nontrivial solutions near $(0, \lambda_0)$: Is there really a "branch" of nontrivial solutions emanating at $(0, \lambda_0)$? The notion "branch" implies that the set is connected in $X \times \mathbb{R}$. This property and more is proved in the **Global Rabinowitz Bifurcation Theorem:**

Theorem II.3.3 *Assume that* $F \in C(X \times \mathbb{R}, X)$, $F(x, \lambda) = x + f(x, \lambda)$, *where* $f : X \times \mathbb{R} \to X$ *is completely continuous, and* $D_x F(0, \cdot) = I + D_x f(0, \cdot) \in C(\mathbb{R}, L(X, X))$. *Let* $\mathcal{S}$ *denote the closure of the set of nontrivial solutions of* $F(x, \lambda) = 0$ *in* $X \times \mathbb{R}$. *Assume that* $D_x F(0, \lambda)$ *has an odd crossing number at* $\lambda = \lambda_0$. *Then* $(0, \lambda_0) \in \mathcal{S}$, *and let* $\mathcal{C}$ *be the (connected) component of* $\mathcal{S}$ *to which* $(0, \lambda_0)$ *belongs. Then*
(i) $\mathcal{C}$ *is unbounded, or*
(ii) $\mathcal{C}$ *contains some* $(0, \lambda_1)$, *where* $\lambda_0 \neq \lambda_1$.

Proof. Assume that the Rabinowitz alternative (i) or (ii) does not hold. Then $\mathcal{C}$ is bounded, and $\mathcal{C}$ contains the only trivial solution $(0, \lambda_0)$. By the complete continuity of $f : X \times \mathbb{R} \to X$ (we tacitly assume in Theorem II.3.3 that F is everywhere defined on $X \times \mathbb{R}$), the set $\mathcal{C}$ is compact: If $(x, \lambda) \in \mathcal{C}$, then $x = -f(x, \lambda)$, so that the projection of $\mathcal{C}$ on X is compact. Its projection on $\mathbb{R}$ is compact, too, since it is bounded.

Let $\mathcal{N}$ be an open neighborhood of $\mathcal{C}$ in $X \times \mathbb{R}$ and let $K = \overline{\mathcal{N}} \cap \mathcal{S}$. Then K is compact and $\partial \mathcal{N} \cap \mathcal{C} = \emptyset$. By a lemma from point-set topology (the so-called Whyburn Lemma [169]) there exist disjoint compact subsets $A, B \subset K$ such that $\mathcal{C} \subset A$, $\partial \mathcal{N} \cap \mathcal{S} \subset B$, and $K = A \cup B$. Let $\tilde{U}$ be an open neighborhood of A such that $\tilde{U} \cap B = \emptyset$. Then $\tilde{U}$ has the properties that $\mathcal{C} \subset \tilde{U}$ and $\partial \tilde{U} \cap \mathcal{S} = \emptyset$. By possibly removing some trivial solutions from $\tilde{U}$, we finally obtain

(II.3.7)
$$a \text{ bounded open set } U \subset X \times \mathbb{R} \text{ such that}$$
$$\mathcal{C} \subset U, \ \partial U \cap \mathcal{S} = \emptyset, \text{ and}$$
$$U \cap \{(0, \lambda) | \lambda \in \mathbb{R}\} = \{0\} \times (\lambda_0 - \tilde{\delta}, \lambda_0 + \tilde{\delta})$$
$$\text{for some arbitrarily small } \tilde{\delta} > 0.$$

By assumption (II.3.3)$_1$, the trivial solution $0 \in X$ of $F(x, \lambda) = 0$ is isolated for all $\lambda \in (\lambda_0 - \delta, \lambda_0) \cup (\lambda_0, \lambda_0 + \delta)$. Thus there exist radii $r(\lambda) > 0$ such that

(II.3.8)
$$\partial B_{r(\lambda)}(0) \cap \mathcal{C} = \emptyset \text{ and}$$
$$i(D_x F(0, \lambda), 0) = d(F(\cdot, \lambda), B_{r(\lambda)}(0), 0)$$
$$\text{for all } \lambda \in (\lambda_0 - \delta, \lambda_0) \cup (\lambda_0, \lambda_0 + \delta).$$

Let $V = \{(x, \lambda) | x \in B_{r(\lambda)}(0), \lambda \in (\lambda_0 - \delta, \lambda_0) \cup (\lambda_0, \lambda_0 + \delta)\}$. If r depends continuously on λ, then $V \subset X \times \mathbb{R}$ is open. We assume that $r(\lambda_0) = 0$ extends $r(\lambda)$ continuously on $(\lambda_0 - \delta, \lambda_0 + \delta)$. Define $W = U \backslash \overline{V}$. If $0 < \tilde{\delta} < \delta$, then the only solution of $F(x, \lambda) = 0$ on ∂W is $(0, \lambda_0)$, and the homotopy invariance (II.2.7) of the Leray–Schauder degree implies that

(II.3.9)
$$\begin{aligned}
&d(F(\cdot, \lambda), W_\lambda, 0) = \text{ const on } (-\infty, \lambda_0) \text{ and} \\
&\text{also on } (\lambda_0, \infty) \text{ and} \\
&d(F(\cdot, \lambda), W_\lambda, 0) = 0 \text{ for all } \lambda \neq \lambda_0, \text{ since} \\
&W_\lambda = \emptyset \text{ if } |\lambda| \text{ is large.}
\end{aligned}$$

By construction of the open set U, the only solutions of $F(x, \lambda) = 0$ on ∂U are $(0, \lambda_0 - \tilde{\delta})$ and $(0, \lambda_0 + \tilde{\delta})$, and therefore by homotopy invariance,

(II.3.10) $d(F(\cdot, \lambda), U_\lambda, 0) = \text{ const for } \lambda \in (\lambda_0 - \tilde{\delta}, \lambda_0 + \tilde{\delta}).$

The additivity of the Leray–Schauder degree finally implies that

(II.3.11)
$$\begin{aligned}
&d(F(\cdot, \lambda), U_\lambda, 0) \\
&= d(F(\cdot, \lambda), V_\lambda, 0) + d(F(\cdot, \lambda), W_\lambda, 0) \\
&= d(F(\cdot, \lambda), B_{r(\lambda)}(0), 0) = i(D_x F(0, \lambda), 0) \\
&\text{for all } \lambda \in (\lambda_0 - \tilde{\delta}, \lambda_0) \cup (\lambda_0, \lambda_0 + \tilde{\delta}).
\end{aligned}$$

By (II.3.10), the index of the trivial solution is constant for $\lambda \in (\lambda_0 - \tilde{\delta}, \lambda_0) \cup (\lambda_0, \lambda_0 + \tilde{\delta})$, contradicting an odd crossing number of $D_x F(0, \lambda)$ at $\lambda = \lambda_0$; cf. (II.3.4).   □

Both alternatives of Theorem II.3.3 are possible. The simplest example of (i) is the linear case $F(x, \lambda) = D_x F(\lambda, 0)x$. Examples of alternative (ii) are given in Section III.6. A refinement of Theorem II.3.3 is given in Theorem II.5.9 below.

The main assumption of Theorems II.3.2 and II.3.3, namely, the odd crossing number of $D_x F(0, \lambda)$ at some $\lambda = \lambda_0$, is difficult to verify, in general, when $f(x, \lambda)$ does not depend linearly on λ. A possibility for a verification is expounded in the next section.

II.4 Odd Crossing Numbers for Fredholm Operators and Local Bifurcation

Let X and Z be real Banach spaces and assume that $X \subset Z$ is continuously embedded. The notion of an odd crossing number defined in Definition II.3.1 is readily extended to families of linear operators $A(\lambda) \in L(X, Z)$ depending continuously on $\lambda \in \mathbb{R}$ such that

(II.4.1) $A(\lambda) : Z \to Z$ with domain of definition $D(A(\lambda)) = X$
                  is closed for each $\lambda \in (\lambda_0 - \delta, \lambda_0 + \delta)$.

An equivalent characterization of (II.4.1) is that the norm in X is equivalent to the graph norm $\|x\|_Z + \|A(\lambda)x\|_Z$. This assumption is reasonable for applications; cf. Section III.1.

Another suitable assumption is that

(II.4.2) $A(\lambda) \in L(X, Z)$ is a Fredholm operator of index zero
for each $\lambda \in (\lambda_0 - \delta, \lambda_0 + \delta)$; cf. Definition I.2.1.

The case considered in Section II.3 in which $A(\lambda) = I + D_x f(0, \lambda) \in L(X, X)$ with compact $D_x f(0, \lambda)$ is clearly a special case: By the Riesz–Schauder Theory, every linear compact perturbation of the identity is a Fredholm operator of index zero.

In this section we assume that

(II.4.3)   0 is an isolated eigenvalue of $A(\lambda_0)$,

which, by the Fredholm property, is an eigenvalue of finite geometric multiplicity: $\dim N(A(\lambda_0)) < \infty$. (As usual, for a spectral theory the real spaces X and Z are complexified as in Section I.8.) The isolatedness of the eigenvalue 0 gives a stronger result (cf. [86], Section IV.5.4): The generalized eigenspace E_{λ_0} of the eigenvalue 0 is finite-dimensional, so that the algebraic multiplicity is finite, too. Furthermore, there is a decomposition into invariant closed subspaces

(II.4.4) $$Z = E_{\lambda_0} \oplus Z_{\lambda_0}, \quad E_{\lambda_0} \subset X \subset Z,$$
$$X = E_{\lambda_0} \oplus X_{\lambda_0}, \quad X_{\lambda_0} = Z_{\lambda_0} \cap X,$$

such that the spectrum of $A(\lambda_0) \in L(E_{\lambda_0}, E_{\lambda_0})$ consists only of the eigenvalue 0, and 0 is in the resolvent set of $A(\lambda_0) \in L(X_{\lambda_0}, Z_{\lambda_0})$; i.e., $A(\lambda_0)$ is an isomorphism from X_{λ_0} onto Z_{λ_0}. The eigenprojection $P(\lambda_0)$ onto E_{λ_0} is in $L(X, X)$ as well as in $L(Z, Z)$. As already mentioned in Section II.3, the generalized eigenspace E_{λ_0} perturbs to E_λ of the same finite dimension, and the eigenprojection $P(\lambda)$ depends continuously on λ in $L(X, X)$ and $L(Z, Z)$ for λ near λ_0. The eigenvalue 0 of $A(\lambda_0)$ perturbs to eigenvalues of $A(\lambda)$ near 0 (the so-called 0-group), which are the eigenvalues of $A(\lambda) \in L(E_\lambda, E_\lambda)$ (which is a finite-dimensional operator). Thus Definition II.3.1 is literally generalized to a family of Fredholm operators satisfying (II.4.1)–(II.4.3):

Definition II.4.1 *Define $\sigma^<(\lambda) = 1$ if there are no negative real eigenvalues in the 0-group of $A(\lambda)$ and $\sigma^<(\lambda) = (-1)^{m_1 + \cdots + m_k}$ if $\mu_1, \ldots, \mu_k$ are all negative real eigenvalues in the 0-group having algebraic multiplicities $m_1, \ldots, m_k$, respectively. If*

(II.4.5) $A(\lambda) \in L(X, Z)$ *is regular (is an isomorphism)*
for $\lambda \in (\lambda_0 - \delta, \lambda_0) \cup (\lambda_0, \lambda_0 + \delta)$ and if
$\sigma^<(\lambda)$ *changes at $\lambda = \lambda_0$,*

then the family $A(\lambda)$ has an odd crossing number at $\lambda = \lambda_0$.

The remarks given after Definition II.3.1 are literally valid also in this generalized case: An odd crossing number means that an odd number of real eigenvalues (counting multiplicities) in the 0-group of $A(\lambda)$ leave the left complex half-plane when λ passes through λ_0.

In this section we show that an odd crossing number of $A(\lambda)$ is detected by the method of Lyapunov–Schmidt and that it implies local bifurcation for Fredholm operators.

We decompose (cf. (I.2.2))

$$
\text{(II.4.6)} \qquad
\begin{aligned}
X &= N \oplus X_0, & N &= N(A(\lambda_0)) \subset E_{\lambda_0}, \\
Z &= R \oplus Z_0, & R &= R(A(\lambda_0)),
\end{aligned}
$$

and $\dim N = \dim Z_0 < \infty$. Since $P(\lambda_0)|_N = I_N$ and $P(\lambda) \in L(X,X)$ is continuous, the projection $P(\lambda)$ is injective on N for all λ near λ_0. We show that $P(\lambda_0) \in L(Z,Z)$ is also injective on Z_0. Decompose $z = r + z_0$ according to (II.4.6)$_2$, where $r = A(\lambda_0)x$. Then $P(\lambda_0)z = A(\lambda_0)P(\lambda_0)x + P(\lambda_0)z_0 = z - (I - P(\lambda_0))z$, whence $z = (I - P(\lambda_0))z + A(\lambda_0)P(\lambda_0)x + P(\lambda_0)z_0$. Since $(I - P(\lambda_0))z \in R$, the assumption $P(\lambda_0)z_0 = 0$ leads to $z \in R$, and by uniqueness of the decomposition, this implies $z_0 = 0$. By the continuous dependence of $P(\lambda) \in L(Z,Z)$ on λ, the injectivity is preserved for $P(\lambda)|_{Z_0}$. Furthermore, the assumption $P(\lambda_0)z_0 \in R$ leads to $z \in R$ as well, and therefore, $z_0 = 0$. Thus $R \cap P(\lambda_0)Z_0 = \{0\}$, and the decompositions (II.4.6) yield the decompositions $X = P(\lambda_0)N \oplus X_0$, $Z = R \oplus P(\lambda_0)Z_0$, which are perturbed to

$$
\text{(II.4.7)} \qquad
\begin{aligned}
X &= P(\tau)N \oplus X_0, \\
Z &= R \oplus P(\tau)Z_0 \quad \text{for } \tau \in (\lambda_0 - \delta, \lambda_0 + \delta).
\end{aligned}
$$

By $P(\lambda_0)N = N$, the space $P(\tau)N$ is a continuous perturbation of N. Next, we construct a continuous transition from the complement Z_0 to $P(\lambda_0)Z_0$ (which is continuously extended to $P(\tau)Z_0$). We define the homotopy $H(t)z = (1-t)z + tP(\lambda_0)z$, which connects I_Z and $P(\lambda_0)$ smoothly in $L(Z,Z)$ for $t \in [0,1]$. If $z = r + z_0 = A(\lambda_0)x + z_0$ according to $Z = R \oplus Z_0$, then we showed above that $z = (I - P(\lambda_0))z + A(\lambda_0)P(\lambda_0)x + P(\lambda_0)z_0$. Since $(I - P(\lambda_0))z \in R$, this gives the decomposition $z = \tilde{r} + \tilde{z}_0$ according to $Z = R \oplus P(\lambda_0)Z_0$ with $\tilde{z}_0 = P(\lambda_0)z_0$. Therefore, $z = (1-t)r + t\tilde{r} + (1-t)z_0 + t\tilde{z}_0 = \hat{r} + H(t)z_0$ with $\hat{r} \in R$ and $z_0 \in Z_0$. If $H(t)z_0 = 0$, then $z \in R$ and $z_0 = 0$. This proves that $H(t)Z_0$ is a complement of R in Z for all $t \in [0,1]$ and $H(0)Z_0 = Z_0$, $H(1) = P(\lambda_0)Z_0$.

To summarize, the complement $P(\tau)Z_0$ in (II.4.7)$_2$ is a continuous perturbation of Z_0 (by inserting the homotopy H between Z_0 and $P(\lambda_0)Z_0$) in the class of complements of R in Z. The decompositions (II.4.7) define projections

$$
\text{(II.4.8)} \qquad
\begin{aligned}
P_\tau &: X \to P(\tau)N \quad \text{along } X_0, \\
Q_\tau &: Z \to P(\tau)Z_0 \quad \text{along } R,
\end{aligned}
$$

so that $P_{\lambda_0} = P : X \to N$ along X_0 is the Lyapunov–Schmidt projection according to (II.4.6)$_1$, and $Q_{\lambda_0} : Z \to P(\lambda_0)Z_0$ along R is homotopic to $Q : Z \to Z_0$ along R, which is the Lyapunov–Schmidt projection according to (II.4.6)$_2$. This homotopy is given by $Q_t : Z \to H(t)Z_0$ along R for $t \in [0,1]$.

Next, we decompose the invariant generalized eigenspace E_τ of $A(\tau)$. Since $P(\tau)N \subset E_\tau$ and $P(\tau)Z_0 \subset E_\tau$, the decompositions (II.4.7) yield

(II.4.9)
$$E_\tau = P(\tau)N \oplus (X_0 \cap E_\tau),$$
$$E_\tau = (R \cap E_\tau) \oplus P(\tau)Z_0 \text{ for } \tau \in (\lambda_0 - \delta, \lambda_0 + \delta).$$

Therefore, $P_{\tau|E_\tau}$ and $Q_{\tau|E_\tau}$ project according to the decompositions (II.4.9). Finally, $P_\tau \in L(X, X)$ and $Q_\tau \in L(Z, Z)$ depend continuously on τ near λ_0.

We represent $A(\lambda) : X \to Z$ according to any decomposition (II.4.7)$_1$ and any decomposition $Z = R \oplus P(\tau)Z_0$ or $Z = R \oplus H(\tau)Z_0$ as a matrix

(II.4.10)
$$A(\lambda) = \begin{pmatrix} (I - \hat{Q})A(\lambda) & (I - \hat{Q})A(\lambda) \\ \hat{Q}A(\lambda) & \hat{Q}A(\lambda) \end{pmatrix}$$

identifying $X = P(\tau)N \oplus X_0$ with $P(\tau)N \times X_0$ (whose elements are written as columns) and $Z = R \oplus P(\tau)Z_0$ or $Z = R \oplus H(t)Z_0$ with the corresponding Cartesian product (whose elements are written as columns as well). The projection $\hat{Q}$ denotes Q_τ or Q_t depending on the choice of the complement $\hat{Q}Z$ of R.

For all λ near λ_0,

(II.4.11) $(I - \hat{Q})A(\lambda)|_{X_0} \in L(X_0, R)$ is an isomorphism,

since this is true for $\lambda = \lambda_0$. We define (with $[(I - \hat{Q})A(\lambda)]^{-1} \in L(R, X_0)$)

$$C(\hat{Q}, \lambda) \equiv \begin{pmatrix} -\hat{Q}A(\lambda)[(I - \hat{Q})A(\lambda)]^{-1} & I_{\hat{Q}Z} \\ [(I - \hat{Q})A(\lambda)]^{-1} & 0 \end{pmatrix},$$
$$C(\hat{Q}, \lambda) : Z = R \times \hat{Q}Z \to \hat{Q}Z \times X_0,$$

(II.4.12) $D(\tau, \hat{Q}, \lambda)$
$$\equiv \begin{pmatrix} \hat{Q}A(\lambda)\{I_{P(\tau)N} - [(I - \hat{Q})A(\lambda)]^{-1}(I - \hat{Q})A(\lambda)\} & 0 \\ [(I - \hat{Q})A(\lambda)]^{-1}(I - \hat{Q})A(\lambda) & I_{X_0} \end{pmatrix},$$

$$D(\tau, \hat{Q}, \lambda) : X = P(\tau)N \times X_0 \to \hat{Q}Z \times X_0.$$

Then an easy computation yields

(II.4.13)
$$C(\hat{Q}, \lambda)A(\lambda) = D(\tau, \hat{Q}, \lambda)$$
for all λ near λ_0, τ near λ_0, and for all $\hat{Q}$.

Since

$$(\text{II.4.14}) \quad C(\hat{Q}, \lambda_0) = \begin{pmatrix} 0 & I_{\hat{Q}Z} \\ A(\lambda_0)^{-1} & 0 \end{pmatrix} \quad \text{with } A(\lambda_0)^{-1} \in L(R, X_0)$$

is an isomorphism, the same holds for all λ near λ_0. This proves by (II.4.13) and by the definition (II.4.12) of $D(\tau, \hat{Q}, \lambda)$ for any λ near λ_0 that

$$(\text{II.4.15}) \quad \begin{aligned} &A(\lambda) \in L(X, Z) \text{ is an isomorphism} \Leftrightarrow \\ &\hat{Q}A(\lambda)\{I_{P(\tau)N} - [(I - \hat{Q})A(\lambda)]^{-1}(I - \hat{Q})A(\lambda)\} \equiv B(\tau, \hat{Q}, \lambda) \\ &\in L(P(\tau)N, \hat{Q}Z) \text{ is regular for all } \tau \text{ near } \lambda_0 \\ &\text{and for all } \hat{Q}. \end{aligned}$$

Next we consider $A(\lambda) \in L(E_\lambda, E_\lambda)$.

By the definitions of P_λ and Q_λ in (II.4.8) and by the fact that $P_\lambda|_{E_\lambda}$ and $Q_\lambda|_{E_\lambda}$ project according to the decompositions (II.4.9) (for $\tau = \lambda$), we see that

$$(\text{II.4.16}) \quad \begin{aligned} &C(Q_\lambda, \lambda), \ D(\lambda, Q_\lambda, \lambda) \in L(E_\lambda, E_\lambda) \text{ and} \\ &C(Q_\lambda, \lambda)A(\lambda) = D(\lambda, Q_\lambda, \lambda) \text{ in } L(E_\lambda, E_\lambda). \end{aligned}$$

Definition II.4.1 of an odd crossing number is equivalent to the following:

$$(\text{II.4.17}) \quad \begin{aligned} &\det(A(\lambda)|_{E_\lambda}) \neq 0 \text{ for } \lambda \in (\lambda_0 - \delta, \lambda_0) \cup (\lambda_0, \lambda_0 + \delta) \\ &\text{and } \det(A(\lambda)|_{E_\lambda}) \text{ changes sign at } \lambda = \lambda_0. \end{aligned}$$

Since $C(Q_\lambda, \lambda)|_{E_\lambda}$ is regular for all $\lambda \in (\lambda_0 - \delta, \lambda_0 + \delta)$, the equality $(\text{II.4.16})_2$ implies that $A(\lambda)$ has an odd crossing number at $\lambda = \lambda_0$ if and only if

$$(\text{II.4.18}) \quad \begin{aligned} &\det(D(\lambda, Q_\lambda, \lambda)|_{E_\lambda}) \neq 0 \text{ for } \lambda \in (\lambda_0 - \delta, \lambda_0) \cup (\lambda_0, \lambda_0 + \delta) \\ &\text{and } \det(D(\lambda, Q_\lambda, \lambda)|_{E_\lambda}) \text{ changes sign at } \lambda = \lambda_0. \end{aligned}$$

According to Definition II.4.1 (or $(\text{II.4.17})_1$) and (II.4.15),

$$(\text{II.4.19}) \quad \begin{aligned} &B(\tau, \hat{Q}, \lambda) \in L(P(\tau)N, \hat{Q}Z) \text{ is regular for every} \\ &\lambda \in (\lambda_0 - \delta, \lambda_0) \cup (\lambda_0, \lambda_0 + \delta), \text{ for all } \tau \in (\lambda_0 - \delta, \lambda_0 + \delta), \\ &\text{and for all } \hat{Q} = Q_\tau \text{ with } \tau \in (\lambda_0 - \delta, \lambda_0 + \delta) \\ &\text{or for all } \hat{Q} = Q_t \text{ with } t \in [0, 1]. \end{aligned}$$

Fix $\lambda \in (\lambda_0 - \delta, \lambda_0) \cup (\lambda_0, \lambda_0 + \delta)$. Then $B(\tau, \hat{Q}, \lambda) \in L(P(\tau)N, \hat{Q}Z)$ is homotopic in the class of regular linear mappings to

$$(\text{II.4.20}) \quad \begin{aligned} &B(\lambda_0, Q, \lambda) = QA(\lambda)\{I_N - [(I - Q)A(\lambda)]^{-1}(I - Q)A(\lambda)\} \\ &\in L(N, Z_0), \text{ where } Q : Z \to Z_0 \text{ along } R; \text{ cf. (II.4.6)}. \end{aligned}$$

(Note that $[(I - Q)A(\lambda)]^{-1} \in L(R, X_0)$.)

In particular, $B(\lambda, Q_\lambda, \lambda) \in L(P(\lambda)N, P(\lambda)Z_0)$ is homotopic to $B(\lambda_0, Q, \lambda)$ $\in L(N, Z_0)$, and by definition (II.4.12) of $D(\lambda, Q_\lambda, \lambda)$, a change of sign of $\det(D(\lambda, Q_\lambda, \lambda)|_{E_\lambda})$ implies a change of sign of $\det B(\lambda, Q_\lambda, \lambda)$ at $\lambda = \lambda_0$. Here we endow $P(\lambda)N$ and $P(\lambda)Z_0$ with bases such that $B(\lambda, Q_\lambda, \lambda)$ is represented by a quadratic matrix and $\det B(\lambda, Q_\lambda, \lambda)$ is the determinant of that matrix. By the homotopy we finally obtain the following theorem:

Theorem II.4.2 *Let $A(\lambda) \in L(X, Z)$ be a family of Fredholm operators satisfying (II.4.1), (II.4.2), and let 0 be an isolated eigenvalue of $A(\lambda_0)$. Choose a Lyapunov–Schmidt decomposition (II.4.6) and fix bases in N and Z_0 such that we can consider the family $B(\lambda_0, Q, \lambda)$ in $L(\mathbb{R}^n, \mathbb{R}^n)$ (see (II.4.20)) for $n = \dim N = \dim Z_0$. Then*

(II.4.21)
$$A(\lambda) \in L(X, Z) \text{ has an odd crossing number}$$
$$\text{at } \lambda = \lambda_0 \Leftrightarrow$$
$$B(\lambda_0, Q, \lambda) \in L(\mathbb{R}^n, \mathbb{R}^n) \text{ has an odd crossing number}$$
$$\text{at } \lambda = \lambda_0 \Leftrightarrow$$
$$\det B(\lambda_0, Q, \lambda) \neq 0 \text{ for } \lambda \in (\lambda_0 - \delta, \lambda_0) \cup \lambda_0, \lambda_0 + \delta)$$
$$\text{and } \det B(\lambda_0, Q, \lambda) \text{ changes sign at } \lambda = \lambda_0.$$

II.4.1 Local Bifurcation via Odd Crossing Numbers

The structure of the family (II.4.20) is perfect for a Lyapunov–Schmidt reduction of a local bifurcation problem. We recall the essential steps.

Let $F : U \times V \to Z$ be a mapping such that $0 \in U \subset X$ and $\lambda_0 \in V \subset \mathbb{R}$ are open. We assume that $F(0, \lambda) = 0$ for all $\lambda \in V$ and that F is a nonlinear Fredholm operator with respect to $x \in U$ and for all $\lambda \in V : F \in C(U \times V, Z)$ and $D_x F \in C(U \times V, L(X, Z))$ such that $A(\lambda) \equiv D_x F(0, \lambda)$ is a family of Fredholm operators satisfying (II.4.1) and (II.4.2). Let 0 be an isolated eigenvalue of $A(\lambda_0)$. In order to find nontrivial bifurcating solutions of $F(x, \lambda) = 0$ near $(0, \lambda_0)$, a Lyapunov–Schmidt reduction as in Section I.2 is adequate. To this end we decompose the Banach spaces X and Z as in (II.4.6), and according to Theorem I.2.3, the problem $F(x, \lambda) = 0$ for (x, λ) locally near $(0, \lambda_0)$ is equivalent to an n-dimensional problem

(II.4.22) $\Phi(v, \lambda) = 0 \quad \text{near } (0, \lambda_0) \in N \times \mathbb{R}.$

Here $\Phi : \tilde{U} \times \tilde{V} \to Z_0$, where $0 \in \tilde{U} \subset N$ and $\lambda_0 \in \tilde{V} \subset \mathbb{R}$, and $\Phi \in C(\tilde{U} \times \tilde{V}, Z_0)$, $D_v \Phi \in C(\tilde{U} \times \tilde{V}, L(N, Z_0))$. To be more precise, $\Phi(v, \lambda) = QF(v + \psi(v, \lambda), \lambda)$, where $Q : Z \to Z_0$ along R is the projection, $v \in \tilde{U} \subset N$, and $\psi : \tilde{U} \times \tilde{V} \to X_0$ solves $(I - Q)F(v + \psi(v, \lambda), \lambda) = 0$ for all $(\lambda, v) \in \tilde{U} \times \tilde{V}$; cf. Section I.2. Differentiating this equation with respect to v yields

(II.4.23) $(I - Q)D_x F(v + \psi(v, \lambda), \lambda)(I_N + D_v \psi(v, \lambda)) \equiv 0.$

Inserting the trivial solution $(v, \lambda) = (0, \lambda)$ into (II.4.23) gives, in view of $\psi(0, \lambda) = 0$,

$$D_v\psi(0,\lambda) = -[(I-Q)D_xF(0,\lambda)]^{-1}(I-Q)D_xF(0,\lambda)$$

(II.4.24) $\quad \in L(N, X_0)$,

where $(I-Q)D_xF(0,\lambda)|_{X_0} \in L(X_0, R)$ is an isomorphism for λ near λ_0 and $(I-Q)D_xF(0,\lambda)|_N \in L(N,R)$.

This gives for $\Phi(v,\lambda) = QF(v+\psi(v,\lambda),\lambda)$ the expression (recall $D_xF(0,\lambda) \equiv A(\lambda)$)

(II.4.25) $\quad D_v\Phi(0,\lambda) = QA(\lambda)\{I_N - [(I-Q)A(\lambda)]^{-1}(I-Q)A(\lambda)\}.$

(Note that $[(I-Q)A(\lambda)]^{-1} \in L(R, X_0)$ and $X_0 = (I-P)X$.)

Thus $D_v\Phi(0,\lambda) = B(\lambda_0, Q, \lambda)$ (cf. (II.4.20)) and we can state the following theorem:

Theorem II.4.3 *Let $\Phi(v,\lambda) = 0$ be an n-dimensional Bifurcation Equation for the problem $F(x,\lambda) = 0$ obtained by the method of Lyapunov–Schmidt as expounded in Section I.2. Assume the trivial solution $F(0,\lambda) = 0$ and therefore $\Phi(0,\lambda) = 0$ for all λ near λ_0. Then*

(II.4.26)
$\quad D_xF(0,\lambda) \in L(X,Z)$ *has an odd crossing number at $\lambda = \lambda_0 \quad \Leftrightarrow$*
$\quad D_v\Phi(0,\lambda) \in L(N, Z_0) \cong L(\mathbb{R}^n, \mathbb{R}^n)$ *has an odd crossing number at $\lambda = \lambda_0 \Leftrightarrow$*
$\quad \det D_v\Phi(0,\lambda)$ *changes its (nonzero) sign at $\lambda = \lambda_0$.*

Note that the decompositions (II.4.6), and therefore the function Φ, are not unique.

An application of a finite-dimensional version of Theorem II.3.3 for $\Phi(v,\lambda) = 0$ finally yields a **Bifurcation Theorem for Fredholm Operators**:

Theorem II.4.4 *Assume for $F(x,\lambda) = 0$ that the derivative $D_xF(0,\lambda) = A(\lambda)$ along the trivial solution satisfies (II.4.1)–(II.4.3). If $D_xF(0,\lambda)$ has an odd crossing number at $\lambda = \lambda_0$, then $(0,\lambda_0)$ is a bifurcation point for $F(x,\lambda) = 0$ in the following sense: The closure of the set of nontrivial solutions near $(0,\lambda_0)$ contains a connected component to which $(0,\lambda_0)$ belongs.*

Indeed, extend Φ to $N \times \mathbb{R} \to Z_0$, apply Theorems II.4.3 and II.3.3 to $\Phi(v,\lambda) = 0$, and recall that $\Phi(v,\lambda) = 0$ gives all solutions of $F(x,\lambda) = 0$ locally near $(0,\lambda_0)$. By its construction, connected solution sets of $\Phi(v,\lambda) = 0$ yield connected solution sets of $F(x,\lambda) = 0$; cf. Section I.2.

In the next section we improve Theorem II.4.4 by a global version in the sense of Rabinowitz's Theorem, Theorem II.3.3.

We give some special cases of Theorem II.4.4.

Case 1

If $\dim N = 1$, i.e., if 0 is a geometrically simple eigenvalue of $A(\lambda_0) = D_x F(0, \lambda_0)$, then $\Phi(v, \lambda) = 0$ is a one-dimensional Bifurcation Equation. An odd crossing number of $D_v \Phi(0, \lambda)$ means simply that the scalar function $D_v \Phi(0, \lambda)$ changes sign at $\lambda = \lambda_0$ if it is expressed with respect to bases in $N = \operatorname{span}[\hat{v}_0]$ and $Z_0 = \operatorname{span}[\hat{v}_0^*]$; cf. (I.4.20)–(I.4.22). Since $D_v \Phi(0, \lambda_0) = 0$ (by Corollary I.2.4), a sufficient condition is $D_{v\lambda}^2 \Phi(0, \lambda_0) \neq 0$, provided that the family $A(\lambda) = D_x F(0, \lambda)$ and therefore $D_v \Phi(0, \lambda)$ is differentiable with respect to λ. In this case, formula (II.4.25) gives (recall that $Q : Z \to Z_0$ along R)

(II.4.27)
$$D_{v\lambda}^2 \Phi(0, \lambda_0) = Q \frac{d}{d\lambda} A(\lambda_0)|_N \in L(N, Z_0) \text{ and}$$

$$D_{v\lambda}^2 \Phi(0, \lambda_0) \neq 0 \Leftrightarrow$$

$$\frac{d}{d\lambda} A(\lambda_0) \hat{v}_0 = D_{x\lambda}^2 F(0, \lambda_0) \hat{v}_0 \notin R(D_x F(0, \lambda_0)) \text{ if}$$

$$N(D_x F(0, \lambda_0)) = \operatorname{span}[\hat{v}_0].$$

Condition $(II.4.27)_3$ is the same as condition (I.5.3) of Theorem I.5.1. In this case, the bifurcating solutions consist of a unique smooth curve.

The nondegeneracy (II.4.27) or (I.5.3), however, is violated in many simple cases of an odd crossing number. Consider the example $D_x F(0, \lambda) = A(\lambda) = A_0 + (\lambda - \lambda_0)I$ and $N(A_0) = \operatorname{span}[\hat{v}_0]$. Then $\frac{d}{d\lambda} A(\lambda_0) \hat{v}_0 = \hat{v}_0 \notin R(A_0)$ if and only if the eigenvalue 0 is not only geometrically but also algebraically simple (usually called simple). On the other hand, by its linear dependence on λ, it is directly seen that the family $A(\lambda) = A_0 + (\lambda - \lambda_0)I$ has an odd crossing number at $\lambda = \lambda_0$ if and only if the eigenvalue 0 has an odd algebraic multiplicity. (In this case, the differentiated bifurcation function $D_v \Phi(0, \lambda)$ given in (II.4.25) is of order $(\lambda - \lambda_0)^m$ if m is the algebraic multiplicity of 0.) We admit that the evaluation of (II.4.25) is not simple, in general, and in many cases the crossing number of the family $A(\lambda)$ is of easier access.

A degenerate change of sign of $D_v \Phi(0, \lambda)$ at $\lambda = \lambda_0$, however, does not necessarily give a unique bifurcating curve as shown in Section I.16 even in the analytic case. In the nonanalytic case, "Bifurcation with a One-Dimensional Kernel" can indeed be as general as stated in Theorem II.4.4.

Remark II.4.5 *It can easily be seen without knowledge of any degree that a change of sign of $D_v \Phi(0, \lambda)$ at $\lambda = \lambda_0$ entails bifurcation from the trivial solution line $\{(0, \lambda)\}$ at $\lambda = \lambda_0$. Namely, in the upper and lower (v, λ) half-plane, the function Φ has opposite signs near the line $\{(0, \lambda)\}$, and these signs change at $\lambda = \lambda_0$; cf. Figure II.4.1.*

By continuity and the mean value theorem, the function Φ has to have a continuum of zeros emanating at $(0, \lambda_0)$, and that continuum exists globally as described in Theorem II.3.3 (provided that Φ is globally defined). (The idea for a proof of that "evident" statement can be taken from the proof of Theorem I.19.2; see also Figure I.19.1.)

In particular, the solution continua of $\Phi(v, \lambda) = 0$, separating domains where Φ is positive and negative, exist in the upper and lower (v, λ) half-planes. This simple observation leads to a refinement of Theorem II.3.3; cf. Theorem II.5.9 below.

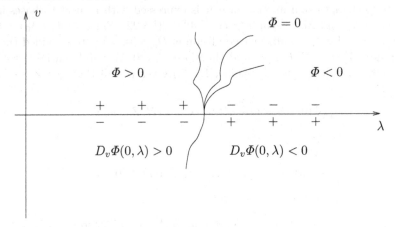

Figure II.4.1

Case 2

Let dim $N = n > 1$ be odd. Assume again that $A(\lambda) = D_x F(0, \lambda)$ is differentiable with respect to λ near $\lambda = \lambda_0$, so that formula $(II.4.27)_1$ holds. Since $D_v \Phi(0, \lambda_0) = 0$ (by Corollary I.2.4), we have

$$D_v \Phi(0, \lambda) = \frac{d}{d\lambda} D_v \Phi(0, \lambda_0)(\lambda - \lambda_0) + R(\lambda) \text{ in } L(N, Z_0)$$

(II.4.28)

$$= Q\frac{d}{d\lambda} A(\lambda_0)|_N (\lambda - \lambda_0) + R(\lambda) \text{ (see } (II.4.27)_1).$$

Here $R(\lambda)$ is a remainder such that $\lim_{\lambda \to \lambda_0} R(\lambda)/(\lambda - \lambda_0) = 0$ in $L(N, Z_0)$. Assume that

$$Q\frac{d}{d\lambda} A(\lambda_0) \in L(N, Z_0) \text{ is regular } \Leftrightarrow$$

$$D_{x\lambda}^2 F(0, \lambda_0)\hat{v} \notin R(D_x F(0, \lambda_0))$$

(II.4.29)   for all $\hat{v} \in N(D_x F(0, \lambda_0))\setminus\{0\} \Leftrightarrow$

$$Z = R(D_x F(0, \lambda_0)) \oplus D_{x\lambda}^2 F(0, \lambda_0)N(D_x F(0, \lambda_0)) \text{ or}$$

$$Z = R \oplus \frac{d}{d\lambda} A(\lambda_0)N.$$

Then, by (II.4.28),

(II.4.30)
$$\det D_v \Phi(0,\lambda) = \det\left(Q\frac{d}{d\lambda}A(\lambda_0)\right)(\lambda - \lambda_0)^n + r(\lambda),$$
so that $\lim_{\lambda \to \lambda_0} r(\lambda)/(\lambda - \lambda_0)^n = 0.$

(Here we identify $L(N, Z_0)$ with $L(\mathbb{R}^n, \mathbb{R}^n)$, so that $\det(Q\frac{d}{d\lambda}A(\lambda_0))$ is defined.) By (II.4.29), $\det(Q\frac{d}{d\lambda}A(\lambda_0)) \neq 0$, and we conclude that

(II.4.31)
if (II.4.29) holds and $\dim N(D_x F(0, \lambda_0)) = n$ is odd, then $D_v \Phi(0, \lambda)$ has an odd crossing number at $\lambda = \lambda_0$. By Theorem II.4.3, $D_x F(0, \lambda)$ has an odd crossing number at $\lambda = \lambda_0$, too, and Theorem II.4.4 guarantees bifurcation for $F(x, \lambda) = 0$ at $\lambda = \lambda_0$.

This bifurcation result is due to [168]. Observe that the number n is the geometric multiplicity of the eigenvalue 0 of $D_x F(0, \lambda_0)$ and that the algebraic multiplicity $m = \dim E_{\lambda_0}$ can be arbitrarily large.

Remark II.4.6 *An odd crossing number in Definitions II.3.1 and II.4.1 plays a crucial role in local and global bifurcation theorems; cf. Theorems II.3.2, II.3.3, II.4.4, II.5.8, and II.5.9. By the simple counterexamples (II.7.18), (II.7.19), it is indispensable for bifurcation, in general. Whereas its definition is transparent, its verification can be complicated. The computation of eigenvalues of (infinite-dimensional) linear operators is not a simple task, and the evaluation of the entire 0-group could be very involved. (Note, however, that for our applications in Chapter III we give cases in which it is easy; cf. (III.2.36), for example.) Theorem II.4.3 tells that the determination of an odd crossing number is reduced to a finite-dimensional problem, to be more precise, to a problem whose dimension is the geometric multiplicity of the critical eigenvalue 0. In view of its form (II.4.25), the evaluation of its determinant could still be a challenge; cf., however, Case 1 of Theorem II.4.4.*

There is a large literature in which one attempts to give a different approach to the crucial property of local and global bifurcation: It is found under the notion of "odd multiplicity." The Krasnosel'skii condition on $D_x F(0, \lambda) = I + \lambda D f(0)$ is an odd algebraic multiplicity of the eigenvalue $-1/\lambda_0$ of $Df(0)$, and by the linear dependence on λ, this means an odd crossing number of $D_x F(0, \lambda)$ at $\lambda = \lambda_0$. In case of a nonlinear dependence on λ, however, the common algebraic multiplicity of an eigenvalue has to be replaced by some multiplicity in which the family $D_x F(0, \lambda)$ apart from $\lambda = \lambda_0$ is involved. Such a definition is rather complicated and not transparent at all. Another drawback is that it applies only to families $D_x F(0, \lambda)$ depending analytically on λ in a neighborhood of $\lambda = \lambda_0$. If it depends only smoothly, it applies only if some derivative with respect to λ does not vanish at $\lambda = \lambda_0$. An advantage is that there are explicit devices for computing odd multiplicities in which the reduction (II.4.25) is used, too. An odd multiplicity is clearly equivalent to an odd crossing number. In view of the complexity of odd mul-

tiplicities, other approaches under fewer regularity assumptions are found in the literature.

For details we refer to [154], [82], [133], [123], [43], [42], [146], [50], [84].

We prefer the notion of an odd crossing number, since it is more natural and, moreover, since it is directly related to a change of Morse index; cf. Definition II.7.1. The Morse index is an important tool in Nonlinear Dynamics, and an odd crossing number links Bifurcation Theory to that wide field; cf. also Section II.7.

All approaches to bifurcation via odd multiplicities or odd crossing numbers are "linear theories" in the respect that the condition for bifurcation is imposed only on the linearization along the trivial solution. The nonlinear remainder is arbitrary in all these theories.

II.5 A Degree for a Class of Proper Fredholm Operators and Global Bifurcation Theorems

In this section, X and Z are both real Banach spaces such that $X \subset Z$ is continuously embedded. We consider mappings $F : U \to Z$, where $U \subset X$ is an open and bounded set. In order to define our admissible class of operators we need some definitions.

Definition II.5.1 *A class of linear operators $A \in L(X, Z)$ is called admissible if the following hold:*
(i) A is a Fredholm operator of index zero.
(ii) $A : Z \to Z$ with domain of definition $D(A) = X$ is closed.
(iii) The spectrum $\sigma(A)$ in a strip $(-\infty, c) \times (-i\varepsilon, i\varepsilon) \subset \mathbb{C}$ for some $c > 0$,
  $\varepsilon > 0$ consists of finitely many eigenvalues of finite algebraic multiplicity. Their total number (counting multiplicities) in that strip is stable under small perturbations in the class of A in $L(X, Z)$.

We remark that property (ii) is a consequence of property (i), so that its requirement could be omitted.

By the perturbation theory for closed operators as expounded in [86], any isolated eigenvalue μ_0 perturbs to its μ_0-group of equal total finite multiplicity. Property (iii) simply excludes that a perturbation of A creates a new eigenvalue as a perturbation of $-\infty$. This phenomenon is certainly excluded for a class of operators $A \in L(X, Z)$ whose spectra are locally uniformly bounded from below. Definition II.5.1 is satisfied by classes of elliptic operators that are locally uniformly sectorial; cf. [52].

Definition II.5.2 *Let $U \subset X$ be open and bounded. An operator $F : \overline{U} \to Z$ is called admissible if*

(i) $F \in C^2(\tilde{U}, Z)$, $\overline{U} \subset \tilde{U}$, *where $\tilde{U}$ is open in X,*
(ii) The class $\{DF(x)|x \in \tilde{U}\} \subset L(X, Z)$ is admissible according to Defini-tion
 II.5.1,
(iii) F is proper; i.e., the inverse image in $\overline{U}$ of a compact set in Z is compact in X.

Definition II.5.2 is motivated by nonlinear differential operators whose Fréchet derivatives are elliptic. For such operators Definition II.5.2 is verified in Section III.5.

If $X = Z$, every compact perturbation of the identity (cf. Definition II.2.1) is admissible if it is a C^2-mapping. Indeed, its derivative is of the form $I + K$, where $K \in L(X, X)$ is compact, and Definition II.5.1 is satisfied by the Riesz–Schauder Theory [170]: There are only finitely many eigenvalues of K in a strip $(-\infty, -1 + c) \times (-i\varepsilon, i\varepsilon)$ if $0 < c < 1$ that are the eigenvalues of $I + K$ in $(-\infty, c) \times (-i\varepsilon, i\varepsilon)$. Since the spectrum of $I + K$ is bounded by $1 + \|K\|$, the total number of eigenvalues in that strip is stable under small compact perturbations of K in $L(X, X)$. Finally, property (iii) of Definition II.5.2 is satisfied for $F = I + f$ if f is completely continuous: If $x_n + f(x_n) = z_n$, the convergence of the sequences (z_n) and $(f(x_n))$ (w.l.o.g.) implies the convergence of (x_n), so that F is proper.

Step 1
Assume that $y \notin F(\partial U)$ is a regular value for F; i.e.,

(II.5.1) $\qquad \begin{aligned} &DF(x) \in L(X, Z) \text{ is regular (is an isomorphism)} \\ &\text{for all } x \in U \text{ with } F(x) = y \quad (x \in F^{-1}(y)). \end{aligned}$

By (II.5.1), the set $F^{-1}(y)$ (possibly empty) consists of isolated points $\{x_j\}$, and by properness it is compact. Therefore, it is a finite set, so that we can make the following definition:

Definition II.5.3 *Assume that F is admissible and that $y \notin F(\partial U)$ is a regular value for F. Then*

(II.5.2) $\qquad \begin{aligned} &d(F, U, y) = \sum_{x \in F^{-1}(y)} i(F, x), \quad \text{where} \\ &i(F, x) = 1 \text{ if } \sigma(DF(x)) \cap (-\infty, 0) = \emptyset, \\ &i(F, x) = (-1)^{m_1 + \cdots + m_\ell} \text{ if} \\ &\sigma(DF(x)) \cap (-\infty, 0) = \{\mu_1, \ldots, \mu_\ell\}, \quad \text{and} \\ &m_k \text{ is the algebraic multiplicity of } \mu_k, k = 1, \ldots, \ell, \end{aligned}$

and $\sum_{\emptyset} = 0$.

As in Definition II.1.1, we call $i(F, x)$ the *index of F at x* (where we omit the dependence on y). Clearly, $d(F, U, y) = d(F - y, U, 0)$.

Definition II.5.3 is motivated by the Leray–Schauder formula (II.2.11), and if $X = Z$ and F is a compact perturbation of the identity (cf. Definition

II.2.1) of class C^1, then Definition II.5.3 indeed gives the Leray–Schauder degree, provided that $y \notin F(\partial U)$ is a regular value for F.

Proposition II.5.4 *The set of regular values $y \notin F(\partial U)$ is open, and the degree (II.5.2) is locally constant with respect to y. To be more precise, there is a $\delta > 0$ such that for all $\tilde{y} \in B_\delta(y) = \{\tilde{y} \in Z | \|\tilde{y} - y\| < \delta\}$, Definition II.5.3 applies and*

$$(\text{II.5.3}) \qquad\qquad d(F, U, \tilde{y}) = d(F, U, y).$$

Proof. By properness, F is a closed map and $y \notin F(\partial U)$ has a positive distance to $F(\partial U)$. Therefore, $B_\delta(y) \cap F(\partial U) = \emptyset$ if $\delta > 0$ is small enough. By (II.5.1), the map F is a diffeormorphism from a neighborhood V_j of any $x_j \in F^{-1}(y)$ in U onto a neighborhood of y. If $\delta > 0$ is possibly decreased, every $\tilde{y} \in B_\delta(y)$ has one preimage $\tilde{x}_j$ in each V_j, and $DF(\tilde{x}_j)$ is regular. Let $V \subset U$ be the union of (finitely many, disjoint) open neighborhoods V_j of $x_j \in F^{-1}(y)$. Then $\overline{U} \backslash V$ is closed, $F(\overline{U} \backslash V)$ is closed, and $y \notin F(\overline{U} \backslash V)$ has a positive distance to $F(\overline{U} \backslash V)$. Therefore, if $\delta > 0$ is small enough, $B_\delta(y) \cap F(\overline{U} \backslash V) = \emptyset$, so that $F^{-1}(B_\delta(y)) \subset V$ and the sets $F^{-1}(y)$ and $F^{-1}(\tilde{y})$ have the same (finite) cardinality for all $\tilde{y} \in B_\delta(y)$. Consider the homotopy $y(\tau) = y + \tau(\tilde{y} - y) \in B_\delta(y)$ for $\tau \in [0, 1]$. Then $F^{-1}(\{y(\tau) | \tau \in [0, 1]\})$ consists of continuously differentiable curves $\{\tilde{x}_j(\tau) | \tau \in [0, 1]\}$ connecting $x_j \in F^{-1}(y)$ to $\tilde{x}_j \in F^{-1}(\tilde{y})$ in each neighborhood V_j of x_j. Then $DF(\tilde{x}_j(\tau))$ is a continuous homotopy from $DF(x_j)$ to $DF(\tilde{x}_j)$ in the open set of regular maps in $L(X, Z)$. By admissibility of F according to Definition II.5.2, the total number of eigenvalues of $DF(\tilde{x}_j(\tau))$ on $(-\infty, 0)$ is finite (counting multiplicities), and eigenvalues of $DF(\tilde{x}_j(\tau))$ can leave the real half-axis $(-\infty, 0)$ only by nonreal complex conjugate pairs. Since by Definition II.5.1 (iii), the total multiplicity of all perturbed eigenvalues is constant, the numbers of all eigenvalues of $DF(\tilde{x}_j(\tau))$ in $(-\infty, 0)$ have the same parity for all $\tau \in [0, 1]$, and (II.5.3) follows directly from Definition II.5.3. $\square$

Step 2

If y is not a regular value for F, we find a sequence $(y_n)_{n \in \mathbb{N}} \subset Z$ with $y_n \to y$ in Z such that $y_n \notin F(\partial U)$ and all y_n are regular values for F. This is the result of Quinn–Sard–Smale [145]. (Originally, it was required that X be separable, so that every open covering of $\overline{U}$ contained a countable subcovering. This stringent assumption is given up in [145].) Since there is no analogue of the Stone–Weierstrass Approximation Theorem in infinite dimensions, we cannot give up the smoothness of F as in the finite-dimensional case. The C^2-differentiability is needed in Theorem II.5.6 below, which states the crucial property of the degree of Definition II.5.3, namely, that it is locally constant when its entries are slightly perturbed in their admissible classes. In other words, the *homotopy invariance* of that degree is valid. This implies that the following definition makes sense:

Definition II.5.5 *Assume that F is admissible and that $y \notin F(\partial U)$. Then for a sequence $(y_n)_{n \in \mathbb{Z}} \subset Z$ of regular values for F such that $y_n \to y$ and $y_n \notin F(\partial U)$,*

(II.5.4)
$$d(F, U, y) = \lim_{n \to \infty} d(F, U, y_n).$$

Clearly, $d(F, U, y) = d(F - y, U, 0)$.

We give the arguments why the degree (II.5.4) is well defined: Let y_0 and y_1 be regular values for F in $Z \backslash F(\partial U)$. We define

(II.5.5)
$$\tilde{F} : [0, 1] \times \overline{U} \to Z \text{ by}$$
$$\tilde{F}(\tau, x) = F(x) - \tau y_1 - (1 - \tau) y_0$$

and clearly,

(II.5.6)
$$d(F, U, y_0) = d(F - y_0, U, 0) = d(\tilde{F}(0, \cdot), U, 0),$$
$$d(F, U, y_1) = d(F - y_1, U, 0) = d(\tilde{F}(1, \cdot), U, 0).$$

The map $\tilde{F}$ is of class C^2, and its derivative $D_{(\tau, x)}\tilde{F} \in L(\mathbb{R} \times X, Z)$ is a Fredholm operator of index 1. By properness of F, the map $\tilde{F}$ is proper, too. According to the result of Quinn–Sard–Smale [145], the set of regular values for $\tilde{F}$ is dense in Z, and by properness, it is also open in Z. (We sketch the argument: Let $y \in Z$ be a regular value for $\tilde{F}$. If not empty, $\tilde{F}^{-1}(y)$ consists of compact continuously differentiable curves in $[0, 1] \times \overline{U}$. Fix some point on such a curve. Then a decomposition of $\mathbb{R} \times X$ into the one-dimensional kernel of $D_{(\tau, x)}\tilde{F}$, which is the tangent space to that curve, and some complementary space $X_0 \subset \mathbb{R} \times X$ yields a regular operator $D_{(\tau, x)}\tilde{F}|_{X_0} \in L(X_0, Z)$. The Implicit Function Theorem and the properness of $\tilde{F}$ imply that the curves in $\tilde{F}^{-1}(y)$ are isolated, that there are only finitely many, that for $\tilde{y} \in B_\delta(y)$ for small $\delta > 0$ the derivative $D_{(\tau, x)}\tilde{F}$ on $\tilde{F}^{-1}(\tilde{y}) \subset [0, 1] \times \overline{U}$ is surjective, and that the set $\tilde{F}^{-1}(\tilde{y})$ consists only of curves that are perturbations of the curves in $\tilde{F}^{-1}(y)$; cf. the arguments in the proof of Proposition II.5.4.)

Since $0 \in Z$ is a regular value for $\tilde{F}(0, \cdot)$ and for $\tilde{F}(1, \cdot)$, according to Proposition II.5.4 we find some $\tilde{y}$ near 0 such that (cf. (II.5.6))

(II.5.7)
$$d(\tilde{F}(0, \cdot), U, \tilde{y}) = d(\tilde{F}(0, \cdot), U, 0) = d(F, U, y_0),$$
$$d(\tilde{F}(1, \cdot), U, \tilde{y}) = d(\tilde{F}(1, \cdot), U, 0) = d(F, U, y_1),$$
and $\tilde{y}$ is also a regular value for $\tilde{F}$
defined in (II.5.5).

We return to the problem whether the degree (II.5.4) is well defined. If $y \notin F(\partial U)$, then it has a positive distance to $F(\partial U)$; i.e., $\|F(x) - y\| \geq d > 0$ for all $x \in \partial U$. Next, we show that if y_0, y_1 are sufficiently close to y and if $\tilde{y}$ is in a small neighborhood of 0, then the endpoints of all curves in $\tilde{F}^{-1}(\tilde{y})$ are in the bottom $\{0\} \times U$ or in the top $\{1\} \times U$ of the cylinder $[0, 1] \times \overline{U}$.

Indeed, let $y_0, y_1 \in B_{d/8}(y)$, $\tilde{y} \in B_{d/8}(0)$. Then

(II.5.8)
$$\begin{aligned} \|\tilde{F}(\tau, x) - \tilde{y}\| &= \|F(x) - y_0 + \tau(y_0 - y_1) - \tilde{y}\| \\ &= \|F(x) - y + y - y_0 + \tau(y_0 - y_1) - \tilde{y}\| \geq \tfrac{d}{2} > 0 \end{aligned}$$
for all $(\tau, x) \in [0, 1] \times \partial U$.

The proof that

(II.5.9)
$$d(\tilde{F}(0, \cdot), U, \tilde{y}) = d(\tilde{F}(1, \cdot), U, \tilde{y}),$$

which implies in view of (II.5.7) that

(II.5.10)
$$d(F, U, y_0) = d(F, U, y_1),$$

is the same as that of the more general Theorem II.5.6 below, where the cylinder $[0, 1] \times \overline{U}$ is replaced by a segment of a noncylindrical domain. This finally proves that Definition II.5.5 makes sense and that (II.5.4) does not depend on the choice of the sequence of regular values $(y_n)_{n \in \mathbb{N}}$.

As mentioned before, if $X = Z$ and F is a compact perturbation of the identity of class C^1, then Definition II.5.3 gives the Leray–Schauder degree, provided that $y \notin F(\partial U)$ is a regular value for F. The homotopy invariance (II.5.9) implies that Definition II.5.5 is valid for the Leray–Schauder degree as well. Thus all proofs in this section apply also for the Leray–Schauder degree for its admissible class $F = I + f$. Note, however, that this approach requires that F be a C^2-mapping, in particular, that smoothness is required in the proofs of Proposition II.5.4 and Theorem II.5.6 below.

Theorem II.5.6 *The degree of Definition II.5.5 is homotopy invariant in the following sense: Let $U \subset \mathbb{R} \times X = \{(\tau, x) | \tau \in \mathbb{R}, x \in X\}$ be open and $U_\tau = \{x \in X | (\tau, x) \in U\}$ be uniformly bounded (possibly empty) for τ in finite intervals of $\mathbb{R}$. We assume that*

(II.5.11)
$$\begin{aligned} &F \in C^2(\tilde{U}, Z), \text{ where } \overline{U} \subset \tilde{U} \text{ is open in } \mathbb{R} \times X, \\ &F : \overline{U} \to Z \text{ is proper,} \\ &\text{the class } \{D_x F(\tau, x) | (\tau, x) \in \tilde{U}\} \subset L(X, Z) \\ &\text{is admissible according to Definition II.5.1, and} \\ &y : \mathbb{R} \to Z \text{ is a } C^2\text{-curve such that} \\ &y(\tau) \neq F(\tau, x) \text{ for all } (\tau, x) \in \partial U. \end{aligned}$$

Then

(II.5.12)
$$d(F(\tau, \cdot), U_\tau, y(\tau)) = \text{ const for all } \tau \in \mathbb{R}$$

and $d(F(\tau, \cdot), U_\tau, y(\tau)) = 0$ if $U_{\tau_0} = \emptyset$ for some $\tau_0 \in \mathbb{R}$.

Proof. Define $\tilde{F}(\tau, x) = F(\tau, x) - y(\tau)$. Then $\tilde{F} \in C^2(\tilde{U}, Z)$, $\tilde{F} : \overline{U} \to Z$ is proper, and $D_{(\tau, x)} \tilde{F}(\tau, x) \in L(\mathbb{R} \times X, Z)$ is a Fredholm operator of index 1. Fix a bounded segment of the noncylindrical domain $U_{a,b} = U \cap ((a, b) \times$

X). Since $F(\partial U)$ is closed, by properness the curve $\{y(\tau)|\tau \in [a,b]\}$ has a positive distance to $F((\partial U)_{a,b})$, where $(\partial U)_{a,b} = \partial U \cap ([a,b] \times X)$. Therefore, $\|\tilde{F}(\tau, x)\| \geq d > 0$ for all $(\tau, x) \in (\partial U)_{a,b}$, and by the Quinn–Sard–Smale Theorem [145] and by the openness of the set of regular values, there exists an element $\tilde{y} \in Z$, $\|\tilde{y}\| < d$, and $\tilde{y}$ is a regular value for $\tilde{F} : U_{a,b} \to Z$, for $\tilde{F}(a, \cdot) : U_a \to Z$, and for $\tilde{F}(b, \cdot) : U_b \to Z$; cf. (II.5.7). By the arguments given before, this implies that $\tilde{F}^{-1}(\tilde{y}) \cap \overline{U}_{a,b}$ consists of finitely many compact one-dimensional manifolds (curves) with boundaries in U_a or in U_b; cf. Figure II.5.1. (Closed curves in $\tilde{F}^{-1}(\tilde{y}) \cap U_{a,b}$ play no role in the sequel. Therefore, Figure II.5.1 does not show closed curves that might clearly exist.)

Pick one curve C in $\tilde{F}^{-1}(\tilde{y})$ starting in U_a or U_b. It ends at a different point in U_a or U_b. By the choice of $\tilde{y}$, the operators

(II.5.13) $D_x\tilde{F}(a, x), D_x\tilde{F}(b, x) \in L(X, Z)$ are regular
when (a, x) or (b, x) are endpoints of C.

Therefore, the degrees $d(\tilde{F}(a, \cdot), U_a, \tilde{y})$ and $d(\tilde{F}(b, \cdot), U_b, \tilde{y})$ are defined as in Definition II.5.3. We show that

(II.5.14) $d(\tilde{F}(a, \cdot), U_a, \tilde{y}) = d(\tilde{F}(b, \cdot), U_b, \tilde{y}),$

where we assume that $U_a \neq \emptyset, U_b \neq \emptyset$; see Figure II.5.1. If $U_a \neq \emptyset, U_b = \emptyset$, for instance, then for some $\tilde{b} < b$ we obtain $U_{\tilde{b}} \neq \emptyset$, but $\tilde{F}(\tilde{b}, \cdot)^{-1}(\tilde{y}) = \emptyset$, so that, according to Definition II.5.3, $d(\tilde{F}(\tilde{b}, \cdot), U_{\tilde{b}}, \tilde{y}) = 0$. In this situation U has a dead end, and it does not reach $\{b\} \times X$; see Figure II.5.1. Thus all curves have to return to U_a, and the subsequent proof shows that the sum of all indices of $\tilde{F}(a, \cdot)$ in U_a is zero, proving that $d(\tilde{F}(a, \cdot), U_a, \tilde{y}) = 0$.

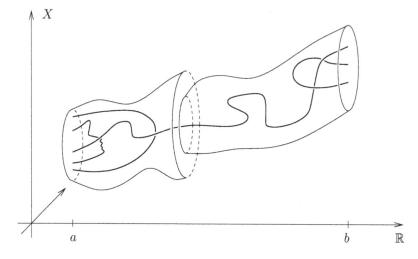

Figure II.5.1

We parameterize the curve C in $\tilde{F}^{-1}(\tilde{y})$ by $\{(\tau(\lambda), x(\lambda)) | \lambda \in [\alpha, \beta]\}$, so that $\tau(\alpha), \tau(\beta) \in \{a, b\}$ and $(\dot{\tau}(\lambda), \dot{x}(\lambda)) \neq (0, 0)$ $(\dot{} = \frac{d}{d\lambda})$. Since $\tilde{y}$ is a regular value for $\tilde{F}$,

$$(\mathrm{II}.5.15) \qquad N(D_{(\tau, x)}\tilde{F}(\tau(\lambda), x(\lambda))) = \mathrm{span}[(\dot{\tau}(\lambda), \dot{x}(\lambda))],$$

which is geometrically the tangent space to C at $(\tau(\lambda), x(\lambda)) \in \mathbb{R} \times X$. Thus

$$(\mathrm{II}.5.16) \qquad \begin{aligned} &D_x \tilde{F}(\tau(\lambda), x(\lambda)) \in L(X, Y) \text{ is regular} \\ &\Leftrightarrow \dot{\tau}(\lambda) \neq 0 \quad \text{for } \lambda \in [\alpha, \beta]. \end{aligned}$$

On all (relatively) open intervals in $[\alpha, \beta]$ where $\dot{\tau}(\lambda) \neq 0$, we see that the index $i(\tilde{F}(\tau(\lambda), \cdot), x(\lambda))$ is constant by the arguments given in the proof of Proposition II.5.4. By the choice of $\tilde{y}$, we know that $\dot{\tau}(\lambda) \neq 0$ for λ near α and λ near β. We investigate the behavior of the index when λ passes the complementary compact set $\Lambda \subset (\alpha, \beta)$, where $\dot{\tau}(\lambda) = 0$. Choose for $\lambda \in \Lambda$ an open interval $(\lambda - \delta_\lambda, \lambda + \delta_\lambda)$ whose length will be appropriately determined below; cf. (II.5.20). Then finitely many such intervals cover Λ, and the union of these consists of one or more (finitely many) intervals in (α, β). We focus on one such open interval $(\tilde{\alpha}, \tilde{\beta})$. By construction, $D_x \tilde{F}(\tau(\tilde{\alpha}), x(\tilde{\alpha}))$ and $D_x \tilde{F}(\tau(\tilde{\beta}), x(\tilde{\beta}))$ are regular, and we determine the indices $i(\tilde{F}(\tau(\tilde{\alpha}), \cdot), x(\tilde{\alpha}))$ and $i(\tilde{F}(\tau(\tilde{\beta}), \cdot), x(\tilde{\beta}))$.

Since $\dot{\tau}(\lambda) = 0$ for $\lambda \in \Lambda$, we have $\dot{x}(\lambda) \neq 0$, and by continuity, this is true for all $\lambda \in [\tilde{\alpha}, \tilde{\beta}]$. By the choice of $\tilde{y}$, the operator $D_{(\tau, x)}\tilde{F}(\tau(\lambda), x(\lambda)) \in L(\mathbb{R} \times X, Z)$ is surjective for all $\lambda \in [\alpha, \beta]$, and therefore $D_\tau \tilde{F}(\tau(\lambda), x(\lambda)) \neq 0$ in Z for all $\lambda \in \Lambda$. By continuity, this is again true for all $\lambda \in [\tilde{\alpha}, \tilde{\beta}]$. We define

$$(\mathrm{II}.5.17) \qquad \begin{aligned} &N(\lambda) = \mathrm{span}[\dot{x}(\lambda)], \quad Z_0(\lambda) = \mathrm{span}[D_\tau \tilde{F}(\tau(\lambda), x(\lambda))], \\ &\text{and decompositions} \\ &X = N(\lambda) \oplus X_0(\lambda), \\ &Z = R(\lambda) \oplus Z_0(\lambda) \quad \text{for } \lambda \in [\tilde{\alpha}, \tilde{\beta}], \text{ where} \\ &R(\lambda) = R(D_x \tilde{F}(\tau(\lambda), x(\lambda))|_{X_0(\lambda)}). \end{aligned}$$

The spaces $X_0(\lambda)$ form a continuous family of complements in the sense that the projections

$$(\mathrm{II}.5.18) \qquad \begin{aligned} &P_\lambda : X \to N(\lambda) \text{ along } X_0(\lambda), \\ &Q_\lambda : Z \to Z_0(\lambda) \text{ along } R(\lambda), \\ &\text{depend continuously on } \lambda \in [\tilde{\alpha}, \tilde{\beta}] \text{ in} \\ &L(X, X), L(Z, Z), \text{ respectively.} \end{aligned}$$

Observe that $N(D_x \tilde{F}(\tau(\lambda), x(\lambda))) \subset N(\lambda)$ with equality only if $\lambda \in \Lambda$. (Using the identity $D_\tau \tilde{F}(\tau(\lambda), x(\lambda))\dot{\tau}(\lambda) + D_x \tilde{F}(\tau(\lambda), x(\lambda))\dot{x}(\lambda) = 0$ it is an easy exercise that $Z_0(\lambda)$ is complementary to $R(\lambda)$ for all $\lambda \in [\tilde{\alpha}, \tilde{\beta}]$.)

We make a Lyapunov–Schmidt reduction according to Section I.2 in a neighborhood of $(\tau(\lambda), x(\lambda))$ that satisfies $\tilde{F}(\tau(\lambda), x(\lambda)) - \tilde{y} = 0$. For this procedure we use the decompositions (II.5.17). (This can be done for all

$\lambda \in [\tilde{\alpha}, \tilde{\beta}]$ for which $N(D_x\tilde{F}(\tau(\lambda), x(\lambda))) = \{0\}$ or $= N(\lambda)$.) We obtain a family of functions

(II.5.19)
$$\tilde{\Phi}_\lambda : (\tau(\lambda) - \tilde{\delta}_\lambda, \tau(\lambda) + \tilde{\delta}_\lambda) \times \tilde{U}_\lambda \to Z_0(\lambda),$$
where $\tilde{U}_\lambda \subset N(\lambda)$ is a neighborhood of $v(\lambda) = P_\lambda x(\lambda)$
and $\tilde{\Phi}_\lambda(\tau, v) = 0 \Leftrightarrow \tilde{F}(\tau, x) = \tilde{y}, \ v = P_\lambda x.$

Thus the zeros of $\tilde{\Phi}_\lambda$ form the local curve $\{(\tau(\tilde{\lambda}), v(\tilde{\lambda})) | \tilde{\lambda} \in (\lambda - \delta_\lambda, \lambda + \delta_\lambda)\}$, $v(\tilde{\lambda}) = P_\lambda x(\tilde{\lambda})$.

Setting $G(x, \lambda) = \tilde{F}(\tau(\lambda), x(\lambda) + x) - \tilde{y}$, we have $G(0, \lambda) = 0, D_xG(0, \lambda) = D_x\tilde{F}(\tau(\lambda), x(\lambda))$, and a Lyapunov–Schmidt reduction near $(0, \lambda)$ using the decompositions (II.5.17) yields a family of functions

(II.5.20)
$$\Phi_\lambda : (\lambda - \delta_\lambda, \lambda + \delta_\lambda) \times U_\lambda \to Z_0(\lambda),$$
where $U_\lambda \subset N(\lambda)$ is a neighborhood of 0 and
$\Phi_\lambda(v, \tilde{\lambda}) = 0 \Leftrightarrow G(x, \tilde{\lambda}) = \tilde{F}(\tau(\tilde{\lambda}), x(\tilde{\lambda}) + x) - \tilde{y} = 0,$
$v = P_\lambda x.$

By uniqueness of the reductions we have

(II.5.21)
$$\Phi_\lambda(v, \tilde{\lambda}) = \tilde{\Phi}_\lambda(\tau(\tilde{\lambda}), P_\lambda x(\tilde{\lambda}) + v) \text{ and}$$
$$D_v\Phi_\lambda(0, \tilde{\lambda}) = D_v\tilde{\Phi}_\lambda(\tau(\tilde{\lambda}), P_\lambda x(\tilde{\lambda})) \text{ for } \tilde{\lambda} \in (\lambda - \delta_\lambda, \lambda + \delta_\lambda).$$

We apply now the results of Section II.4 to $A(\tilde{\lambda}) = D_xG(0, \tilde{\lambda})$. Let $\lambda \in \Lambda$. Then, for $\tilde{\lambda} \in (\lambda - \delta_\lambda, \lambda + \delta_\lambda)$,

(II.5.22)
$$D_x\tilde{F}(\tau(\tilde{\lambda}), x(\tilde{\lambda})) \in L(X, Z) \text{ is regular}$$
$$\Leftrightarrow D_v\tilde{\Phi}_\lambda(\tau(\tilde{\lambda}), P_\lambda x(\tilde{\lambda})) \neq 0, \text{ and in this case,}$$
$$i(\tilde{F}(\tau(\tilde{\lambda}), \cdot), x(\tilde{\lambda})) = S(\tilde{\lambda})\text{sign}D_v\tilde{\Phi}_\lambda(\tau(\tilde{\lambda}), P_\lambda x(\tilde{\lambda})).$$

Here $S(\tilde{\lambda}) \in \{1, -1\}$, and formulas (II.4.13), (II.4.14), (II.4.25) show that $S(\tilde{\lambda})$ depends continuously on $\tilde{\lambda}$ in $(\lambda - \delta_\lambda, \lambda + \delta_\lambda)$, and therefore, it is constant. (Recall that $\hat{Q}$ belongs to a continuous family of projections in $L(Z, Z)$.)

If $\lambda \in [\tilde{\alpha}, \tilde{\beta}] \backslash \Lambda$, then we choose $\delta_\lambda > 0$ such that $(\lambda - \delta_\lambda, \lambda + \delta_\lambda) \cap \Lambda = \emptyset$ and (II.5.22)$_1$ is valid for all $\tilde{\lambda} \in (\lambda - \delta_\lambda, \lambda + \delta_\lambda)$. Choosing $\hat{Q} = Q_\lambda$ in (II.4.13), we obtain also (II.5.22)$_2$ with a continuous and therefore constant function $S(\tilde{\lambda})$ on $(\lambda - \delta_\lambda, \lambda + \delta_\lambda)$.

Thus $S(\tilde{\lambda})$ is locally constant on the entire interval $[\tilde{\alpha}, \tilde{\beta}]$, which means that $S(\tilde{\lambda}) \equiv S_0 \in \{1, -1\}$ for all $\tilde{\lambda} \in [\tilde{\alpha}, \tilde{\beta}]$, and (II.5.22) implies that

(II.5.23)
$$\text{if } D_v\tilde{\Phi}_\lambda(\tau(\lambda), P_\lambda x(\lambda)) \neq 0 \text{ for some } \lambda \in [\tilde{\alpha}, \tilde{\beta}], \text{ then}$$
$$i(\tilde{F}(\tau(\lambda), \cdot), x(\lambda)) = S_0\text{sign}D_v\tilde{\Phi}_\lambda(\tau(\lambda), P_\lambda x(\lambda)).$$

By $\tilde{\Phi}_\lambda(\tau(\tilde{\lambda}), v(\tilde{\lambda})) = 0$ for all $\tilde{\lambda} \in (\lambda - \delta_\lambda, \lambda + \delta_\lambda)$ (where $v(\tilde{\lambda}) = P_\lambda x(\tilde{\lambda})$; cf. (II.5.19)), we obtain, by differentiation with respect to $\tilde{\lambda}$ at $\tilde{\lambda} = \lambda$,

$$(\text{II}.5.24) \qquad D_\tau \tilde{\Phi}_\lambda(\tau(\lambda), v(\lambda))\dot{\tau}(\lambda) + D_v \tilde{\Phi}_\lambda(\tau(\lambda), v(\lambda))\dot{v}(\lambda) = 0.$$

By $N(\lambda) = \text{span}[\dot{x}(\lambda)]$, see (II.5.17), we have $\dot{v}(\lambda) = P_\lambda \dot{x}(\lambda) = \dot{x}(\lambda)$, and by $Z_0(\lambda) = \text{span}[D_\tau \tilde{F}(\tau(\lambda), x(\lambda))]$, we get in view of the construction of the bifurcation function $\tilde{\Phi}_\lambda$ in Section I.2 $D_\tau \tilde{\Phi}_\lambda(\tau(\lambda), x(\lambda)) = Q_\lambda D_\tau \tilde{F}(\tau(\lambda), x(\lambda))$ $= D_\tau \tilde{F}(\tau(\lambda), x(\lambda))$, where we use also the definition (II.5.18) of Q_λ. In order to define the sign of $D_v \tilde{\Phi}_\lambda(\tau(\lambda), v(\lambda))$, we have to identify $L(N(\lambda), Z_0(\lambda))$ with $L(\mathbb{R}, \mathbb{R}) \cong \mathbb{R}$. This is done by the choice of bases $\{\dot{v}(\lambda)\}$ in $N(\lambda)$ and $\{D_\tau \tilde{F}(\tau(\lambda), x(\lambda))\}$ in $Z_0(\lambda)$. Thus (II.5.24) gives

$$(\text{II}.5.25) \qquad \begin{aligned} &\text{sign} D_v \tilde{\Phi}_\lambda(\tau(\lambda), v(\lambda)) = -\text{sign}\dot{\tau}(\lambda) \in \{-1, 0, 1\} \\ &\text{for all } \lambda \in [\tilde{\alpha}, \tilde{\beta}]. \end{aligned}$$

A combination of (II.5.23) and (II.5.25) leads to the following result:

$$(\text{II}.5.26) \qquad \begin{aligned} &\text{Let } (\tilde{\alpha}, \tilde{\beta}) \text{ be any open interval in the covering} \\ &\text{of } \Lambda = \{\lambda \in (\alpha, \beta) | \dot{\tau}(\lambda) = 0\}. \text{ Then for all} \\ &\lambda \in [\tilde{\alpha}, \tilde{\beta}] \text{ such that } \dot{\tau}(\lambda) \neq 0, \text{ in particular for} \\ &\lambda = \tilde{\alpha} \text{ and } \lambda = \tilde{\beta}, \text{ the index of} \\ &\tilde{F}(\tau(\lambda), \cdot) \text{ at } x(\lambda) \text{ exists and} \\ &i(\tilde{F}(\tau(\lambda), \cdot), x(\lambda))\text{sign}\dot{\tau}(\lambda) \equiv -S_0. \end{aligned}$$

Since, as stated before, the index $i(\tilde{F}(\tau(\lambda), \cdot), x((\lambda))$ is constant in any interval of $[\alpha, \beta] \backslash \Lambda$, the behavior (II.5.26) extends to all $\lambda \in [\alpha, \beta]$, i.e., to the entire curve C:

$$(\text{II}.5.27) \qquad \begin{aligned} &i(\tilde{F}(\tau(\lambda), \cdot), x(\lambda))\text{sign}\dot{\tau}(\lambda) \equiv -S_0 \in \{1, -1\} \\ &\text{for all } \lambda \in [\alpha, \beta] \text{ with } \dot{\tau}(\lambda) \neq 0. \end{aligned}$$

Note that the index exists if $\dot{\tau}(\lambda) \neq 0$.

A simple but crucial consequence of (II.5.27) is the following (see also Figure II.5.1): Since $\text{sign}\dot{\tau}(\alpha) = -\text{sign}\dot{\tau}(\beta) \neq 0$ if the endpoints of C are both in U_a or both in U_b and $\text{sign}\dot{\tau}(\alpha) = \text{sign}\dot{\tau}(\beta) \neq 0$ if the endpoints of C are in U_a and in U_b,

$$(\text{II}.5.28) \qquad \begin{aligned} &\text{the indices of } \tilde{F}(\tau, \cdot) \text{ at the endpoints} \\ &\text{of a solution curve } C \text{ are opposite (equal)} \\ &\text{if the endpoints are in the same set} \\ &U_a \text{ or } U_b \text{ (in different sets } U_a \text{ and } U_b\text{).} \end{aligned}$$

In the regular case, the degree is the sum of the indices, and (II.5.28) proves (II.5.14). This completes the proof that Definition II.5.5 makes sense; cf. (II.5.9). Since the regular value $\tilde{y}$ is arbitrarily close to 0, Definition II.5.5 implies

$$(\text{II.5.29}) \quad \begin{aligned} d(F(a,\cdot),U_a,y(a)) &= d(\tilde{F}(a,\cdot),U_a,0) \\ &= d(\tilde{F}(a,\cdot),U_a,\tilde{y}) = d(\tilde{F}(b,\cdot),U_b,\tilde{y}) \\ &= d(\tilde{F}(b,\cdot),U_b,0) = d(F(b,\cdot),U_b,y(b)), \end{aligned}$$

which finally proves (II.5.12). The last statement of Theorem II.5.6 follows from $d(\tilde{F}(a,\cdot),U_a,\tilde{y}) = 0$ if $U_b = \emptyset$; cf. the arguments after (II.5.14). $\square$

Remark II.5.7 *Property (II.5.27) is in complete coincidence with the index along a plane solution curve of a function $\Phi : \mathbb{R} \times \mathbb{R} \to \mathbb{R}$. Let 0 be a regular value for Φ such that the solutions of $\Phi(\tau,x) = 0$ form smooth curves in $\mathbb{R} \times \mathbb{R}$. The index $i(\Phi(\tau,\cdot),x)$, if it exists, is simply $\mathrm{sign}D_x\Phi(\tau,x)$ if $\Phi(\tau,x) = 0$. Since 0 is a regular value for Φ, any curve of zeros of Φ separates regions in the plane $\mathbb{R} \times \mathbb{R}$ where Φ is positive and where Φ is negative. Therefore, if $\dot{\tau}(\lambda) \neq 0$ (for a suitable parameterization $\{(\tau(\lambda),x(\lambda))\}$ of such a curve), the derivative $D_x\Phi(\tau(\lambda),x(\lambda))$ is positive (negative) if Φ is negative (positive) below and positive (negative) above that curve. As visualized in Figure II.5.2, the signs of Φ below and above a curve of zeros change at every turning point, i.e., if $\mathrm{sign}\dot{\tau}(\lambda)$ changes. Observe that for such behavior it is not required that the points where $\dot{\tau}(\lambda) = 0$ be isolated.*

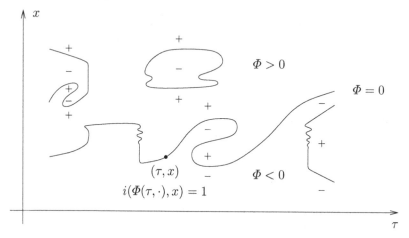

Figure II.5.2

We summarize the essential properties of the degree defined in Definition II.5.5:

(1) The *normalization* (II.1.10) holds.
(2) The *additivity* (II.1.11) holds.
(3) The *homotopy invariance* holds as described in Theorem II.5.6.
(4) If $y_n \to y$, $y,y_n \notin F(\partial U)$, and $F(x_n) = y_n$ for some $x_n \in U$, then properness implies the existence of some $x \in U$ such that $F(x) = y$. Then by Definitions II.5.3 and II.5.5,

$$\text{(II.5.30)} \qquad \begin{aligned} &\text{if } d(F, U, y) \neq 0, \text{ then there is some} \\ &x \in U \text{ such that } F(x) = y. \end{aligned}$$

(5) The homotopy invariance shows that the degree depends only on the values of F of the boundary ∂U.

(6) The *excision* (II.1.16) holds.

Finally, the notion of the *index* given in Definition II.5.3 is extended to any isolated solution $x_0 \in U$ of $F(x) = y$ as in (II.1.17). Thus the degree of Definition II.5.5 is the sum of its indices if there are only finitely many solutions of $F(x) = y$ in U.

To summarize, the degree for our class of proper Fredholm operators has precisely the same properties as the Brouwer degree or the Leray–Schauder degree for their classes of operators.

II.5.1 Global Bifurcation via Odd Crossing Numbers

The degree is a good tool in the bifurcation theory for parameter-dependent problems as described in Section II.3. We consider

$$\text{(II.5.31)} \qquad \begin{aligned} &F : U \times V \to Z \text{ with open sets} \\ &0 \in U \subset X, \ V \subset \mathbb{R}, \end{aligned}$$

and we assume the "trivial solution" $F(0, \lambda) = 0$ for all $\lambda \in V$. For our global bifurcation result below, we simply assume that $U = X$ and $V = \mathbb{R}$; i.e., $F : X \times \mathbb{R} \to Z$ is everywhere defined. Furthermore,

$$\text{(II.5.32)} \qquad \begin{aligned} &F \in C^2(X \times \mathbb{R}, Z), \\ &\text{the class } \{D_x F(x, \lambda) | (x, \lambda) \in X \times \mathbb{R}\} \subset L(X, Z) \\ &\text{is admissible according to Definition II.5.1, and} \\ &F \text{ is proper on every closed and bounded} \\ &\text{subset of } X \times \mathbb{R} \text{ according to Definition II.5.2 (iii).} \end{aligned}$$

Thus for an open and bounded set $U \subset X$, the degree $d(F(\cdot, \lambda), U, y)$ is defined by Definition II.5.5, provided that $y \notin F(\partial U, \lambda)$.

A necessary condition for bifurcation from the trivial solution line $\{(0, \lambda)\}$ at some $(0, \lambda_0) \in X \times \mathbb{R}$ is that

$$\text{(II.5.33)} \qquad 0 \text{ be an eigenvalue of } D_x F(0, \lambda_0).$$

By admissibility of $A(\lambda) \equiv D_x F(0, \lambda)$, conditions (II.4.1), (II.4.2), and (II.4.3) are satisfied, so that the notion of an odd crossing number according to Definition II.4.1 is well defined. By (II.4.5)$_1$, the index $i(F(\cdot, \lambda), 0) = i(D_x F(0, \lambda), 0)$ exists locally for $\lambda \neq \lambda_0$, and by Definition II.5.3 and by property (iii) of admissibility in Definition II.5.1 (see also the arguments given in the proof of Proposition II.5.4),

(II.5.34)   an odd crossing number of $D_x F(0, \lambda)$ at $\lambda = \lambda_0$
means that the index $i(F(\cdot, \lambda), 0)$ changes at $\lambda = \lambda_0$.

Since all ingredients of Rabinowitz's Theorem (Theorem II.3.3) are available, we obtain the following **Global Bifurcation Theorem for Fredholm Operators** (cf. Theorem II.4.4):

Theorem II.5.8 *Let S denote the closure of the set of nontrivial solutions of $F(x, \lambda) = 0$ in $X \times \mathbb{R}$, where $F : X \times \mathbb{R} \to Z$ satisfies (II.5.32). Assume that $D_x F(0, \lambda)$ has an odd crossing number at $\lambda = \lambda_0$. Then $(0, \lambda_0) \in S$, and let C be the (connected) component of S to which $(0, \lambda_0)$ belongs. Then*
(i) C is unbounded, or
(ii) C contains some $(0, \lambda_1)$, where $\lambda_0 \neq \lambda_1$.

The proof is the same as that of Theorem II.3.3. Observe simply that if (i) and (ii) do not hold, the component C is compact by properness.

For applications, both global Theorems II.3.3 and II.5.8 are valuable. To prove global bifurcation for quasilinear elliptic problems, for instance, both theorems are applicable as expounded in Section III.5. Theorem II.5.8 needs more differentiability of F, but for the application of Theorem II.3.3 the mapping F has to be put into the form of a compact perturbation of the identity. Such a transformation usually requires the inversion of linear operators having a compact resolvent, and this inversion might cause problems, as it does for the examples mentioned in Remark III.5.2. But even if it is possible, the verification of an odd crossing number might be hard after F has been transformed into a compact perturbation of the identity. Example (III.5.6) shows the problems as explained after it. To summarize, if F has enough differentiability, a conversion into a compact perturbation of the identity in order to apply the Leray–Schauder degree is not useful even if possible. Rather, we recommend that one apply the degree for proper Fredholm operators to F.

II.5.2 Global Bifurcation with One-Dimensional Kernel

Next, we give a refinement of Theorem II.3.3 if $\dim N(D_x F(0, \lambda_0)) = 1$. In Section II.3, the mapping $F(\cdot, \lambda)$ is assumed to be a compact perturbation of the identity; i.e., $X = Z$ and $F(x, \lambda) = x + f(x, \lambda)$, where f is completely continuous. Therefore, $D_x F(0, \lambda) = I + D_x f(0, \lambda)$ and $D_x f(0, \lambda) \in L(X, X)$ is compact. As mentioned already in Sections II.4 and II.5, the mappings considered in Section II.3 fit completely into the framework of Section II.4, and all results for (nonlinear) Fredholm operators presented in Section II.4 are clearly also true for compact perturbations of the identity.

Locally near the bifurcation point $(0, \lambda_0)$, all solutions of $F(x, \lambda) = 0$ are given by the solutions of $\Phi(v, \lambda) = 0$, which is the bifurcation equation obtained by the method of Lyapunov–Schmidt. We use the decompositions (II.4.6), where $N = N(D_x F(0, \lambda_0))$, $R = R(D_x F(0, \lambda_0))$, and $X = Z$. If $\dim N = \operatorname{codim} R = 1$, then $\Phi(v, \lambda) = 0$ is a scalar bifurcation equation. As discussed in Section II.4, Case 1 of Theorem II.4.4, and in particular in Remark II.4.5, a change of sign of $D_v \Phi(0, \lambda)$ at $\lambda = \lambda_0$ (which is equivalent to an odd crossing number of $D_x F(0, \lambda)$ at $\lambda = \lambda_0$) causes a change of sign of the function Φ in the upper and lower (v, λ)-planes near the line $\{(0, \lambda)\}$; cf. Figure II.4.1. By continuity of Φ, a continuum of zeros emanates at $(0, \lambda_0)$ existing globally in a neighborhood of $(0, \lambda_0) \in N \times \mathbb{R}$ where Φ is defined. We make this more precise. Let

$$(\text{II.5.35}) \qquad N(D_x F(0, \lambda_0)) = \operatorname{span}[\hat{v}_0], \quad \hat{v}_0 \in X, \quad \|\hat{v}_0\| = 1.$$

As proved in Section I.2,

$$(\text{II.5.36}) \qquad \begin{aligned} &F(x, \lambda) = 0 \text{ for } (x, \lambda) \in B_r(0) \times (\lambda_0 - \delta, \lambda_0 + \delta) \Leftrightarrow \\ &x = s\hat{v}_0 + \psi(s\hat{v}_0, \lambda) \text{ and } \Phi(s\hat{v}_0, \lambda) = 0. \end{aligned}$$

If the ball $B_r(0)$ and the interval $(\lambda_0 - \delta, \lambda_0 + \delta)$ are small enough, all continua for $s > 0$ and for $s < 0$ reach the boundary of $B_r(0) \times (\lambda_0 - \delta, \lambda_0 + \delta)$. Moreover, since $\psi(0, \lambda) = 0$ and $D_v \psi(0, \lambda_0) = 0 \in L(N, X_0)$, cf. Corollary I.2.4, all solutions of $F(x, \lambda) = 0$ in $B_r(0) \times (\lambda_0 - \delta, \lambda_0 + \delta)$ satisfy

$$(\text{II.5.37}) \qquad \begin{aligned} &x = s\hat{v}_0 + x_0, \text{ where } x_0 = s \int_0^1 D_v \psi(ts\hat{v}_0, \lambda) dt \hat{v}_0 \in X_0 \\ &\text{and } \|x_0\| \leq \frac{1}{2}|s| \text{ (for small } |s| \text{ such that } x \in B_r(0)). \end{aligned}$$

Thus all local nontrivial solution continua are in closed cones of $X \times \mathbb{R}$ defined as follows:

$$(\text{II.5.38}) \qquad \begin{aligned} &\text{Let } x \in X \text{ be decomposed as } x = s\hat{v}_0 + x_0, \text{ where} \\ &s \in \mathbb{R} \text{ and } x_0 \in X_0, \text{ cf. (II.4.6) and (II.5.35). Then} \\ &K_\delta^+ = \left\{ (x, \lambda) \,\middle|\, s \geq 0, \|x_0\| \leq \frac{1}{2}s, |\lambda - \lambda_0| \leq \delta \right\}, \\ &K_\delta^- = \left\{ (x, \lambda) \,\middle|\, s \leq 0, \|x_0\| \leq -\frac{1}{2}s, |\lambda - \lambda_0| \leq \delta \right\}. \end{aligned}$$

Clearly, $K_\delta^+ \cap K_\delta^- = \{0\} \times [\lambda_0 - \delta, \lambda_0 + \delta]$.

The component $\mathcal{C}$ in $\mathcal{S}$ (= the closure of all nontrivial solutions of $F(x, \lambda) = 0$ in $X \times \mathbb{R}$) to which $(0, \lambda_0)$ belongs decomposes locally as

$$\mathcal{C} \cap (B_r(0) \times (\lambda_0 - \delta, \lambda_0 + \delta)) = \mathcal{C}_{loc}^+ \cup \mathcal{C}_{loc}^-, \text{ where}$$

(II.5.39) $\mathcal{C}_{loc}^+ \subset K_\delta^+$, $\mathcal{C}_{loc}^- \subset K_\delta^-$ and $\mathcal{C}_{loc}^+ \cap \mathcal{C}_{loc}^- = \{(0, \lambda_0)\}$.

Furthermore, $\mathcal{C}_{loc}^\pm \cap \partial(B_r(0) \times (\lambda_0 - \delta, \lambda_0 + \delta)) \neq \emptyset$.

Both subcontinua $\mathcal{C}_{loc}^+$ and $\mathcal{C}_{loc}^-$ of $\mathcal{C}$ have global extensions. We let

(II.5.40) $\mathcal{C}^\pm$ denote the maximal components of $\mathcal{C}_{loc}^\pm \backslash \{(0, \lambda_0)\}$, respectively, in $\mathcal{C} \backslash \{(0, \lambda_0)\}$.

By definition, $\mathcal{C} = \mathcal{C}^+ \cup \mathcal{C}^- \cup \{(0, \lambda_0)\}$, but $\mathcal{C}^+ = \mathcal{C}^-$ is not excluded. The following result refines Theorem II.3.3.

Theorem II.5.9 *Assume the hypotheses of Theorem II.3.3 and that* $\dim N(D_x F(0, \lambda_0)) = 1$. *Then each of* $\mathcal{C}^+$ *and* $\mathcal{C}^-$ *either satisfies the alternatives (i) and (ii) of Theorem II.3.3 or*
(iii) contains a pair of points $(x, \lambda), (-x, \lambda)$, *where* $x \neq 0$.

Proof. At least one of $\mathcal{C}^+$ and $\mathcal{C}^-$ must satisfy the alternatives of Theorem II.3.3. Suppose $\mathcal{C}^+$ does not satisfy any of (i)–(iii). We define

(II.5.41) $$\hat{\mathcal{C}}^- = \{(x, \lambda) | (-x, \lambda) \in \mathcal{C}^+\}.$$

Then $\mathcal{C}^+ \cap \hat{\mathcal{C}}^- = \emptyset$, and $\hat{\mathcal{C}} \equiv \mathcal{C}^+ \cup \hat{\mathcal{C}}^- \cup \{(0, \lambda_0)\}$ is compact. Let

(II.5.42) U be an open neighborhood of $\hat{\mathcal{C}}$ in $X \times \mathbb{R}$ such that $(-x, \lambda) \in U$ whenever $(x, \lambda) \in U$.

Since $\hat{\mathcal{C}}$ is generated by $\mathcal{C}^+$ via the involution $(x, \lambda) \mapsto (-x, \lambda)$, we can assume for U that

(II.5.43) $U = U_+ \cup U_- \cup (B_r(0) \times (\lambda_0 - \delta, \lambda_0 + \delta))$,
$U_+ \cap U_- = \emptyset$, $U_- = \{(x, \lambda) | (-x, \lambda) \in U_+\}$, and
$\mathcal{C}^+ \subset U_+ \cup (B_r(0) \times (\lambda_0 - \delta, \lambda_0 + \delta))$,
$\hat{\mathcal{C}}^- \subset U_- \cup (B_r(0) \times (\lambda_0 - \delta, \lambda_0 + \delta))$.

We define a new mapping on the closure $\overline{U}$ as follows:

$$(\text{II.5.44}) \quad \hat{F}(x, \lambda) = \begin{cases} F(x, \lambda) \text{ if} \\ (x, \lambda) \in K_\delta^+ \cap (\overline{B}_r(0) \times [\lambda_0 - \delta, \lambda_0 + \delta]), \\ \\ \dfrac{s}{2\|x_0\|} F(2\|x_0\|\hat{v}_0 + x_0, \lambda) \text{ if} \\ x = s\hat{v}_0 + x_0, \ 0 \le s < 2\|x_0\|, \text{ and} \\ (x, \lambda) \in \overline{B}_r(0) \times [\lambda_0 - \delta, \lambda_0 + \delta], \\ \\ -\hat{F}(-x, \lambda) \text{ if} \\ x = s\hat{v}_0 + x_0, \ s \le 0, \text{ and} \\ (x, \lambda) \in \overline{B}_r(0) \times [\lambda_0 - \delta, \lambda_0 + \delta], \\ \\ F(x, \lambda) \text{ if} \\ (x, \lambda) \in \overline{U}_+ \backslash (B_r(0) \times (\lambda_0 - \delta, \lambda_0 + \delta)), \\ \\ -\hat{F}(-x, \lambda) \text{ if} \\ (x, \lambda) \in \overline{U}_- \backslash (B_r(0) \times (\lambda_0 - \delta, \lambda_0 + \delta)). \end{cases}$$

The goal of Definition (II.5.44) is that

$$(\text{II.5.45}) \quad \begin{aligned} &\hat{F} \text{ is continuous on } \overline{U}, \\ &\hat{F} \text{ is a compact perturbation of the identity, i.e.,} \\ &\hat{F}(x, \lambda) = x + \hat{f}(x, \lambda) \text{ and} \\ &\hat{f} : \overline{U} \to X \text{ is completely continuous,} \\ &D_x \hat{F}(0, \lambda) = D_x F(0, \lambda) \text{ has an odd crossing} \\ &\text{number at } \lambda = \lambda_0, \text{ and finally,} \\ &\hat{F}(\cdot, \lambda) \text{ is an odd mapping.} \end{aligned}$$

Furthermore, $\hat{F} = F$ on $(K_\delta^+ \cap (\overline{B}_r(0) \times [\lambda_0 - \delta, \lambda_0 + \delta])) \cup \overline{U}_+ \backslash (B_r(0) \times (\lambda_0 - \delta, \lambda_0 + \delta))$, which contains $\mathcal{C}^+$ by (II.5.39) and (II.5.43). By the oddness of $\hat{F}(\cdot, \lambda)$,

$$(\text{II.5.46}) \quad \begin{aligned} &\hat{\mathcal{C}} = \mathcal{C}^+ \cup \hat{\mathcal{C}}^- \cup \{(0, \lambda_0)\} \text{ is the component of} \\ &\hat{S} \text{ (which is the closure of the set of nontrivial} \\ &\text{solutions of } \hat{F}(x, \lambda) = 0 \text{ in } U) \text{ to which} \\ &(0, \lambda_0) \text{ belongs.} \end{aligned}$$

Since $\hat{\mathcal{C}}$ is compact, we obtain, as in the proof of Theorem II.3.3,

$$(\text{II.5.47}) \quad \begin{aligned} &\text{a bounded open set } \hat{U} \subset X \times \mathbb{R} \text{ such that} \\ &\hat{\mathcal{C}} \subset \hat{U}, \ \partial \hat{U} \cap S = \emptyset, \text{ and} \\ &\hat{U} \cap \{(0, \lambda) | \lambda \in \mathbb{R}\} = \{0\} \times (\lambda_0 - \hat{\delta}, \lambda_0 + \hat{\delta}) \\ &\text{for some arbitrarily small } \hat{\delta} > 0. \end{aligned}$$

This leads to a contradiction as in the proof of Theorem II.3.3. □

Since the Global Bifurcation Theorem for Fredholm operators, Theorem II.5.8, has the same proof as Theorem II.3.3, one might expect the refinement as stated in Theorem II.5.9 also for nonlinear Fredholm operators. The obstruction, however, is in the Definition (II.5.44) of the odd mapping $\hat{F}$, which is only continuous and not of class C^2. However, if $F : X \times \mathbb{R} \to Z$ is admissible as in (II.5.32) and if $F(\cdot, \lambda)$ is odd, we do not need Definition (II.5.44), and we obtain immediately the following:

Corollary II.5.10 *Assume the hypotheses of Theorem II.3.3 or of Theorem II.5.8 and that* $\dim N(D_x F(0, \lambda_0)) = 1$. *If* $F(\cdot, \lambda)$ *is odd, then the component* $\mathcal{C}$ *of* $\mathcal{S}$ *to which* $(0, \lambda_0)$ *belongs decomposes into* $\mathcal{C} = \mathcal{C}^+ \cup \mathcal{C}^- \cup \{(0, \lambda_0)\}$, *so that* $\mathcal{C}^- = \{(x, \lambda) | (-x, \lambda) \in \mathcal{C}^+\}$. *Then each of* $\mathcal{C}^+$ *and* $\mathcal{C}^-$ *satisfies the alternatives of Theorem II.3.3 or of Theorem II.5.8.*

This does not exclude alternative (iii) of Theorem II.5.9.

If $\dim N(D_x F(0, \lambda_0)) > 1$, then for odd $F(\cdot, \lambda)$, the solution set is "odd" in the sense that it is invariant under the involution $(x, \lambda) \mapsto (-x, \lambda)$. The definition of $\mathcal{C}^+$ and $\mathcal{C}^-$, however, does not make sense. In particular, a local decomposition as in (II.5.39) has no analogue, in general, if $\dim N(D_x F(0, \lambda_0)) > 1$.

II.6 A Global Implicit Function Theorem

We consider a continuous mapping $F \colon X \times \mathbb{R} \to Z$, where $X \subset Z$ is continuously embedded, and we assume a solution

$$(\text{II.6.1}) \qquad\qquad F(x_0, \lambda_0) = 0.$$

Apart from the assumptions (I.1.2) and (I.1.3) for the local Implicit Function Theorem, Theorem I.1.1, we need a setting such that a degree for $F(\cdot, \lambda)$ can be defined.

The Leray–Schauder degree is applicable if $X = Z$, $F(x, \lambda) = x + f(x, \lambda)$, and $f : X \times \mathbb{R} \to X$ is completely continuous. The degree for Fredholm operators can be used if $F \in C^2(X \times \mathbb{R}, Z)$, $D_x F(x, \lambda) \in L(X, Z)$ is admissible according to Definition II.5.1, and F is proper on every closed and bounded subset of $X \times \mathbb{R}$ according to Definition II.5.2(iii).

A **Global Implicit Function Theorem** then reads as follows:

Theorem II.6.1 *Assume the preceding properties of* $F : X \times \mathbb{R} \to Z$ *and (II.6.1) such that*

$$(\text{II.6.2}) \qquad \begin{aligned} &D_x F(x_0, \lambda_0) \in L(X, Z) \text{ is bijective and} \\ &D_x F \in C(B_r(x_0) \times (\lambda_0 - \rho, \lambda_0 + \rho), L(X, Z)) \text{ for some } r, \rho > 0. \end{aligned}$$

Let S denote the set of all solutions of $F(x, \lambda) = 0$ in $X \times \mathbb{R}$ and let C be the (connected) component of S that contains the local solution curve $\{(x(\lambda), \lambda)|\lambda \in (\lambda_0 - \delta, \lambda_0 + \delta)\}$ through $(x_0, \lambda_0) = (x(\lambda_0), \lambda_0)$ given by Theorem I.1.1. Then

(i) $C = \{(x_0, \lambda_0)\} \cup C^+ \cup C^-$, $C^+ \cap C^- = \emptyset$, and C^+, C^- are each unbounded,

   or

(ii) $C \backslash \{(x_0, \lambda_0)\}$ is connected.

Proof. Assume that $C \backslash \{(x_0, \lambda_0)\}$ is not connected. Then $C = \{(x_0, \lambda_0)\} \cup C^+ \cup C^-$, where $C^+ \cap C^- = \emptyset$, and let C^+ denote the component of $\{(x(\lambda), \lambda)|\lambda \in (\lambda_0, \lambda_0 + \delta)\}$ in $S \backslash \{(x_0, \lambda_0)\}$.

Assume that C^+ is bounded. As shown in the proofs of Theorem II.3.3 and Theorem II.5.8, the bounded and closed set $C^+ \cup \{(x_0, \lambda_0)\} = \overline{C}^+$ is compact. As in the proof of Theorem II.3.3, we can construct a bounded open set $U \subset X \times \mathbb{R}$ such that

$$(\text{II.6.3}) \qquad C^+ \subset U \text{ and } \partial U \cap S = \{(x_0, \lambda_0)\}.$$

Setting as before $U_\lambda = \{x \in X | (x, \lambda) \in U\}$, we can also assume that

$$(\text{II.6.4}) \qquad \begin{aligned} &\overline{U}_{\lambda_0} \cap \overline{B}_r(x_0) = \emptyset \text{ and} \\ &\overline{U}_{\lambda_0} \cup \overline{B}_r(x_0) = (\overline{U})_{\lambda_0}, \end{aligned}$$

where $r > 0$ is so small that $x = x_0$ is an isolated solution of $F(x, \lambda_0) = 0$ in $\overline{B}_r(x_0)$. (Note that in $(\text{II.6.4})_1$ $\overline{U}_{\lambda_0}$ denotes the closure of the fiber U_{λ_0} in X, whereas $(\text{II.6.4})_2$ denotes the fiber of the closure $\overline{U}$ in $X \times \mathbb{R}$.) Then the additivity and the homotopy invariance of the respective degree imply

$$(\text{II.6.5}) \qquad \begin{aligned} &d(F(\cdot, \lambda_0), B_r(x_0), 0) + d(F(\cdot, \lambda_0), U_{\lambda_0}, 0) \\ &= d(F(\cdot, \lambda), U_\lambda, 0) \text{ for } \lambda \geq \lambda_0, \\ &= 0, \text{ since } U_\lambda = \emptyset \text{ for large } \lambda > \lambda_0. \end{aligned}$$

On the other hand, if $U_{\lambda_0} \neq \emptyset$,

$$(\text{II.6.6}) \qquad \begin{aligned} &d(F(\cdot, \lambda_0), U_{\lambda_0}, 0) \\ &= d(F(\cdot, \lambda), U_\lambda, 0) \text{ for } \lambda \leq \lambda_0, \\ &= 0, \text{ since } U_\lambda = \emptyset \text{ for large } \lambda < \lambda_0. \end{aligned}$$

This proves

$$(\text{II.6.7}) \qquad d(F(\cdot, \lambda_0), B_r(x_0), 0) = 0.$$

But $D_x F(x_0, \lambda_0) \in L(X, Z)$ is bijective, and the local degree (II.6.7) is the index $i(F(\cdot, \lambda_0), x_0) \in \{-1, 1\}$; cf. (II.2.10), (II.2.11), or (II.5.2). This contradiction proves that C^+ is unbounded, and the unboundedness of C^- is proved in the same way. $\qquad \square$

Remark II.6.2 *The proof of Theorem II.6.1 shows that the assumption (II.6.2) can be reduced to $(\text{II.6.2})_1$. Note that the local Implicit Function The-*

*orem does not hold without assumption (II.6.2)$_2$, which means that there is
not necessarily a unique local curve of solutions $\{(x(\lambda), \lambda)\}$ through (x_0, λ_0).
But the nonzero local degree (II.6.7) and the homotopy invariance of the degree imply that the solution (x_0, λ_0) is continued for $\lambda \in (\lambda_0 - \delta, \lambda_0 + \delta)$. Let
C denote the component in S containing (x_0, λ_0). Then the same alternative
(i), (ii) holds, and in any case, $C \backslash \{(x_0, \lambda_0)\} \neq \emptyset$.*

Remark II.6.3 *The possibility of a global extension of the local solution
curve given by the Implicit Function Theorem is also called* **Global Continuation.** *It gives solutions of $F(x, \lambda) = 0$ for all $\lambda \in \mathbb{R}$, provided that (x_0, λ_0)
is the only solution for $\lambda = \lambda_0$ and that there is an a priori estimate for
solutions x for all λ in finite intervals of $\mathbb{R}$. This possibility motivated Leray
and Schauder to extend the Brouwer degree to infinite dimensions in order
to solve nonlinear elliptic partial differential equations; cf. Section III.5.*

II.7 Change of Morse Index and Local Bifurcation for Potential Operators

Zero is called a hyperbolic equilibrium of a linear operator $A \in L(X, Z), X \subset Z$, if after a natural complexification of X and Z (cf. Section I.8), there is no
spectral point of A on the imaginary axis of $\mathbb{C}$. If A has only a point spectrum,
the number of eigenvalues (counting multiplicities) in the right half-plane of
$\mathbb{C}$ is called the Morse index of A. For general operators, however, there might
be infinitely many such eigenvalues, even if A is a Fredholm operator. Under
the assumptions of Section II.4, however, a local Morse index is defined as
follows.

   Assume for a family of linear operators $A(\lambda) \in L(X, Z)$ for $\lambda \in \mathbb{R}$ the
properties (II.4.1), (II.4.2), and (II.4.3). Then the generalized eigenspace E_{λ_0}
of the eigenvalue 0 of $A(\lambda_0)$ is finite-dimensional, and that eigenspace is
perturbed to an invariant space E_λ for $A(\lambda)$ of the same dimension for λ
near λ_0. The eigenvalue 0 of $A(\lambda_0)$ perturbs to eigenvalues of $A(\lambda)$ near 0
(the so-called 0-group) that are the eigenvalues of $A(\lambda) \in L(E_\lambda, E_\lambda)$; cf. [86],
Chapters II and III.

Definition II.7.1 *Assume that zero is a locally hyperbolic equilibrium of
$A(\lambda) \in L(X, Z)$ for all $\lambda \in (\lambda_0 - \delta, \lambda_0) \cup (\lambda_0, \lambda_0 + \delta)$; i.e., there is no eigenvalue
in the 0-group of $A(\lambda)$ on the imaginary axis. Let $n^>(\lambda)$ be the number of
all eigenvalues in the 0-group of $A(\lambda)$ (counting multiplicities) in the positive
complex half-plane. This number is constant for $\lambda \in (\lambda_0 - \delta, \lambda_0)$ and for
$\lambda \in (\lambda_0, \lambda_0 + \delta)$, and it is called the local Morse index of $A(\lambda)$ at 0. The
number*

(II.7.1) $\chi(A(\lambda), \lambda_0) = n^>(\lambda_0 - \varepsilon) - n^>(\lambda_0 + \varepsilon), \quad 0 < \varepsilon < \delta,$

is the crossing number of the family $A(\lambda)$ at $\lambda = \lambda_0$ through 0.

Clearly, if $\chi(A(\lambda), \lambda_0)$ is an odd number, then the family $A(\lambda)$ has an odd crossing number at $\lambda = \lambda_0$ according to Definition II.4.1. Observe, however, that Definition II.4.1 does not require a local hyperbolicity of the equilibrium zero of $A(\lambda)$. A nonzero crossing number (II.7.1) means that the local Morse index of $A(\lambda)$ changes at $\lambda = \lambda_0$. If $\chi(A(\lambda), \lambda_0)$ is nonzero but even, then the family $A(\lambda)$ does not have an odd crossing number at $\lambda = \lambda_0$ according to Definition II.4.1.

As in Section II.4, we show that a change of the local Morse index is detected by the method of Lyapunov–Schmidt, provided that the family $A(\lambda)$ is symmetric in the sense of Section I.3.

For the sake of convenience, we restate the assumptions. On the Banach space Z a continuous and definite scalar product (I.3.1) is defined. We assume that

(II.7.2) $\begin{aligned} &(A(\lambda)x_1, x_2) = (x_1, A(\lambda)x_2) \\ &\text{for all } x_1, x_2 \in X \subset Z, \quad \lambda \in (\lambda_0 - \delta, \lambda_0 + \delta). \end{aligned}$

Let E_λ be the perturbed finite-dimensional invariant space for $A(\lambda)$, where E_{λ_0} is the generalized eigenspace of the eigenvalue 0 of $A(\lambda_0)$. Then $E_\lambda \subset X \subset Z$, so that E_λ is endowed with a scalar product and (II.7.2) holds on E_λ. Thus $A(\lambda) \in L(E_\lambda, E_\lambda)$ is symmetric, and therefore

(II.7.3) $\begin{aligned} &E_\lambda \text{ possesses a basis of eigenvectors of } A(\lambda), \\ &\text{and all eigenvalues of } A(\lambda) \text{ are real.} \end{aligned}$

Furthermore, the algebraic multiplicity $\dim E_\lambda = \dim E_{\lambda_0}$ of the eigenvalue 0 of $A(\lambda_0)$ coincides with its geometric multiplicity.

Thus $N(A(\lambda_0)) = E_{\lambda_0}$, and there are invariant decompositions of X and Z as follows (cf. (II.4.4)):

(II.7.4) $\begin{aligned} X &= N \oplus X_0, \quad N = N(A(\lambda_0)), \quad X_0 = R(A(\lambda_0)) \cap X, \\ Z &= R \oplus Z_0, \quad R = R(A(\lambda_0)), \quad Z_0 = N. \end{aligned}$

By the symmetry (II.7.2), the spaces N and X_0 as well as R and Z_0 are orthogonal with respect to the scalar product $(\ , \ )$. We also make use of the following decompositions, which are continuous perturbations of (II.7.4) for τ near λ_0:

(II.7.5) $\begin{aligned} X &= E_\tau \oplus X_\tau, \quad X_\tau = (I - P(\tau))X, \\ Z &= R_\tau \oplus E_\tau, \quad R_\tau = (I - P(\tau))Z. \end{aligned}$

Here $P(\tau)$ is the eigenprojection of $A(\tau)$ onto the eigenspace $E_\tau \subset X \subset Z$, which depends continuously on τ in $L(X, X)$ and in $L(Z, Z)$ for τ near λ_0. Since $A(\tau)$ is symmetric with respect to the scalar product $(\ , \ )$, the eigenprojection $P(\tau)$ is symmetric as well, so that the decompositions (II.7.5) are orthogonal with respect to $(\ ,\ )$. Clearly, (II.7.4) is embedded into (II.7.5) for $\tau = \lambda_0$. The Lyapunov–Schmidt projections defined by (II.7.5) are $Q_\tau = P(\tau) : Z \to E_\tau$ along R_τ and $P_\tau = Q_\tau|_X : X \to E_\tau$ along X_τ, and $P(\tau)N = P(\tau)E_{\lambda_0} = E_\tau$.

We represent $A(\lambda) : X \to Z$ according to a decomposition (II.7.5) as a matrix

(II.7.6) $$A(\lambda) = \begin{pmatrix} (I - Q_\tau)A(\lambda) & (I - Q_\tau)A(\lambda) \\ Q_\tau A(\lambda) & Q_\tau A(\lambda) \end{pmatrix},$$

identifying the direct sums in (II.7.5) with Cartesian products whose elements are written as columns; cf. (II.4.10).

Defining $C(Q_\tau, \lambda)$ and $D(\tau, Q_\tau, \lambda)$ as in (II.4.12), we obtain, as in (II.4.13),

(II.7.7) $$C(Q_\tau, \lambda)A(\lambda) = D(\tau, Q_\tau, \lambda).$$

Following (II.4.15) with $[(I - Q_\tau)A(\lambda)]^{-1} \in L(R_\tau, X_\tau)$ and

(II.7.8) $$\begin{aligned} B(\tau, Q_\tau, \lambda) &= Q_\tau A(\lambda)\{I_{E_\tau} - [(I - Q_\tau)A(\lambda)]^{-1}(I - Q_\tau)A(\lambda)\} \\ &\in L(E_\tau, E_\tau), \end{aligned}$$

we see that the symmetry of Q_τ and $A(\lambda)$ with respect to the scalar product $(\ ,\ )$ implies the symmetry of $B(\tau, Q_\tau, \lambda) \in L(E_\tau, E_\tau)$. Therefore, all eigenvalues of $B(\tau, Q_\tau, \lambda)$ are real for all τ and λ near λ_0.

By definition (II.4.12), $C(Q_\lambda, \lambda)|_{E_\lambda} = I_{E_\lambda}$, and $B(\lambda, Q_\lambda, \lambda) = A(\lambda)|_{E_\lambda}$. Furthermore, $D(\tau, Q_\tau, \lambda)|_{E_\tau} = B(\tau, Q_\tau, \lambda)$, and therefore (II.7.7) defines a homotopy in $L(E_\tau, E_\tau)$:

(II.7.9) $$C(Q_\tau, \lambda)A(\lambda)|_{E_\tau} = \begin{cases} A(\lambda) \in L(E_\lambda, E_\lambda) & \text{for } \tau = \lambda, \\ B(\lambda_0, Q_{\lambda_0}, \lambda) \in L(N, N) & \text{for } \tau = \lambda_0. \end{cases}$$

(Recall that $N = E_{\lambda_0}$.) Since $C(Q_\tau, \lambda)$ is an isomorphism for all τ and λ near λ_0, cf. (II.4.14), $B(\tau, Q_\tau, \lambda) \in L(E_\tau, E_\tau)$ is regular for all $\lambda \in (\lambda_0 - \delta, \lambda_0) \cup (\lambda_0, \lambda_0 + \delta)$ and for all $\tau \in (\lambda_0 - \delta, \lambda_0 + \delta)$. This, in turn, implies that for all τ between λ and λ_0, zero is a hyperbolic equilibrium of $B(\tau, Q_\tau, \lambda)$ (all eigenvalues are real) and that the Morse index of $B(\tau, Q_\tau, \lambda)$ is constant. According to Definition II.7.1, this proves the following theorem:

Theorem II.7.2 Let $A(\lambda) \in L(X, Z)$ be a family of Fredholm operators satisfying (II.4.1), (II.4.2), and that is symmetric in the sense of (II.7.2). Let 0 be an isolated eigenvalue of $A(\lambda_0)$ and assume that zero is a locally hyperbolic equilibrium of $A(\lambda)$ for $\lambda \in (\lambda_0 - \delta, \lambda_0) \cup (\lambda_0, \lambda_0 + \delta)$. Choose the Lyapunov–Schmidt decomposition (II.7.4) with projection $Q = Q_{\lambda_0} : Z \to N$ along R. Then zero is a hyperbolic equilibrium of $B(\lambda_0, Q, \lambda) \in L(N, N)$ for all $\lambda \in (\lambda_0 - \delta, \lambda_0) \cup (\lambda_0, \lambda_0 + \delta)$, and the families $A(\lambda) \in L(X, Z)$ and $B(\lambda_0, Q, \lambda) \in L(N, N)$ have the same crossing numbers at $\lambda = \lambda_0$ through 0:

(II.7.10) $$\chi(A(\lambda), \lambda_0) = \chi(B(\lambda_0, Q, \lambda), \lambda_0).$$

We apply Theorem II.7.2 for the proof of a local bifurcation theorem for a family of potential operators.

II.7.1 Local Bifurcation for Potential Operators

Let $F : U \times V \to Z$ be a mapping such that $0 \in U \subset X$ and $\lambda_0 \in V \subset \mathbb{R}$ are open. We assume that $F(0, \lambda) = 0$ for all $\lambda \in V$ and that F is a nonlinear Fredholm operator with respect to $x \in U$ and for all $\lambda \in V$: To be more precise, $F \in C(U \times V, Z)$ and $D_x F \in C(U \times V, L(X, Z))$ such that $A(\lambda) \equiv D_x F(0, \lambda)$ is a family of Fredholm operators satisfying (II.4.1), (II.4.2). Moreover, we assume that Z is endowed with a scalar product $(\ , \ )$ satisfying (I.3.1) and that

(II.7.11)
$\qquad F(\cdot, \lambda)$ is a potential operator from U into Z
$\qquad$ according to Definition I.3.1 for all $\lambda \in V$.

A consequence of (II.7.11) is that $A(\lambda) = D_x F(0, \lambda) \in L(X, Z)$ is a family of symmetric operators satisfying (II.7.2); cf. Proposition I.3.2.

We prove next a **Bifurcation Theorem for Potential Operators**:

Theorem II.7.3 *Let $F(\cdot, \lambda)$ be a family of potential operators satisfying the hypotheses summarized above in this Section. Let 0 be an isolated eigenvalue of $A(\lambda_0) = D_x F(0, \lambda_0)$. If zero is a locally hyperbolic equilibrium of $A(\lambda) = D_x F(0, \lambda)$ for $\lambda \in (\lambda_0 - \delta, \lambda_0) \cup (\lambda_0, \lambda_0 + \delta)$ according to Definition II.7.1 and if the crossing number $\chi(A(\lambda), \lambda_0)$ of the family $A(\lambda)$ at $\lambda = \lambda_0$ through 0 is nonzero, then $(0, \lambda_0)$ is a bifurcation point of $F(x, \lambda) = 0$ in the following sense: $(0, \lambda_0)$ is a cluster point of nontrivial solutions $(x, \lambda) \in X \times \mathbb{R}$, $x \neq 0$, of $F(x, \lambda) = 0$.*

In a few words: Any change of the local Morse index of $A(\lambda) = D_x F(0, \lambda)$ at $\lambda = \lambda_0$ implies bifurcation of $F(x, \lambda) = 0$ at $\lambda = \lambda_0$.

Proof. In order to find nontrivial solutions of $F(x, \lambda) = 0$ near $(0, \lambda_0)$ a Lyapunov–Schmidt reduction as in Section I.2 is adequate. To this end, we decompose the Banach spaces X and Z as in (II.7.4), and according to Theorem I.2.3, the problem $F(x, \lambda) = 0$ for (x, λ) near $(0, \lambda_0)$ is equivalent to a finite-dimensional problem $\Phi(v, \lambda) = 0$ near $(0, \lambda_0)$ in $N \times \mathbb{R}$. Clearly, $\Phi(0, \lambda) = 0$, and as shown in Section II.4, formula (II.4.25),

(II.7.12)
$$\begin{aligned} D_v \Phi(0, \lambda) &= B(\lambda_0, Q, \lambda) \\ &= QA(\lambda)\{I_N - [(I - Q)A(\lambda)]^{-1}(I - Q)A(\lambda)\} \in L(N, N). \end{aligned}$$

By Corollary I.2.4, $D_v \Phi(0, \lambda_0) = 0 \in L(N, N)$, but by Theorem II.7.2, zero is a hyperbolic equilibrium of $D_v \Phi(0, \lambda)$ for all $\lambda \in (\lambda_0 - \delta, \lambda_0) \cup (\lambda_0, \lambda_0 + \delta)$, and the crossing number $\chi(D_v \Phi(0, \lambda), \lambda_0)$ is nonzero. If we consider the dynamical system in N,

(II.7.13)
$$\frac{dv}{dt} = \Phi(v, \lambda),$$

this means, by definition, that

(II.7.14)   zero is a hyperbolic equilibrium of the dynamical
system (II.7.13) for $\lambda \in (\lambda_0 - \delta, \lambda_0) \cup (\lambda_0, \lambda_0 + \delta)$
whose Morse index changes at $\lambda = \lambda_0$.

Using Conley's index theory [22], the change of index (II.7.14) has the follow-ing consequences for (II.7.13): $\{0\} \subset N$ is an isolated invariant set of (II.7.13) for all $\lambda \in (\lambda_0 - \delta, \lambda_0) \cup (\lambda_0, \lambda_0 + \delta)$ and let $N_\varepsilon \subset \overline{B}_\varepsilon(0) \subset N$ be a compact isolating neighborhood of $\{0\}$ for (II.7.13) with $\lambda = \lambda_0 \pm \varepsilon, 0 < \varepsilon < \delta$. Then there is some $\tilde{\lambda} \in (\lambda_0 - \varepsilon, \lambda_0 + \varepsilon)$ such that (II.7.13) has a global solution whose trajectory is in N_ε for all $t \in (-\infty, \infty)$ and touches the boundary of N_ε at some t. If there were no such $\tilde{\lambda} \in (\lambda_0 - \varepsilon, \lambda_0 + \varepsilon)$, then N_ε would define a continuation from $\lambda_0 - \varepsilon$ to $\lambda_0 + \varepsilon$, and the Morse indices would be the same, contradicting (II.7.14). The union of all such bounded trajectories forms a nontrivial bounded invariant set in $N_\varepsilon \subset \overline{B}_\varepsilon(0)$ for (II.7.13) with $\tilde{\lambda} \in (\lambda_0 - \varepsilon, \lambda_0 + \varepsilon)$. Since $0 < \varepsilon < \delta$ is arbitrary, we can state that

(II.7.15) $(0, \lambda_0)$ is a bifurcation point of nontrivial
bounded invariant sets of (II.7.13).

By Theorem I.3.4, $\Phi(\cdot, \lambda) : \tilde{U} \to N, \tilde{U} \subset N$, is a family of potential operators for all $\lambda \in (\lambda_0 - \delta, \lambda_0 + \delta)$: There exists a function $\varphi : \tilde{U} \times (\lambda_0 - \delta, \lambda_0 + \delta) \to \mathbb{R}$ such that

(II.7.16) $D_v\varphi(v, \lambda)h = (\Phi(v, \lambda), h)$ for all
$(v, \lambda) \in \tilde{U} \times (\lambda_0 - \delta, \lambda_0 + \delta) \subset N \times \mathbb{R}, \quad h \in N.$

Here the scalar product $(\ ,\ )$ on N is induced by the scalar product on Z. Let $\{v(t) | t \in (-\infty, \infty)\}$ be a global nontrivial bounded trajectory of (II.7.13) with $\lambda = \tilde{\lambda}$. Then

(II.7.17) $\dfrac{d}{dt}\varphi(v(t), \tilde{\lambda}) = D_v\varphi(v(t), \tilde{\lambda})\dot{v}(t)$
$= \|\Phi(v(t), \tilde{\lambda})\|^2 \geq 0$ for all $t \in (-\infty, \infty),$

where $\|v\|^2 = (v, v), \ v \in N$. In this sense, $\varphi(\cdot, \tilde{\lambda})$ is a Lyapunov func-tion for (II.7.13) with $\lambda = \tilde{\lambda}$ whose orbital derivative (II.7.17) is nonneg-ative and vanishes only at equilibria of (II.7.13) with $\lambda = \tilde{\lambda}$. Furthermore, $\{\varphi(v(t), \tilde{\lambda}) | t \in (-\infty, \infty)\}$ is a bounded monotonic function such that the lim-its $\lim_{t \to \pm\infty} \varphi(v(t), \tilde{\lambda})$ exist and $\lim_{t \to \pm\infty} \frac{d}{dt}\varphi(v(t), \tilde{\lambda}) = 0$. The α-(or ω-)limit set of $\{v(t)\}$ is not empty. Let v_0 be any element of the ω-limit set, say. Then $\lim_{t \to \infty} \varphi(v(t), \tilde{\lambda}) = \varphi(v_0, \tilde{\lambda}), \lim_{t \to \infty} \frac{d}{dt}\varphi(v(t), \tilde{\lambda}) = \|\Phi(v_0, \tilde{\lambda})\|^2 = 0$, and v_0 is an equilibrium of (II.7.13) with $\lambda = \tilde{\lambda}$. This proves the well-known fact that the α- and ω-limit sets of a bounded trajectory of a dynamical system having a Lyapunov function $\varphi(\cdot, \tilde{\lambda})$ as in (II.7.17) consist only of equilibria, i.e., of solutions of $\Phi(v, \tilde{\lambda}) = 0$. If the α- and ω-limit sets were both $\{0\}$, then by the monotonicity, $\varphi(v(t), \tilde{\lambda}) \equiv \varphi(0, \tilde{\lambda})$ or $\frac{d}{dt}\varphi(v(t), \tilde{\lambda}) = \|\Phi(v(t), \tilde{\lambda})\|^2 \equiv 0$, which means that $v(t) \equiv 0$ or v is trivial. Thus at least one of the α- or ω-limit sets of a nontrivial bounded trajectory v is nontrivial. Clearly, the nontrivial equilibrium $(v_0, \tilde{\lambda})$ is in $N_\varepsilon \times (\lambda_0 - \varepsilon, \lambda_0 + \varepsilon)$. Since $0 < \varepsilon < \delta$ is arbitrary, every neighborhood of $(0, \lambda_0)$ contains a nontrivial solution of $\Phi(v, \lambda) = 0$, which proves the theorem.   □

Remark II.7.4 *The proof of Theorem II.7.3 reveals its validity also for $F(\cdot, \lambda)$, which is not a family of potential operators: For the application of Theorem II.7.2 one needs the symmetry of $D_x F(0, \lambda) = A(\lambda)$ in the sense of (II.7.2). If the dynamical system (II.7.13) has a Lyapunov function as in (II.7.17), then the α- and ω-limit sets consist only of equilibria. If these properties for F and Φ can be verified, a nonzero crossing number of the family $A(\lambda)$ at $\lambda = \lambda_0$ through 0 implies the same conclusion for $F(x, \lambda) = 0$ as given in Theorem II.7.3.*

The set of nontrivial solutions $\{(x, \lambda)\}$ of $F(x, \lambda) = 0$ near $(0, \lambda_0)$ does not form a curve or a continuum in general (provided that $\chi(A(\lambda), \lambda_0)$ is not odd; cf. Theorem II.4.4). A counterexample can be found in [14].

The example in [14] is a counterexample to Theorem I.21.2. We show that Theorem I.21.2 is a special case of Theorem II.7.3:

Assumption (I.21.2) on F implies (II.7.11) for $F(x) - \lambda x$ with potential $f(x) - \frac{\lambda}{2}(x, x)$. Here $A(\lambda) = DF(0) - \lambda I = A_0 - \lambda I$ satisfies (II.4.1), (II.4.2) for $\lambda \in (\lambda_0 - \delta, \lambda_0 - \delta)$ according to (I.21.3), and the 0-group of $A(\lambda)$ consists of $\{\lambda_0 - \lambda\}$ if λ_0 is an isolated eigenvalue of A_0. Since $E_{\lambda_0} = N(A_0 - \lambda_0 I)$ by assumption (I.21.3), we obtain $E_\lambda = N(A_0 - \lambda_0 I)$ as well, and counting the multiplicity of the positive eigenvalue $\lambda_0 - \lambda$ of $A_0 - \lambda I$ for $\lambda \in (\lambda_0 - \delta, \lambda_0)$, we obtain $\chi(A(\lambda), \lambda_0) = n > 0$ if $n = \dim N(A_0 - \lambda_0 I)$. Therefore, the assumptions for Theorem I.21.2 allow us also to apply Theorem II.7.3, and the example in [14] is a counterexample to Theorem II.7.3.

We close this section by a remark that odd crossing numbers in general and nonzero crossing numbers in case of a family of potential operators are indispensable for bifurcation. This is illustrated by the following simple example in $\mathbb{R}^2$:

$$(II.7.18) \qquad F(x, \lambda) = \lambda \begin{pmatrix} x_1 \\ x_2 \end{pmatrix} + \begin{pmatrix} x_2^3 \\ -x_1^3 \end{pmatrix} = 0, \qquad x = \begin{pmatrix} x_1 \\ x_2 \end{pmatrix} \in \mathbb{R}^2.$$

Here $F(0, \lambda) = 0$, $A(\lambda) = D_x F(0, \lambda) = \lambda I$ such that $\chi(A(\lambda), 0) = -2$. Nonetheless, $x = (x_1, x_2) = (0, 0)$ is the only solution of (II.7.18) for all $\lambda \in \mathbb{R}$. Thus even and nonzero crossing numbers do not imply bifurcation in general unless $F(\cdot, \lambda)$ has a potential. Finally, a zero crossing number does not entail bifurcation even if $F(\cdot, \lambda)$ has a potential. This is seen by the following example in $\mathbb{R}$:

$$(II.7.19) \qquad F(x, \lambda) = \lambda^2 x + x^3 = 0, \quad x \in \mathbb{R}.$$

Again, $F(0, \lambda) = 0$, $A(\lambda) = D_x F(0, \lambda) = \lambda^2 I$, $\chi(A(\lambda), 0) = 0$, and $x = 0$ is the only solution of (II.7.19) for all $\lambda \in \mathbb{R}$.

Remark II.7.5 *According to Conley's bifurcation result (II.7.15) under the hypotheses (II.7.14), the point $(0, \lambda_0) = (0, 0)$ is a bifurcation point of non-trivial bounded invariant sets of the dynamical system $\dot{x} = F(x, \lambda)$, where F is given in (II.7.18).*

Indeed, in view of $\frac{d}{dt}(x_1^4 + x_2^4) = 4(x_1^3 \dot{x}_1 + x_2^3 \dot{x}_2) = 4\lambda(x_1^4 + x_2^4)$, the sets $T_r = \{(x_1, x_2) | x_1^4 + x_2^4 = r^4\}$ are invariant for the system $\dot{x} = F(x, 0)$ for all $r > 0$, i.e., the bounded invariant sets T_r bifurcate vertically. The sets T_r are closed curves and therefore they represent periodic solutions with periods P_r.

By Green's formula the area of $\Omega_r = \{(x_1, x_2) | x_1^4 + x_2^4 < r^4\}$ is given by $|\Omega_r| = 2 \int_{T_r} x_1 dx_2 - x_2 dx_1 = 2 \int_0^{P_r} x_1^4 + x_2^4 dt = 2r^4 P_r$, whence $P_r = |\Omega_r|/2r^4 = |\Omega_1|/2r^2$. The periods of the bifurcating periodic solutions tend to infinity as the amplitudes tend to zero.

Remark II.7.6 *The local Theorem II.7.3 is included in Chapter II about a global theory, since the methods to prove it differ considerably from the analytic methods expounded in Chaper I about a local theory. The Brouwer or Leray–Schauder degree as well as Conley's index are topological and global tools. Furthermore, the condition for bifurcation in terms of a nonzero crossing number fits perfectly into the framework of Chapter II.*

II.8 Notes and Remarks to Chapter II

Degree theories have been developed for various classes of mappings, not all of which are mentioned here. A degree for Fredholm operators was suggested by Smale [160], and after this, using a concept of orientable Fredholm structures, such a degree was introduced in [41]. In [50], [51] that concept was replaced by a notion of orientability of Fredholm maps via their Fréchet derivatives, yielding a degree in the usual way. For a subclass of bounded Fredholm operators whose Fréchet derivatives have bounded and finite spectra on the negative real axis, thus having an orientation in a natural way, a degree was defined in [45], [31], [40], [128]. The extension to unbounded Fredholm operators whose Fréchet derivatives have again finite spectra on the negative real axis is given in [46] and independently in [98]. For relatively compact perturbations of linear Fredholm mappings the coincidence degree was defined in [136]; cf. also [54], [137]. A preliminary synopsis of these degree theories is given in [13].

The application of the degree in Bifurcation Theory goes back to Krasnosel'skii [118]. Here the sufficient (and in some sense also necessary) condition of an odd multiplicity was introduced. For a linear dependence on the
parameter, it is the algebraic multiplicity of an eigenvalue, whereas for a nonlinear dependence on the parameter, the notion of a multiplicity is much more
involved; cf. Remark II.4.6. We mention [154], [82], [133], [123], [98], [43], [42],
[146], [50], [84]. In most articles, the different notions of multiplicity are compared and proved to be equivalent (up to regularity of the mappings with
respect to the parameter).

In all theories an odd multiplicity means that the index of the linearization
along the trivial solution line changes. We call this an odd crossing number.
Bifurcation then takes place for any nonlinear remainder. In this sense these
are linear theories. For special nonlinearities, however, local and global bifurcation can also be proved for even multiplicities, i.e., for even crossing
numbers; see [130].

The bifurcation results for Fredholm operators are taken from [98]. Here
the relation between an odd crossing number and (local and global) bifurcation was established.

The Global Rabinowitz Bifurcation Theorem, including the case of a one-
dimensional kernel, was published in [147]. For the structure of global continua we refer also to [30], [84], where Theorem II.5.9 is sharpened.

Bifurcation for Potential Operators has a long history. Usually, the bifurcation parameter plays the role of a Lagrange multiplier; cf. Section I.21 and
the remarks at the end of Section I.22. This means that it appears linearly
in the equation. A nonlinear dependence on the parameter was first allowed
in [20], and after this, in [100], [146]. The approach here is taken from [100].

Chapter III
Applications

III.1 The Fredholm Property of Elliptic Operators

The Fredholm property is a leitmotiv, since it was assumed in all sections of Chapters I and II. Clearly, all finite-dimensional linear operators have that property, but we have infinite-dimensional applications in mind. A prototype of a Fredholm operator, playing also a central role in applications, is an elliptic operator over a bounded domain. We confine ourselves to operators of second order acting on scalar functions.

Let $\Omega \subset \mathbb{R}^n$ be a bounded domain with sufficiently smooth boundary $\partial\Omega$ such that the elliptic regularity theory is valid. For $u : \Omega \to \mathbb{R}$ (in $C^2(\Omega)$, say), the linear operator

$$Lu = \sum_{i,j=1}^n a_{ij}(x)u_{x_i x_j} + \sum_{i=1}^n b_i(x)u_{x_i} + c(x)u$$

with smooth and bounded coefficients
$a_{ij}, b_i, c : \overline{\Omega} \to \mathbb{R}$ is called elliptic if

(III.1.1) $\qquad \sum_{i,j=1}^n a_{ij}(x)\xi_i\xi_j \geq d\|\xi\|^2$ for all $x \in \Omega$, $\xi \in \mathbb{R}^n$,

with some uniform constant $d > 0$ and the
Euclidean norm $\|\ \ \|$ on $\mathbb{R}^n$.
Without loss of generality, we can assume that
$a_{ij}(x) = a_{ji}(x)$ for $i, j = 1, \ldots, n$.

In the literature, the nomenclature includes "uniformly elliptic" and "strongly elliptic," but we call it simply elliptic. The indices x_i denote partial derivatives with respect to x_i.

Closely related to the operator L is its bilinear form B of first order, obtained by an integration by parts if u is a test function in $C_0^\infty(\Omega)$:

(III.1.2)
$$B(u, v) = (Lu, v)_0 = (u, L^*v)_0$$
for all $u, v \in C_0^\infty(\Omega)$,
where $(\ , \ )_0$ is the scalar product in $L^2(\Omega)$.
L^* is called the formally adjoint operator.

If $L = L^*$, then L is formally self-adjoint, and in this case,

(III.1.3)
$$Lu = \sum_{i,j=1}^n (a_{ij}(x)u_{x_i})_{x_j} + c(x)u,$$
with $a_{ij}(x) = a_{ji}(x)$ for $i, j = 1, \dots, n$; i.e.,
L is of divergence form.

The bilinear form is symmetric, i.e., $B(u, v) = B(v, u)$, if and only if $L = L^*$.

In the sequel we define the operator L on function spaces that include boundary conditions on the functions u, v such that (III.1.2) holds (the regularity and boundary behavior of test functions is too much). For general L, *homogeneous Dirichlet boundary conditions*

(III.1.4)
$$u = 0 \quad \text{on } \partial\Omega,$$

and for formally self-adjoint L, so-called *natural boundary conditions*

(III.1.5)
$$\sum_{i,j=1}^n a_{ij}(x)u_{x_i}\nu_j = 0 \text{ for } x \in \partial\Omega,$$
where $\nu = (\nu_1, \dots, \nu_n)$ is the outward
unit normal vector field on $\partial\Omega$,

guarantee $B(u, v) = (Lu, v)_0 = (u, L^*v)_0$ for all $u, v \in C^2(\overline{\Omega})$ satisfying (III.1.4) or (III.1.5). In particular, if $L = \Delta$, then the natural boundary conditions are *homogeneous Neumann boundary conditions*. They play an important role in the calculus of variations, since (local) minimizers of $B(u, u) - 2(f, u)_0$ satisfy the natural boundary conditions if no boundary conditions on u are required a priori; cf. our remarks after (III.1.20).

Remark III.1.1 *Other homogeneous boundary conditions that provide an elliptic operator with all properties summarized in (III.1.6) below are possible. We confine ourselves to Dirichlet and natural boundary conditions, since on the one hand, they simplify our presentation, and on the other hand, they are important in applications.*

For our functional analysis, the linear operator L is defined in a Banach space (or Hilbert space) with an appropriate domain of definition. Then L is unbounded but closed, and when its domain of definition is given the graph norm (or an equivalent norm), then L is a bounded operator from one Banach space into another Banach space. In our abstract setting we use both aspects; cf. (II.4.1), (II.4.2): Let X and Z be real Banach spaces such that $X \subset Z$ is continuously embedded. Then we need the following properties of the linear operator L:

(III.1.6)
$$L : Z \to Z \text{ with domain of definition}$$
$$D(L) = X \subset Z \text{ is closed and}$$
$$L : X \to Z \text{ is bounded; i.e.,}$$
$$L \in L(X, Z) \text{ and}$$
$$L \text{ is a Fredholm operator of index zero.}$$

The first setting is needed for a spectral theory of L, whereas the last (and more subtle) property is crucial for a Lyapunov–Schmidt reduction, the universal tool of our abstract chapters. We verify (III.1.6) for different classes of spaces $X \subset Z$.

This can be done (and it is well known) because an elliptic operator L has a regularizing property with respect to various scales of smoothness. Due to this crucial property, it suffices to establish its Fredholm property in a Hilbert space, which is the most convenient space for linear problems: The geometry of a Hilbert space is Euclidean, and apart from local compactness, it is the same as the geometry of $\mathbb{R}^n$ or $\mathbb{C}^n$. If the domain of definition of a linear operator is dense, its adjoint is defined and acts in the same space.

In view of these advantages of a Hilbert space, one is tempted to formulate all applications in a Hilbert space. However, the benefits of a Hilbert space for linear problems turn eventually into obstructions for nonlinear problems that cannot be overcome. Nonlinear operators, in general, are defined only on Banach spaces of continuous functions (with continuous derivatives, etc.), which from a functional-analytic point of view are "bad," since they are not reflexive.

Therefore, when analyzing a nonlinear problem, we recommend the following flexibility: Use the Hilbert space for its linear aspects and switch to the Banach space to study its nonlinearity.

An elliptic operator (III.1.1) with homogeneous Dirichlet boundary conditions (III.1.4) defines an operator

(III.1.7)
$$L : L^2(\Omega) \to L^2(\Omega) \text{ with domain of definition}$$
$$D(L) = H^2(\Omega) \cap H^1_0(\Omega).$$

We use the following notation for the Sobolev spaces: $W^{k,p}(\Omega)$, $W^{k,p}_0(\Omega)$ for $p \in [1, \infty]$, $k \in \mathbb{N} \cup \{0\}$, $W^{k,2}(\Omega) = H^k(\Omega)$, $W^{k,2}_0(\Omega) = H^k_0(\Omega)$, and $H^0(\Omega) = H^0_0(\Omega) = L^2(\Omega)$. The respective norms are denoted by $\| \ \|_{k,p}$ and $\| \ \|_{k,2} = \| \ \|_k$, which are generated by scalar products $(\ , \ )_k$. (A good reference for Sobolev spaces is [1], and in [24], [52], [56], [122], [126] one finds all we need for elliptic operators.)

By ellipticity, Poincaré's inequality, and Gårding's inequality the bilinear form $B(u, u) - c\|u\|^2_0$ is negative definite on $H^1_0(\Omega)$ for some constant $c \geq 0$ such that the Lax–Milgram Theorem gives a unique weak solution $u \in H^1_0(\Omega)$ of $Lu - cu = f$ for every $f \in H^0(\Omega)$. The above-mentioned regularity of weak solutions implies that $u \in H^2(\Omega) \cap H^1_0(\Omega)$, and therefore,

$$\text{(III.1.8)} \qquad \begin{array}{l} L - cI : H^2(\Omega) \cap H_0^1(\Omega) \to H^0(\Omega) \\ \text{is bounded and bijective.} \end{array}$$

Here $D(L) = H^2(\Omega) \cap H_0^1(\Omega)$ is given the norm $\| \ \|_2$, which turns it into a Hilbert space. By Banach's Theorem, $(L - cI)^{-1} : H^0(\Omega) \to D(L)$ is continuous, which implies that the operator L defined in (III.1.7) is closed.

Furthermore,

$$\text{(III.1.9)} \qquad \begin{array}{l} (L - cI)^{-1} \equiv K_c \in L(H^0(\Omega), H^0(\Omega)) \\ \text{is compact,} \end{array}$$

by the compact embedding $D(L) \subset H^0(\Omega)$. Then for $f \in H^0(\Omega)$,

$$\text{(III.1.10)} \qquad \begin{array}{l} Lu = f, \quad u \in D(L) \Leftrightarrow \\ (I + cK_c)u = K_c f, \ u \in H^0(\Omega). \end{array}$$

The Riesz–Schauder Theory implies that $I + cK_c$ is a Fredholm operator of index zero. The equivalence (III.1.10) proves directly that

$$\text{(III.1.11)} \qquad \begin{array}{l} \dim N(L) = \dim N(I + cK_c) = n < \infty, \\ f \in R(L) \Leftrightarrow K_c f \in R(I + cK_c), \text{ and} \\ R(L) \text{ is closed in } H^0(\Omega). \end{array}$$

The trivial decomposition $f = (I + cK_c)f - cK_c f$ shows that $H^0(\Omega) = R(I + cK_c) + R(K_c)$. Then $H^0(\Omega) = R(I + cK_c) \oplus K_c(Z_0)$ for some n-dimensional space $Z_0 \subset H^0(\Omega)$ with $R(L) \cap Z_0 = \{0\}$. On the other hand, if $\tilde{Z}_0$ is any complementary space in the sense that $R(L) \oplus \tilde{Z}_0 = H^0(\Omega)$ (choose the orthogonal complement $\tilde{Z}_0 = R(L)^\perp$, e.g.), then (III.1.11) implies also that $K_c(\tilde{Z}_0) \cap R(I + cK_c) = \{0\}$. This proves $n \leq \text{codim} R(L) \leq n$, and in view of (III.1.11),

$$\text{(III.1.12)} \qquad \begin{array}{l} L : D(L) = H^2(\Omega) \cap H_0^1(\Omega) \to H^0(\Omega) \\ \text{is a Fredholm operator of index zero.} \end{array}$$

Considering L as an operator in $H^0(\Omega)$ as in (III.1.7), it is closed, densely defined, and in view of (III.1.12), the Closed Range Theorem in its Hilbert space version is applicable:

$$\text{(III.1.13)} \quad R(L) = \{f \in H^0(\Omega) | (f, u)_0 = 0 \quad \text{for all } u \in N(L^*)\}.$$

Here $L^* : H^0(\Omega) \to H^0(\Omega)$ is the adjoint of L with $D(L^*) \subset H^0(\Omega)$. Recall that the definition of the adjoint includes the definition of $D(L^*)$. If the formal adjoint defined in (III.1.2) is given the domain of definition $H^2(\Omega) \cap H_0^1(\Omega)$, then the same arguments as for (III.1.8) imply that $L^* - cI : H^2(\Omega) \cap H_0^1(\Omega) \to H^0(\Omega)$ is bijective, which proves that

(III.1.14) $\quad$ $L^* : L^2(\Omega) \to L^2(\Omega)$ with domain of definition
$D(L^*) = H^2(\Omega) \cap H_0^1(\Omega)$ is the adjoint of L.

(See [170] for a definition of L^* and its properties.) Interchanging the roles of L and L^* (clearly, $L^{**} = L$), we obtain the decompositions

(III.1.15) $\quad$
$$H^0(\Omega) = N(L) \oplus R(L^*),$$
$$H^0(\Omega) = R(L) \oplus N(L^*),$$

which are also orthogonal with respect to $(\ , \ )_0$. For applications the following representation of the corresponding orthogonal projections is useful:

(III.1.16)

Let $\{\hat{v}_1, \ldots, \hat{v}_n\} \subset N(L)$ be an orthonormal basis.

Then $Pu = \sum_{k=1}^n (u, \hat{v}_k)_0 \hat{v}_k$ is an orthogonal projection

$P : H^0(\Omega) \to N(L)$ along $R(L^*)$.

Let $\{\hat{v}_1^*, \ldots, \hat{v}_n^*\} \subset N(L^*)$ be an orthonormal basis.

Then $Qu = \sum_{k=1}^n (u, \hat{v}_k^*)_0 \hat{v}_k^*$ is an orthogonal projection

$Q : H^0(\Omega) \to N(L^*)$ along $R(L)$.

Since $N(L) \subset D(L) = X$ and $X \subset Z = H^0(\Omega)$, we obtain Lyapunov–Schmidt decompositions

(III.1.17)
$$X = N(L) \oplus (R(L^*) \cap X),$$
$$Z = R(L) \oplus N(L^*),$$
with projections (III.1.16); note that
$$P \in L(X, X) \text{ and } Q \in L(Z, Z).$$

If $L = L^*$ (formally), then L with homogeneous Dirichlet boundary conditions is self-adjoint, and (III.1.15)–(III.1.17) hold with $L = L^*$.

Remark III.1.2 *Note that the Lyapunov–Schmidt decomposition is not unique, and in particular, it is not necessarily orthogonal with respect to the scalar product $(\ , \ )_0$. In case of a semisimple eigenvalue 0 of L, one might choose $X = N(L) \oplus (R(L) \cap X)$, $Z = R(L) \oplus N(L)$, so that $P = Q|_X$; cf. (I.18.2)–(I.18.4).*

If $L = L^*$ (formally), the Dirichlet boundary conditions can be replaced by the natural boundary conditions (III.1.5). Accordingly,

(III.1.18) $\quad$ $L : L^2(\Omega) \to L^2(\Omega)$ with domain of definition
$D(L) = H^2(\Omega) \cap \{u | \sum_{i,j=1}^n a_{ij}(x) u_{x_i} \nu_j = 0 \text{ for } x \in \partial\Omega\}$,

where the boundary conditions hold in the generalized sense of the trace of functions in $H^2(\Omega)$ on $\partial\Omega$. By $B(u, v) = (Lu, v)_0 = (u, Lv)_0$ for all $u, v \in D(L)$, all conclusions drawn for $D(L) = H^2(\Omega) \cap H_0^1(\Omega)$ hold literally also for $D(L)$ as defined in (III.1.18), since the regularity of weak solutions is also true in this case. In particular, a bijectivity as in (III.1.8) holds, which implies by the steps (III.1.9)–(III.1.12) that

(III.1.19) $L : D(L) \to H^0(\Omega)$, where $D(L)$ as defined in (III.1.18) is given the norm $\| \quad \|_2$, is a Fredholm operator of index zero.

Since $L = L^*$ (formally), we obtain for the same reasons as for (III.1.14) that the operator L defined in (III.1.18) is self-adjoint, and the Closed Range Theorem gives

(III.1.20)
$$H^0(\Omega) = N(L) \oplus R(L),$$
yielding Lyapunov–Schmidt decompositions
$$X = D(L) = N(L) \oplus (R(L) \cap X),$$
$$Z = H^0(\Omega) = R(L) \oplus N(L).$$

In this case, $P = Q|_X$.

The peculiarity of the natural boundary conditions can be seen from the following: Consider the associated bilinear form $B : H^1(\Omega) \times H^1(\Omega) \to \mathbb{R}$, which is negative definite if $c(x) \leq -c_0 < 0$ for all $x \in \Omega$; cf. (III.1.3). Then the Riesz Representation Theorem gives a weak solution $u \in H^1(\Omega)$ of $Lu = f$ for every $f \in H^0(\Omega)$, i.e., $B(u, v) = (f, v)_0$ for all $v \in H^1(\Omega)$. By elliptic regularity, $u \in H^2(\Omega)$ and the Divergence Theorem (in its generalized form) then give $u \in D(L)$ as defined in (III.1.18). This observation is restated as follows: If $u \in H^1(\Omega)$ is a solution of the Euler–Lagrange equation for $B(u, u) - 2(f, u)_0$ in its weak form, which is $B(u, v) = (f, v)_0$ for all $v \in H^1(\Omega)$, then u satisfies the natural boundary conditions.

We emphasize that L has the Fredholm property also in other spaces. Assume that a Banach space $Z \subset L^2(\Omega)$ is continuously embedded and that there is a domain of definition $X \subset Z$ for L such that

(III.1.21) $L : X \to Z$ is continuous when $X = D(L|_Z)$ is given a norm that turns it into a Banach space.

We assume an *elliptic regularity* in the following sense:

(III.1.22) $Lu = f$ for $u \in D(L), f \in Z \Rightarrow u \in X$.

(Note that $D(L) = H^2(\Omega) \cap H_0^1(\Omega)$.) Then

(III.1.23)
$$N(L) = N(L|_Z) \subset X, \text{ and}$$
$$R(L) \cap Z = R(L|_Z) \text{ is closed in } Z,$$

since $Z \subset L^2(\Omega)$ is continuously embedded and $R(L)$ is closed in $L^2(\Omega)$. The ellipticity of L^* implies (III.1.22) also for L^* (with $D(L^*) = D(L)$), so that

(III.1.24) $$N(L^*) \subset X.$$

The decomposition $(III.1.15)_2$ implies that for every $z \in Z$,

(III.1.25)
$$
\begin{aligned}
&z = Lu + u^*, \quad \text{where } u \in D(L), u^* \in N(L^*),\\
&Lu = z - u^* \in Z \Rightarrow u \in X, \text{ whence}\\
&Z = R(L|_Z) \oplus N(L^*).
\end{aligned}
$$

Since $\dim N(L|_Z) = \dim N(L) = \dim N(L^*)$,

(III.1.26)
$$
\begin{aligned}
&L : X \to Z, \ X = D(L|_Z),\\
&\text{is a Fredholm operator of index zero.}
\end{aligned}
$$

Finally, the decomposition $(III.1.15)_1$ implies, by $(III.1.23)_1$,

(III.1.27) $$X = N(L|_Z) \oplus (R(L^*) \cap X),$$

so that the projections (III.1.16) also define Lyapunov–Schmidt projections

(III.1.28)
$$
\begin{aligned}
&P|_X : X \to N(L|_Z) \text{ along } (R(L^*) \cap X),\\
&Q|_Z : Z \to N(L^*) \text{ along } R(L|_Z),\\
&P|_X \in L(X, X), \ Q|_Z \in L(Z, Z).
\end{aligned}
$$

The crucial assumptions (III.1.21) and (III.1.22) are true for the following spaces:

(III.1.29)
$$
\begin{aligned}
&Z = L^p(\Omega), \ X = W^{2,p}(\Omega) \cap W_0^{1,p}(\Omega),\\
&\text{with norms } \|\ \|_{0,p} \text{ and } \|\ \|_{2,p}, \text{ and}\\
&Z = C^\alpha(\overline{\Omega}), \ X = C^{2,\alpha}(\overline{\Omega}) \cap \{u|u = 0 \text{ on } \partial\Omega\},\\
&\text{with norms } \|\ \|_{0,\alpha} \text{ and } \|\ \|_{2,\alpha},
\end{aligned}
$$

where for the Sobolev spaces $2 \le p < \infty$ and where for the Hölder spaces $0 < \alpha < 1$. *In all cases, the elliptic operator $L : X \to Z$ is a Fredholm operator of index zero.* For applications it is very useful that for all settings the same projections (III.1.16) can be used; cf. (III.1.28).

Replacing the homogeneous Dirichlet boundary conditions by the natural boundary conditions for $L = L^*$, we see that the elliptic regularity (III.1.22) holds as well when in $Z = L^p(\Omega)$ or $Z = C^\alpha(\overline{\Omega})$ the domain of definition X for L is defined as in (III.1.18), where $H^2(\Omega)$ is replaced by $W^{2,p}(\Omega)$ or $C^{2,\alpha}(\overline{\Omega})$, respectively. The decomposition (III.1.20) then yields

(III.1.30)
$$
\begin{aligned}
&X = N(L|_Z) \oplus (R(L) \cap X),\\
&Z = R(L|_Z) \oplus N(L),
\end{aligned}
$$

where L denotes the operator (III.1.18). Again, $N(L|_Z) = N(L)$ and $R(L|_Z) = R(L) \cap Z$, so that $Q|_X$ and $Q|_Z$ are continuous projections according to the decompositions (III.1.30) when Q is the projection defined in (III.1.16). Finally, $P = Q|_X$.

For one-dimensional domains, i.e., for intervals $\Omega = (a, b) \subset \mathbb{R}$, we can also choose the spaces $Z = C[a, b]$ and $X = C^2[a, b] \cap \{u | u(a) = u(b) = 0\}$ endowed with the usual maximum norm and the sum of the maximum norms of u and its first and second derivatives, respectively. In this case ellipticity means $a_{11}(x) \geq d > 0$ for all $x \in [a, b]$ and the regularity (III.1.22) follows from a continuous embedding $H^2(a, b) \subset C^1[a, b]$ and from the classical fundamental theorem of the calculus of variations due to DuBois–Reymond ([16], [109], e.g.): A weak derivative of a continuous function that is continuous is in fact a classical derivative.

Another peculiarity of a one-dimensional linear elliptic differential operator of second order is the fact that a multiplication by a suitable positive function ρ transforms it into a formally self-adjoint form. In this case the natural boundary conditions for an elliptic operator are $u'(a) = u'(b) = 0$ in the classical sense since $u \in C^1[a, b]$. The regularity (III.1.22) holds also in case of these or mixed boundary conditions, i.e., in case of $u(a) = u'(b) = 0$ or vice versa. Consequently, L is a Fredholm operator of index zero.

The theory for so-called Sturm–Liouville boundary value and eigenvalue problems yields that the kernel of L is at most one-dimensional and that 0 is a simple eigenvalue in this case. Therefore Remark III.1.2 holds. In particular, $Qu = (u, \rho\hat{v}_0)_0\hat{v}_0$, where $N(L) = span[\hat{v}_0]$ and $(\hat{v}_0, \rho\hat{v}_0)_0 = 1$. Indeed, $f \in R(L|_Z) = R(L) \cap Z \Leftrightarrow \rho f \in R(\rho L|_Z) = R(\rho L) \cap Z \Leftrightarrow (\rho f, \hat{v}_0)_0 = (f, \rho\hat{v}_0)_0 = 0$ for $\hat{v}_0 \in N(\rho L) = N(L) = N(L|_Z)$.

III.1.1 Elliptic Operators on a Lattice

Recall that the elliptic regularity (III.1.22) for the spaces (III.1.29) requires a smooth boundary $\partial\Omega$. For regular domains with corners, however, such as rectangles, squares, and certain triangles, we argue as follows: For

$$\Omega = (0, a) \times (0, b) \subset \mathbb{R}^2, \text{ we define}$$
$$X_D = \{u : \mathbb{R}^2 \to \mathbb{R} | u(x_1, x_2) = -u(-x_1, x_2) = -u(x_1, -x_2)$$
(III.1.31)
$$= u(x_1 + 2a, x_2) = u(x_1, x_2 + 2b) \text{ for } (x_1, x_2) \in \mathbb{R}^2\},$$
$$X_N = \{u : \mathbb{R}^2 \to \mathbb{R} | u(x_1, x_2) = u(-x_1, x_2) = u(x_1, -x_2)$$
$$= u(x_1 + 2a, x_2) = u(x_1, x_2 + 2b) \text{ for } (x_1, x_2) \in \mathbb{R}^2\}.$$

Then all functions $u \in C(\mathbb{R}^2) \cap X_D$ satisfy homogeneous Dirichlet boundary conditions $u = 0$ on $\partial\Omega$.

All functions $u \in C^1(\mathbb{R}^2) \cap X_N$ satisfy homogeneous Neumann boundary conditions $(\nabla u, \nu) = 0$ on $\partial\Omega$.

Let L be an elliptic operator (III.1.1) with smooth coefficients defined on $\mathbb{R}^2$ such that

$$a_{11}, a_{22} \in X_N, \ a_{12}, a_{21} \in X_D, \ c \in X_N,$$

(III.1.32)
$$
\begin{aligned}
b_1 &\in \{u : \mathbb{R}^2 \to \mathbb{R} \,|\, u(x_1, x_2) = -u(-x_1, x_2) = u(x_1, -x_2) \\
&= u(x_1 + 2a, x_2) = u(x_1, x_2 + 2b) \text{ for } (x_1, x_2) \in \mathbb{R}^2\}, \\
b_2 &\in \{u : \mathbb{R}^2 \to \mathbb{R} \,|\, u(x_1, x_2) = u(-x_1, x_2) = -u(x_1, -x_2) \\
&= u(x_1 + 2a, x_2) = u(x_1, x_2 + 2b) \text{ for } (x_1, x_2) \in \mathbb{R}^2\}.
\end{aligned}
$$

(If $L = L^*$ is of divergence form (III.1.3), then (III.1.32)$_1$ implies automatically the assumptions on b_1 and b_2.)

Define $\Omega_{-2,2} = (-2a, 2a) \times (-2b, 2b)$, $\Omega_{-2,0} = (-2a, 0) \times (-2b, 2b)$, and $\Omega_{0,2} = (0, 2a) \times (-2b, 2b)$. In dealing with weak solutions, the boundary plays no role. Choose $c \geq 0$ such that $B(u, u) - c\|u\|_0^2$ is negative definite over $H_0^1(\Omega_{-2,2})$ and that by the Lax–Milgram Theorem there is a unique weak solution $u \in H_0^1(\Omega_{-2,2})$ of $Lu - cu = f$ for every $f \in L^2(\Omega_{-2,2})$. Choose $f \in L^2(\Omega_{-2,2}) \cap X_D$. By symmetry of $\Omega_{-2,2}$, the function $\tilde{u}(x_1, x_2) \equiv u(-x_1, x_2)$ is also in $H_0^1(\Omega_{-2,2})$. The assumptions (III.1.32) imply that $\tilde{u}$ is a weak solution of $Lu - cu = \tilde{f} = -f$, so that by uniqueness, $\tilde{u} = -u$. This oddness of u with respect to x_1 implies that $u(0, x_2) = 0$, and by approximation by smooth functions (using the odd part of approximating sequences), $u \in H_0^1(\Omega_{-2,0}) \cap H_0^1(\Omega_{0,2})$. Therefore, $\hat{u}(x_1, x_2) \equiv u(x_1 + 2a, x_2) \in H_0^1(\Omega_{-2,0})$, too, and the assumptions (III.1.32) imply that $\hat{u}$ is a weak solution of $Lu - cu = \hat{f} = f$. Since $B(u, u) - c\|u\|_0^2$ is also negative definite over $H_0^1(\Omega_{-2,0})$, uniqueness of weak solutions in $H_0^1(\Omega_{-2,0})$ implies $u = \hat{u}$. In view of the oddness of u, we obtain then $u(-a, x_2) = u(a, x_2) = -u(a, x_2)$, and as before, $u(-a, x_2) = u(a, x_2) = 0$ implies $u \in H_0^1(\Omega_{-1,0}) \cap H_0^1(\Omega_{0,1})$, where $\Omega_{-1,0} = (-a, 0) \times (-2b, 2b)$ and $\Omega_{0,1} = (0, a) \times (-2b, 2b)$.

Using the same arguments for an inverse reflection and a $2b$-shift in the direction of x_2, we have proved that the weak solution $u \in H_0^1(\Omega_{-2,2})$ of $Lu - cu = f$ for $f \in L^2(\Omega_{-2,2}) \cap X_D$ is in $H_0^1(\Omega_{-2,2}) \cap X_D$, and in particular, $u \in H_0^1(\Omega)$ for $\Omega = (0, a) \times (0, b)$.

By *interior* regularity of weak solutions in $H_0^1(\Omega_{-2,2})$ we obtain $u \in H^2(\Omega) \cap H_0^1(\Omega)$. Since there is a one-to-one correspondence between $f \in L^2(\Omega)$ and $f \in L^2(\Omega_{-2,2}) \cap X_D$ (by extension via inverse reflections and periodicities), we have proved (III.1.8) also for $\Omega = (0, a) \times (0, b)$.

Consequently, (III.1.12) holds, and since the corresponding coefficients of L^* satisfy (III.1.32) as well, the statements (III.1.14) and (III.1.15) are proved analogously.

Let $Lu = f$ for $u \in H^2(\Omega) \cap H_0^1(\Omega)$ and $f \in C^\alpha(\mathbb{R}^2) \cap X_D$. Extending u via inverse reflections and periodicities to a function in X_D, the properties (III.1.32) of the coefficients of L imply $Lu = f$ (almost everywhere) on $\mathbb{R}^2$. By interior regularity, $u \in C^{2,\alpha}(\mathbb{R}^2) \cap X_D$, which proves (III.1.22) for $Z = C^\alpha(\mathbb{R}^2) \cap X_D$ and $X = C^{2,\alpha}(\mathbb{R}^2) \cap X_D$. Consequently,

$$L : C^{2,\alpha}(\mathbb{R}^2) \cap X_D \to C^{\alpha}(\mathbb{R}^2) \cap X_D$$

(III.1.33)   is a Fredholm operator of index zero,

$$C^{2,\alpha}(\mathbb{R}^2) \cap X_D = N(L) \oplus (R(L^*) \cap X),$$
$$C^{\alpha}(\mathbb{R}^2) \cap X_D = R(L) \oplus N(L^*).$$

Observe that the formal adjoint in (III.1.33) is also considered as an operator $L^* : X \to Z$.

The homogeneous Neumann boundary conditions are the natural boundary conditions for $L = \Delta + cI = L^*$. For negative $c \in X_N$, the corresponding bilinear form $B(u,u) + c\|u\|_0^2$ (where $B(u,u)$ is the negative Dirichlet integral) is negative definite over $H^1(\Omega_{-2,2})$, so that by Riesz's Theorem there is a unique weak solution $u \in H^1(\Omega_{-2,2})$ of $\Delta u + cu = f$ for every $f \in L^2(\Omega_{-2,2})$. Proceeding as before, if $f \in L^2(\Omega_{-2,2}) \cap X_N$, then $u \in H^1(\Omega_{-2,2}) \cap X_N$. (Here the arguments are easier, since we do not have to take care about boundary conditions.) By interior regularity of weak solutions in $H^1(\Omega_{-2,2})$ we obtain $u \in H^2(\Omega)$. We have two choices in proving the homogeneous Neumann boundary conditions for u: As mentioned before, weak solutions of $\Delta u + cu = f$ in $H^2(\Omega)$ satisfy $(\nabla u, \nu) = 0$ on $\partial\Omega$ in a generalized sense by the Divergence Theorem. On the other hand, if $u \in H^2(\tilde{\Omega})$ for $\overline{\Omega} \subset \tilde{\Omega} \subset (\Omega_{-2,2})$ and $u \in X_N$, then the symmetries and periodicities of functions in $C^1(\tilde{\Omega}) \cap X_N$ imply $(\nabla u, \nu) = 0$ on $\partial\Omega$ in the classical sense, which is extended to functions in $H^2(\tilde{\Omega}) \cap X_N$ by approximation. In any case, we end up with the statement that for negative $c \in X_N$,

(III.1.34) $$\Delta + cI : H^2(\Omega) \cap \{u | (\nabla u, \nu) = 0 \text{ on } \partial\Omega\} \to H^0(\Omega)$$
              is bounded and bijective; cf. (III.1.8).

This, in turn, implies (III.1.19) and (III.1.20) for $L = \Delta + cI$ for every (smooth) $c \in X_N$.

Let $Lu = f$ for $u \in D(L)$ (defined in (III.1.34)) and $f \in C^{\alpha}(\mathbb{R}^2) \cap X_N$. As before, replacing inverse reflections by reflections, we obtain by interior regularity $u \in C^{2,\alpha}(\mathbb{R}^2) \cap X_N$, proving (III.1.22) for $Z = C^{\alpha}(\mathbb{R}^2) \cap X_N$ and $X = C^{2,\alpha}(\mathbb{R}^2) \cap X_N$. This implies, by (III.1.20), for any (smooth) $c \in X_N$, that

$$L = \Delta + cI : C^{2,\alpha}(\mathbb{R}^2) \cap X_N \to C^{\alpha}(\mathbb{R}^2) \cap X_N$$

(III.1.35)   is a Fredholm operator of index zero,

$$C^{2,\alpha}(\mathbb{R}^2) \cap X_N = N(L) \oplus (R(L) \cap X),$$
$$C^{\alpha}(\mathbb{R}^2) \cap X_N = R(L) \oplus N(L),$$

where the decompositions are orthogonal with respect to the scalar product $(\ ,\ )_0$ in $L^2(\Omega)$.

The previous analysis for the rectangle $\Omega = (0,a) \times (0,b)$ and for a rectangular lattice in $\mathbb{R}^2$ is easily generalized to higher dimensions.

For a square lattice, i.e., when $a = b$, we can impose another (inverse) reflection across a diagonal, yielding

$$X_D^1 = \{u : \mathbb{R}^2 \to \mathbb{R} | u \in X_D, u(x_1, x_2) = -u(-x_2 + a, -x_1 + a)\},$$
$$X_N^1 = \{u : \mathbb{R}^2 \to \mathbb{R} | u \in X_N, u(x_1, x_2) = u(-x_2 + a, -x_1 + a)\}.$$

(III.1.36)

For $u \in C(\mathbb{R}^2) \cap X_D^1$, the nodal set $\{(x_1, x_2) | u(x_1, x_2) = 0\}$ contains $\{(x_1, x_2) | x_1 = ka, x_2 = ka, x_1 + x_2 = (2k+1)a, -x_1 + x_2 = (2k+1)a$ for all $k \in \mathbb{Z}\}$. Therefore, homogeneous Dirichlet boundary conditions are satisfied for the "tile" $\Omega = \{(x_1, x_2) | 0 \le x_1, 0 \le x_2, x_1 + x_2 \le a\}$, which is the triangle with corners $(0,0), (a,0), (0,a)$. Every other tile obtained by reflections and translations defining X_D^1 has clearly the same property. We sketch the lattices of X_D and X_D^1 in Figure III.1.1.

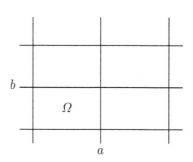

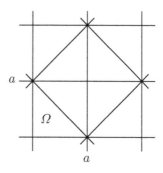

Lattice of X_D, X_N Lattice of X_D^1, X_N^1

Figure III.1.1

In addition to (III.1.32), we assume that

(III.1.37)   the coefficients a_{ij} and c have a reflection symmetry and all b_j have an inverse reflection symmetry across the line $\{x_1 + x_2 = a\}$.

Then L and L^* have the "equivariance" mapping $C^2(\mathbb{R}^2) \cap X_D^1$ into $C(\mathbb{R}^2) \cap X_D^1$. By the previous arguments, $L - cI : H^2(\Omega) \cap H_0^1(\Omega) \to H^0(\Omega)$ is bijective for the square $\Omega = (0, a) \times (0, a)$. If $f \in L^2(\Omega) \cap X_D^1$, in view of the equivariance of $L - cI$, the inverse reflection symmetry of f across the diagonal of Ω is inherited by the unique solution u of $Lu - cu = f$. As shown earlier, this implies the Fredholm property and the decompositions (III.1.33) where X_D is replaced by X_D^1.

The arguments for (III.1.35) apply also when X_N is replaced by X_N^1. Every $u \in C^{2,\alpha}(\mathbb{R}^2) \cap X_N^1$ satisfies homogeneous Neumann boundary conditions on the boundary of the triangle with corners $(0,0), (a,0), (0,a)$, and of every other triangular tile of the lattice of X_N^1; cf. Figure III.1.1.

Adding to X_D^1, X_N^1 another (inverse) reflection across the second diagonal $\{x_1 = x_2\}$ of the square, we obtain the lattice shown in Figure III.1.2.

Note that the square lattices in Figure III.1.1 and Figure III.1.2 are not essentially distinct, since they are transformed into each other by an affine mapping.

Another class of triangles for which our analysis applies is given by hexagonal lattices shown in Figure III.1.3. The double periodicity of u in both cases is $u(x_1 + a, x_2) = u(x_1, x_2) = u(x_1 + \frac{1}{2}a, x_2 + \frac{1}{2}\sqrt{3}a)$.

Assume first an inverse reflection across each line of a lattice shown in Figure III.1.3, which, together with the two periodicities, defines the "isotropy group" of the respective lattice. If the coefficients of the elliptic operator L on $\mathbb{R}^2$ have the double periodicity and all (inverse) reflection symmetries by analogy with (III.1.32), then L is "equivariant" with respect to the "isotropy group" of the lattice. (We leave it to the reader to give the precise conditions on L. A sufficient condition is found in (III.6.15), (III.6.16).)

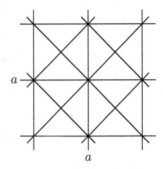

Figure III.1.2

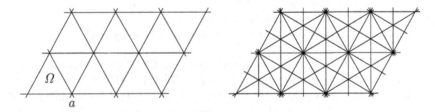

Figure III.1.3

Let u be the unique weak solution of $Lu - cu = f$ with weak homogeneous Dirichlet boundary conditions on a hexagon of side length $2a$ in the lattice. We sketch the proof that u inherits the isotropy of f, which is the isotropy of the lattice shown in the left picture of Figure III.1.3. (For details, see the proof for the rectangular lattice.) By uniqueness, u has an inverse reflection symmetry across all diagonals of the hexagon. Therefore, u satisfies weak homogeneous Dirichlet boundary conditions on the boundaries of each of the six equilateral triangles of side length $2a$. Shift u along each diagonal by an amount of a to the middle of the hexagon. The shifted triangles do not match any other of the six triangles but form triangles with three sides in the interior and one corner on one side of the hexagon. By the periodicities of f and the equivariance of L, the shifted u are weak solutions of $Lu - cu = f$ with weak homogeneous Dirichlet boundary conditions on each shifted triangle. On the other hand, the original u is also a weak solution of $Lu - cu = f$ on each of these triangles, but on two sides of the triangles, which are not on a diagonal of the hexagon, weak homogeneous Dirichlet boundary conditions are not necessarily satisfied.

Here is a new argument: Let Ω_1 denote one of these triangles. Then the bilinear form $B(u, u) - c\|u\|_0^2$ is negative definite on $H^1(\Omega_1)$, too, if $c > 0$ is large enough. Therefore, by uniqueness of weak solutions in $H^1(\Omega_1)$, the solutions u and the shifted u coincide, proving $u \in H_0^1(\Omega_1)$. A closer look at the geometry of a hexagon of side length $2a$ reveals the following: u has weak homogeneous Dirichlet boundary conditions on the boundaries of all 12 equilateral triangles of side length $2a$ and therefore also on all 24 equilateral triangles of side length a; u has double periodicity; and u has also inverse reflection symmetry across every line shown in the left picture of Figure III.1.3, provided that it is in the chosen hexagon of side length $2a$.

By interior regularity, $u \in H^2(\Omega) \cap H_0^1(\Omega)$ and $L - cI : H^2(\Omega) \cap H_0^1(\Omega) \to H^0(\Omega)$ is bijective for every tile Ω of the left lattice of Figure III.1.3 that is an equilateral triangle of side length a. If f has the additional three inverse reflection symmetries shown in the right picture of Figure III.1.3, then again by uniqueness and assumed equivariance of L, these symmetries are inherited by the solution of $Lu - cu = f$, too.

If $f \in C^\alpha(\mathbb{R}^2)$ has the isotropy of one of the lattices shown in Figure III.1.3, then every solution $u \in H^2(\Omega) \cap H_0^1(\Omega)$ of $Lu = f$ is extended to $\mathbb{R}^2$ having that isotropy, too, and by interior regularity, $u \in C^{2,\alpha}(\mathbb{R}^2)$.

Consequently, (III.1.33) holds if the isotropy of the function space X_D defined in (III.1.31) is replaced by a "fixed-point space" of any of the isotropy groups of the lattices shown in Figure III.1.1–Figure III.1.3, provided that L has the equivariance with respect to that isotropy group. Recall that each line of the lattices is an inverse reflection line and therefore a nodal line of every u in their respective "fixed-point spaces." Thus every u in a fixed-point space satisfies homogeneous Dirichlet boundary conditions on the boundary of each tile of the respective lattice.

Finally, if each line of the lattices is a reflection line and therefore a line where u satisfies homogeneous Neumann conditions, then (III.1.35) holds if the isotropy of X_N defined in (III.1.31) is replaced by any of the isotropies of the lattices shown in Figure III.1.1–Figure III.1.3.

III.1.2 Spectral Properties of Elliptic Operators

At the end of this section we make some useful remarks on the spectrum of an elliptic operator L, which is closely connected to its Fredholm property.

As usual, for a spectral theory, the underlying real function spaces are complexified in a natural way; cf. Section I.8. For the operator L defined in (III.1.7), property (III.1.8) means that the number $c \geq 0$ is in the resolvent set $\rho(L)$ of L and that the resolvent $(L - cI)^{-1} = K_c$ is compact; cf. (III.1.9). This allows the application of the Theorem on Compact Resolvents in [86]:

> The spectrum $\sigma(L)$ of L defined in (III.1.7)
> consists of isolated eigenvalues of finite algebraic
> multiplicities, and all resolvents are compact.
> If $c \in \rho(L)$, then for $(L - cI)^{-1} = K_c$,
>
> (III.1.38) $\quad \mu \in \rho(K_c) \Leftrightarrow \dfrac{1}{\mu} + c \in \rho(L),$
>
> and the geometric and algebraic multiplicities
> of an eigenvalue μ_0 of K_c and of
> the eigenvalue $\dfrac{1}{\mu_0} + c$ of L are the same.

(Observe that $0 \notin \rho(K_c)$ and 0 is not an eigenvalue of K_c.) Defining

(III.1.39)
$$L : Z \to Z \text{ with } D(L|_Z) = X \subset Z$$
$$\text{with } Z \text{ and } X \text{ as in (III.1.29)},$$

then the elliptic regularity (III.1.22) implies that

> the geometric and algebraic eigenspaces
> of L for an eigenvalue μ are in X,
> (III.1.40)   so that the spectrum and the geometric and algebraic
> multiplicities of an eigenvalue of L
> are the same for all settings (III.1.39).

This means that not only the Lyapunov–Schmidt projections (III.1.16) but also necessary spectral information on L (or on a family of elliptic operators) can be taken from its Hilbert space realization (III.1.7).

The results (III.1.38) and (III.1.40) hold accordingly for $L = L^*$ defined in (III.1.18) and for its restriction (III.1.39) when $Z = L^p(\Omega)$ or $Z = C^\alpha(\overline{\Omega})$ and X is defined as in (III.1.18), where $H^2(\Omega)$ is replaced by $W^{2,p}(\Omega)$ or $C^{2,\alpha}(\overline{\Omega})$, respectively. It is well known that a (formally) self-adjoint elliptic operator L has only real eigenvalues.

If $\Omega = (a, b) \subset \mathbb{R}$ is a finite interval, then, as mentioned in Section III.1, a multiplication by a suitable positive function transforms L into a (formal) self-adjoint form. Therefore one-dimensional elliptic operators with homogeneous Dirichlet, Neumann, or mixed boundary conditions have only real eigenvalues. A closer analysis of those so-called Sturm–Liouville eigenvalue problems yields that these eigenvalues are simple, which means that their geometric and algebraic multiplicities are one; cf. [16], [109], for example.

Finally, the spectral properties (III.1.38) and (III.1.40) hold also for equivariant L if Ω is a tile in one of the lattices shown in Figures III.1.1–III.1.3. The alternative setting here is (III.1.39) with $Z = C^\alpha(\mathbb{R}^2) \cap X_D$ and $X = C^{2,\alpha}(\mathbb{R}^2) \cap X_D$ when X_D is the fixed-point space of the isotropy group of one of the lattices. The subscript D means "inverse reflections" across the lines of the lattice. For $L = \Delta + cI$ we can also choose $Z = C^\alpha(\mathbb{R}^2) \cap X_N$ and $X = C^{2,\alpha}(\mathbb{R}^2) \cap X_N$, where the subscript N means "reflection" across the lines of the respective lattice.

Remark III.1.3 *Properties (III.1.38) and (III.1.40) provide another proof of the Fredholm property of L in all settings (III.1.39): If $0 \in \rho(L)$, then $N(L) = \{0\}$ and $R(L) = Z$, so that the Fredholm property of L is trivial. If $0 \in \sigma(L)$, it is an isolated eigenvalue of finite algebraic multiplicity. By the results in [86], [170] on isolated eigenvalues, the Banach space Z decomposes into $Z = E_0 \oplus Z_0$, where E_0 is the finite-dimensional generalized eigenspace of L for the eigenvalue 0. Both spaces E_0 and Z_0 are invariant for L, $L \in L(E_0, E_0)$, and $0 \in \rho(L|_{Z_0})$, so that $N(L) \subset E_0$ and $Z_0 \subset R(L)$. From finite-dimensional linear algebra we know that $E_0 = R(L|_{E_0}) \oplus \tilde{E}_0$ with $\dim \tilde{E}_0 = \dim N(L)$. Therefore, $Z = R(L|_{E_0}) \oplus \tilde{E}_0 \oplus Z_0 = R(L) \oplus \tilde{E}_0$, proving the Fredholm property of L.*

III.2 Local Bifurcation for Elliptic Problems

Let $\Omega \subset \mathbb{R}^n$ be a bounded domain with a boundary $\partial\Omega$ such that linear elliptic operators over Ω have the properties provided in Section III.1. A fully nonlinear elliptic problem with homogeneous Dirichlet boundary conditions is of the form

(III.2.1)
$$G(\nabla^2 u, \nabla u, u, x, \lambda) = 0 \quad \text{in } \partial\Omega, \ \lambda \in \mathbb{R},$$
$$u = 0 \quad \text{on } \partial\Omega,$$

where we use the following notation:

(III.2.2)
∇u is the gradient of u with components u_{x_i},
$\nabla^2 u$ is the second gradient or Hessian of u
with components $u_{x_i x_j}, i, j = 1, \ldots, n$,
$u : \Omega \to \mathbb{R}$ is in $C^2(\Omega)$, say, and $x \in \Omega$.

If $G \in C^{k+1}(\mathbb{R}^{n \times n}_{sym} \times \mathbb{R}^n \times \mathbb{R} \times \overline{\Omega} \times \mathbb{R}, \mathbb{R})$, then

$$(III.2.3) \qquad \begin{aligned} &F \in C^k(C^{2,\alpha}(\overline{\Omega}) \times \mathbb{R}, C^\alpha(\overline{\Omega})), \text{ where} \\ &F(u, \lambda)(x) \equiv G(\nabla^2 u(x), \nabla u(x), u(x), x, \lambda). \end{aligned}$$

Assuming $G(\mathbf{0}, \mathbf{0}, 0, x, \lambda) = 0$ for all $(x, \lambda) \in \Omega \times \mathbb{R}$, we have $F(0, \lambda) = 0$ and the trivial solution $(0, \lambda)$ for all $\lambda \in \mathbb{R}$. The Fréchet derivative of F with respect to u along the trivial solution is given by

$$D_u F(0, \lambda)h =$$

$$(III.2.4) \qquad \begin{aligned} &\sum_{i,j=1}^n G_{w_{ij}}(\mathbf{0}, \mathbf{0}, 0, x, \lambda)h_{x_i x_j} + \sum_{i=1}^n G_{v_i}(./.)h_{x_i} + G_u(./.)h \\ &\text{for } h \in C^{2,\alpha}(\overline{\Omega}), \text{ where the variables of} \\ &G : \mathbb{R}^{n \times n}_{sym} \times \mathbb{R}^n \times \mathbb{R} \times \overline{\Omega} \times \mathbb{R} \to \mathbb{R} \text{ are} \\ &\text{denoted by } (\mathbf{W}, \mathbf{v}, u, x, \lambda), \ \mathbf{W} = (w_{ij}), \ \mathbf{v} = (v_i). \end{aligned}$$

If for some $\lambda = \lambda_0$ the operator $D_u F(0, \lambda_0)$ is **elliptic** in the sense of (III.1.1), then for $X = C^{2,\alpha}(\overline{\Omega}) \cap \{u | u = 0 \text{ on } \partial\Omega\}$ and $Z = C^\alpha(\overline{\Omega})$,

$$(III.2.5) \qquad \begin{aligned} &F \in C^k(X \times \mathbb{R}, Z), \text{ and} \\ &D_u F(0, \lambda_0) \text{ is a Fredholm operator of index zero;} \end{aligned}$$

see (III.1.26), (III.1.29). (For the local analysis of this section we need smoothness of F only in a neighborhood of the bifurcation point $(0, \lambda_0)$. Accordingly, the global smoothness of G can be reduced to $G \in C^{k+1}(W \times V \times U \times \overline{\Omega} \times (\lambda_0 - \delta, \lambda_0 + \delta), \mathbb{R})$, where $W \times V \times U$ is an open neighborhood of $(\mathbf{0}, \mathbf{0}, 0) \in \mathbb{R}^{n \times n}_{sym} \times \mathbb{R}^n \times \mathbb{R}$.)

In many applications the operator in (III.2.1) is of a more special form,

$$(III.2.6) \qquad \begin{aligned} &G(\nabla^2 u, \nabla u, u, x, \lambda) \equiv \\ &\sum_{i,j=1}^n a_{ij}(\nabla u, u, x, \lambda)u_{x_i x_j} + g(\nabla u, u, x, \lambda), \end{aligned}$$

called *quasilinear*, since the highest-order derivatives of u appear linearly. Euler–Lagrange equations of first-order variational problems are quasilinear, which explains the importance of this class of problems. Moreover, for an Euler–Lagrange equation we obtain automatically $a_{ij} = a_{ji}$ in (III.2.6), and when linearized about $u = 0$, an Euler–Lagrange equation is of divergence form (III.1.3) and therefore self-adjoint. Its ellipticity is directly related to the convexity of the underlying functional, which explains the importance of ellipticity. For such problems, not only Dirichlet but also natural boundary conditions are of interest; cf. the example at the end of this section. In Section III.1 we provide the Fredholm property (III.2.5) also for this class of problems.

If $\Omega = (a, b) \subset \mathbb{R}$ is a (finite) interval, then $F(u, \lambda)(x) \equiv G(u''(x), u'(x), u(x), x, \lambda)$ defines a map $F \in C^k(C^2[a, b] \times \mathbb{R}, C[a, b])$, provided that $G \in C^k(\mathbb{R} \times \mathbb{R} \times \mathbb{R} \times [a, b], \mathbb{R})$. Therefore, for one-dimensional domains we can save one order of differentiability. The Fréchet derivative $D_u F(0, \lambda)h = G_w(0, 0, 0, x, \lambda)h'' + G_v(./.)h' + G_u(./.)h$ is elliptic if $G_w(0, 0, 0, x, \lambda) \neq 0$ for all $x \in [a, b]$. As pointed out in Section III.1, it defines a Fredholm operator of index zero when it is considered as an operator in $L(X, Z)$ for $X = C^2[a, b] \cap \{u|u(a) = u(b) = 0\}$ and $Z = C[a, b]$. Instead of Dirichlet boundary conditions we can take Neumann boundary conditions, i.e., $u'(a) = u'(b) = 0$, or mixed boundary conditions.

III.2.1 Bifurcation with a One-Dimensional Kernel

We start with the application of Theorem I.5.1. We set

$$D_u F(0, \lambda)h =$$

$$(III.2.7) \; L(\lambda)h = \sum_{i,j=1}^{n} a_{ij}(x, \lambda)h_{x_i x_j} + \sum_{i=1}^{n} b_i(x, \lambda)h_{x_i} + c(x, \lambda)h,$$

with coefficients given in (III.2.4).

Then the hypotheses of Theorem I.5.1 are the following:

$$L(\lambda_0) \text{ is elliptic,}$$

$$N(L(\lambda_0)) = \text{span}[\hat{v}_0], \; \|\hat{v}_0\|_0 = 1,$$

(III.2.8) $\qquad N(L(\lambda_0)^*) = \text{span}[\hat{v}_0^*], \; \|\hat{v}_0^*\|_0 = 1,$

$$(L_\lambda(\lambda_0)\hat{v}_0, \hat{v}_0^*)_0 \neq 0, \text{ where}$$

$$L_\lambda(\lambda_0)h = \frac{\partial}{\partial \lambda} L(\lambda)h|_{\lambda=\lambda_0} = D_{u\lambda}^2 F(0, \lambda_0)h.$$

Here $L(\lambda_0)^*$ is the adjoint operator according to (III.1.2), and $\| \quad \|_0, (\quad , \quad)_0$ are norm and scalar product in $L^2(\Omega)$. Observe that in view of the results of Section III.1, the properties $(III.2.8)_{2,3}$ are independent of the setting of $L(\lambda_0) : X \to Z$, and for the nondegeneracy (I.5.3) we choose the projection Q of (III.1.16); cf. also (III.1.17).

If (III.2.8) is satisfied, there exists a smooth (depending on $k \geq 2$ in (III.2.3)) nontrivial curve of solutions $\{(u(s), \lambda(s))|s \in (-\delta, \delta)\}$ of $F(u, \lambda) = 0$ through $(u(0), \lambda(0)) = (0, \lambda_0)$ in $X \times \mathbb{R}$, and by (I.5.16), (I.5.17), $u(s) = s\hat{v}_0 + o(s)$ in X.

The Bifurcation Formulas of Section I.6 determine the bifurcation diagram for (III.2.1) near $(u, \lambda) = (0, \lambda_0)$. The formula for $\dot{\lambda}(0)$ (cf. (I.6.3)) reads in this case as

(III.2.9) $\qquad \dot{\lambda}(0) = -\frac{1}{2} \frac{(D_{uu}^2 F(0, \lambda_0)[\hat{v}_0, \hat{v}_0], \hat{v}_0^*)_0}{(L_\lambda(\lambda_0)\hat{v}_0, \hat{v}_0^*)_0},$

where F is defined in (III.2.3). The derivatives of F with respect to u are expressed by derivatives of G with respect to its variables (w_{ij}, v_i, u), cf. (III.2.4).

In the so-called *semilinear* case,

(III.2.10) $G(\nabla^2 u, \nabla u, u, x, \lambda) \equiv \sum_{i,j=1}^{n} a_{ij}(x, \lambda)u_{x_i x_j} + g(\nabla u, u, x, \lambda),$

the formula for $D^2_{uu}F(0, \lambda_0)$ uses only derivatives of g, and in the simplest case, in which $g = g(u, x, \lambda)$, then $D^2_{uu}F(0, \lambda_0)[\hat{v}_0, \hat{v}_0] = g_{uu}(0, x, \lambda_0)\hat{v}_0^2$.

If $\dot{\lambda}(0) \neq 0$, then we have a transcritical bifurcation as sketched in Figure I.6.1. If $\dot{\lambda}(0) = 0$, the computation of $\ddot{\lambda}(0)$ given in (I.6.11) is more involved, in particular if in (I.6.9) the second derivatives do not vanish. We give its computation in this case for a particular example; cf. (III.2.89). Since this example serves as a paradigm for more phenomena in local and global bifurcation theory, we postpone it until the end of this section.

If 0 is a simple eigenvalue of $D_u F(0, \lambda_0) = L(\lambda_0)$ in the sense of (I.7.4) then the Principle of Exchange of Stability expounded in Section I.7 is applicable.

The **buckling of the Euler rod** mentioned in the Introduction is an important historical example. Therefore we discuss it here although it is modeled by an ODE. An incompressible but elastic rod of length ℓ is clamped at one end and free at the other end. Due to its incompressibility the rod is not deformed if it is subject to an axial load P. However, if the load exceeds a crititical value, this "trivial state" becomes unstable and the rod deflects from the straight state, i.e., it "buckles." This phenomenon is a paradigm for bifurcation.

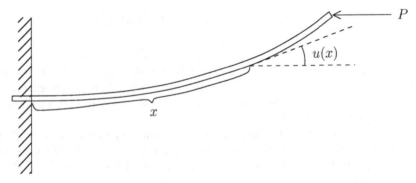

Figure III.2.1

The function $u(x)$ for $x \in [0, \ell]$ describes the angle as sketched in Figure III.2.1. As developed in [153], it satisfies the following boundary value problem:

(III.2.11)
$$u'' + \lambda \sin u = 0 \quad \text{in } [0, \ell],$$
$$u(0) = u'(\ell) = 0.$$

The real parameter λ is proportional to the load P. Setting $F(u, \lambda) = u'' + \lambda \sin u$ for $u \in X = C^2[0, \ell] \cap \{u | u(0) = u'(\ell) = 0\}$, we obtain $F : X \times \mathbb{R} \to Z$ with $Z = C[0, \ell]$. We have the trivial solution $F(0, \lambda) = 0$ for all $\lambda \in \mathbb{R}$, and $D_u F(0, \lambda)h = L(\lambda)h = h'' + \lambda h$ is an elliptic operator, which is a Fredholm operator of index zero with the mixed boundary conditions (III.2.11)$_2$; cf. Section III.1. Moreover, as an operator in $L^2(0, \ell)$ with domain of definition $D(L(\lambda)) = H^2(0, \ell) \cap \{u | u(0) = u'(\ell) = 0\}$ it is self-adjoint, i.e., $L(\lambda) = L(\lambda)^*$. (Since $H^2(0, \ell) \subset C^1[0, \ell]$, the boundary conditions can be imposed in the classical sense.) For the following eigenvalues λ_k it has a one-dimensional kernel spanned by $\hat{v}_k$:

$$(\text{III.2.12}) \quad \lambda_k = \left((2k+1)\frac{\pi}{2\ell}\right)^2, \ \hat{v}_k(x) = c_k \sin \sqrt{\lambda_k} x, \ k \in \mathbb{N} \cup \{0\},$$

where c_k normalizes to $\|\hat{v}_k\|_0 = 1$. Since $L_\lambda(\lambda_k)h = h$ and $\hat{v}_k^* = \hat{v}_k$, condition (III.2.8) is fulfilled for all $k \in \mathbb{N} \cup \{0\}$. By Theorem I.5.1 there exist smooth nontrivial curves of solutions $\{(u_k(s), \lambda_k(s)) | s \in (-\delta_k, \delta_k)\} \subset X \times \mathbb{R}$ of $F(u, \lambda) = 0$ through $(u_k(0), \lambda_k(0)) = (0, \lambda_k)$ for all $k \in \mathbb{N} \cup \{0\}$. Since $D_{uu}^2 F(0, \lambda_k)[h, h] = 0$ for all $h \in X$, formula (III.2.9) gives $\dot{\lambda}_k(0) = 0$ for all $k \in \mathbb{N} \cup \{0\}$, and formula (I.6.11) yields

$$(\text{III.2.13}) \quad \begin{aligned} \ddot{\lambda}_k(0) &= -\frac{1}{3} \frac{(D_{uuu}^3 F(0, \lambda_k)[\hat{v}_k, \hat{v}_k, \hat{v}_k], \hat{v}_k)_0}{(D_{u\lambda}^2 F(0, \lambda_k)\hat{v}_k, \hat{v}_k)_0} \\ &= \frac{1}{3}\lambda_k \int_0^\ell \hat{v}_k^4 dx \ > 0 \quad \text{for all } k \in \mathbb{N} \cup \{0\}. \end{aligned}$$

All bifurcations are supercritical pitchfork bifurcations. The oddness of F with respect to u, i.e., $F(-u, \lambda) = -F(u, \lambda)$, implies for the bifurcating curves the following properties (we omit the index k): With the notation of (I.2.7), (I.2.8) we have

$$(\text{III.2.14}) \quad \begin{aligned} (I-Q)F(-v - \psi(v, \lambda), \lambda) &= 0 \quad \text{and} \\ (I-Q)F(-v + \psi(-v, \lambda), \lambda) &= 0. \end{aligned}$$

By uniqueness guaranteed by the Implicit Function Theorem we see that ψ is odd with respect to v, i.e., $\psi(-v, \lambda) = -\psi(v, \lambda)$. This, in turn, implies the oddness of the bifurcation function Φ with respect to v; cf. (I.2.9). By definition of $\tilde{\Phi}$ in (I.5.10) we have $\Phi(s\hat{v}_0, \lambda) = s\tilde{\Phi}(s, \lambda)$, and therefore $\tilde{\Phi}$ is even with respect to s, i.e., $\tilde{\Phi}(-s, \lambda) = \tilde{\Phi}(s, \lambda)$. This gives in (I.5.14)

$$(\text{III.2.15}) \quad \begin{aligned} \tilde{\Phi}(-s, \lambda(s)) &= 0 \quad \text{and} \\ \tilde{\Phi}(-s, \lambda(-s)) &= 0. \end{aligned}$$

Again by uniqueness we see that λ is even in s, i.e., $\lambda(-s) = \lambda(s)$ for all $s \in (-\delta, \delta)$ and that u is odd in s, i.e., $u(-s) = -u(s)$; cf. (I.5.16). This provides another proof of $\dot{\lambda}(0) = 0$ and it implies the symmetry of all pitchforks with respect to the λ-axis.

Since 0 is a simple eigenvalue of $D_uF(0, \lambda_0)$, we can apply the Principle of Exchange of Stability expounded in Section I.7. By $D_uF(0, \lambda)\hat{v}_k = \hat{v}_k'' + \lambda\hat{v}_k = \hat{v}_k'' + \lambda_k\hat{v}_k + (\lambda - \lambda_k)\hat{v}_k = (\lambda - \lambda_k)\hat{v}_k$ the trivial solution line $\{(0, \lambda)|\lambda \in \mathbb{R}\}$ is stable for $\lambda < \lambda_0$: all (simple) eigenvalues $\lambda - \lambda_k$ of $D_uF(0, \lambda)$ are negative for $k \in \mathbb{N} \cup \{0\}$. For $\lambda > \lambda_0$ the greatest eigenvalue $\lambda - \lambda_0$ of $D_uF(0, \lambda)$ is positive and therefore the trivial solution line becomes unstable. In Section I.7.3 we prove that a loss of stability implies that the supercritical pitchfork is stable; see also Figure I.7.3. The local bifurcation scenario is sketched in Figure III.2.2.

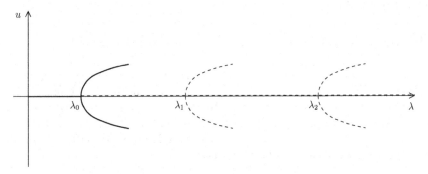

Figure III.2.2

Each half of the pitchforks has a global extension; see Section III.5, Theorem III.5.1, and Corollary II.5.10. A celebrated result of Crandall and Rabinowitz [26] for nonlinear Sturm–Liouville eigenvalue problems is the following: Each half of all pitchforks is unbounded in $X \times \mathbb{R}$, they are all globally separated, and they never intersect the trivial solution line a second time. A key ingredient of their proof, besides the Leray–Schauder degree, is the demonstration that the number of zeros of a solution in the interval $(0, \ell)$ is preserved along each nontrivial global branch. That number is inherited from the function $\hat{v}_k$ and therefore it is different for each global pitchfork. We come back to these ideas in Section III.6.

The properties of the elliptic operator (III.2.7) expounded in Section III.1 allow us also to apply Theorem II.4.4.

In Case 1, i.e., when $\dim N(L(\lambda_0)) = 1$, an odd crossing number of the family $L(\lambda)$ at $\lambda = \lambda_0$ is equivalent to a change of sign of the one-dimensional bifurcation function $D_v\Phi(0, \lambda)$ given in (II.4.25) at $\lambda = \lambda_0$. This generalizes the nondegeneracy (III.2.8)$_4$, which is equivalent to $D_{v\lambda}^2\Phi(0, \lambda_0) \neq 0$; cf. (II.4.27). Consider the example (III.2.17) below, where the coefficients a_{ij} and b_i of $L(\lambda)$ given by (III.2.4) do not depend on λ, and the coefficient c does not depend on x. Then (III.2.8)$_4$ reduces to

$$(\text{III.2.16}) \qquad c'(\lambda_0)(\hat{v}_0, \hat{v}_0^*)_0 \neq 0, \quad ' = \frac{d}{d\lambda}.$$

The condition $(\hat{v}_0, \hat{v}_0^*)_0 \neq 0$, however, requires that 0 be an (algebraically) simple eigenvalue of $L(\lambda_0)$; cf. (I.7.4), (I.16.4). On the other hand, the family

(III.2.17) $L(\lambda)h = \sum_{i,j=1}^{n} a_{ij}(x)h_{x_i x_j} + \sum_{i=1}^{n} b_i(x)h_{x_i} + c(\lambda)h$

has an odd crossing number at $\lambda = \lambda_0$ if

(III.2.18)
> the eigenvalue 0 of $L(\lambda_0)$ has an
> odd algebraic multiplicity and if
> $c(\lambda)$ is strictly monotonic near $\lambda = \lambda_0$.

This clearly allows $c'(\lambda_0) = 0$. Note that condition (III.2.18) for local bifurcation applies to a fully nonlinear elliptic problem of the form

(III.2.19)
$$L(\lambda)u + g(\nabla^2 u, \nabla u, u, x, \lambda) = 0 \quad \text{in} \quad \Omega,$$
$$u = 0 \quad \text{on} \quad \partial\Omega,$$

where $L(\lambda)$ is of the form (III.2.17) and $g_{w_{ij}}(\mathbf{0},\mathbf{0},0,x,\lambda) = g_{v_i}(./.) = g_u(./.) = 0$. The famous result of Krasnosel'skii about "odd algebraic multiplicities" relies on the Leray–Schauder degree and does not apply to fully nonlinear problems. (The one-dimensional kernel is clearly not necessary for the application of Theorem II.4.4, as we note below in (III.2.36).)

The price one has to pay if the nondegeneracy (III.2.8)$_4$ is given up is that the bifurcation diagram does not necessarily consist of only one nontrivial curve; see also Remark II.4.5. This is explicitly seen by the example (III.2.20) below for Degenerate Bifurcation discussed in Section I.16. In this case, 0 is an algebraically simple eigenvalue of $L(\lambda)$ (cf. (I.16.4)); i.e., $(\hat{v}_0, \hat{v}_0^*)_0 \neq 0$. Nonetheless, (III.2.16) is violated by $c'(\lambda_0) = 0$.

Remark III.2.1 *If the lowest-order coefficient of an elliptic operator L as in (III.1.1) is nonpositive, $c(x) \leq 0$, then the elliptic maximum principle and Hopf's boundary lemma are valid. Consequently, for all settings discussed in Section III.1, $L : X \rightarrow Z$ is bijective, $L^{-1} \in L(Z, Z)$ is compact, and $-L^{-1}$ is strictly positive in the sense of ordered Banach spaces. This means in particular that every solution $u \in X$ of $-Lu = f \in Z$ is nonnegative, provided that f is nonnegative. The Krein–Rutman Theorem (cf. [171]) then states that the eigenvalue $\mu_0 \in \sigma(L)$ of smallest modulus is unique, real, negative, and algebraically simple. The corresponding eigenfunction $\hat{v}_0$ is positive (or negative) in Ω.*

The statement for the dual operator implies that $\mu_0 \in \sigma(L^*)$ is simple and the eigenfunction $\hat{v}_0^*$ is positive in Ω as well. Here L^* is the adjoint operator in the sense of (III.1.2), (III.1.14).

For an arbitrary elliptic operator L we have $c(x) \leq \gamma \in \mathbb{R}$, so that the operator $(-L + \gamma I)^{-1}$ is strictly positive in the above sense. Therefore, $L - \gamma I$ has a simple eigenvalue $\mu_0 < 0$ with positive eigenfunction $\hat{v}_0$. Then $\mu_0 + \gamma \in \mathbb{R}$ is a simple eigenvalue of L, or 0 is a simple eigenvalue of $L - (\mu_0 + \gamma)I$ with a positive eigenfunction $\hat{v}_0$. Accordingly, 0 is a simple eigenvalue of the adjoint $(L - (\mu_0 + \gamma)I)^* = L^* - (\mu_0 + \gamma)I$.

If L is self-adjoint, $L = L^*$, then all eigenvalues in $\sigma(L)$ are real, and μ_0 is the largest negative eigenvalue, called the **principal** eigenvalue. It was known long before the Krein–Rutman Theorem, in particular it was known to Courant, Hilbert, Fischer, and Weyl, that the maximum of the negative definite Rayleigh quotient $B(u, u)/\|u\|_0^2$ in $H_0^1(\Omega)\backslash\{0\}$ is attained for some positive (or negative) $\hat{v}_0 \in H_0^1(\Omega)$ and that its value $\mu_0 < 0$ is the largest eigenvalue. Its simplicity follows directly from the positivity of the eigenfunction $\hat{v}_0$: Every eigenfunction that is orthogonal to $\hat{v}_0$ in $L^2(\Omega)$ changes its sign. On the other hand, the maximum μ_0 is attained only for some positive (or negative) function.

The significance of the principal eigenvalue of a parameter-dependent family $L(\lambda)$ is that it is the first eigenvalue that might cross the imaginary axis through 0 at $\lambda = \lambda_0$, causing instability in the sense discussed in Section I.7. The benefits for bifurcation "created by that instability" is that the kernel of $L(\lambda_0)$ is one-dimensional and that due to the positivity of the eigenfunction $\hat{v}_0$, it is easily seen which terms in bifurcation formulas do not vanish. Last but not least, by the simplicity of the eigenvalue 0 of $L(\lambda_0)$, the Principle of Exchange of Stability is valid; cf. Section I.7.

As an example of Degenerate Bifurcation expounded in Section I.16 we consider the semilinear boundary value problem

$$(\text{III}.2.20) \qquad \begin{aligned} L_0 u + \lambda^7 u - \lambda^4 u^2 + \lambda^2 u^4 - u^7 &= 0 \quad \text{in} \quad \Omega, \\ u &= 0 \quad \text{on} \quad \partial\Omega, \end{aligned}$$

where 0 is an algebraically simple eigenvalue of some elliptic operator L_0 with positive eigenfunction $\hat{v}_0$. According to Remark III.2.1, the adjoint L_0^* has a simple eigenvalue 0 with positive eigenfunction $\hat{v}_0^*$, too.

In view of the simplicity of the eigenvalue 0, we can use the Lyapunov–Schmidt decomposition

$$(\text{III}.2.21) \qquad \begin{aligned} X &= N(L_0) \oplus (R(L_0) \cap X), \\ Z &= R(L_0) \oplus N(L_0), \end{aligned}$$

with eigenprojection $Qu = (u, \hat{v}_0^*)_0 \hat{v}_0$ on $N(L_0)$ along $R(L_0)$ when we normalize $(\hat{v}_0, \hat{v}_0^*)_0 = 1$; cf. (I.16.4)–(I.16.6). (Observe that (III.2.21) is orthogonal with respect to $(\ ,\ )_0$ only if $\hat{v}_0 = \hat{v}_0^*$ or $L_0 = L_0^*$. In contrast to the orthogonal projections (III.1.16) valid for (III.1.17) or (III.1.27), we need here only one projection Q, since $P = Q|_X$. Note that the Lyapunov–Schmidt reduction is not unique and that bifurcation formulas depend clearly on the chosen reduction. Their vanishing or nonvanishing, however, or a change of sign does not depend on the chosen reduction.)

For $L(\lambda) = L_0 + \lambda^7 I$ we obtain

$$(\text{III}.2.22) \qquad\qquad (L(\lambda)\hat{v}_0, \hat{v}_0^*)_0 = \lambda^7,$$

so that the nondegeneracy $(\text{III.2.8})_4$ for $\lambda_0 = 0$ is violated (in accordance with
(I.7.36), since the simple eigenvalue perturbation is trivially $\mu(\lambda) = \lambda^7$, so
that $\mu'(0) = 0$). The degenerate condition (III.2.18), however, is satisfied, and
the method of Section I.16 provides the following: Identifying $v \in N(L(0)) =$
$N(L_0)$ with its coordinate $s = (v, \hat{v}_0^*)_0$ and $\Phi \in N(L_0)$ with its coordinate
$(\Phi, \hat{v}_0^*)_0$ with respect to $\hat{v}_0$, respectively, we see that the scalar bifurcation
function $\tilde{\Phi}(s, \lambda) = \Phi(s, \lambda)/s$ given in (I.16.23) is of the following form:

(III.2.23)
$$\tilde{\Phi}(s, \lambda) = c_{70}\lambda^7 + c_{41}\lambda^4 s + c_{23}\lambda^2 s^3 + c_{06}s^6 + \text{h.o.t.},$$
where $c_{70} = 1$ by $\mu(\lambda) = \lambda^7$ and Theorem I.16.3,
$c_{41} = -(\hat{v}_0^2, \hat{v}_0^*)_0, \ c_{23} = (\hat{v}_0^4, \hat{v}_0^*)_0, \ c_{06} = -(\hat{v}_0^7, \hat{v}_0^*)_0.$

By positivity of the eigenfunctions $\hat{v}_0, \hat{v}_0^*$ we have $c_{70} = 1 > 0, \ c_{41} < 0, \ c_{23} >$
$0, \ c_{06} < 0$, so that the Newton Polygon Method described in Section I.15
yields the bifurcation diagram shown in Figure I.16.3; see also (I.16.55),
(I.16.56).

As mentioned at the end of Remark III.2.1, the Principle of Exchange of
Stability is valid; see Theorem I.16.8. If $L_0 = L_0^*$ is self-adjoint, then $\mu(\lambda) =$
λ^7 is the principal eigenvalue of $L(\lambda) = L_0 + \lambda^7 I$, and the trivial solution
$(u, \lambda) = (0, \lambda)$ of (III.2.20) indeed loses stability at $\lambda_0 = 0$. Consequently, the
bifurcating branches have the stability properties that are marked in Figure
I.16.3.

Another example is

(III.2.24)
$$L_0 u + (\lambda^2 - \varepsilon)u - u^3 = 0 \quad \text{in} \quad \Omega,$$
$$u = 0 \quad \text{on} \quad \partial\Omega,$$

with the same operator L_0 as before in (III.2.20). All solutions of (III.2.24)
near $(u, \lambda, \varepsilon) = (0, 0, 0)$ are obtained by solving $\Phi(s, \lambda^2 - \varepsilon) = s\tilde{\Phi}(s, \lambda^2 - \varepsilon) =$
0, where $\Phi(s, \tilde{\lambda})$ is the bifurcation function for (III.2.24) when we simply
substitute $\lambda^2 - \varepsilon = \tilde{\lambda}$. As in (III.2.23), we obtain

(III.2.25)
$$\tilde{\Phi}(s, \lambda^2 - \varepsilon) = \lambda^2 - \varepsilon + c_{02}s^2 + \text{h.o.t.},$$
where $c_{02} = -(\hat{v}_0^3, \hat{v}_0^*)_0 < 0.$

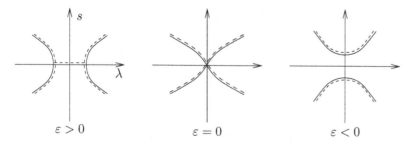

$\varepsilon > 0$ $\qquad\qquad\qquad\qquad$ $\varepsilon = 0$ $\qquad\qquad\qquad\qquad$ $\varepsilon < 0$

Figure III.2.3

The solution set $\{(s, \lambda)\}$ of $\Phi(s, \lambda^2 - \varepsilon) = 0$ is sketched in Figure III.2.3 for different values of ε near 0.

If $L_0 = L_0^*$, then 0 is the principal eigenvalue of L_0, so that the solutions have the stability indicated in Figure III.2.3.

Remark III.2.2 *The "generic" simple bifurcations at* $\lambda = \pm\sqrt{\varepsilon}$ *for* $\varepsilon > 0$ *become a degenerate bifurcation for* $\varepsilon = 0$, *and for* $\varepsilon < 0$ *no bifurcation takes place at all. One might believe that the eigenvalue 0 of* $L_0 + (\lambda^2 - \varepsilon)I$ *for* $\lambda = \varepsilon = 0$ *is no longer simple, since two eigenvalues "collide" at 0. This is a misunderstanding: The eigenvalue* $\lambda^2 - \varepsilon$ *of* $L_0 + (\lambda^2 - \varepsilon)I$ *is throughout simple, and bifurcation takes place at those values of* λ *for which* $\lambda^2 - \varepsilon = 0$. *This is completely different from the Hamiltonian Hopf Bifurcation, cf. Section I.11, where for fixed* λ *the operator* $A(\lambda)$ *has indeed two different eigenvalues that collide for* $\lambda = \lambda_0$. *As mentioned in Remark I.17.7, a degenerate bifurcation for* $\varepsilon = 0$ *is always perturbed to "generic" bifurcations, but those perturbations are best understood from the degenerate diagram. We recommend a study of the bifurcations for* $L_0 + \lambda(\lambda - \varepsilon)I$ *in (III.2.24), too. For more examples see* [96].

III.2.2 Bifurcation with a Two-Dimensional Kernel

In this section we apply the method of Section I.19.1 to the following example:

$$
\text{(III.2.26)} \qquad
\begin{aligned}
\mathbf{\Delta u} + \mathbf{f}(\mathbf{u}, \lambda) &= \mathbf{0} \quad \text{in } \Omega \subset \mathbb{R}^n, \\
\mathbf{u} &= \mathbf{0} \quad \text{on } \partial\Omega,
\end{aligned}
$$

where $\mathbf{u} = (u_1, u_2)$, $\mathbf{\Delta u} = (\Delta u_1, \Delta u_2)$, and Δ denotes the scalar Laplacian. We assume that the vector field $\mathbf{f}$ is of the following form:

$$\mathbf{f} : \mathbb{R}^2 \times \mathbb{R} \to \mathbb{R}^2 \quad \text{is in } C^{k+1}(\mathbb{R}^2 \times \mathbb{R}, \mathbb{R}^2), \ k \geq 2,$$

$$\mathbf{f}(\mathbf{u}, \lambda) = \lambda\mathbf{u} + \mathbf{f}_{0k_0}(\mathbf{u}) + \mathbf{R}(\mathbf{u}, \lambda)$$

for some homogeneous polynomial

(III.2.27) $\mathbf{f}_{0k_0} : \mathbb{R}^2 \to \mathbb{R}^2$ of order $k_0 \geq 2$, and where

$\mathbf{R}(\mathbf{u}, \lambda)$ contains all terms of higher order in $(\mathbf{u}, \lambda - \lambda_0)$,

i.e., of order $|\lambda - \lambda_0|^j \|\mathbf{u}\|^k$, where

$k > k_0$ for $j = 0$, $k > 1$ for $j = 1$, and $k \geq 1$ for $j > 1$.

(For the value of $\lambda_0 \in \mathbb{R}$ see (III.2.29).) If the domain $\Omega \subset \mathbb{R}^n$ is bounded and the boundary $\partial\Omega$ is sufficiently smooth, we know by Section III.1 that $\Delta + \lambda I : X \to Z$ with $X = C^{2,\alpha}(\overline{\Omega}) \cap \{u | u = 0 \text{ on } \partial\Omega\}$ and $Z = C^\alpha(\overline{\Omega})$ is a Fredholm operator of index zero. Then the same holds for $\mathbf{\Delta} + \lambda \mathbf{I} : X^2 \to Z^2$ and

$$\mathbf{F}(\mathbf{u}, \lambda)(x) = \mathbf{\Delta u}(x) + \mathbf{f}(\mathbf{u}(x), \lambda)$$

(III.2.28)   defines a mapping

$$\mathbf{F} : X^2 \times \mathbb{R} \to Z^2 \quad \text{of class } C^k(X^2 \times \mathbb{R}, Z^2).$$

By $\mathbf{F}(\mathbf{0}, \lambda) = \mathbf{0}$ we have the trivial solution $(\mathbf{0}, \lambda)$ for all $\lambda \in \mathbb{R}$. We find a two-dimensional kernel of $D_\mathbf{u}\mathbf{F}(\mathbf{0}, \lambda) = \mathbf{\Delta} + \lambda\mathbf{I}$ to the first (principal) eigenvalue λ_0 of $-\Delta$, i.e.,

(III.2.29)
$$\begin{aligned} &N(D_\mathbf{u}\mathbf{F}(\mathbf{0}, \lambda_0)) = \text{span}[\hat{\mathbf{v}}_1, \hat{\mathbf{v}}_2], \quad \text{where} \\ &\hat{\mathbf{v}}_1 = (\hat{v}_0, 0) \quad \text{and} \quad \hat{\mathbf{v}}_2 = (0, \hat{v}_0) \text{ and} \\ &\hat{v}_0 \quad \text{is the positive eigenfunction of} \\ &-\Delta : X \to Z \quad \text{corresponding to } \lambda_0 \\ &\text{that is normalized to} \quad \|\hat{v}_0\|_0 = 1. \end{aligned}$$

Since $\mathbf{\Delta} + \lambda_0\mathbf{I}$ is (formally) self-adjoint in $(L^2(\Omega))^2$ with scalar product $(\mathbf{u}, \mathbf{v})_0 = (u_1, v_1)_0 + (u_2, v_2)_0$, we obtain the following decomposition and projections:

$$Z^2 = R(D_\mathbf{u}\mathbf{F}(\mathbf{0}, \lambda_0)) \oplus N(D_\mathbf{u}\mathbf{F}(\mathbf{0}, \lambda_0)),$$

(III.2.30) $Q : Z^2 \to N \quad \text{along } R, \; P : X^2 \to N \text{ along } R \cap X^2,$

$$Q\mathbf{u} = (\mathbf{u}, \hat{\mathbf{v}}_1)_0\hat{\mathbf{v}}_1 + (\mathbf{u}, \hat{\mathbf{v}}_2)_0\hat{\mathbf{v}}_2, \; P = Q|_{X^2};$$

cf. (III.1.16), (III.1.30). In view of assumption (III.2.27) we obtain by (I.19.25), (I.19.26) a bifurcation function of the form (I.19.40), i.e.,

(III.2.31)
$$\begin{aligned} &\Phi(v, \lambda) = \Phi_{0k_0}(v) + (\lambda - \lambda_0)\Phi_{11}v + R(v, \lambda), \quad \text{where} \\ &\Phi_{0k_0}(v) = Q\mathbf{f}_{0k_0}(v), \; \Phi_{11}v = v \quad \text{for} \\ &v = (z_1, z_2)\hat{v}_0 \in N, \; \mathbf{z} = (z_1, z_2) \in \mathbb{R}^2. \end{aligned}$$

Since assumption (I.19.44) is fulfilled by (III.2.31)$_2$ ($Z_0 = N$ and $\hat{v}_k^* = \hat{v}_k$, $k = 1, 2$, in this case), the crucial condition in Theorem I.19.2 is (I.19.45). For $v = \mathbf{z}\hat{v}_0 \in N$ we obtain by definition (I.19.43) and with $\mathbf{f}_{0k_0} = (f_{0k_0}^1, f_{0k_0}^2)$,

(III.2.32)
$$\begin{aligned} &(\Phi_{0k_0}(v), R_{\pi/2}\Phi_{11}v)_0 \\ &= -z_2 \int_\Omega f_{0k_0}^1(\mathbf{z}\hat{v}_0)\hat{v}_0 dx + z_1 \int_\Omega f_{0k_0}^2(\mathbf{z}\hat{v}_0)\hat{v}_0 dx \\ &= (-f_{0k_0}^1(\mathbf{z})z_2 + f_{0k_0}^2(\mathbf{z})z_1) \int_\Omega \hat{v}_0^{k_0+1} dx \\ &= \mathbf{f}_{0k_0}(\mathbf{z}) \cdot R_{\pi/2}\mathbf{z} \int_\Omega \hat{v}_0^{k_0+1} dx, \end{aligned}$$

where "$\cdot$" is the Euclidean scalar product in $\mathbb{R}^2$ and $R_{\pi/2}$ is a rotation about $\pi/2$ in $\mathbb{R}^2$. Since $\hat{v}_0 > 0$ in Ω, (I.19.45) is satisfied in the following cases:

(III.2.33)

If k_0 is odd, assume the existence of
$\mathbf{z}_1, \mathbf{z}_2 \in \mathbb{R}^2$ with $\|\mathbf{z}_1\| = \|\mathbf{z}_2\| = 1$ and
$\mathbf{f}_{0k_0}(\mathbf{z}_1) \cdot \mathbf{R}_{\pi/2}\mathbf{z}_1 < 0$,
$\mathbf{f}_{0k_0}(\mathbf{z}_2) \cdot \mathbf{R}_{\pi/2}\mathbf{z}_2 > 0$.

If k_0 is even, assume the existence of
a $\mathbf{z} \in \mathbb{R}^2$ with $\|\mathbf{z}\| = 1$ and
$\mathbf{f}_{0k_0}(\mathbf{z}) \cdot \mathbf{R}_{\pi/2}\mathbf{z} \neq 0$.

In both cases Corollary I.19.3 guarantees the existence of at least two local continua or one continuum, respectively, of nontrivial solutions of $\mathbf{F}(\mathbf{u}, \lambda) = 0$ through $(\mathbf{0}, \lambda_0)$ in $X^2 \times \mathbb{R}$.

Choosing the parameterization $\mathbf{z}(t) = (\cos t, \sin t)$ for $t \in [0, 2\pi]$ of $S_1 \subset \mathbb{R}^2$, Corollary I.19.4 applies if there exists a $\mathbf{z} \in S_1$ with $\mathbf{f}_{0k_0}(\mathbf{z}) \cdot \mathbf{R}_{\pi/2}\mathbf{z} \neq 0$ and if $\int_0^{2\pi} \mathbf{f}_{0k_0}(\mathbf{z}(t)) \cdot \mathbf{R}_{\pi/2}\mathbf{z}(t)dt = 0$. Simple examples are

$$\mathbf{f}_{0k_0}(\mathbf{z}) = \begin{pmatrix} \alpha & \beta \\ \gamma & \delta \end{pmatrix} \begin{pmatrix} z_1^{k_0} \\ z_2^{k_0} \end{pmatrix} \quad \text{with}$$

(III.2.34) $(\alpha, \beta, \gamma, \delta) \neq (0, 0, 0, 0)$ for even k_0,
and the additional condition $\beta = \gamma$ for odd k_0;

$$\mathbf{f}_{02}(\mathbf{z}) = \begin{pmatrix} z_1^2 - z_2^2 \\ z_1 z_2 \end{pmatrix}.$$

The last example has the peculiarity that the two zeros of $g(\tilde{v}, 0) = (\Phi_{02}(\tilde{v}), R_{\pi/2}\tilde{v})_0 = z_2^3 \int_\Omega \hat{v}_0^3 dx$ on $S_1 \subset N$, i.e., for $z_1^2 + z_2^2 = 1$, are degenerate such that the Implicit Function Theorem is not applicable to solve (I.19.51).

We remark that the computation (III.2.32) is also valid for any other simple eigenvalue λ_k of $-\Delta$ with an eigenfunction $\hat{v}_k, k \geq 1$. Then (I.19.45) is satisfied under the assumptions (III.2.33), provided $\int_\Omega \hat{v}_k^{k_0+1} dx \neq 0$. This is always true if k_0 is odd.

However, when considered over a square $\Omega \subset \mathbb{R}^2$, the second eigenvalue λ_1 of $-\Delta$ subject to homogeneous Dirichlet boundary conditions is no longer simple but has multiplicity two. Consequently, $\dim N(D_\mathbf{u}\mathbf{F}(\mathbf{0}, \lambda_1)) = \dim N(\Delta + \lambda_1 \mathbf{I}) = 4$ and the method of Section I.19.1 is not applicable. (For the functional analytic setting of (III.2.26) over a square see Section III.1.1.) In order to reduce the dimension of the kernel to two, we restrict the mapping $\mathbf{F}(\cdot, \lambda)$ in (III.2.28) to fixed-point spaces of certain symmetry groups of the square. For this purpose we need an equivariance of the vector field $\mathbf{f}$ in (III.2.27) with respect to these symmetries. Note that Δ is equivariant under the complete orthogonal group of the plane. One of the benefits of equivariance is that fixed-point spaces are invariant under the nonlinear mapping $\mathbf{F}(\cdot, \lambda)$. Furthermore, the dimension four of the kernel $N(D_\mathbf{u}\mathbf{F}(\mathbf{0}, \lambda_1))$ reduces to two in a fixed-point space such that under suitable additional conditions, Theorem I.19.2 and Corollary I.19.3 are applicable, yielding bifurcating nontrivial continua also at $(\mathbf{0}, \lambda_1)$. All details can be found in [119].

Finally, we study in [119] bifurcation functions of type (III.2.31)$_1$ having the equivariance with respect to the symmetry group of a regular n-gon, $n \geq 3$.

III.2.3 Bifurcation with High-Dimensional Kernels

The most general local result (in view of counterexamples) for fully nonlinear elliptic problems is given by the application of Theorem II.4.4. With $D_u F(0, \lambda) = L(\lambda)$ with $L(\lambda)$ as in (III.2.7), as previously, the hypotheses of Theorem II.4.4 are the following:

(III.2.35)
$$\begin{aligned} &L(\lambda) \text{ is elliptic for } \lambda \in (\lambda_0 - \delta, \lambda_0 + \delta), \\ &0 \text{ is an eigenvalue of } L(\lambda_0), \text{ and} \\ &L(\lambda) \text{ has an odd crossing number at } \lambda = \lambda_0. \end{aligned}$$

The last statement means that an odd number of eigenvalues (counting multiplicities) in the 0-group of $L(\lambda)$ leave the left complex half-plane when λ passes through λ_0 (see Definition II.4.1). As proved in Theorem II.4.3, assumption (III.2.35) implies that $\det(D_v \Phi(0, \lambda))$ changes sign at $\lambda = \lambda_0$ for every bifurcation function $\Phi(v, \lambda)$ obtained by the method of Lyapunov–Schmidt.

The verification of (III.2.35)$_3$ is not easy, in general, but for the special case that $L(\lambda)$ is given by (III.2.17) with a coefficient $c(\lambda)$ satisfying (III.2.18) it is simple. We summarize this result for convenience briefly as follows:

(III.2.36)
$$\begin{aligned} &\text{If } 0 \text{ is an eigenvalue of odd algebraic multiplicity} \\ &\text{of } L(\lambda_0) \text{ that is of the form (III.2.17)} \\ &\text{with strictly monotonic coefficient } c(\lambda), \text{ then} \\ &\text{a continuum of nontrivial solutions of the} \\ &\text{fully nonlinear elliptic problem (III.2.1) bifurcates} \\ &\text{at } (0, \lambda_0) \text{ in } X \times \mathbb{R}, \text{ where } X = C^{2,\alpha}(\overline{\Omega}) \cap \{u | u = 0 \text{ on } \partial\Omega\}. \end{aligned}$$

Note that the hypotheses (III.2.35) are more general, and the statement of Theorem II.4.4 is sharper than that of Krasnosel'skii's Theorem in [118].

Another type of bifurcation theorem is given by (II.4.31), which is summarized thus:

(III.2.37)
$$\begin{aligned} &L(\lambda) \text{ is elliptic for } \lambda \in (\lambda_0 - \delta, \lambda_0 + \delta), \\ &\dim N(L(\lambda_0)) \text{ is odd,} \\ &(L_\lambda(\lambda_0)\hat{v}, \hat{v}^*)_0 \neq 0 \text{ for all} \\ &\hat{v} \in N(L(\lambda_0))\backslash\{0\}, \ \hat{v}^* \in N(L(\lambda_0)^*)\backslash\{0\}. \end{aligned}$$

We recall that the spectral properties of an elliptic family $L(\lambda)$ are the same for all settings $L(\lambda) : Z \to Z$ with $D(L) = X \subset Z$; cf. (III.1.39), (III.1.40).

III.2.4 Variational Methods I

We give an application of Theorem I.21.2 to a so-called nonlinear eigenvalue problem

(III.2.38)
$$Lu = \mu g(x, u) \text{ in } \Omega, \ \mu \in \mathbb{R},$$
$$u = 0 \text{ on } \partial\Omega,$$

where $L = L^*$ is of divergence form (III.1.3) and elliptic. Assuming $c(x) \leq 0$, then the bilinear and symmetric form $B(u, v)$ is negative definite on $H_0^1(\Omega)$; cf. (III.1.8) (use Poincaré's inequality). This means that $(\ , \ )_1$ and $-B(\ , \ )$ are equivalent scalar products on $H_0^1(\Omega)$.

Weak Solutions for Functions g with at Most Critical Growth

For the function $g : \overline{\Omega} \times \mathbb{R} \to \mathbb{R}$, we assume that the partial derivative $g_u : \overline{\Omega} \times \mathbb{R} \to \mathbb{R}$ exists, that $g, g_u \in C(\overline{\Omega} \times \mathbb{R}, \mathbb{R})$, and that

(III.2.39)
$$|g(x, u)| \leq c_1 + c_2 |u|^r,$$
$$|g_u(x, u)| \leq c_3 + c_4 |u|^{r-1} \text{ for all } (x, u) \in \Omega \times \mathbb{R},$$
$$1 \leq r \leq \frac{n+2}{n-2} \text{ for } n > 2, \ 1 \leq r < \infty \text{ for } n = 2,$$

and no restriction on g for $n = 1$. Here n is the dimension of the domain Ω, i.e., $\Omega \subset \mathbb{R}^n$.

Defining the primitive

(III.2.40)
$$G(x, u) = \int_0^u g(x, s)ds \quad \text{for } (x, u) \in \Omega \times \mathbb{R} \text{ and}$$
$$f(u) = \int_\Omega G(x, u(x))dx \quad \text{for } u \in H_0^1(\Omega),$$

then the following is well known; see [118], for example. The function $f : H_0^1(\Omega) \to \mathbb{R}$ satisfies

(III.2.41)
$$f \in C^2(H_0^1(\Omega), \mathbb{R}) \text{ and}$$
$$Df(u)h = \int_\Omega g(x, u(x))h(x)dx \quad \text{for } u, h \in H_0^1(\Omega).$$

Setting $\hat{g}(u)(x) = g(x, u(x))$, we see that the function $\hat{g}(u)$ is in $L^q(\Omega)$ with $q = 2n/(n+2)$, by assumption (III.2.39)$_1$ and by continuous embedding $H_0^1(\Omega) \subset L^p(\Omega)$ for $p = 2n(n-2)$, where we assume $n > 2$; the cases $n \leq 2$ are left to the reader. Since $1/p + 1/q = 1$, Hölder's inequality proves that $Df(u) \in L(H_0^1(\Omega), \mathbb{R})$, which is the dual space of $H_0^1(\Omega)$. Furthermore,

$$(\text{III.2.42}) \qquad D^2 f(u)[h_1, h_2] = \int_\Omega g_u(x, u(x)) h_1(x) h_2(x) dx$$
$$\text{for } u, h_1, h_2 \in H_0^1(\Omega),$$

and the same arguments using $(\text{III.2.39})_2$ prove that $D^2 f(u) : H_0^1(\Omega) \times H_0^1(\Omega) \to \mathbb{R}$ is bilinear and continuous; i.e., $D^2 f(u) \in L_2(H_0^1(\Omega), \mathbb{R})$. The continuity of f, Df, and $D^2 f$ with respect to u in the norms of $H_0^1(\Omega), L(H_0^1(\Omega), \mathbb{R}), L_2(H_0^1(\Omega), \mathbb{R})$, respectively, is more subtle and is proved in [118]. By Riesz's Representation Theorem, for each $u \in H_0^1(\Omega)$ there is a unique

$$(\text{III.2.43}) \qquad \nabla_L f(u) \in H_0^1(\Omega) \text{ such that}$$

$$Df(u)h = -B(\nabla_L f(u), h) \text{ for all } u, h \in H_0^1(\Omega).$$

According to Definition I.3.1, the mapping $\nabla_L f$ is the gradient of f with respect to the scalar product $-B(\ ,\ )$, and the subscript L denotes its dependence on L. In the notation of Theorem I.21.2 we have $X = Z = H_0^1(\Omega)$. Staying with the notation of Section III.1, where X and Z have a different meaning, we set $H_0^1(\Omega) = X_1$, and we define

$$F \equiv -\nabla_L f : X_1 \to X_1,$$

$$F(0) = 0 \text{ if } g(x, 0) = 0 \text{ for all } x \in \Omega,$$

$$(\text{III.2.44}) \qquad F \in C^1(X_1, X_1), \text{ by (III.2.40)--(III.2.43), and}$$

$$B(DF(u)h_1, h_2) = \int_\Omega \hat{g}_u(u) h_1 h_2 dx,$$

where as before, $\hat{g}_u(u)(x) = g_u(x, u(x))$. For $A_0 = DF(0)$ we verify hypothesis (I.21.3) of Theorem I.21.2.

Remark III.2.3 *Let $r = r(x)$ be any function $r \in C(\overline{\Omega})$. Then the operator $A_0 \in L(X_1, X_1)$, defined by*

$$(\text{III.2.45}) \qquad B(A_0 v, h) = (rv, h)_0 \quad \text{for all } v, h \in X_1 = H_0^1(\Omega),$$

is symmetric (self-adjoint) with respect to the scalar product $-B(\ ,\ )$ and compact: By continuity, the linear operator A_0 maps weakly convergent sequences in X_1 onto weakly convergent sequences in X_1, which, by compact embedding $H_0^1(\Omega) \subset L^2(\Omega)$, converge strongly in $L^2(\Omega)$. Choosing $h = A_0 v$, we see that the defining equation (III.2.45) proves that A_0 maps weakly convergent sequences in X_1 onto strongly convergent sequences in X_1, which means compactness of A_0.

Therefore, by the Riesz–Schauder Theory, the spectrum $\sigma(A_0)$ consists of real nonzero eigenvalues of finite (algebraic) multiplicities and of 0, which is the only possible cluster point of $\sigma(A_0)$. If $A_0 \neq 0$, then its spectral radius $\|A_0\|_{L(X_1 X_1)}$ (with the operator norm generated by $-B(\ ,\ )$) is po-

sitive. Therefore, there exists an eigenvalue λ_0 with $|\lambda_0| = \|A_0\| > 0$. Furthermore, it is known that the following completeness of orthonormal systems of eigenfunctions holds: If $\lambda_0, \lambda_1, \ldots$ are all nonzero eigenvalues of A_0 with orthonormal eigenfunctions $\hat{v}_0, \hat{v}_1, \ldots$ (with respect to $-B(\ ,\ ))$, then $X_1 = N(A_0) \oplus \overline{\operatorname{span}[\hat{v}_0, \hat{v}_1, \ldots]}$.

A closer look reveals more: Let $r(x) > 0$ for x in an open subset $\Omega_+ \subset \Omega$. Then $\dim\{v \in X_1 | \operatorname{supp}(v) \subset \Omega_+\} = \infty$ (where "supp" denotes the support), so that the complement of $N(A_0)$ in X_1 is infinite-dimensional. Let $v \neq 0$, $\operatorname{supp}(v) \subset \Omega_+$, and $v = \sum \alpha_k \hat{v}_k$ be its Fourier series. Then $0 < (rv, v)_0 = B(A_0 v, v) = -\sum \alpha_k^2 \lambda_k$. If there were only finitely many eigenvalues $\lambda_k < 0$, then there would exist such a $v \neq 0$ with Fourier coefficients $\alpha_k = 0$ for all $\lambda_k < 0$ such that $0 < (rv, v)_0 = -\sum \alpha_k^2 \lambda_k \leq 0$. This contradiction proves that there exist infinitely many negative eigenvalues $\lambda_k < 0$ of A_0 with $\lambda_k \to 0$ as $k \to \infty$. The same argument yields infinitely many positive eigenvalues $\lambda_\ell > 0$ with $\lambda_\ell \to 0$ as $\ell \to \infty$ if $r(x) < 0$ for $x \in \Omega_- \subset \Omega$.

The eigenvalue problem for A_0, i.e.,

$$(\text{III.2.46}) \quad B(A_0 v, h) = \lambda B(v, h) = (rv, h)_0 \text{ for all } h \in H_0^1(\Omega),$$

is the weak formulation of the elliptic eigenvalue problem

$$(\text{III.2.47}) \qquad \begin{aligned} Lv &= \mu rv \quad \text{in} \quad \Omega, \ \mu = \lambda^{-1}, \\ v &= 0 \quad \text{on} \quad \partial\Omega, \end{aligned}$$

with possibly indefinite weight function $r \in C(\overline{\Omega})$. The preceding results about the eigenvalues $\lambda \neq 0$ of A_0 yield the following: If $\Omega_+ = \{x \in \Omega | r(x) > 0\} \neq \emptyset$ (or $\Omega_- = \{x \in \Omega | r(x) < 0\} \neq \emptyset$), then there exist infinitely many negative eigenvalues $\mu_k < 0$ with $\mu_k \to -\infty$ as $k \to \infty$ (or infinitely many positive eigenvalues $\mu_\ell > 0$ with $\mu_\ell \to +\infty$ as $\ell \to \infty$) of the weak eigenvalue problem (III.2.47). Furthermore, there exists a largest negative (or smallest positive) eigenvalue, and these possibly two "principal" eigenvalues are the maximum (or minimum) of $\{-B(v, v)/(rv, v)_0 | v \in H_0^1(\Omega), (rv, v)_0 > 0 \text{ (or } (rv, v)_0 < 0)\}$ (respectively). These extrema are attained at some positive (or negative) function $\hat{v}_0 \in H_0^1(\Omega)$; cf. Remark III.2.1.

If r changes sign on Ω, the simplicity of a principal eigenvalue μ_0 is not as obvious as in the definite case. Furthermore, all other eigenfunctions of (III.2.47) that do not belong to a principal eigenvalue change sign in Ω. For details we refer to [77], [108].

If the boundary of Ω allows an elliptic regularity theory, then the weak formulation (III.2.46) for $v \in H_0^1(\Omega)$ and the strong version (III.2.47) for $v \in D(L) = H^2(\Omega) \cap H_0^1(\Omega)$ are equivalent. If the weight function r belongs to $C^\alpha(\overline{\Omega})$, then elliptic regularity implies that $v \in C^{2,\alpha}(\overline{\Omega}) \cap \{u | u = 0 \text{ on } \partial\Omega\}$ for every eigenfunction v of (III.2.47); cf. (III.1.40). Conversely, all eigenfunctions of A_0 with eigenvalues $\lambda_k = \mu_k^{-1}$ are in $C^{2,\alpha}(\overline{\Omega}) \cap \{u | u = 0 \text{ on } \partial\Omega\}$ if the data of the problem are smooth enough.

We apply the results described in Remark III.2.3 to the weight $r = \hat{g}_u(0) \in C(\overline{\Omega})$. If $\hat{g}_u(0) \neq 0$ (i.e., if $g_u(x,0) \not\equiv 0$), then *all* eigenvalues $\lambda_k \neq 0$ of A_0 are candidates for the application of Theorem I.21.2. (Since $A_0 \in L(X_1, X_1)$ is self-adjoint, we have clearly (I.21.3)$_4$, and Remark I.21.1 is trivially satisfied.)

Let $\lambda_0 \neq 0$ be one of the eigenvalues of A_0. Then for every sufficiently small $\varepsilon > 0$ there are at least two solutions $(u, \lambda) = (u(\varepsilon), \lambda(\varepsilon)) \in H_0^1(\Omega) \times \mathbb{R}$ of

$$(III.2.48) \qquad \begin{aligned} F(u) = \lambda u, \quad -B(u,u) = \varepsilon^2, \\ \text{and } \lambda(\varepsilon) \to \lambda_0 \text{ as } \varepsilon \searrow 0. \end{aligned}$$

By definitions (III.2.43), (III.2.44), and (III.2.40), this means that

$$(III.2.49) \qquad \begin{aligned} B(u,h) &= \mu \int_\Omega g(x,u) h \, dx, \quad \mu = \lambda^{-1} \\ &\text{for all } h \in H_0^1(\Omega) \text{ and } \mu = \mu(\varepsilon) \to \mu_0 = \lambda_0^{-1}, \\ &\text{where } \mu_0 \text{ is an eigenvalue of (III.2.47)} \\ &\text{with } r = \hat{g}_u(0). \end{aligned}$$

In other words, $(u, \mu) \in H_0^1(\Omega) \times \mathbb{R}$ are nontrivial **weak solutions** of (III.2.38) clustering at $(0, \mu_0)$.

For $u \in H_0^1(\Omega)$, assumption (III.2.39)$_1$ implies that $\hat{g}(u) \in L^q(\Omega)$ with $q = 2n/(n+2)$. If the data, in particular the boundary $\partial\Omega$, allow an elliptic regularity theory, then (III.2.49) implies $u \in W^{2,q}(\Omega)$. Since this does not give more regularity for $\hat{g}(u)$, that is $\hat{g}(u) \in L^q(\Omega)$ with $q = 2n/(n+2)$, no further gain of regularity via a so-called "bootstrapping" is possible, in general. (It is possible, however, if the growth in (III.2.39)$_1$ is "subcritical," i.e., if $r < \frac{n+2}{n-2}$.)

Some comments on this result are in order: Theorem I.21.2 provides weak solutions of (III.2.38) in the sense of (III.2.48) or (III.2.49) under rather weak regularity conditions on the data of the problem. The coefficients of L of the form (III.1.3) have to satisfy only $a_{ij} \in C^1(\Omega), c \in C(\overline{\Omega})$, and for the nonlinearity it suffices that $g, g_u \in C(\overline{\Omega} \times \mathbb{R}, \mathbb{R})$. Then all hypotheses on $F \in C^1(H_0^1(\Omega), H_0^1(\Omega))$ and on $A_0 = DF(0) \in L(H_0^1(\Omega), H_0^1(\Omega))$ to apply Theorem I.21.2 are satisfied irrespective of the regularity of the boundary $\partial\Omega$. We obtain bifurcation of weak solutions at every eigenvalue of the weak eigenvalue problem (III.2.38) *for every bounded domain Ω with no condition on its boundary $\partial\Omega$*.

Apart from the lack of regularity of the weak solutions, the Hilbert space approach for (III.2.38) clearly also has the drawback mentioned in Section III.1: The Hilbert space setting might impose obstructions to nonlinear problems, which in the case in question are the growth conditions (III.2.39) on the nonlinearity g. As recommended in Section III.1, we overcome this drawback by a Banach space approach.

Strong Solutions for Arbitrary Functions g

Now the boundary $\partial\Omega$ and the coefficients of L are again smooth enough to ensure the properties of the elliptic operator L as expounded in Section III.1. Let $g : \overline{\Omega} \times \mathbb{R} \to \mathbb{R}$ be *any* function in $C^2(\overline{\Omega} \times \mathbb{R}, \mathbb{R})$ such that $g(x, 0) = 0$ for all $x \in \Omega$. Then, for G as defined in (III.2.40)$_1$, the functional f given in (III.2.40)$_2$ is defined on $X_{1,\alpha} \equiv C^{1,\alpha}(\overline{\Omega}) \cap \{u | u = 0 \text{ on } \partial\Omega\}$. Furthermore, $f \in C^2(X_{1,\alpha}, \mathbb{R})$, and its derivative $Df(u)$ is given by (III.2.41)$_2$ for $u, h \in X_{1,\alpha}$. By $\hat{g}(u), \hat{g}_u(u) \in C^\alpha(\overline{\Omega}) = Z \subset L^2(\Omega)$, we can state (III.2.41)$_2$ and (III.2.42) as

$$Df(u)h = (\hat{g}(u), h)_0 \text{ for } u, h \in X_{1,\alpha},$$

(III.2.50)

$$D^2 f(u)[h_1, h_2] = (\hat{g}_u(u)h_1, h_2)_0 \text{ for } u, h_1, h_2 \in X_{1,\alpha}.$$

By the assumption that $c(x) \le 0$, the elliptic operator L from (III.1.3) is bijective via $L : H^2(\Omega) \cap H_0^1(\Omega) \to H^0(\Omega)$. Therefore, L is also bijective as a mapping $L : X \equiv C^{2,\alpha}(\overline{\Omega}) \cap \{u | u = 0 \text{ on } \partial\Omega\} \to Z$; cf. Section III.1, in particular (III.1.40). Defining

(III.2.51)
$$F(u) \equiv L^{-1}\hat{g}(u) \text{ for } u \in X_{1,\alpha}, \text{ then}$$
$$B(F(u), h) = (\hat{g}(u), h)_0 = Df(u)h$$
$$\text{for all } u, h \in X_{1,\alpha},$$

so that $-f \in C^2(X_{1,\alpha}, \mathbb{R})$ is a potential for $F \in C^1(X_{1,\alpha}, X_{1,\alpha})$ with respect to the scalar product $-B(\ ,\ )$ on $X_{1,\alpha} \subset H_0^1(\Omega) = X_1$. Furthermore,

(III.2.52) $$DF(u)h = L^{-1}(\hat{g}_u(u)h) \quad \text{for} \quad u, h \in X_{1,\alpha},$$

which implies for $A_0 = DF(0)$ the equivalence

(III.2.53)
$$(A_0 - \lambda I)v = 0 \quad \text{for } v \in X_{1,\alpha}, \ \lambda \ne 0 \Leftrightarrow$$
$$Lv = \mu \hat{g}_u(0)v \quad \text{for } v \in X, \ \mu = \lambda^{-1}.$$

As discussed at the end of Remark III.2.3, for $r = \hat{g}_u(0) \in C^\alpha(\overline{\Omega})$, every eigenfunction v of (III.2.47) in $H^2(\Omega) \cap H_0^1(\Omega)$ is also in X, so that all results for (III.2.47) summarized in Remark III.2.3 hold for the eigenvalue problem (III.2.53)$_2$ as well. Therefore, the eigenvalue problems (III.2.46) and (III.2.53)$_1$ are equivalent, too.

Let $\lambda_0 \ne 0$ be one of the eigenvalues of A_0. Then the decomposition $X_1 = N(A_0 - \lambda_0 I) \oplus R(A_0 - \lambda_0 I)$ for $A_0 \in L(X_1, X_1)$, defined in (III.2.45) for $r = \hat{g}_u(0)$, implies for $X_{1,\alpha} \subset H_0^1(\Omega) = X_1$ the decomposition $X_{1,\alpha} = N(A_0 - \lambda_0 I) \oplus R(A_0 - \lambda_0 I)$ for $A_0 \in L(X_{1,\alpha}, X_{1,\alpha})$ defined by $A_0 = L^{-1} \circ \hat{g}_u(0)I$. (The same notation A_0 for different settings should not be confusing.)

Thus all hypotheses of Theorem I.21.2 are satisfied for $F \in C^1(X_{1,\alpha}, X_{1,\alpha})$ defined in (III.2.51), and every nontrivial pair solving $F(u) = \lambda u$ yields a **classical solution** $(u, \mu) \in X \times \mathbb{R}$ of (III.2.38) for $\mu = \lambda^{-1}$.

The two approaches for problem (III.2.38) show that weak solutions under weak regularity assumptions are obtained via a Hilbert space setting that requires growth conditions on g, whereas strong solutions under strong regularity of all data are given by a Banach space approach having the advantage that it applies *for every smooth nonlinearity g, irrespective of any growth at infinity.*

In general, we know only the properties (III.2.48) of the nontrivial solutions. However, if $\dim N(L - \mu_0 \hat{g}_u(0)I) = 1$, then Theorem I.5.1 applies: For the family $L(\mu) = L - \mu \hat{g}_u(0)I$, the nondegeneracy (III.2.8)$_4$ is satisfied by

$$
\begin{aligned}
\text{(III.2.54)} \qquad (L_\mu(\mu_0)\hat{v}_0, \hat{v}_0)_0 &= (\hat{g}_u(0)\hat{v}_0, \hat{v}_0)_0 \\
&= \lambda_0 B(\hat{v}_0, \hat{v}_0) \neq 0 \qquad \text{for } \lambda_0 = \mu_0^{-1},
\end{aligned}
$$

so that the bifurcating solution set is a smooth curve $\{(u(s), \mu(s)) | s \in (-\delta, \delta)\}$ in $X \times \mathbb{R}$ through $(0, \mu_0)$. This is true, in particular, if μ_0 is a principal eigenvalue of $Lv = \mu \hat{g}_u(0)v$; cf. (III.2.47) and the comments after it.

The application of Corollary I.21.3 gives many more solutions in the Hilbert space as well as in the Banach space approach:

(III.2.55) If $g(x, -u) = -g(x, u)$ for all $(x, u) \in \Omega \times \mathbb{R}$, then at $(0, \mu_0)$ at least n pairs of solutions $(\pm u, \mu)$ of (III.2.38) bifurcate, where $n = \dim N(L - \mu_0 \hat{g}_u(0)I)$.

III.2.5 Variational Methods II

There is still a drawback to overcome: The method used to prove Theorem I.21.2 allows only a linear dependence on the parameter λ, which means that the parameter μ in (III.2.38) appears linearly, too. Application of Theorem II.7.3, however, allows an arbitrary dependence on the parameter, which we denote again by λ. Although we could stay with the foregoing setting, we change it in order to avoid an inversion of L.

Let $L(\lambda) = L(\lambda)^*$ be a family of operators of divergence form (III.1.3) with coefficients $a_{ij} \in C(\mathbb{R}, C^{1,\alpha}(\overline{\Omega})), c \in C(\mathbb{R}, C^\alpha(\overline{\Omega}))$. (This means that $\lambda \mapsto a_{ij}(\cdot, \lambda)$, $\lambda \mapsto c(\cdot, \lambda)$ are continuous from $\mathbb{R}$ into the Banach spaces $C^{1,\alpha}(\overline{\Omega}), C^\alpha(\overline{\Omega})$, respectively.) The boundary $\partial\Omega$ is smooth enough to apply all properties of elliptic operators summarized in Section III.1. Assuming that

$$
\text{(III.2.56)} \qquad
\begin{aligned}
&L(\lambda_0) \text{ is elliptic and that} \\
&0 \text{ is an eigenvalue of } L(\lambda_0),
\end{aligned}
$$

then the continuous family $L(\lambda) : X \to Z$ satisfies (II.4.1), (II.4.2), and (II.4.3) for $X = C^{2,\alpha}(\overline{\Omega}) \cap \{u | u = 0 \text{ on } \partial\Omega\}$ and $Z = C^\alpha(\overline{\Omega})$. We consider the semilinear problem

$$L(\lambda)u + g(u, x, \lambda) = 0 \quad \text{in} \quad \Omega,$$
(III.2.57)
$$u = 0 \quad \text{on} \quad \partial\Omega,$$

where $g \in C^2(\mathbb{R} \times \overline{\Omega} \times \mathbb{R}, \mathbb{R})$ satisfies $g(0, x, \lambda) = g_u(0, x, \lambda) = 0$ for all $(x, \lambda) \in \overline{\Omega} \times \mathbb{R}$. (The first condition gives the trivial solution $(u, \lambda) = (0, \lambda)$ and the second is imposed without loss of generality by adding $g_u(0, x, \lambda)$ to the coefficients $c(x, \lambda)$ of $L(\lambda)$.) Then

(III.2.58)
$$
\begin{aligned}
&F : X \times \mathbb{R} \to Z \text{ is in } C(X \times \mathbb{R}, Z), \text{ where} \\
&F(u, \lambda)(x) = (L(\lambda)u)(x) + g(u(x), x, \lambda), \\
&D_u F \in C(X \times \mathbb{R}, L(X, Z)), \text{ and} \\
&D_u F(0, \lambda)h = L(\lambda)h.
\end{aligned}
$$

If we endow Z with the scalar product $(\ ,\ )_0$, then $F(\cdot, \lambda)$ is a potential operator from X into Z according to Definition I.3.1. We give the potential:

$$f(u, \lambda) = \frac{1}{2}(L(\lambda)u, u)_0 + \int_\Omega G(u(x), x, \lambda)dx$$
(III.2.59)   for $(u, \lambda) \in X \times \mathbb{R}$, where
$$G(u, x, \lambda) = \int_0^u g(s, x, \lambda)ds \text{ for } (u, x, \lambda) \in \mathbb{R} \times \overline{\Omega} \times \mathbb{R}.$$

Then Theorem II.7.3 provides nontrivial solutions of $F(u, \lambda) = 0$ in $X \times \mathbb{R}$, i.e., of (III.2.57), which cluster at $(0, \lambda_0)$, if

(III.2.60)   the crossing number $\chi(L(\lambda), \lambda_0)$ of $L(\lambda)$ at $\lambda = \lambda_0$ is nonzero.

Since all eigenvalues of $L(\lambda)$ are real, local hyperbolicity is true if 0 is not an eigenvalue of $L(\lambda)$ for $\lambda \in (\lambda_0 - \delta, \lambda_0) \cup (\lambda_0, \lambda_0 + \delta)$. By Definition II.7.1, a nonzero crossing number means that a nonzero number of eigenvalues (counting multiplicities) leaves or enters the positive real axis when λ passes through λ_0. We give a simple sufficient condition that implies (III.2.60):

$$L(\lambda)h = \sum_{i,j=1}^n (a_{ij}(x)h_{x_i})_{x_j} + c(\lambda)h,$$
(III.2.61)
and $c(\lambda)$ is strictly monotonic near $\lambda = \lambda_0$;

cf. (III.2.17), (III.2.18). Note that in contrast to (III.2.36), the multiplicity of the eigenvalue 0 of $L(\lambda_0)$ is arbitrary in the variational case. Therefore, for $F(u, \lambda) = \Delta u + \lambda g(u) = 0$ with $g(0) = 0$, $g'(0) > 0$, for example, *every* eigenvalue $\mu_n > 0$ of $-\Delta$ provides a bifurcation point $(0, \lambda_n)$, where $\mu_n = \lambda_n g'(0)$; cf. Section III.7.

Since Theorem II.7.3 generalizes Theorem I.21.2, our result on (III.2.57) is our most general for elliptic problems of variational structure.

III.2.6 An Example

This example, which is continued in Sections III.5 and III.6, is presented because it serves as a paradigm for the following techniques:

- By an exploitation of symmetry, only nondegenerate bifurcations with one-dimensional kernels occur.
- An evaluation of bifurcation formulas reveals sub- and supercritical pitch-fork bifurcations and determines, in turn, their stability.
- The global extensions of the local bifurcating curves satisfy the *second* Rabinowitz alternative. The proof includes the following steps:
- A combination of symmetry and of the elliptic maximum principle applied to a *differentiated equation* proves that the maxima and minima of all solutions on a global branch have a fixed location.
- This qualitative property separates all global branches and helps at the same time to prove an a priori estimate excluding the first Rabinowitz alternative.
- In view of the separation of branches, there is only one possibility left to meet the trivial solution line a second time.
- This establishes the global bifurcation diagram.

We start now to discuss the following model: Minimize or find critical points of the energy

$$E_\varepsilon(u) = \int_\Omega (\frac{\varepsilon}{2}\|\nabla u\|^2 + W(u))dx, \ \varepsilon \geq 0,$$

(III.2.62)   over $\Omega = (0,1) \times (0,1)$

under the constraint $\int_\Omega u\, dx = m.$

This functional is commonly called the **Cahn–Hilliard energy**, describing the total energy of a binary alloy of mass m in Ω with $(\varepsilon > 0)$ or without $(\varepsilon = 0)$ interfacial energy. According to the two components of the alloy, the free energy potential $W: \mathbb{R} \to \mathbb{R}$ is a so-called two-well potential having two minima; cf. Figure III.2.4. We assume that W is sufficiently smooth (C^4 is enough).

The Euler–Lagrange equation for (III.2.62) is

$$-\varepsilon\Delta u + W'(u) = \lambda \ \text{ in } \Omega, \ \lambda = \text{Lagrange multiplier},$$

(III.2.63) $\sum_{i=1}^2 u_{x_i}\nu_i = 0 \ \text{ on } \partial\Omega,$

$$\int_\Omega u\, dx = m,$$

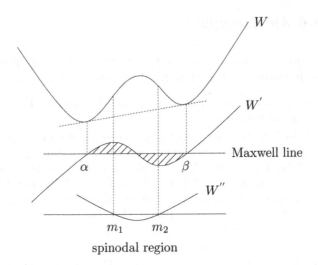

spinodal region

Figure III.2.4

where ν is the outer normal; i.e., (III.2.63)$_2$ are the natural boundary conditions for the self-adjoint operator Δ called homogeneous Neumann boundary conditions; cf. (III.1.5). For $\varepsilon > 0$, (III.2.63) is a semilinear elliptic problem with a nonlocal term, namely,

$$(\text{III.2.64}) \qquad\qquad \lambda = \int_\Omega W'(u)dx.$$

In order to incorporate the constraint (III.2.63)$_3$ into a function space, we make the substitution

$$(\text{III.2.65}) \qquad\qquad u = m + v, \quad \text{where} \quad \int_\Omega v\,dx = 0$$

(recall that $|\Omega| = 1$), and we set

$$(\text{III.2.66}) \quad F_\varepsilon(v, m) = -\varepsilon\Delta v + W'(m + v) - \int_\Omega W'(m + v)dx.$$

In the subsequent analysis we fix $\varepsilon > 0$, and we use the mass m as a bifurcation parameter. We have the trivial solutions $(v, m) = (0, m)$ of $F_\varepsilon(v, m) = 0$ for all $m \in \mathbb{R}$ (describing a homogeneous mixture), and before choosing the adequate function spaces, we study its linearization along the trivial solution. We obtain from (III.2.66)

$$(\text{III.2.67}) \qquad \begin{aligned} -\varepsilon\Delta v + W''(m)v &= 0 \quad \text{in} \quad \Omega, \\ (\nabla v, \nu) &= 0 \quad \text{on} \quad \partial\Omega, \\ \int_\Omega v\,dx &= 0, \end{aligned}$$

having the following special classes of eigenfunctions:

(III.2.68)
$$\hat{v}(x_1, x_2) = \alpha_1 \cos n\pi x_1 + \alpha_2 \cos n\pi x_2, \ \alpha_1, \alpha_2 \in \mathbb{R},$$
$$\text{provided that } W''(m) = -\varepsilon n^2 \pi^2, n \in \mathbb{N},$$

$$\hat{v}_{\ell n}(x_1, x_2) = \alpha \cos \ell\pi x_1 \cos n\pi x_2, \ \alpha \in \mathbb{R},$$
$$\text{provided that } W''(m) = -\varepsilon(\ell^2 + n^2)\pi^2, \ \ell, n \in \mathbb{N}.$$

The kernels are not necessarily one-dimensional: Whereas this is obvious in the first case, it is possible also in the second case if $\tilde{n}^2 = \ell^2 + n^2$ or $\tilde{\ell}^2 + \tilde{n}^2 = \ell^2 + n^2$ for different integers.

Since higher-dimensional kernels are not convenient in bifurcation theory, we can pursue two tracks: Restrict the problem to a one-dimensional one, i.e., allow the dependence on only one variable x_1, say (in which case the Euler–Lagrange equation is an ODE), or impose symmetry constraints that reduce the dimension of the kernels to one. We choose the second possibility, since the first one is simpler and is treated analogously.

From the first class we pick the eigenfunction

(III.2.69) $$\hat{v}_n(x_1, x_2) = \cos n\pi x_1 + \cos n\pi x_2,$$

which belongs to the following symmetry class:

(III.2.70)
$$X_n = \left\{ v : \mathbb{R}^2 \to \mathbb{R} \middle| v(x_1, x_2) = v(-x_1, x_2) = v(x_1, -x_2) \right.$$
$$\left. = v(x_2, x_1) = v\left(x_1 + \frac{2}{n}, x_2\right) = v\left(x_1, x_2 + \frac{2}{n}\right) \right\};$$

cf. (III.1.31)$_2$. We change the notation of (III.1.31): The subscript n refers to the period, and we have an additional symmetry with respect to the diagonal $\{x_1 = x_2\}$. The symmetry lattice is shown in Figure III.1.1 for $a = \frac{1}{n}$ when it is shifted by the vector $(\frac{1}{n}, 0)$.

The eigenfunctions of the second class have the symmetry

(III.2.71) $$v(x_1, x_2) = v\left(\frac{n}{\ell}x_2, \frac{\ell}{n}x_1\right),$$

which is not orthogonal if $\ell \neq n$. Therefore, the operator Δ is not equivariant with respect to the oblique symmetry, so that its fixed-point space is not invariant; cf. the discussion in Section III.1. For $\ell = n$, however, we set

(III.2.72) $$\hat{v}_{nn}(x_1, x_2) = 2 \cos n\pi x_1 \cos n\pi x_2,$$

which belongs to the symmetry class

(III.2.73) $$X_{nn} = \left\{ v \in X_n \middle| v(x_1, x_2) = v\left(-x_2 + \frac{1}{n}, -x_1 + \frac{1}{n}\right) \right\},$$

whose lattice is shown in Figure III.1.2. As remarked in Section III.1, the lattices for X_n and X_{nn} are not essentially distinct, since they are transformed into each other by an affine mapping. We confine our analysis to the symmetry class X_n and state simply that the modifications to the class X_{nn} are simple enough to be left to the reader. (For details see [106].)

The classes X_n are appropriate for homogeneous Neumann boundary conditions on $\Omega = (0,1) \times (0,1)$ for all $n \in \mathbb{N}$. Accordingly, we define

(III.2.74)
$$X_n^{2,\alpha} = C^{2,\alpha}(\mathbb{R}^2) \cap X_n \cap \left\{ v \Big| \int_\Omega v \, dx = 0 \right\},$$
X_n^α analogously,

and the mapping defined in (III.2.66) satisfies

(III.2.75)
$$F_\varepsilon : X_n^{2,\alpha} \times \mathbb{R} \to X_n^\alpha,$$

since by the boundary conditions, $\int_\Omega F_\varepsilon(v,m)dx = 0$. Clearly, all solutions $(v,m) \in X_n^{2,\alpha} \times \mathbb{R}$ of $F_\varepsilon(v,m) = 0$ give via $u = v + m$ solutions of the Euler–Lagrange equation (III.2.63) satisfying the constraint (III.2.63)$_3$.

By the analysis of Section III.1, in particular by (III.1.35),

(III.2.76)
$$D_v F_\varepsilon(0,m) = -\varepsilon\Delta + W''(m)I : X_n^{2,\alpha} \to X_n^\alpha$$
is a Fredholm operator of index zero.

Note that the Fredholm property remains valid in the subspaces of functions with mean value zero in Ω. (Constant functions span $N(\Delta)$ and are complementary to $R(\Delta)$. In this case constant functions vanish.) As seen in (III.2.68),

(III.2.77)
$$\dim N(D_v F_\varepsilon(0,m)) = 1, \text{ provided that}$$
$$W''(m) = -\varepsilon n^2 \pi^2.$$

By the assumed shape of W'' shown in Figure III.2.4, there are two solutions m_n^1, m_n^2 of the characteristic equation (III.2.77)$_2$ for $n = 1, \ldots, N(\varepsilon)$ if $\varepsilon > 0$ is small enough. These candidates for bifurcation points are in the spinodal region (m_1, m_2).

By formal self-adjointness of $L = -\varepsilon\Delta + W''(m)I$, we have by (III.1.35) the decomposition

(III.2.78)
$$X_n^\alpha = R(L) \oplus N(L),$$

where the spaces $R(L)$ and $N(L)$ are orthogonal with respect to the scalar product $(\ ,\ )_0$ in $L^2(\Omega)$. This yields the Lyapunov–Schmidt projections

(III.2.79)
$$Qv = (v, \hat{v}_n)_0 \hat{v}_n, \ Q : X_n^\alpha \to N(L) \text{ along } R(L),$$
$$P = Q|_{X_n^{2,\alpha}} : X_n^{2,\alpha} \to N(L) \text{ along } R(L) \cap X_n^{2,\alpha}.$$

Observe that $\|\hat{v}_n\|_0 = 1$ for all $n \in \mathbb{N}$.

Again by the assumption on W'', cf. Figure III.2.4, the nondegeneracy (I.5.3) of Theorem I.5.1 is satisfied:

(III.2.80)
$$(D^2_{vm} F_\varepsilon(0, m)\hat{v}_n, \hat{v}_n)_0 = W'''(m) \neq 0$$

if $W''(m) = -\varepsilon n^2 \pi^2$ has two solutions m^1_n, m^2_n.

Therefore, there exist nontrivial curves of solutions $\{(v(s), m(s)) | s \in (-\delta, \delta)\}$ of $F_\varepsilon(v, m) = 0$ through $(v(0), m(0)) = (0, m^i_n)$, $i = 1, 2$, and by (I.5.16), (I.5.17),

(III.2.81)
$$v(s) = s\hat{v}_n + o(s) \quad \text{in} \quad X^{2,\alpha}_n.$$

If $(v, m) \in X^{2,\alpha}_n \times \mathbb{R}$ is a solution of $F_\varepsilon(v, m)$, then the reversion

(III.2.82)
$$\tilde{v}(x_1, x_2) = v\left(\frac{1}{n} - x_1, \frac{1}{n} - x_2\right)$$

defines a solution $(\tilde{v}, m) \in X^{2,\alpha}_n \times \mathbb{R}$, too. Since the reversion of the eigenfunction $\hat{v}_n$ is its negative $-\hat{v}_n$, the uniqueness of the bifurcating curve implies that the two components $\{(v(s), m(s)) | s \in [0, \delta)\}$ and $\{(v(s), m(s)) | s \in (-\delta, 0]\}$ are transformed into each other by the reversion (III.2.82). This implies, in particular, $m(-s) = m(s)$ and $\dot{m}(0) = 0$. We verify this by evaluating the Bifurcation Formula of Section I.6. Sub- or supercriticality of the pitchfork is then determined by $\ddot{m}(0)$, provided that it is nonzero.

By definition (III.2.66),

(III.2.83)
$$D^2_{vv} F_\varepsilon(v, m)[\hat{v}_n, \hat{v}_n] = W'''(m + v)\hat{v}^2_n - \int_\Omega W'''(m + v)\hat{v}^2_n dx,$$
$$D^3_{vvv} F_\varepsilon(0, m)[\hat{v}_n, \hat{v}_n, \hat{v}_n] = W^{(4)}(m)\hat{v}^3_n \quad \text{by} \int_\Omega \hat{v}^3_n dx = 0.$$

Using the projection (III.2.79), we see that the numerator of formula (I.6.3) for $\dot{m}(0)$ (see also (III.2.9)) is

(III.2.84)
$$(D^2_{vv} F_\varepsilon(0, m)[\hat{v}_n, \hat{v}_n], \hat{v}_n)_0$$
$$= W'''(m) \int_\Omega \hat{v}^3_n dx - W'''(m)\|\hat{v}_n\|^2_0 \int_\Omega \hat{v}_n dx = 0.$$

Therefore, $\dot{m}(0) = 0$. Formula (I.6.11) for $\ddot{m}(0)$ is more involved. We evaluate the numerator (I.6.9) step by step:

(III.2.85)
$$(D^3_{vvv} F_\varepsilon(0, m)[\hat{v}_n, \hat{v}_n, \hat{v}_n], \hat{v}_n)_0 = W^{(4)}(m) \int_\Omega \hat{v}^4_n dx = \frac{9}{4} W^{(4)}(m),$$
$$D^2_{vv} F_\varepsilon(0, m)[\hat{v}_n, \hat{v}_n] = W'''(m)(\hat{v}^2_n - 1),$$
$$Q D^2_{vv} F_\varepsilon(0, m)[\hat{v}_n, \hat{v}_n] = 0.$$

Next we solve

(III.2.86)
$$D_v F_\varepsilon(0, m)v = W'''(m)(\hat{v}_n^2 - 1)$$
$$\text{for } v \in (I - P)X_n^{2,\alpha}; \text{ i.e., } (v, \hat{v}_n)_0 = 0.$$

By $\hat{v}_n^2 - 1 = \frac{1}{2}(\cos 2n\pi x_1 + \cos 2n\pi x_2) + 2\cos n\pi x_1 \cos n\pi x_2$ we obtain, in view of (III.2.76) and $W''(m) = -\varepsilon n^2 \pi^2$ for the solution of (III.2.86),

(III.2.87) $v = \dfrac{W'''(m)}{\varepsilon n^2 \pi^2}\left(\dfrac{1}{6}(\cos 2n\pi x_1 + \cos 2n\pi x_2) + 2\cos n\pi x_1 \cos n\pi x_2\right).$

Finally,

(III.2.88)
$$(D_{vv}^2 F_\varepsilon(0, m)[\hat{v}_n, v], \hat{v}_n)_0 = W'''(m)\int_\Omega v\hat{v}_n^2 dx$$
$$= \frac{13}{12}\frac{W'''(m)^2}{\varepsilon n^2 \pi^2}, \quad \text{since } (v, \hat{v}_n)_0 = 0.$$

For the denominator of (I.6.11) we use (III.2.80), and we obtain for $m = m(0)$,

(III.2.89) $\ddot{m}(0) = -\dfrac{1}{3}\left\{\dfrac{9}{4}\dfrac{W^{(4)}(m)}{W'''(m)} - \dfrac{13W'''(m)}{4\varepsilon n^2 \pi^2}\right\}.$

By the assumed shape of W'' we have $W^{(4)}(m) > 0$ and $W'''(m_n^1) < 0$, $W'''(m_n^2) > 0$. Therefore, for small $\varepsilon > 0$,

(III.2.90) $\ddot{m}(0) < 0$ at $(0, m_n^1)$, $\ddot{m}(0) > 0$ at $(0, m_n^2)$,

yielding the bifurcation diagrams sketched in Figure III.2.5.

In Sections III.5, III.6 we see how the two local curves shown in Figure III.2.5 are extended globally; see Figure III.6.2.

In Figure III.2.5 the instability of the solutions (m, v) is marked by a hatched line. We define the stability of a conditionally critical point of the energy $E_\varepsilon(u) = E_\varepsilon(m+v)$ given in (III.2.62) by its stability as an equilibrium of the negative gradient flow, i.e., of

(III.2.91) $\dfrac{dv}{dt} = -F_\varepsilon(v, m);$ see (III.2.66).

(The evolution equation (III.2.91) is not what is called the dynamical Cahn–Hilliard model. In that model the preservation of mass is ensured by the application of the Laplacian together with an additional boundary condition. It does not contain a nonlocal term, but it is a parabolic PDE of fourth order.)

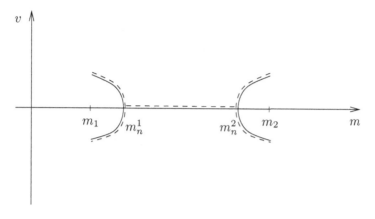

Figure III.2.5

In the symmetry class X_n, the value $W''(m_n^i) = -\varepsilon n^2 \pi^2$ is the principal, i.e., greatest negative, eigenvalue of $\varepsilon \Delta$; cf. Remark III.2.1. In other words, 0 is the principal eigenvalue of $-D_v F_\varepsilon(0, m_n^i)$, $i = 1, 2$. The assumed shape of the graph of W'' shown in Figure III.2.4 implies that the eigenvalue 0 becomes a negative eigenvalue of $D_v F_\varepsilon(0, m)$ for $m < m_n^1$, $m > m_n^2$, and a positive eigenvalue for $m \in (m_n^1, m_n^2)$. This proves the stability properties of the trivial solutions $(0, m)$, as indicated in Figure III.2.5.

By the Principle of Exchange of Stability proved in Section I.7, both bifurcating pitchforks sketched in Figure III.2.5 are therefore unstable; cf. Figure I.7.3. Presumably the branches regain stability at the next turning point; cf. Figure I.7.1 and Figure III.6.2.

Remark III.2.4 *One of the best-known phenomena in applied bifurcation theory is the appearance of Taylor vortices in the so-called* **Couette–Taylor** **model.** *Since there exists a vast literature about that model, we do not give all details. We refer to the monographs [17], [80], and to the survey [115], where the problem is thoroughly discussed and where many historical and important references are given.*

The gap Ω between two coaxial cylinders of infinite length with radii $0 < R_1 < R_2$ is filled with a viscous and incompressible fluid; see Figure III.2.6.

There are no external forces, and a rotation of the interior cylinder causes a flow due to the adhesion of the fluid to the cylinders. A flow is described by its velocity $\mathbf{u} = (u_1, u_2, u_3)$ at time t and at a point $\mathbf{x} = (x_1, x_2, x_3)$ in the gap Ω.

One introduces dimensionless reference quantities for length, velocity, time, and pressure, which are all directly adapted to the model, and one defines $r_i = R_i/(R_2 - R_1)$, $i = 1, 2$, and the Reynolds number $\lambda = R_1 \omega (R_2 - R_1)/\nu$, where ω is the angular velocity of the interior cylinder and where ν denotes the kinematic viscosity. Then the dimensionless velocity $\mathbf{u}$ satisfies the Navier–Stokes system

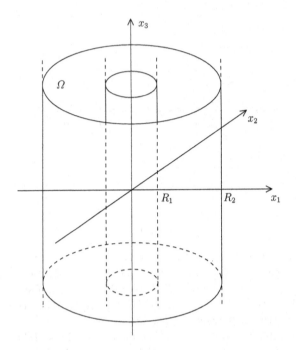

Figure III.2.6

(III.2.92)

$$\frac{\partial \mathbf{u}}{\partial t} - \mathbf{\Delta u} + \lambda(\mathbf{u} \cdot \nabla)\mathbf{u} + \lambda \nabla p = \mathbf{0} \quad in \quad \Omega,$$

$$\mathrm{div}\mathbf{u} = 0 \quad in \quad \Omega,$$

$$\mathbf{u}(x_1, x_2, x_3) = \frac{1}{r_1}\begin{pmatrix} -x_2 \\ x_1 \\ 0 \end{pmatrix} \quad for \ \sqrt{x_1^2 + x_2^2} = r_1,$$

$$\mathbf{u}(x_1, x_2, x_3) = \mathbf{0} \quad for \ \sqrt{x_1^2 + x_2^2} = r_2.$$

Here p denotes the (unknown and dimensionless) pressure, and $\mathbf{\Delta u} = (\Delta u_1, \Delta u_2, \Delta u_3)$, where Δ is the scalar Laplacian, and the operator $\mathbf{u} \cdot \nabla = u_1\frac{\partial}{\partial x_1} + u_2\frac{\partial}{\partial x_2} + u_3\frac{\partial}{\partial x_3}$ acts on each component of $\mathbf{u}$.

There is a stationary solution $(\mathbf{u}_0, p_0)$ for all $\lambda > 0$, the so-called Couette flow, given by

$$\mathbf{u}_0(x_1, x_2, x_3) = \frac{r_1}{r_2^2 - r_1^2}\left(\frac{r_2^2}{r^2} - 1\right)\begin{pmatrix} -x_2 \\ x_1 \\ 0 \end{pmatrix}, \quad r = \sqrt{x_1^2 + x_2^2},$$

(III.2.93)

$$p_0(x_1, x_2, x_3) = \left(\frac{r_1}{r_2^2 - r_1^2}\right)^2 \int_{r_1}^{\sqrt{x_1^2+x_2^2}} \frac{(r_2^2 - \rho^2)^2}{\rho^3}\,d\rho.$$

That flow does not depend on the axial variable x_3 and it is rotationally symmetric in the following sense: Let

(III.2.94) $\quad \mathbf{T}_\varphi = \begin{pmatrix} \cos\varphi & -\sin\varphi & 0 \\ \sin\varphi & \cos\varphi & 0 \\ 0 & 0 & 1 \end{pmatrix}, \qquad \varphi \in [0, 2\pi],$

denote the rotation in the (x_1, x_2)-plane about the angle φ.
Then

(III.2.95) $\mathbf{u}_0(\mathbf{T}_\varphi\mathbf{x}) = \mathbf{T}_\varphi\mathbf{u}_0(\mathbf{x}), \quad p_0(\mathbf{T}_\varphi\mathbf{x}) = p_0(\mathbf{x}) \quad$ *for all $\mathbf{T}_\varphi$.*

The velocity profile decays monotonically from $(III.2.92)_3$ at $r = r_1$ to zero at $r = r_2$. It is what one expects. (System $(III.2.92)$ and its solution $(III.2.93)$ are usually given in cylindrical coordinates.)

Experiments, however, show that for angular velocities ω greater than some critical value ω_0, i.e., for Reynolds numbers λ greater than some λ_0, the Couette flow is no longer observed but a new stationary flow appears. This flow is also rotationally symmetric in the sense of $(III.2.95)$, but it depends also on the axial direction x_3. As a matter of fact, it is periodic in that direction with a well-defined period. It is called the Taylor vortex flow. The trajectory of any single particle is a closed spiral and all spirals form tori stacked one on top of the other, with opposite orientations.

From a mathematical point of view this organization of a new flow can be described as a bifurcation. To be more precise, it is a "symmetry-breaking bifurcation," since the symmetry (=constancy) in direction x_3 is broken. Furthermore, an exchange of stability takes place, which means that the Couette flow loses its stability, whereas the Taylor vortices are stable, at least for angular velocities $\omega > \omega_0$ that are near ω_0.

In order to prove this scenario one makes the ansatz $\mathbf{u} = \mathbf{u}_0 + \mathbf{v}$, $p = p_0 + q$ in $(III.2.92)$, yielding for stationary $\mathbf{v}$,

(III.2.96)
$$-\boldsymbol{\Delta}\mathbf{v} + \lambda[(\mathbf{v}\cdot\nabla)\mathbf{u}_0 + (\mathbf{u}_0\cdot\nabla)\mathbf{v}] + \lambda(\mathbf{v}\cdot\nabla)\mathbf{v} + \lambda\nabla q = \mathbf{0} \quad in \ \ \Omega,$$
$$\mathrm{div}\,\mathbf{v} = 0 \quad in \ \ \Omega,$$
$$\mathbf{v} = \mathbf{0} \quad for \ r = r_i,$$

$i = 1, 2$, where $r = \sqrt{x_1^2 + x_2^2}$. Furthermore one requires a rotational symmetry, i.e.,

(III.2.97) $\mathbf{v}(\mathbf{T}_\varphi \mathbf{x}) = \mathbf{T}_\varphi \mathbf{v}(\mathbf{x})$, $q(\mathbf{T}_\varphi \mathbf{x}) = q(\mathbf{x})$ *for all* $\mathbf{T}_\varphi$ *in (III.2.94)*.

The restriction (III.2.97) is motivated by experiments and mathematically justified by the following: The Navier–Stokes system (III.2.92)$_{1,2}$ is equivariant with respect to the entire orthogonal group $O(3)$, whence it allows symmetric solutions in the sense of (III.2.95), (III.2.97).

Obviously system (III.2.96) has the trivial solution $\mathbf{v} = \mathbf{0}$ and $q = 0$ for all parameters λ, and it is the goal to detect a bifurcation from the trivial solution line. However, there is an obstruction to the application of all common bifurcation theorems, namely the unboundedness of the domain Ω. The linearization of the operator (III.2.96)$_{1,2}$ at $(\mathbf{v}, q, \lambda) = (\mathbf{0}, 0, \lambda)$ has the properties of an elliptic operator including its drawbacks caused by an unbounded domain: The crucial Fredholm property is no longer granted and the spectrum is no longer necessarily discrete.

Experiments suggest to restrict problem (III.2.96) to vector fields $(\mathbf{v}, q)$ that are periodic in the axial direction x_3. Then the domain Ω can be restricted to a section D having the height p_3 of one period, i.e., the problem is reduced to a bounded domain.

A crucial question arises at this point: Which period is physically correct? Recall that the period of the Taylor vortices is not imposed but organized by itself. We discuss that deep question later and we describe briefly the steps to prove bifurcation.

The starting point is an orthogonal decomposition of the Hilbert space $L^2(D)^3$ into the closure of divergence-free (solenoidal) vector fields $\mathbf{Z}$ and the gradient of scalar functions: $L^2(D)^3 = \mathbf{Z} \oplus \mathbf{Z}^\perp$. Let $\mathbf{P} : L^2(D)^3 \to \mathbf{Z}$ denote the orthogonal projection along $\mathbf{Z}^\perp$. If $\mathbf{P}$ is applied to (III.2.96)$_1$, then the gradient of the pressure is eliminated, since $\mathbf{P}\nabla q = \mathbf{0}$. On the other hand, if $\mathbf{Pf} = \mathbf{0}$, then $\mathbf{f} = \nabla q$ for some $q \in H^1(D)$. We define

(III.2.98) $\mathbf{F}(\mathbf{v}, \lambda) = -\mathbf{P}\Delta\mathbf{v} + \lambda\mathbf{P}[(\mathbf{v} \cdot \nabla)\mathbf{u}_0 + (\mathbf{u}_0 \cdot \nabla)\mathbf{v}] + \lambda\mathbf{P}(\mathbf{v} \cdot \nabla)\mathbf{v}.$

The operator $\mathbf{F}(\cdot, \lambda)$ is given a domain $\mathbf{X} \subset \mathbf{Z}$, which is a closed subspace of the Hilbert space $H^2(D)^3$. It consists of solenoidal vector fields satisfying homogeneous Dirichlet boundary conditions on the cylinders. Furthermore, all vector fields in $\mathbf{X}$ have the symmetry (III.2.97) and they are restrictions of vector fields on Ω that are periodic in the axial direction x_3 with a period p_3, which is the height of the segment D. Then

(III.2.99)
$$
\begin{aligned}
&\mathbf{F} : \mathbf{X} \times \mathbb{R} \to \mathbf{Z} \quad \textit{is continuous,} \\
&\mathbf{F}(\mathbf{0}, \lambda) = \mathbf{0} \quad \textit{for all } \lambda \in \mathbb{R}, \textit{ and} \\
&\textit{problem (III.2.96) is reduced to} \\
&\mathbf{F}(\mathbf{v}, \lambda) = \mathbf{0} \quad \textit{for} \quad (\mathbf{v}, \lambda) \in \mathbf{X} \times \mathbb{R}.
\end{aligned}
$$

It is known that $-\mathbf{P}\Delta : \mathbf{X} \to \mathbf{Z}$ is bijective and continuous, whence it has a continuous inverse $\mathbf{K} : \mathbf{Z} \to \mathbf{X}$. Since the linear operator $\mathbf{L}(\mathbf{u}_0)\mathbf{v} = (\mathbf{v}\cdot\nabla)\mathbf{u}_0 + (\mathbf{u}_0 \cdot \nabla)\mathbf{v}$ is continuous from $H^1(D)^3$ into $L^2(D)^3$, and since the embedding

$\mathbf{X} \subset H^1(D)^3$ *is compact, the operator* $\mathbf{KPL}(\mathbf{u}_0) : H^1(D)^3 \rightarrow H^1(D)^3$ *is compact. For* $D_{\mathbf{v}}\mathbf{F}(\mathbf{0}, \lambda) = -\mathbf{P}\boldsymbol{\Delta} + \lambda\mathbf{PL}(\mathbf{u}_0)$ *the relation* $D_{\mathbf{v}}\mathbf{F}(\mathbf{0}, \lambda)\mathbf{v} = \mathbf{f}$ *for* $\mathbf{f} \in \mathbf{Z}$, $\mathbf{v} \in \mathbf{X}$ *is equivalent to* $(\mathbf{I} + \lambda\mathbf{KPL}(\mathbf{u}_0))\mathbf{v} = \mathbf{Kf}$ *for* $\mathbf{v} \in H^1(D)^3$. *The proof following (III.1.10) shows that*

$$(\text{III.2.100}) \qquad \begin{aligned} & D_{\mathbf{v}}\mathbf{F}(\mathbf{0}, \lambda) : \mathbf{X} \rightarrow \mathbf{Z} \\ & \textit{is a Fredholm operator of index zero.} \end{aligned}$$

Thus problem (III.2.99)$_4$ allows the application of the (local) bifurcation theorems presented in this book. First of all, one has to find some $\lambda_0 > 0$ *such that* $D_{\mathbf{v}}\mathbf{F}(\mathbf{0}, \lambda_0)$ *has a nontrivial kernel. This amounts to the solution of the linear problem*

$$
\begin{aligned}
-\boldsymbol{\Delta}\mathbf{v} + \lambda_0[(\mathbf{v} \cdot \nabla)\mathbf{u}_0 + (\mathbf{u}_0 \cdot \nabla)\mathbf{v}] + \lambda_0\nabla q &= \mathbf{0} && in && \Omega, \\
\mathrm{div}\,\mathbf{v} &= \mathbf{0} && in && \Omega, \\
\mathbf{v} &= \mathbf{0} && for\ r = r_i,\ i = 1, 2,
\end{aligned}
$$

$$(\text{III.2.101})$$
$$
\begin{aligned}
\mathbf{v}(\mathbf{T}_\varphi\mathbf{x}) &= \mathbf{T}_\varphi\mathbf{v}(\mathbf{x}),\ q(\mathbf{T}_\varphi\mathbf{x}) = q(\mathbf{x})\ \textit{for all}\ \mathbf{T}_\varphi\ \textit{in (III.2.94)}, \\
\mathbf{v}(x_1, x_2, x_3 + p_3) &= \mathbf{v}(x_1, x_2, x_3), \\
q(x_1, x_2, x_3 + p_3) &= q(x_1, x_2, x_3).
\end{aligned}
$$

If $(\mathbf{v}, q)(x_1, x_2, x_3)$ *is a (nontrivial) solution of (III.2.96), (III.2.97), or (III.2.101), then for all numbers* $a \in \mathbb{R}$ *the function* $(\mathbf{v}, q)(x_1, x_2, x_3 + a)$ *is a solution, too. In order to obtain a one-dimensional kernel one fixes the "phase" in the axial direction by the following additional restrictions (which can be satisfied by solutions of (III.2.96), (III.2.97), and (III.2.101) due to the corresponding equivariance of the systems):*

$$(\text{III.2.102}) \qquad
\begin{aligned}
v_1(x_1, x_2, -x_3) &= v_1(x_1, x_2, x_3), \\
v_2(x_1, x_2, -x_3) &= v_2(x_1, x_2, x_3), \\
v_3(x_1, x_2, -x_3) &= -v_3(x_1, x_2, x_3), \\
q(x_1, x_2, -x_3) &= q(x_1, x_2, x_3).
\end{aligned}
$$

An analysis of (III.2.101) yields a one-dimensional kernel of $D_{\mathbf{v}}\mathbf{F}(\mathbf{0}, \lambda_0)$ *for some* $\lambda_0 > 0$ *depending on the chosen period* p_3. *To be more precise, there are infinitely many such parameters* λ_j, $j \in \mathbb{N} \cup \{0\}$, *but for physical reasons only the smallest value* λ_0 *is of interest. Furthermore, excluding a countable set of periods* p_3 *for which degeneracies could occur, zero is not only a geometrically simple eigenvalue of* $D_{\mathbf{v}}\mathbf{F}(\mathbf{0}, \lambda_0)$, *but also "simple" in the following sense: Define* $\mathbf{A} = -\mathbf{P}\boldsymbol{\Delta}$, *which is self-adjoint and positive definite as an operator* $\mathbf{A} : \mathbf{Z} \rightarrow \mathbf{Z}$ *with domain* $D(\mathbf{A}) = \mathbf{X}$, *and define* $\mathbf{M} = \mathbf{PL}(\mathbf{u}_0)$ *and its formal adjoint* $\mathbf{M}^*$ *as operators* $\mathbf{M}, \mathbf{M}^* : \mathbf{Z} \rightarrow \mathbf{Z}$ *with domains* $\mathbf{X}$, *respectively. If for* $\hat{\mathbf{v}}_0 \in \mathbf{X}$, $\hat{\mathbf{w}}_0^* \in \mathbf{Z}$,

$$(\text{III.2.103}) \qquad
\begin{aligned}
N(\mathbf{I} + \lambda_0\mathbf{A}^{-1}\mathbf{M}) &= \mathrm{span}[\hat{\mathbf{v}}_0], \\
N(\mathbf{I} + \lambda_0\mathbf{M}^*\mathbf{A}^{-1}) &= \mathrm{span}[\hat{\mathbf{w}}_0^*],\quad then \\
(\hat{\mathbf{v}}_0, \hat{\mathbf{w}}_0^*)_0 &\neq 0.
\end{aligned}
$$

By $(\mathbf{A}^{-1}\mathbf{Mv}, \mathbf{w})_0 = (\mathbf{v}, \mathbf{M}^*\mathbf{A}^{-1}\mathbf{w})_0$ *for all* $\mathbf{v} \in X$, $\mathbf{w} \in Z$, *the vector* $\hat{\mathbf{v}}_0$ *is not in the range* $R(\mathbf{I} + \lambda_0\mathbf{A}^{-1}\mathbf{M})$.

The "simplicity" (III.2.103) implies the nondegeneracy (I.5.3)$_2$ of Theorem I.5.1: Let

$$
\begin{aligned}
&\hat{\mathbf{v}}_0^* = \mathbf{A}^{-1}\hat{\mathbf{w}}_0^* \in X. \quad \text{Then} \\
&N(D_{\mathbf{v}}\mathbf{F}(\mathbf{0}, \lambda_0)) = N(\mathbf{A} + \lambda_0\mathbf{M}) = \text{span}[\hat{\mathbf{v}}_0], \\
&N((D_{\mathbf{v}}\mathbf{F}(\mathbf{0}, \lambda_0))^*) = N(\mathbf{A} + \lambda_0\mathbf{M}^*) = \text{span}[\hat{\mathbf{v}}_0^*], \\
&0 \neq (\hat{\mathbf{v}}_0, \hat{\mathbf{w}}_0^*)_0 = -\lambda_0(\mathbf{A}^{-1}\mathbf{M}\hat{\mathbf{v}}_0, \hat{\mathbf{w}}_0^*)_0 \\
&= -\lambda_0(\mathbf{M}\hat{\mathbf{v}}_0, \mathbf{A}^{-1}\hat{\mathbf{w}}_0^*)_0 \\
&= -\lambda_0(D_{\mathbf{v}\lambda}^2\mathbf{F}(\mathbf{0}, \lambda_0)\hat{\mathbf{v}}_0, \hat{\mathbf{v}}_0^*)_0,
\end{aligned}
$$

(III.2.104)

whence $D_{\mathbf{v}\lambda}^2\mathbf{F}(\mathbf{0}, \lambda_0)\hat{\mathbf{v}}_0 \notin R(D_{\mathbf{v}}\mathbf{F}(\mathbf{0}, \lambda_0))$.

Thus Theorem I.5.1 is applicable to $\mathbf{F}(\mathbf{v}, \lambda) = 0$ at $(\mathbf{0}, \lambda_0)$, and by some numerical evidence a supercritical pitchfork bifurcation takes place at $(\mathbf{0}, \lambda_0)$. The bifurcating curve $\{(\mathbf{v}(s), \lambda(s))\}$ through $(\mathbf{v}(0), \lambda(0)) = (\mathbf{0}, \lambda_0)$ yields via $\mathbf{u}(s) = \mathbf{u}_0 + \mathbf{v}(s)$ Taylor vortices for Reynolds numbers $\lambda(s) > \lambda_0$. The two parts of the pitchfork give vortices having opposite orientations.

As mentioned before, experiments suggest a "Principle of Exchange of Stability," which means that the Couette flow is stable for Reynolds numbers $\lambda < \lambda_0$, that it is unstable for $\lambda > \lambda_0$, and that the supercritically bifurcating vortex flow is stable.

Linearized stability of a stationary solution $(\mathbf{v}, \lambda)$ is true if the spectrum of $D_{\mathbf{v}}\mathbf{F}(\mathbf{v}, \lambda)$ is in the right complex half-plane. On the other hand, if $D_{\mathbf{v}}\mathbf{F}(\mathbf{v}, \lambda)$ has an eigenvalue in the left complex half-plane, then $(\mathbf{v}, \lambda)$ is unstable; cf. [115]. (Note that the time-dependent Navier–Stokes system is of the form $\frac{d\mathbf{v}}{dt} = -\mathbf{F}(\mathbf{v}, \lambda)$. By the definition (III.2.98) of $\mathbf{F}$, the spectral properties of $D_{\mathbf{v}}\mathbf{F}(\mathbf{v}, \lambda)$ imply also the stability properties of $\mathbf{u} = \mathbf{u}_0 + \mathbf{v}$ as a solution of (III.2.92).) The Couette flow $\mathbf{u}_0$ is stable for $0 < \lambda < \lambda_0$ if the real parts of all eigenvalues of $D_{\mathbf{v}}\mathbf{F}(\mathbf{0}, \lambda)$ are positive for all $0 < \lambda < \lambda_0$. Assuming that zero is the only eigenvalue of $D_{\mathbf{v}}\mathbf{F}(\mathbf{0}, \lambda_0)$ with vanishing real part, the stability of the Couette flow and of the vortex flow for $\lambda > \lambda_0$ is determined by the critical eigenvalue perturbation of $D_{\mathbf{v}}\mathbf{F}(\mathbf{0}, \lambda)$ and of $D_{\mathbf{v}}\mathbf{F}(\mathbf{v}(s), \lambda(s))$, respectively. These perturbations are studied in Section I.7 under the assumption that zero is a simple eigenvalue of $D_{\mathbf{v}}\mathbf{F}(\mathbf{0}, \lambda_0) = \mathbf{A} + \lambda_0\mathbf{M}$. This means that $(\hat{\mathbf{v}}_0, \hat{\mathbf{v}}_0^*)_0 \neq 0$, where the vectors $\hat{\mathbf{v}}_0$ and $\hat{\mathbf{v}}_0^*$ are defined in (III.2.104). However, to our knowledge, only the simplicity of the eigenvalue zero of $\mathbf{A}^{-1}D_{\mathbf{v}}\mathbf{F}(\mathbf{0}, \lambda_0) = \mathbf{I} + \lambda_0\mathbf{A}^{-1}\mathbf{M}$, i.e., $(\hat{\mathbf{v}}_0, \hat{\mathbf{w}}_0^*)_0 = (\hat{\mathbf{v}}_0, \mathbf{A}\hat{\mathbf{v}}_0^*)_0 \neq 0$, has been shown in the literature; cf. (III.2.103).

Let $\mu(\lambda)$ denote the simple eigenvalue perturbation of $D_{\mathbf{v}}\mathbf{F}(\mathbf{0}, \lambda)$ for λ near λ_0. Then $\mu(\lambda_0) = 0$, and by (I.7.34) and (III.2.104) we obtain $\mu'(\lambda_0)(\hat{\mathbf{v}}_0, \hat{\mathbf{v}}_0^*)_0 = -(\hat{\mathbf{v}}_0, \mathbf{A}\hat{\mathbf{v}}_0^*)_0/\lambda_0 \neq 0$. (This does not prove $(\hat{\mathbf{v}}_0, \hat{\mathbf{v}}_0^*)_0 \neq 0$, since the eigenvalue perturbation is valid only in case of $(\hat{\mathbf{v}}_0, \hat{\mathbf{v}}_0^*)_0 \neq 0$.)

If $\text{sign}(\hat{\mathbf{v}}_0, \hat{\mathbf{v}}_0^*)_0 = \text{sign}(\hat{\mathbf{v}}_0, \mathbf{A}\hat{\mathbf{v}}_0^*)_0$, then the Couette flow becomes unstable for $\lambda > \lambda_0$, and in view of the results of Section I.7 the supercritically bifurcating *vortex flow is stable; see Figure I.7.3. However, this exchange of stability has not yet been rigorously proved.*

Some more remarks have to be made:

Stability in this context means stability under perturbations in the class of velocity fields having the same symmetries and the same periodicity in the axial direction x_3. This notion of stability is more restrictive than physical stability under any perturbation.

The crucial open problem, however, is the choice of the axial period p_3 of the Taylor vortices. Whereas a fixed period is imposed in the mathematical analysis, the period is organized by itself in reality. Thus the question of the physically relevant period cannot be answered by the approach described previously. However, there is some plausible attempt at a solution: Since the bifurcation point $(\mathbf{0}, \lambda_0)$ depends on the period p_3, i.e., $\lambda_0 = \lambda_0(p_3)$, one assumes that the period $p_{3,c}$, where $\lambda_{0,c} = \lambda_0(p_{3,c})$ is minimal, is a correct period, at least at the bifurcation point. Note, however, that the function $\lambda_0(p_3)$ is obtained only in the class of vector fields $(\mathbf{v}, q)$ having rotational symmetry and the prescribed period p_3; cf. (III.2.101)$_{4,5}$.

A different approach to understanding the pattern formation of the Couette–Taylor model is an analysis of the Navier–Stokes system (III.2.92) in the unbounded domain Ω without a fixed periodicity in axial direction x_3. A new pattern appears only if the Couette flow $\mathbf{u}_0$ loses its stability as a stationary solution of (III.2.92). The critical threshold $\lambda = \lambda_{cr}$, where a loss of stability takes place, is determined by the Principle of Linearized Stability, which means that for $\lambda = \lambda_{cr}$ the linearization at $\mathbf{u}_0$ has eigenvalues on the imaginary axis and that the rest of the spectrum is in the stable (positive) half-plane. A Fourier analysis in the unbounded axial direction x_3 determines the critical unstable modes. Assume that the loss of stability is caused only by an eigenvalue zero with eigenfunctions defining one single critical mode k_{cr} or a critical period $p_{cr} = 2\pi/k_{cr}$ in the x_3-direction. (Obviously $\lambda_{cr} \leq \lambda_{0,c}$, where $\lambda_{0,c}$ is the minimal value of $\lambda_0(p_3)$ described above. If $\lambda_{cr} = \lambda_{0,c}$ then $p_{cr} = p_{3,c}$ under the foregoing assumption.) Under more generic assumptions one finds for $\lambda > \lambda_{cr}$ (nontrivial) stationary solutions of (III.2.92) near $(\mathbf{0}, \lambda_{cr})$, which have axial periods in intervals around p_{cr}. These intervals shrink to p_{cr} as λ tends to λ_{cr} and they are determined by the unstable modes for $\lambda > \lambda_{cr}$. These solutions represent Taylor vortex flows, but it seems to be hard to select a single period in these intervals that is physically relevant. This is another open stability problem.

The method is briefly described as follows: Since system (III.2.92) for stationary fields $\mathbf{u}$ is "reversible," the x_3-axis can play the role of a "time axis." Using a "center manifold," system (III.2.92) for stationary $\mathbf{u}$ with λ near λ_{cr} is reduced to a four-dimensional system of "amplitude equations," which are put into "normal form." This method detects also classes of bounded stationary solutions, which are not periodic in the direction x_3, but, for instance, are quasiperiodic or "homoclinic." For details we refer to [17] and to the litera-

ture cited here. We mention only the origins [114] and [139], where elliptic problems in unbounded cylinders are treated in this way. Somehow related to this method is the use of the "Ginzburg-Landau Equation" as a "Modulation Equation;" cf. the survey [140].

In experiments, Taylor vortices are observed only in a specific regime of Reynolds numbers beyond a critical value λ_{cr}. If that parameter, i.e., if the angular velocity of the interior cylinder, reaches another critical value, a secondary bifurcation to a so-called wavy vortex flow takes place. That flow is no longer stationary but time-periodic such that the bifurcation is of the type of a Hopf bifurcation. However, due to the complexity of the mathematical model, this bifurcation has not yet been rigorously proved.

This is not the end of the story: One has experimentally found a large "zoo" of different flows, one bifurcating from another at characteristic values of the Reynolds number, which, in general, depends also on the angular velocity of the exterior cylinder. Opposite directions of rotations of the interior and exterior cylinders yield interesting flows. In particular, a wavy vortex flow can bifurcate directly from the Couette flow. The experimental results are listed in [17], [80], for instance. Many of the flows are mathematically understood to a certain extent.

There is another prominent model in fluid dynamics, namely the so-called Bénard model. A viscous fluid in the gap between horizontal planes moves by convection when the lower plane is heated. Experiments show that the convection creates flows forming regular patterns: one observes, for instance, strips (rolls) and hexagons. These and more patterns are mathematically detected as stationary flows bifurcating from a basic trivial state.

III.3 Free Nonlinear Vibrations

Another class of nonlinear problems to which many of our local bifurcation theorems apply is given by the nonlinear wave equation. We discuss the one-dimensional case in detail and give the extensions to higher dimensions in Remark III.3.4.

The one-dimensional wave equation for scalar $u = u(t, x)$ with $(t, x) \in \mathbb{R} \times \mathbb{R}$ is

(III.3.1) $$u_{tt} - u_{xx} = g(u),$$

where $g : \mathbb{R} \to \mathbb{R}$ is some smooth function. As usual, we assume $g(0) = 0$, so that we have the trivial solution $u = 0$. Its linearization at $u = 0$,

(III.3.2) $$v_{tt} - v_{xx} - g'(0)v = 0,$$

has nontrivial solutions

(III.3.3)
$$v(t, x) = \genfrac{}{}{0pt}{}{\text{Re}}{\text{Im}} e^{i(\omega t \pm kx)}, \text{ provided that}$$

$$k^2 - \omega^2 - g'(0) = 0.$$

These solutions are periodic in t and in x. We try to answer the following question:

Do any of these doubly periodic solutions persist in a perturbed form for the nonlinear equation, at least for small amplitudes?

Following P. Rabinowitz [152], we call nontrivial solutions of (III.3.1) *free vibrations.* We treat this as a bifurcation problem.

We assume that the periods in time and space are locked and that they are perturbations of the periods of the linearized equation: After a rescaling $t \mapsto t/\lambda$, $x \mapsto x/\lambda$, we obtain the problem of small-amplitude solutions of

(III.3.4)
$$u_{tt} - u_{xx} = \lambda^2 g(u) \quad \text{for } \lambda \text{ near } 1$$
with fixed periods $2\pi/w$ in time and $2\pi/k$ in space.

The locked periods serve as a hidden bifurcation parameter λ for (III.3.1).

Actually, we consider the following generalization (after another rescaling, we can assume w.l.o.g. that the period in space is 2π):

(III.3.5)
$$\begin{aligned} u_{tt} - u_{xx} - c(\lambda)u &= g(t, x, u, \lambda), \\ u(t + P, x) &= u(t, x), \\ u(t, x + 2\pi) &= u(t, x), \end{aligned}$$

or instead of periodicity in space, we impose homogeneous Dirichlet or Neumann boundary conditions,

(III.3.6)
$$\begin{aligned} u(t, 0) &= u(t, \pi) = 0, \quad \text{or} \\ u_x(t, 0) &= u_x(t, \pi) = 0. \end{aligned}$$

The period P in time is a period of the linearized problem (a *linear period*) and will be specified later. The function $c : \mathbb{R} \to \mathbb{R}$ is in $C^1(\mathbb{R}, \mathbb{R})$, the function $g : \mathbb{R}^4 \to \mathbb{R}$ is sufficiently smooth and satisfies

(III.3.7)
$$\begin{aligned} g(t, x, 0, \lambda) &= g_u(t, x, 0, \lambda) = 0, \\ g(t + P, x, u, \lambda) &= g(t, x, u, \lambda), \\ g(t, x + 2\pi, u, \lambda) &= g(t, x, u, \lambda), \end{aligned}$$

and in case of homogeneous Neumann boundary conditions (III.3.6)$_2$,

(III.3.8)
$$g_x(t, 0, u, \lambda) = g_x(t, \pi, u, \lambda) = 0$$
for all $(t, x, u, \lambda) \in \mathbb{R}^4$;

cf. also (III.3.20) below. We define

(III.3.9) $\hat{g}(u, \lambda)(t, x) = g(t, x, u(t, x), \lambda)$

and consider
(III.3.10) $u_{tt} - u_{xx} - c(\lambda)u - \hat{g}(u, \lambda) = 0$

in an appropriate function space that incorporates the periodicities and boundary conditions.

For the special case (III.3.4) we assume simply $g(0) = 0$, $g'(0) \neq 0$, and we write it as $u_{tt} - u_{xx} - \lambda^2 g'(0)u = \lambda^2(g(u) - g'(0)u)$. Then all assumptions (III.3.7), (III.3.8) are satisfied and $c(\lambda) = \lambda^2 g'(0)$ is nontrivial. Note that a rescaling of solutions of (III.3.4) yields solutions of (III.3.1).

Remark III.3.1 *When written as a first-order system,*

$$(III.3.11) \qquad \begin{aligned} u_t &= v, \\ v_t &= u_{xx} + c(\lambda)u + \hat{g}(u, \lambda), \end{aligned}$$

the nonlinear hyperbolic PDE (III.3.10) can be considered to be an infinite-dimensional Hamiltonian system, provided that g does not depend on t.

We give the formal arguments, confining ourselves to the boundary conditions (III.3.6)$_1$. Let $X = H^2(0, \pi) \cap H_0^1(0, \pi)$ and $Z = L^2(0, \pi)$. Then a solution (u, v) of (III.3.11) defines a trajectory $(u, v)(t)$ in $X \times X$ via $[(u, v)(t)](x) = (u(t, x), v(t, x))$, solving the parameter-dependent evolution equation

$$\frac{d}{dt}\begin{pmatrix} u \\ v \end{pmatrix} = \begin{pmatrix} v \\ L(\lambda)u + \hat{g}(u, \lambda) \end{pmatrix} \equiv F(u, v, \lambda)$$
(III.3.12)

in $Z \times Z$ with $L(\lambda)u = u_{xx} + c(\lambda)u$.

We define a Hamiltonian

$$H : X \times X \times \mathbb{R} \to \mathbb{R} \qquad by$$

$$H(u, v, \lambda) = \frac{1}{2}(L(\lambda)u, u)_0 + \int_0^\pi \hat{G}(u, \lambda)dx - \frac{1}{2}(v, v)_0,$$
(III.3.13)
where $(\ ,\ )_0$ is the scalar product in $Z = L^2(0, \pi)$,

G is a primitive such that $G_u(x, u, \lambda) = g(x, u, \lambda)$, and $\hat{G}(u, \lambda)(x) = G(x, u(x), \lambda)$.

Then the gradient of H with respect to the scalar product $(\ ,\ )_0$ on $Z \times Z$ (cf. Definition I.3.1) is

$$\nabla_{(u,v)} H(u,v,\lambda) = \begin{pmatrix} L(\lambda)u + \hat{g}(u,\lambda) \\ -v \end{pmatrix} \in Z \times Z$$

for $(u,v,\lambda) \in X \times X \times \mathbb{R}$, *and*

(III.3.14)

$$F(u,v,\lambda) = \begin{pmatrix} 0 & -I \\ I & 0 \end{pmatrix} \nabla_{(u,v)} H(u,v,\lambda) \text{ or } F = J\nabla H,$$

$$D_{(u,v)} F(0,0,\lambda_0) = \begin{pmatrix} 0 & -I \\ I & 0 \end{pmatrix} \begin{pmatrix} L(\lambda_0) & 0 \\ 0 & -I \end{pmatrix} \equiv JB_0,$$

by $g_u(x,0,\lambda) = 0$; *cf.* *(III.3.7)*$_1$. *Now,* $L(\lambda_0) : Z \to Z$ *with domain of definition* $D(L(\lambda_0)) = X$ *has the eigenvalues* $c(\lambda_0) - n^2$, $n \in \mathbb{N}$, *so that*

(III.3.15) $A_0 = D_{(u,v)} F(0,0,\lambda_0) = JB_0$ *has*
the eigenvalues $\mu_n = \pm\sqrt{c(\lambda_0) - n^2}$, $n \in \mathbb{N}$.

Therefore, A_0 *has infinitely many (discrete) eigenvalues on the imaginary axis, and* A_0 *does not generate a holomorphic semigroup. Apart from that obstruction, the infinitely many purely imaginary eigenvalues could be "almost resonant," a phenomenon that gives rise to "small-divisor problems"; cf. Remark III.3.2 below. Therefore, a Lyapunov Center Theorem like Theorem I.11.4 cannot be proved for the nonlinear wave equation (III.3.10) or (III.3.11). Nonetheless, some of the problems are overcome by methods of KAM Theory in [25], for example, providing time-periodic solutions of (III.3.10) of small amplitude for a frozen parameter* λ, *i.e., in the spirit of a Lyapunov Center Theorem.*

The appropriate function spaces for treating (III.3.10) with the periodicities (III.3.5) or boundary conditions (III.3.6) are the following:

$$X = \{u(t,x) = \sum_{k=-\infty}^{\infty} \sum_{n=-\infty}^{\infty} c_{kn} e^{ikx} e^{i\frac{2\pi}{P} nt}, c_{-k,-n} = \bar{c}_{kn},$$

$$\|u\|_X^2 = \sum_{k=-\infty}^{\infty} \sum_{n=-\infty}^{\infty} |c_{kn}|^2 (k^2 + n^2) < \infty\},$$

(III.3.16) $$X = \{u(t,x) = \sum_{k=1}^{\infty} \sum_{n=-\infty}^{\infty} c_{kn} \sin kx e^{i\frac{2\pi}{P} nt}, c_{k,-n} = \bar{c}_{kn},$$

$$\|u\|_X^2 \text{ as above }\}, \text{ or}$$

$$X = \{u(t,x) = \sum_{k=0}^{\infty} \sum_{n=-\infty}^{\infty} c_{kn} \cos kx e^{i\frac{2\pi}{P} nt}, c_{k,-n} = \bar{c}_{kn},$$

$$\|u\|_X^2 \text{ as above }\}.$$

Using Fourier analysis, it is not difficult to prove the alternative definitions

$$X = \{u \in W^{2,2}((0,P) \times (0,2\pi))|u(0,x) = u(P,x),$$
$$u_t(0,x) = u_t(P,x), u(t,0) = u(t,2\pi), u_x(t,0) = u_x(t,2\pi)\}$$
with equivalent norms $\|u\|_X, \|u\|_{2,2},$

(III.3.17)
$$X = \{u \in W^{2,2}((0,P) \times (0,\pi))|u(0,x) = u(P,x),$$
$$u_t(0,x) = u_t(P,x), u(t,0) = u(t,\pi) = 0\}, \text{ or}$$

$$X = \{u \in W^{2,2}((0,P) \times (0,\pi))|u(0,x) = u(P,x),$$
$$u_t(0,x) = u_t(P,x), u_x(t,0) = u_x(t,\pi) = 0\},$$

for all $t \in (0,P), x \in (0,2\pi)$, or $x \in (0,\pi)$, respectively.

All spaces X are Hilbert spaces with norm $\| \ \|_X$ or $\| \ \|_{2,2}$. (Since $C^2([0,P] \times [0,2\pi])$ and $C^2([0,P] \times [0,\pi])$ are dense in the spaces $W^{2,2}((0,P) \times (0,2\pi))$ and $W^{2,2}((0,P) \times (0,\pi))$, respectively, the conditions on u on the boundary of the respective rectangles are extended from C^2-functions to $W^{2,2}$-functions by approximation. One could also say that the boundary condition on the Lipschitz boundary holds in the sense of the "*trace*" of $W^{2,2}$-functions on the boundary.)

The wave operator acts as follows:

(III.3.18)
$$Au \equiv u_{tt} - u_{xx} = \sum\sum c_{kn}(k^2 - \tfrac{4\pi^2}{P^2}n^2)\varphi_k(x)e^{i\frac{2\pi}{P}nt}$$

for $u \in X$ with $\varphi_k(x) = e^{ikx}$, $\sin kx$, or $\cos kx$.

We define

(III.3.19)
$$A : X \to X \text{ with } D(A) = \{u \in X|Au \in X\} \equiv Y$$
and A acting as in (III.3.18).

Since $A : X \to L^2(Q)$ is continuous and $X \subset L^2(Q)$ for $Q = (0,P) \times (0,2\pi)$ or $Q = (0,P) \times (0,\pi)$, it follows immediately that A as defined in (III.3.19) is a closed operator.

Endowing $Y = D(A)$ with the graph norm $(\|u\|_X^2 + \|Au\|_X^2)^{1/2}$, we obtain a Hilbert space Y, and $A : Y \to X$ is obviously continuous.

Next, we show that the nonlinear mapping $\hat{g}$ (see (III.3.9)) can be defined on $X \times \mathbb{R}$. Since the rectangle Q is in $\mathbb{R}^2$, the Sobolev space $W^{2,2}(Q)$ is a Banach algebra: Let $u \in W^{2,2}(Q) \subset C(\overline{Q})$ and $D = \frac{\partial}{\partial t}$ or $D = \frac{\partial}{\partial x}$. By $u \in C(\overline{Q})$, the element $\hat{g}(u,\lambda)$ is in $C(\overline{Q}) \subset L^2(Q)$ and $D\hat{g}(u,\lambda) = \hat{g}_t(u,\lambda) + \hat{g}_u(u,\lambda)Du$ or $D\hat{g}(u,\lambda) = \hat{g}_x(u,\lambda) + \hat{g}_u(u,\lambda)Du \in L^2(Q)$, too, since $\hat{g}_t(u,\lambda), \hat{g}_x(u,\lambda), \hat{g}_u(u,\lambda) \in C(\overline{Q})$ and $Du \in L^2(Q)$. For each second derivative we obtain $D^2\hat{g}(u,\lambda) = \hat{g}_{tt}(u,\lambda) + 2\hat{g}_{tu}(u,\lambda)Du + \hat{g}_u(u,\lambda)D^2u$ or the same expression with t replaced by x. Since $\hat{g}_{tt}(u,\lambda)$, $\hat{g}_{tu}(u,\lambda)$, $\hat{g}_{xx}(u,\lambda)$, $\hat{g}_{xu}(u,\lambda)$, $\hat{g}_u(u,\lambda) \in C(\overline{Q})$, and $Du, D^2u \in L^2(Q)$, we end up with $D^2\hat{g}(u,\lambda) \in L^2(Q)$. This proves $\hat{g}(u,\lambda) \in W^{2,2}(Q)$, provided that $u \in W^{2,2}(Q)$. Furthermore, $(u,\lambda) \mapsto \hat{g}(u,\lambda)$ is a continuous mapping from $W^{2,2}(Q) \times \mathbb{R}$ into $W^{2,2}(Q)$.

The space X is also defined via periodicities and boundary conditions; cf. (III.3.17). By assumptions $(III.3.7)_{2,3}$, the periodicities of u in time and space are preserved for $\hat{g}(u, \lambda)$. By $(III.3.7)_1$, the homogeneous Dirichlet boundary conditions of u imply the same boundary conditions for $\hat{g}(u, \lambda)$. Finally, for $D_x \hat{g}(u, \lambda) = \hat{g}_x(u, \lambda) + \hat{g}_u(u, \lambda) D_x u$, assumption (III.3.8) implies that $\hat{g}(u, \lambda)$ satisfies homogeneous Neumann boundary conditions, provided that u satisfies them. This proves that $(u, \lambda) \mapsto \hat{g}(u, \lambda)$ is a continuous mapping from $X \times \mathbb{R}$ into X for all three cases (III.3.17).

(Since spaces of C^2-functions satisfying the Dirichlet or Neumann boundary conditions in space are dense in X, all the above proofs can be carried out for classical derivatives. The extension to weak derivatives follows then by approximation.)

Finally, $D_u \hat{g}(u, \lambda) v = \hat{g}_u(u, \lambda) v$ for all $u, v \in X$, and by assumption $(III.3.7)_1$,

(III.3.20)
$$\begin{aligned} &\hat{g} : X \times \mathbb{R} \to X \text{ is in } C^1(X \times \mathbb{R}, X), \\ &D_u \hat{g}(0, \lambda) = 0 \text{ for all } \lambda \in \mathbb{R}, \text{ if} \\ &\hat{g}(u, \lambda) \text{ is defined by (III.3.9) and if} \\ &g, g_u \in C^2(\mathbb{R}^4, \mathbb{R}). \end{aligned}$$

The functional-analytic setting of problems (III.3.5), (III.3.6) is the following: Solve

$$F(u, \lambda) = 0 \text{ for } (u, \lambda) \in Y \times \mathbb{R}, \text{ where}$$

$$F : Y \times \mathbb{R} \to X \text{ is defined by}$$

(III.3.21)
$$F(u, \lambda) = (A - c(\lambda)I)u - \hat{g}(u, \lambda),$$

$$F \in C(Y \times \mathbb{R}, X), D_u F \in C(Y \times \mathbb{R}, L(Y, X)),$$

$$D_u F(0, \lambda) = A - c(\lambda)I \equiv A(\lambda).$$

For all three realizations of X, the mapping $F(\cdot, \lambda) : Y \to X$ defines a potential operator with respect to the scalar product $(\ , \ )_0$ of $L^2(Q)$; cf. Definition I.3.1. We give the potential:

$$f(u, \lambda) = \frac{1}{2}((A - c(\lambda))u, u)_0 - \int_Q \hat{G}(u, \lambda) dt\, dx,$$

(III.3.22) where $G_u(t, x, u, \lambda) = g(t, x, u, \lambda)$ and
$$\hat{G}(u, \lambda)(t, x) = G(t, x, u(t, x), \lambda) \text{ for } (t, x) \in Q,$$
$$(u, \lambda) \in Y \times \mathbb{R}, \ Q = (0, P) \times (0, 2\pi) \text{ or } Q = (0, P) \times (0, \pi).$$

The Fredholm property of $D_u F(0, \lambda) = A - c(\lambda)I$ is crucial. In contrast to families of elliptic operators $L(\lambda)$ studied in Sections III.1, III.2, the hyperbolic family $A(\lambda) = A - c(\lambda)I$ has the Fredholm property only for specific values of $c(\lambda)$. For $c(\lambda)$ in a complementary set the properties of $A - c(\lambda)I$ might change dramatically; cf. Remark III.3.2 below.

In view of (III.3.18), we define for fixed $\lambda_0 \in \mathbb{R}$,

$$(\text{III.3.23}) \qquad S = \left\{ (k,n) \,\middle|\, k^2 - \frac{4\pi^2}{P^2} n^2 - c(\lambda_0) = 0 \right\}$$
$$\subset \mathbb{Z} \times \mathbb{Z}, \ \mathbb{N} \times \mathbb{Z}, \ \text{or} \ (\mathbb{N} \cup \{0\}) \times \mathbb{Z};$$

cf. (III.3.16). Then

$$(\text{III.3.24}) \qquad N(D_u F(0, \lambda_0)) = \left\{ u(t,x) = \sum_{(k,n) \in S} c_{kn} \varphi_k(x) e^{i \frac{2\pi}{P} nt} \right\}$$
$$\text{and } 0 < \dim N(D_u F(0, \lambda_0)) < \infty \Leftrightarrow$$
$$S \neq \emptyset \text{ is a finite set.}$$

First of all,

$$(\text{III.3.25}) \qquad S \neq \emptyset \Leftrightarrow k_0^2 - \frac{4\pi^2}{P^2} n_0^2 - c(\lambda_0) = 0 \text{ for some } (k_0, n_0)$$
$$\Leftrightarrow P = \frac{2\pi n_0}{\sqrt{k_0^2 - c(\lambda_0)}} \equiv P_0 \text{ is a } linear \ period.$$

Here we exclude the case $n_0 = 0$ or $k_0^2 - c(\lambda_0) = 0$. If g does not depend on t, then a special class of solutions of (III.3.5), (III.3.6) is given by stationary solutions of

$$(\text{III.3.26}) \qquad \begin{aligned} -u_{xx} - c(\lambda)u &= g(x, u, \lambda), \\ u(x + 2\pi) &= u(x) \quad \text{or} \\ u(0) = u(\pi) &= 0 \quad \text{or} \\ u_x(0) = u_x(\pi) &= 0, \end{aligned}$$

which are one-dimensional elliptic problems and are treated with methods described in Section III.2. Parameters λ_0 where bifurcation for (III.3.26) is possible are characterized by $k_0^2 - c(\lambda_0) = 0$ for some $k_0 \in \mathbb{N} \cup \{0\}$. Here we assume that the linear period P_0 in (III.3.25) is well defined by a nonstationary element $u_0 \in N(D_u F(0, \lambda_0))$; i.e., $n_0 \neq 0$ and $k_0^2 - c(\lambda_0) > 0$.

Fixing the period $P = P_0$, we see that the so-called *characteristic equation* defining the set S is

$$(\text{III.3.27}) \qquad n_0^2 k^2 - (k_0^2 - c(\lambda_0))n^2 - n_0^2 c(\lambda_0) = 0.$$

The (nonempty) solution set S of (III.3.27) depends on $c(\lambda_0)$ in a sensitive way: If $c(\lambda_0)$ is irrational, then $S = \{(\pm k_0, \pm n_0)\}$ or $S = \{(k_0, \pm n_0)\}$. In this case, the kernel $N(D_u F(0, \lambda_0))$ is 4- or 2-dimensional, respectively, but the behavior of $D_u F(0, \lambda_0)$ on its complement is hard to control; cf. Remark III.3.2. In particular, we cannot prove the Fredholm property of $D_u F(0, \lambda_0)$ if $c(\lambda_0)$ is irrational.

Therefore, we assume that $c(\lambda_0) = \frac{p}{q} \in \mathbb{Q}$, $q \in \mathbb{N}$, $p \in \mathbb{Z} \backslash \{0\}$.

In this case, the characteristic equation becomes a Diophantine equation

$$(\text{III.3.28}) \qquad n_0^2 q^2 k^2 - (q^2 k_0^2 - pq)n^2 = n_0^2 pq \neq 0,$$

whose solution set S is characterized as follows:

(III.3.29)
$$S \neq \emptyset \text{ is a finite set} \Leftrightarrow$$
$$q^2 k_0^2 - pq = r^2 \text{ for some } r \in \mathbb{N}.$$

For a proof we refer to most books on elementary number theory, where one can find (III.3.29) in sections about Pell's equation.

It is of interest that the set

(III.3.30)
$$\Lambda = \left\{ \frac{p}{q} \mid q^2 k_0^2 - pq = r^2 \text{ for some } (k_0, r) \in \mathbb{N}_0 \times \mathbb{N} \right\}$$
is dense in $\mathbb{R}$ $\quad (\mathbb{N}_0 = \mathbb{N} \cup \{0\} \text{ or } \mathbb{N})$,

which means that the set of rational values of $c(\lambda_0)$ such that the kernel $N(D_u F(0, \lambda_0)) = N(A - c(\lambda_0)I)$ is finite-dimensional *for some linear period* P_0 is dense in $\mathbb{R}$.

We call a rational number in Λ *admissible*. The linear periods for which a rational number is admissible are rational multiples of 2π. In this sense they are locked to the period in space, (III.3.5)$_3$, or to the boundary conditions (III.3.6).

Let $c(\lambda_0) \neq 0$ be admissible for a period P_0 and let $(k, n) \notin S$. Then

(III.3.31)
$$\left| k^2 - \frac{4\pi^2}{P_0^2} n^2 - c(\lambda_0) \right|$$
$$= \frac{1}{n_0^2 q^2} \left| n_0^2 q^2 k^2 - (q^2 k_0^2 - pq) n^2 - n_0^2 pq \right|$$
$$= \frac{1}{n_0^2 q^2} \left| n_0^2 q^2 k^2 - r^2 n^2 - n_0^2 pq \right| \geq \frac{1}{n_0^2 q^2},$$

and for

(III.3.32)
$$f(t, x) = \sum_{(k,n) \notin S} d_{kn} \varphi_k(x) e^{i \frac{2\pi}{P_0} nt} \in X,$$
$$u(t, x) = \sum_{(k,n) \notin S} c_{kn} \varphi_k(x) e^{i \frac{2\pi}{P_0} nt} \text{ with}$$
$$c_{kn} = d_{kn} / \left(k^2 - \frac{4\pi^2}{P_0^2} n^2 - c(\lambda_0) \right) \text{ is in } X,$$
it solves $Au - c(\lambda_0)u = f$, and by

$$Au = c(\lambda_0)u + f \in X \text{ we obtain } u \in Y = D(A).$$

This proves that the closed orthogonal complement of $N(D_u F(0, \lambda_0))$ with respect to the scalar product $(\ ,\ )_0$ of $L^2(Q)$ is the range of $D_u F(0, \lambda_0)$, i.e.,

(III.3.33)
$$X = N(D_u F(0, \lambda_0)) \oplus R(D_u F(0, \lambda_0)).$$

In order to complete the list of all hypotheses of our various bifurcation theorems we prove that

(III.3.34)
$$0 \text{ is an isolated eigenvalue of } D_u F(0, \lambda_0)$$
$$\text{with algebraic multiplicity } n = \dim N(D_u F(0, \lambda_0)).$$

By (III.3.31) we obtain for all $(k, n) \in \mathbb{Z} \times \mathbb{Z}$ (or $\mathbb{N} \times \mathbb{Z}$ or $(\mathbb{N} \cup \{0\}) \times \mathbb{Z}$) and for $\mu \in \mathbb{C}$,

(III.3.35)
$$\left| k^2 - \frac{4\pi^2}{P_0^2} n^2 - c(\lambda_0) - \mu \right|$$
$$\geq \min \left\{ |\mu|, \frac{1}{n_0^2 q^2} - |\mu| \right\} \text{ for } 0 < |\mu| < \frac{1}{n_0^2 q^2},$$

so that the argument of (III.3.32) proves that

(III.3.36)
$$\left\{ \mu \in \mathbb{C} \Big| 0 < |\mu| < \frac{1}{n_0^2 q^2} \right\} \text{ is in the}$$
$$\text{resolvent set of } A - c(\lambda_0) I = D_u F(0, \lambda_0).$$

Corresponding to the isolated eigenvalue 0 of $D_u F(0, \lambda_0) = A_0$ there exists a generalized closed eigenspace $E_0 \subset X$ such that $A_0 \in L(E_0, E_0)$ with spectrum $\sigma(A_0|_{E_0}) = \{0\}$. Since the operator $A_0 = A - c(\lambda_0) I$ is symmetric with respect to the scalar product $(\ ,\ )_X$ defining the norm $\| \ \|_X$ in (III.3.16), the spectral radius of the bounded operator, $A_0|_{E_0}$ is given by its norm. This implies $A_0|_{E_0} = 0$ and $E_0 \subset N(A_0) \subset E_0$, which proves (III.3.34) (see also (III.3.33)).

We assume that the function

(III.3.37) $c(\lambda)$ is strictly monotonic near $\lambda = \lambda_0$.

Then the eigenvalue 0 of $A_0 = D_u F(0, \lambda_0) = A - c(\lambda_0) I$ perturbs to the eigenvalue $\mu(\lambda) = c(\lambda_0) - c(\lambda) \neq 0$ of $D_u F(0, \lambda_0) = A - c(\lambda) I$ for $\lambda \neq \lambda_0$, so that according to Definition II.7.1, zero is a locally hyperbolic equilibrium of $D_u F(0, \lambda_0)$ for all $\lambda \in (\lambda_0 - \delta, \lambda_0) \cup (\lambda_0, \lambda_0 + \delta)$, and the crossing number of the family $D_u F(0, \lambda_0)$ at $\lambda = \lambda_0$ through 0 is $\pm \dim N(D_u F(0, \lambda_0))$, which is *nonzero*. Note, however, that the *crossing number* is always even.

Remark III.3.2 *By (III.3.20) the operator F defined in (III.3.21) also satisfies*

(III.3.38)
$$F : X \times \mathbb{R} \to Z = L^2(Q),$$
$$F \in C(X \times \mathbb{R}, Z), \quad D_u F \in C(X \times \mathbb{R}, L(X, Z)).$$

One might ask why we lift it to $F : Y \times \mathbb{R} \to X$. The reason is that $D_u F(0, \lambda_0) \in L(X, Z)$ is not a Fredholm operator even if $c(\lambda_0)$ is admissible for some P_0. We show that

(III.3.39)
$$Z = N(D_u F(0, \lambda_0)) \oplus \overline{R(D_u F(0, \lambda_0))},$$
$$\text{but } R(D_u F(0, \lambda_0)) \text{ is not closed.}$$

Define $f \in N(D_u F(0, \lambda_0))^{\perp} \subset Z$ as in (III.3.32)$_1$. If the Fourier series of f is finite, then the function u defined in (III.3.32)$_2$ is in X, and $D_u F(0, \lambda_0)u = f \in R(D_u F(0, \lambda_0))$. The set $\tilde{S} = \{(k, n) | n_0^2 q^2 k^2 - r^2 n^2 = 0\}$ is infinite, and $S \cap \tilde{S} = \emptyset$. Define $f \in N(D_u F(0, \lambda_0))^{\perp} \subset Z$ by a Fourier series with infinitely many modes $(k, n) \in \tilde{S}$. The function u defined in (III.3.32)$_2$ is then $u = -c(\lambda_0)^{-1}f$, $u \notin X$ if $f \notin X$, and in this case $f \notin R(D_u F(0, \lambda_0))$.

If $c(\lambda_0)$ is not admissible but rational, we have an infinite-dimensional kernel $N(D_u F(0, \lambda_0))$. None of our abstract bifurcation theorems allows an infinite-dimensional kernel, so that we cannot treat that case. If $c(\lambda_0)$ is irrational, we have a 4- or 2-dimensional kernel, and the analysis of (III.3.32) yields formal solutions of $D_u F(0, \lambda_0)u = f$ for all $f \in N(D_u F(0, \lambda_0))^{\perp}$. Although the denominators in (III.3.32)$_3$ do not vanish, they cannot be controlled as in (III.3.31). As a matter of fact, depending on a "degree of irrationality" of $c(\lambda_0)$, the absolute values of the divisors (III.3.32)$_3$ become arbitrarily small, and $Au - c(\lambda)u = f$ is a typical "small-divisor problem." In KAM Theory, specific values for $c(\lambda_0)$ are distinguished for which the divisor can become arbitrarily small but in a controlled way. A "rapidly convergent" Fourier series can then overcome the deficiency caused by the divisors. The set Λ of (III.3.30) can be considered as a subset of admissible sets in KAM Theory.

In order to solve $F(u, \lambda) = 0$ defined in (III.3.21), we apply various of our abstract bifurcation theorems. Before doing this, we mention that another "lift" of F provides classical solutions of (III.3.5), (III.3.6): Define for X as in (III.3.17),

$$(\text{III.3.40}) \quad X_4 \equiv X \cap \left\{ u \in W^{4,2}(Q) | \frac{\partial^j}{\partial t^j} u(0, x) = \frac{\partial^j}{\partial t^j} u(P, x), j = 2, 3 \right\}$$

endowed with the Sobolev norm $\| \ \ \|_{4,2}$.

Then the wave operator A of (III.3.18) defines a closed operator $A : X_4 \to X_4$ with $D(A) = \{u \in X_4 | Au \in X_4\} \equiv Y_4$, and $\hat{g}$ of (III.3.20) satisfies $\hat{g} : X_4 \times \mathbb{R} \to X_4$ with all properties listed in (III.3.20), where X is replaced by X_4, provided that $g, g_u \in C^4(\mathbb{R}^4, \mathbb{R})$. Consequently, $F(u, \lambda) = (A - c(\lambda)I)u - \hat{g}(u, \lambda)$ defines a mapping $F : Y_4 \times \mathbb{R} \to X_4$ that has the same properties as $F : Y \times \mathbb{R} \to X$. By embedding $X_4 \subset C^2(\overline{Q})$, all solutions $(u, \lambda) \in Y_4 \times \mathbb{R}$ of $F(u, \lambda) = 0$ are classical solutions of (III.3.5), (III.3.6). For simplicity, we stay with the setting $F : Y \times \mathbb{R} \to X$. (Note that this "lifting" to gain regularity differs from a "bootstrapping" to prove regularity for elliptic problems.)

III.3.1 Variational Methods

Our most general results are given by the application of Theorem II.7.3. Note that all its hypotheses are satisfied for $F : Y \times \mathbb{R} \to X$.

Theorem III.3.3 *If $c(\lambda_0) \neq 0$ is admissible for a linear period P_0, and if (III.3.37) holds, then $(0, \lambda_0) \in Y \times \mathbb{R}$ is a cluster point of nontrivial solutions (u, λ) of (III.3.5), (III.3.6).*

Together with the observation (III.3.30), we obtain the following result:

(III.3.41)
    If $c : \mathbb{R} \to \mathbb{R}$ is globally strictly monotonic,
    then the cluster points $\{(0, \lambda_0)\} \subset Y \times \mathbb{R}$
    of nontrivial solutions (u, λ) of (III.3.5), (III.3.6)
    for some period $P = P_0$ depending on λ_0
    form a dense set in $\{0\} \times \mathbb{R}$.

We mention that we cannot apply Theorem II.4.4 in order to provide connected sets of solutions: Since the cardinality of the solution set S of (III.3.28) is 0 (mod 4) or 0 (mod 2), the dimension of $N(D_u F(0, \lambda_0))$ always yields even crossing numbers by (III.3.37). (Note, however, the results (III.3.46), (III.3.47), below.)

For $c(\lambda) = \lambda$ and for nonlinearities g that do not depend on λ, Theorem I.21.2 and Corollary I.21.3 are applicable: If $\lambda_0 \neq 0$ is admissible for a linear period P_0, then there exist at least two solutions $(u(\varepsilon), \lambda(\varepsilon)) \in Y \times \mathbb{R}$ with $\|u(\varepsilon)\|_0 = \varepsilon$ and $\lambda(\varepsilon) \to \lambda_0$ as $\varepsilon \searrow 0$. (Here $\| \quad \|_0$ is the norm in $L^2(Q)$.)

If $g(t, x, -u) = -g(t, x, u)$, Corollary I.21.3 gives at least n pairs $(\pm u(\varepsilon), \lambda(\varepsilon)) \in Y \times \mathbb{R}$ of solutions with $\|u(\varepsilon)\|_0 = \varepsilon$ and $\lambda(\varepsilon) \to \lambda_0$ as $\varepsilon \searrow 0$, where $n = \dim N(D_u F(0, \lambda_0))$.

If (III.3.5) is autonomous, i.e., if g does not depend on t, then with each solution $u(t, x)$, all phase-shifted functions $S_\theta u(t, x) = u(t + \theta, x)$ are again solutions. Let $g(x, -u) = -g(x, u)$. Then speaking of n pairs of solutions does not make much sense, since a phase shift creates a continuum of infinitely many solutions. Let us consider the "orbit of u" $\{\pm S_\theta u | \theta \in \mathbb{R}\}$ to be *one* element. How many solutions has (III.3.5), (III.3.6) with an odd nonlinearity $g = g(x, u)$?

The critical points of the potential f (cf. (III.3.22) for $\lambda = \lambda_0$) on the manifold M_ε (cf. (I.21.22) in the proof of Theorem I.21.2) are obtained via a "minimax method" using families of subsets of M_ε whose "genera" are bounded from below. (We cannot go into the details here, but we refer to [118], [150], [110].) The minimax method gives critical values, and different critical values yield clearly different critical points. If $r + 1$, say, critical values are equal, then a theorem about multiplicities tells that the genus of the set of critical points at that level is at least $r + 1$. The genus of an orbit is 1 or 2, depending on whether it has two components or one. Using the subadditivity of the genus, we obtain the following result: Let $\lambda_0 \neq 0$ be admissible for the period P_0. Then

(III.3.42)
$$u_{tt} - u_{xx} - \lambda u = g(x, u),$$
$$u(t + P_0, x) = u(t, x),$$
with boundary conditions (III.3.5)$_3$ or (III.3.6)
and $g(x, -u) = -g(x, u)$,
has at least m different solution orbits $(\{\pm S_\theta u(\varepsilon)\}, \lambda(\varepsilon))$
with $\|u(\varepsilon)\|_0 = \varepsilon$ and $\lambda(\varepsilon) \to \lambda_0$ as $\varepsilon \searrow 0$,
where $2m = \dim N(D_u F(0, \lambda_0)) = \dim N(A - \lambda_0 I)$;
cf. (III.3.18).

III.3.2 Bifurcation with a One-Dimensional Kernel

The high-dimensionality of the kernels $N(D_u F(0, \lambda_0))$ can be reduced to dimension one as follows: Let $c(\lambda_0) = \frac{p}{q} \neq k^2$ for all $k \in \mathbb{N} \cup \{0\}$ be admissible in the sense of (III.3.30); i.e., $c(\lambda_0) \in \Lambda$. Then it is admissible for all periods $2\pi n_0 q/r = P_0$ such that $pq = q^2 k_0^2 - r^2$ for $(k_0, r) \in \mathbb{N}_0 \times \mathbb{N}$. Obviously, there are only finitely many such pairs $(k_0, r) \in \mathbb{N}_0 \times \mathbb{N}$. Choose r_1 maximal among all solutions of $pq = q^2 k_0^2 - r^2$. Then $P_1 = 2\pi q/r_1$ is the *minimal period* for which $c(\lambda_0) = \frac{p}{q}$ is admissible and

(III.3.43)
$$S_1 = \left\{ (k, n) | k^2 - \frac{4\pi^2}{P_1^2} n^2 - c(\lambda_0) = 0 \right\}$$
$$= \{(\pm k_0, \pm 1)\} \text{ or } \{(k_0, \pm 1)\}.$$

By definition of P_1, the characteristic equation in (III.3.43) is $q^2 k^2 - r_1^2 n^2 = pq$. By assumption on $c(\lambda_0) = \frac{p}{q} \neq k^2$, a solution $(k, 0)$ is excluded. If there were a solution (k, n) with $n > 1$, then $r = nr_1 > r_1$ and $pq = q^2 k^2 - r^2$ would contradict the maximal choice of r_1. This means, in view of (III.3.24), that

(III.3.44)
if P_1 is the minimal period
for which $c(\lambda_0) = \frac{p}{q} \neq k^2$ is admissible, then
$\dim N(D_u F(0, \lambda_0)) = 4$ or 2.

The kernel is 4-dimensional if $k_0 \neq 0$ and if we impose periodic boundary conditions (III.3.5)$_3$. In the cases of the homogeneous Dirichlet or Neumann boundary conditions (III.3.6), the kernel is 2-dimensional.

The reduction to one-dimensional kernels is possible if

(III.3.45)
$$g(-t, x, u, \lambda) = g(t, x, u, \lambda),$$
and in case of periodic boundary conditions, additionally
$$g(t, -x, u, \lambda) = g(t, x, u, \lambda)$$
for all $(t, x, u, \lambda) \in \mathbb{R}^4$.

When the wave equation (III.3.5)$_1$ with the boundary conditions (III.3.6) is restricted to the function space $\{u \in Y | u(-t, x) = u(t, x)\}$, then

(III.3.46)
$$N(D_u F(0, \lambda_0)) = \text{span}[\sin k_0 x \cos \tfrac{2\pi}{P_1} t] \text{ or}$$
$$N(D_u F(0, \lambda_0)) = \text{span}[\cos k_0 x \cos \tfrac{2\pi}{P_1} t].$$

In case of the boundary condition (III.3.5)$_3$, i.e., in case of periodicity in space, we restrict problem (III.3.5) to the subspace $\{u \in Y | u(-t, x) = u(t, x), u(t, -x) = u(t, x)\}$, and then

(III.3.47) $N(D_u F(0, \lambda_0)) = \text{span}[\cos k_0 x \cos \tfrac{2\pi}{P_1} t].$

By (III.3.33), the number 0 is an algebraically simple eigenvalue of the operator $D_u F(0, \lambda_0) = A - c(\lambda_0) I$, and the nondegeneracy (I.5.3) of Theorem I.5.1 is satisfied if $c'(\lambda_0) \neq 0$. We obtain a nontrivial solution curve $\{(u(s), \lambda(s)) | s \in (-\delta, \delta)\}$ through $(u(0), \lambda(0)) = (0, \lambda_0)$ and

(III.3.48)
$$u(s)(t, x) = s \varphi_{k_0}(x) \cos \tfrac{2\pi}{P_1} t + o(s),$$
$$\text{where } \varphi_{k_0}(x) = \cos k_0 x \text{ or } \sin k_0 x.$$

The Bifurcation Formulas of Section I.6 are applicable with projection $Qu = (u, \hat{v}_0,)_0 \hat{v}_0$, where $\hat{v}_0(t, x) = \alpha \varphi_{k_0}(x) \cos \tfrac{2\pi}{P_1} t$, so that $\|\hat{v}_0\|_0 = 1$. Here $(\ ,\ )_0$ and $\|\ \|_0$ denote the scalar product and the norm in $L^2(Q)$ with $Q = (0, P_1) \times (0, 2\pi)$ or $Q = (0, P_1) \times (0, \pi)$, respectively.

We give an example: Consider

(III.3.49) $u_{tt} - u_{xx} - \lambda u = u^\ell$ with $\ell = 2$ or 3,
periodicity (III.3.5)$_3$, or boundary conditions (III.3.6).

Then the value $\lambda_0 = 24$ is admissible for the linear periods $P_0 = 2\pi$ and $P_0 = 2\pi/5$. Since a function with period $2\pi/5$ also has period 2π, the solutions of the characteristic equation for $P_0 = 2\pi/5$ also yield solutions of the characteristic equation for $P_0 = 2\pi$, that is, $(\pm 5, \pm 1), (\pm 7, \pm 5)$ or $(5, \pm 1), (7, \pm 5)$. These solutions give 8- or 4-dimensional kernels, respectively, and for $\ell = 2$, Theorem I.21.2 provides at least two different 2π-periodic orbits $(\{S_\theta u\}, \lambda)$ near $(0, 24)$. For $\ell = 3$, Corollary I.21.3 guarantees at least four different 2π-periodic orbits $(\{\pm S_\theta u\}, \lambda)$ near $(0, 24)$ in case of periodic boundary conditions in space.

When we restrict the period to the minimal period $2\pi/5$, then the characteristic equation has only the solutions $(\pm 7, \pm 1)$ or $(7, \pm 1)$. In the spaces of even functions in time, and also in space in case of periodic boundary conditions, Theorem I.5.1 provides a curve of solutions $\{(u(s), \lambda(s)) | s \in (-\delta, \delta)\}$ of (III.3.49), where $u(s)(t, x) = s \cos 7x \cos 5t + o(s)$ or $u(s)(t, x) = s \sin 7x \cos 5t + o(s)$. Note that in view of the autonomy of equation (III.3.49), all phase-shifted functions $u(s)(t + \theta, x)$ are solutions, too, and in case of periodic boundary conditions, we can also allow any phase shift in space, namely, $u(s)(t, x + \xi)$. Note that for $\ell = 3$, Corollary I.21.3 does not give more than two different $2\pi/5$-periodic orbits.

Finally, the Bifurcation Formulas of Section I.6 determine the local shape of the bifurcating curve $\{(u(s), \lambda(s)) | s \in (-\delta, \delta)\}$ of (III.3.49). By

(III.3.50)
$$\int_0^{2\pi/5} \cos^3 5t \, dt = 0, \quad \text{we obtain for } \ell = 2,$$
$$(D_{uu}^2 F(0, \lambda_0)[\hat{v}_0, \hat{v}_0], \hat{v}_0)_0 = 0, \quad \text{whence } \dot{\lambda}(0) = 0.$$
For $\ell = 3$, we obtain $\dot{\lambda}(0) = 0$ by $D_{uu}^2 F(0, \lambda_0) = 0$.

Note that (III.3.50) holds in the cases of periodic, Dirichlet, or Neumann boundary conditions. Evaluation of formula (I.6.11) for $\ddot{\lambda}(0)$ is more involved. We give only the results:

(III.3.51)
$$\text{For } \ell = 2, \text{ we obtain } \ddot{\lambda}(0) > 0;$$
$$\text{for } \ell = 3, \text{ we have } \ddot{\lambda}(0) < 0,$$

so that in all cases, (III.3.49) gives rise to pitchfork bifurcations at $\lambda_0 = 24$ that are supercritical for $\ell = 2$ and subcritical for $\ell = 3$; cf. Figure I.6.1. Whereas by oddness, the pitchfork is obvious for $\ell = 3$, it is surprising for $\ell = 2$, since one would expect a transcritical bifurcation. (In case of Dirichlet boundary conditions, the computation of $\ddot{\lambda}(0)$ for $\ell = 2$ is not easy.)

By (III.3.34), the algebraic multiplicity of the eigenvalue 0 of $D_u F(0, \lambda_0)$ is not larger than its geometric multiplicity, which means that 0 is semisimple. In case (III.3.44), when the assumptions (III.3.45) allow us to reduce the bifurcation of P_1-periodic solutions of (III.3.5), (III.3.6) to a one-dimensional kernel, the eigenvalue 0 of $D_u F(0, \lambda_0)$ is algebraically simple. Therefore, the analysis of Section I.16 about Degenerate Bifurcation is applicable. We give an example:

(III.3.52)
$$u_{tt} - u_{xx} - \lambda_0 u + (\lambda^2 - \varepsilon)u = u^3$$
$$\text{with periodicity (III.3.5)}_3,$$
$$\text{or boundary conditions (III.3.6).}$$

Let $\lambda_0 \neq k^2$ be admissible, i.e., $\lambda_0 \in \Lambda$ (cf. (III.3.30)), and let P_1 be the minimal period for which λ_0 is admissible. Then the analysis of P_1-periodic solutions for (III.3.52) is precisely the same as for (III.2.24): The solutions $\{(u, \lambda)\}$ of (III.3.52) near $(u, \lambda) = (0, 0)$ are sketched in Figure III.2.1 for different values of ε near 0. (By Lyapunov–Schmidt reduction, every solution (u, λ) near $(0, 0)$ is of the form (III.3.48), and there is a one-to-one correspondence between solutions (u, λ) of (III.3.52) near $(0, 0)$ and solutions (s, λ) of $\Phi(s, \lambda)$ near $(0, 0)$, and Figure III.2.1 shows the solution sets $\{(s, \lambda)\}$.)

If 0 is an algebraically simple eigenvalue of $D_u F(0, \lambda_0)$, then it perturbs to an eigenvalue $\mu(s)$ of $D_u F(u(s), \lambda(s))$ according to the *Principle of Exchange of Stability* stated in Theorem I.7.4. However, the sign of $\mu(s)$ has nothing to do with the stability of $u(s)$ considered as a time-periodic solution of an evolution equation, namely, the nonlinear wave equation (III.3.5)$_1$. The "energy" of (III.3.5)$_1$ with boundary conditions (III.3.6), e.g.,

$$(\text{III.3.53}) \qquad \frac{1}{2}\int_0^\pi u_t^2 + u_x^2 - c(\lambda)u^2 dx - \int_0^\pi \hat{G}(u,\lambda)dx,$$

for autonomous $\hat{G}(u,\lambda)(x) = G(x,u(x),\lambda)$ with $G_u(x,u,\lambda) = g(x,u,\lambda)$, is constant along any solution of (III.3.5)$_1$, (III.3.6). This provides nonlinear stability for *stationary* solutions u_0 if u_0 is a strict local minimizer of the functional

$$(\text{III.3.54}) \qquad \frac{1}{2}\int_0^\pi u_x^2 - c(\lambda)u^2 dx - \int_0^\pi \hat{G}(u,\lambda)dx,$$

i.e., if its second derivative (its "Hessian") is positive definite (cf. the remarks in Section I.7 after (I.7.3)). If u_0 is a nonstationary periodic solution, however, this stability analysis fails.

Remark III.3.4 *It is interesting and instructive to extend the analysis of this section to the two-dimensional analogue of the one-dimensional wave equation, namely,*

$$(\text{III.3.55}) \qquad u_{tt} - \Delta u - c(\lambda)u = g(t,x,u,\lambda),$$

where $u(t,x)$ is defined on $\mathbb{R}\times\Omega$, $\Omega\subset\mathbb{R}^2$. The Fourier analysis is completely analogous if $\Omega = (0,\pi)\times(0,\pi)$, and we end up with a characteristic equation of the form

$$(\text{III.3.56}) \qquad A_0^2(k^2 + \ell^2) - B_0 n^2 = C_0,$$

where $A_0, B_0,$ and C_0 are integers with $A_0 \neq 0, B_0 > 0$. Classical results from number theory show that an indefinite quadratic ternary form like (III.3.56) has either no solution or infinitely many solutions [12]. This means that there is no admissible set (III.3.30) with the property (III.3.29) in this case. (For $c(\lambda_0) \notin \mathbb{Q}$, the same obstructions arise as for the one-dimensional wave equation.) Consequently, we are not able to treat (III.3.55) over the square $\Omega = (0,\pi)\times(0,\pi)$.

However, if we replace the square by the unit sphere $S^2 \subset \mathbb{R}^3$ and "Δ" by the Laplace–Beltrami operator, the analysis of this section is applicable: Here the characteristic equation is again a binary form

$$(\text{III.3.57}) \qquad A_0^2 k(k+1) - B_0 n^2 = C_0,$$

for which an admissible set Λ with the property (III.3.29) can be defined. Moreover, for the nonlinear wave equation (III.3.55) on the sphere, and when $c(\lambda_0)$ is an integer, not only do we find small-amplitude solutions, but we obtain **global and unbounded branches of free vibrations** *(in the sense of Theorem II.5.8).*

There is another new feature of the wave equation on the sphere: By the rich symmetry of S^2, we can distinguish solutions by their precise spatiotemporal patterns. In order to discover these patterns systematically, they are classified via subgroups of $0(2) \times 0(3)$, which leaves $[0, P_0] \times S^2$ invariant (by periodicity, $[0, P_0]$ is identified with S^1). In this way we discover local and global branches of so-called **standing waves,** *which are characterized by a fixed spatial symmetry while oscillating periodically in time. We find* **rotating waves,** *which correspond to rigidly rotating patterns. Finally, we get* **discrete rotating waves:** *Unlike rotating waves, these patterns do not rotate rigidly but rather reappear in rotated form at regular fractions of the period. All these results are found in [61], where all types of waves are also illustrated by their nodal lines. For convenience, we give some examples in Figure III.3.1–Figure III.3.3.*

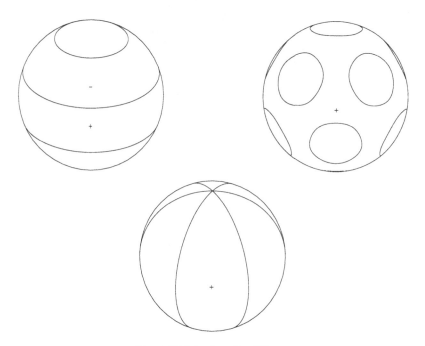

Figure III.3.1   Standing Waves.

(These methods apply also to nonlinear wave equations on the spheres S^n for $n \geq 3$. The eigenvalues of the Laplace–Beltrami operator are given by $k(k + n - 1)$, so that the characteristic equation is always a binary quadratic form. An exploitation of symmetry, however, for $0(2) \times 0(n)$ is much more involved.)

However, for $\Omega \subset \mathbb{R}^2$ a square and an equilateral triangle, the nonlinear plate equation

(III.3.58) $\qquad u_{tt} + \Delta^2 u - c(\lambda)u = g(t, x, u, \lambda)$

on $\mathbb{R} \times \Omega$ is amenable to our method; see [72], [73]. Here the characteristic equations are of the form

(III.3.59)
$$A_0^2(k^2 + \ell^2)^2 - B_0 n^2 = C_0 \text{ for the square,}$$
$$A_0^2(k^2 + \ell^2 - k\ell)^2 - B_0 n^2 = C_0 \text{ for the triangle,}$$

allowing us to define admissible sets Λ having the property (III.3.29). Accordingly, we find free vibrations bifurcating from $(0, \lambda_0)$ whenever $c(\lambda_0) \in \Lambda$. Due to the symmetries of Ω, we distinguish solutions by their spatiotemporal symmetry of any subgroup of $0(2) \times D_4$ and $0(2) \times D_3$. We find **standing waves** and **discrete rotating waves** having the properties described previously for waves on the sphere. On the triangle we discover a new family of solutions that we call **spatiotemporal reflection waves**: A rigid pattern reappears in reflected form after half the period. In [72], [73] all types of waves are sketched by their nodal lines.

The analysis for the plate equation on the square as well as on the equilateral triangle depends on the fact that the eigenvalues and the eigenfunctions of the Laplacian with homogeneous Dirichlet boundary conditions are well known. It is worth mentioning that the embedding of the square into the square lattice and the embedding of the triangle into the hexagonal lattice as expounded in Section III.1 is crucial for that analysis.

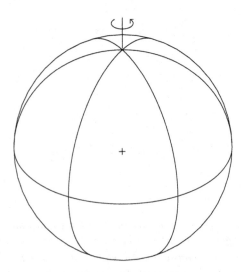

Figure III.3.2   Rotating Wave.

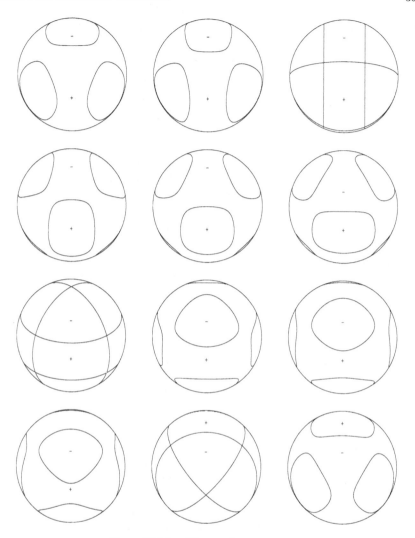

Figure III.3.3   Discrete Rotating Wave.

Finally, rotating waves having spatial symmetries of the sphere, and discrete rotating waves having spatial symmetries of the square and equilateral triangle, cannot be solutions of an evolution equation that is of first order in t. If a periodic solution has some spatial symmetry for some fixed time, then the fixed-point space of its isotropy subgroup (of the entire isotropy group of the domain) is invariant for all times t. Therefore, the spatial pattern cannot rotate.

III.4 Hopf Bifurcation for Parabolic Problems

A fully nonlinear parabolic problem with homogeneous Dirichlet boundary
conditions is of the form

(III.4.1)
$$u_t = G(\nabla^2 u, \nabla u, u, x, \lambda) \quad \text{in } \mathbb{R} \times \Omega, \ \lambda \in \mathbb{R},$$
$$u = 0 \quad \text{on } \mathbb{R} \times \partial\Omega,$$

where $u = u(t, x)$ is a scalar function of time $t \in \mathbb{R}$ and of space $x \in \Omega \subset \mathbb{R}^n$.
Problem (III.4.1) is well posed only if initial values $u(0, x)$ are prescribed, i.e.,
if it is an initial–boundary value problem. For mathematical and physical
reasons, solutions of (III.4.1) exist only for positive time, in general, but
special classes, such as periodic solutions, are defined for all times $t \in \mathbb{R}$.

If $G(\mathbf{0}, \mathbf{0}, 0, x, \lambda) = 0$ for all $(x, \lambda) \in \Omega \times \mathbb{R}$, we have the trivial solution
$u = 0$ for all $\lambda \in \mathbb{R}$, and we provide nontrivial periodic solutions via Hopf
Bifurcation.

If $G \in C^{k+1}(\mathbb{R}_{sym}^{n \times n} \times \mathbb{R}^n \times \mathbb{R} \times \overline{\Omega} \times \mathbb{R}, \mathbb{R})$ for $k \geq 1$, then for

(III.4.2) $$F(u, \lambda)(x) \equiv G(\nabla^2 u(x), \nabla u(x), u(x), x, \lambda),$$

cf. (III.2.3), the Fréchet derivative at $(0, \lambda)$ is given by

$$D_u F(0, \lambda)h =$$

(III.4.3) $$\sum_{i,j=1}^n a_{ij}(x, \lambda)h_{x_i x_j} + \sum_{i=1}^n b_{ij}(x, \lambda)h_{x_i} + c(x, \lambda)h$$

$$\equiv L(\lambda)h; \quad \text{cf. (III.2.7).}$$

Assume that for some $\lambda = \lambda_0$ the operator $L(\lambda_0)$ is **elliptic** in the sense
of (III.1.1). When considered as an operator $L(\lambda_0) : L^2(\Omega) \to L^2(\Omega)$ with
domain of definition $D(L(\lambda_0)) = H^2(\Omega) \cap H_0^1(\Omega)$, it is a Fredholm operator
of index zero having a discrete spectrum of eigenvalues of finite (algebraic)
multiplicity; cf. (III.1.12), (III.1.38). (We assume that $\Omega \subset \mathbb{R}^n$ is bounded
and that $\partial\Omega$ has the properties such that the results of Section III.1 are
applicable. Furthermore, for a reasonable spectral analysis the spaces of this
section are complexified, though all operators are real.)

The *bilinear* pairing of $Z = L^2(\Omega) = Z'$ is explicitly given by

(III.4.4) $$\langle u, v \rangle = \int_\Omega uv \, dx,$$

and the conditions (I.8.5)–(I.8.7) for Hopf Bifurcation are satisfied if

(III.4.5)

$i\kappa_0(\neq 0)$ is a simple eigenvalue of $L(\lambda_0)$ and of its adjoint $L(\lambda_0)^*$ according to (III.1.2), with eigenfunctions $\varphi_0, \varphi_0^* \in D(L(\lambda_0))$, respectively,

$$\int_\Omega \varphi_0 \varphi_0^* \, dx = 1, \text{ and}$$

$$\mathrm{Re} \int_\Omega (L_\lambda(\lambda_0)\varphi_0)\varphi_0^* \, dx \neq 0; \text{ cf. (I.8.44)},$$

where $L_\lambda(\lambda_0)$ is defined in (III.2.8)$_5$.

By the results of Section III.1, in particular (III.1.40), these spectral conditions are the same for the operators

(III.4.6)

$$L(\lambda_0) : L^p(\Omega) \to L^p(\Omega),$$
$$D(L(\lambda_0)) = W^{2,p}(\Omega) \cap W_0^{1,p}(\Omega), \ 2 \le p < \infty,$$

$$L(\lambda_0) : C^\alpha(\overline{\Omega}) \to C^\alpha(\overline{\Omega}),$$
$$D(L(\lambda_0)) = C^{2,\alpha}(\overline{\Omega}) \cap \{u | u = 0 \text{ on } \partial\Omega\}.$$

In other words, it is sufficient to check the spectral properties (III.4.5) for $L(\lambda)$ in its Hilbert space representation in order to know them for all representations (III.4.6).

Setting $L_0 = L(\lambda_0) = D_u F(0, \lambda_0)$, we see that the crucial property for proving Hopf Bifurcation is that

(III.4.7)

$$J_0 \equiv \kappa_0 \frac{d}{dt} - L_0 : Y \cap E \to W$$

is a continuous Fredholm operator of index zero,

provided that the nonresonance condition (I.8.14) holds. The spaces Y, E, and W are defined in (I.8.11) for $Z = L^p(\Omega)$ or $Z = C^\alpha(\overline{\Omega})$ and $X = D(L_0) = W^{2,p}(\Omega) \cap W_0^{1,p}(\Omega)$ or $X = C^{2,\alpha}(\overline{\Omega}) \cap \{u | u = 0 \text{ on } \partial\Omega\}$, respectively.

The Fredholm property (III.4.7) is proved in Proposition I.8.1 under the assumption that L_0 generates a compact holomorphic semigroup. We give now some details about the generation of semigroups of the operators (III.4.6).

For some $c \ge 0$, the operator $L_0 - cI : H^2(\Omega) \cap H_0^1(\Omega) \to L^2(\Omega)$ is bijective; cf. (III.1.8). This means that $c \in \mathbb{R}$ is in the resolvent set of $L_0 : L^2(\Omega) \to L^2(\Omega)$, and by (III.1.40), it is also in the resolvent set of $L_0 : L^p(\Omega) \to L^p(\Omega)$ and of $L_0 : C^\alpha(\overline{\Omega}) \to C^\alpha(\overline{\Omega})$. This observation implies the a priori estimates

(III.4.8)

$$\|u\|_{2,p} \le c_p(\|L_0 u\|_{0,p} + \|u\|_{0,p})$$
for all $u \in W^{2,p}(\Omega) \cap W_0^{1,p}(\Omega), c_p > 0, \ 2 \le p < \infty,$

$$\|u\|_{2,\alpha} \le c_\alpha(\|L_0 u\|_{0,\alpha} + \|u\|_{0,\alpha})$$
for all $u \in C^{2,\alpha}(\overline{\Omega}) \cap \{u | u = 0 \text{ on } \partial\Omega\}, c_\alpha > 0.$

By the famous trick of Agmon [2], the elliptic a priori estimates in $L^p(\Omega)$ imply the following estimates of the resolvent of L_0:

(III.4.9)

$$\text{For } z \in \Sigma = \{z \in \mathbb{C} | |\arg z| \leq \tfrac{\pi}{2} + \delta, |z| \geq R\}$$
$$\text{with some suitable } \delta > 0, R > 0,$$

$$|z| \|u\|_{0,p} \leq C_p \|(L_0 - zI)u\|_{0,p}$$
$$\text{for all } u \in W^{2,p}(\Omega) \cap W_0^{1,p}(\Omega).$$

When the same trick is applied in $C^\alpha(\overline{\Omega})$, one obtains (see [166], [88], [90]) that

(III.4.10)

$$\text{for } z \in \Sigma \text{ and } u \in C^{2,\alpha}(\overline{\Omega}) \cap \{u | u = 0 \text{ on } \partial\Omega\},$$

$$|z| \|u\|_{0,0} + |z|^{1-(\alpha/2)} \|u\|_{0,\alpha} \leq C_\alpha \|(L_0 - zI)u\|_{0,\alpha},$$

where $\| \ \|_{0,0}$ denotes the maximum norm in $C(\overline{\Omega})$. Counterexamples show that (III.4.10) cannot be improved; cf. [166]. In any case, the sector Σ is in the resolvent set of L_0, and the estimates (III.4.9), (III.4.10) imply the existence of the family of operators

(III.4.11)

$$e^{L_0 t} = \frac{1}{2\pi i} \int_{\partial\Sigma} e^{zt}(zI - L_0)^{-1} dz \quad \text{for } t > 0,$$

$$\text{in } L(Z, Z) \text{ for } Z = L^p(\Omega) \text{ and } Z = C^\alpha(\overline{\Omega}).$$

Whereas $e^{L_0 t}$ is a holomorphic semigroup in $L^p(\Omega)$ generated by L_0 (see [86], [142], [170], for example), it is not a holomorphic semigroup in $C^\alpha(\overline{\Omega})$ in its usual sense. Note that $D(L_0) = C^{2,\alpha}(\overline{\Omega}) \cap \{u | u = 0 \text{ on } \partial\Omega\}$ is not dense in $C^\alpha(\overline{\Omega})$. Nonetheless, the family $e^{L_0 t} \in L(C^\alpha(\overline{\Omega}), C^\alpha(\overline{\Omega}))$ defined by (III.4.11) has similar properties to those of a holomorphic semigroup for $t > 0$, but it has a singularity at $t = 0$:

(III.4.12)

$$\|e^{L_0 t} u\|_{0,\alpha} \leq M_\alpha e^{dt} t^{-\alpha/2} \|u\|_{0,\alpha} \text{ for } t > 0,$$

$$\|e^{L_0 t} u\|_{0,0} \leq M_\alpha e^{dt} \|u\|_{0,\alpha},$$

$$\text{with some } M_\alpha > 0, d \in \mathbb{R}, \text{ and for all } u \in C^\alpha(\overline{\Omega});$$

cf. [166], [88], [90]. In both cases, definition (III.4.11) implies that $e^{L_0 t} u \in D(L_0)$ for $t > 0$ and $u \in Z$. Since $D(L_0)$ is compactly embedded into Z, the semigroup $e^{L_0 t}$ is compact for $t > 0$.

Therefore, Proposition I.8.1 is applicable to $L_0 : Z \to Z$ with $Z = L^p(\Omega)$ and $D(L_0) = W^{2,p}(\Omega) \cap W^{1,p}(\Omega)$, which means that under the assumptions (III.4.5) and (I.8.14), the Fredholm property (III.4.7) holds.

For $Z = C^\alpha(\overline{\Omega})$ and $D(L_0) = C^{2,\alpha}(\overline{\Omega}) \cap \{u | u = 0 \text{ on } \partial\Omega\}$, however, the proof of Proposition I.8.1 fails for the following reason: If $f \in C_{2\pi}^\beta(\mathbb{R}, Z)$ for some $\beta > \alpha/2$, then the solution of $J_0 u = f$ given by (I.8.23) is only in $C^{\beta-(\alpha/2)}(\mathbb{R}, X) \cap C^{1+\beta-(\alpha/2)}(\mathbb{R}, Z)$, in general (the Hölder exponents β in time and α in space are not necessarily identical). Nonetheless, we make use of semigroups in $C^\alpha(\overline{\Omega})$ in proving the regularity of bifurcating periodic solutions; cf. (III.4.20)–(III.4.22) below. See also Remark III.4.2 for a different treatment of a fully nonlinear parabolic problem.

For an L^p-theory we have to confine ourselves to quasilinear parabolic problems; i.e., the right-hand side of (III.4.1) is of the form (III.2.6). Since $W^{2,p}(\Omega) \subset C^1(\overline{\Omega})$ for $p > n$, the quasilinear differential operator

$$G(\nabla^2 u, \nabla u, u, x, \lambda) \equiv$$

(III.4.13)
$$\sum_{i,j=1}^n a_{ij}(\nabla u, u, x, \lambda) u_{x_i x_j} + g(\nabla u, u, x, \lambda)$$

defines via (III.4.2), and by our assumptions on G, an operator

$$F \in C^{k+1}((W^{2,p}(\Omega) \cap W_0^{1,p}(\Omega)) \times \mathbb{R}, L^p(\Omega)).$$

(For the local analysis of this section we need smoothness of F only in a neighborhood of the bifurcation point $(0, \lambda_0)$. Accordingly, the global smoothness of G can be reduced to $a_{ij}, g \in C^{k+1}(V \times U \times \overline{\Omega} \times (\lambda_0 - \delta, \lambda_0 + \delta), \mathbb{R})$, where $V \times U$ is an open neighborhood of $(\mathbf{0}, 0) \in \mathbb{R}^n \times \mathbb{R}$.)

Thus, for $k \geq 2$, assumption (III.4.5) and the nonresonance condition (I.8.14) imply the applicability of the Hopf Bifurcation Theorem (Theorem I.8.2) to the evolution equation

(III.4.14) $$\frac{du}{dt} = F(u, \lambda) \quad \text{in } L^p(\Omega) \text{ for } p > n.$$

Solutions of (III.4.14), periodic or not periodic, are, by definition, not classical solutions of the parabolic problem (III.4.1). Note that d/dt means the derivative of the trajectory $u(t)(x) = u(t, x)$ with respect to the norm in $L^p(\Omega)$. We show now that bifurcating $2\pi/\kappa$-periodic solutions of (III.4.14) are indeed classical solutions of the parabolic partial differential equation with homogeneous Dirichlet boundary conditions.

In the sequel we fix λ near λ_0 and consider a small-amplitude solution $u \in C_{2\pi/\kappa}^{1+\beta}(\mathbb{R}, L^p(\Omega)) \cap C_{2\pi/\kappa}^\beta(\mathbb{R}, W^{2,p}(\Omega) \cap W_0^{1,p}(\Omega))$ of (III.4.14) for some $\beta \in (0, 1]$ and $p > n$.

By a continuous embedding $W^{2,p}(\Omega) \cap W_0^{1,p}(\Omega) \subset (C^{1,\alpha}(\overline{\Omega}) \cap \{u | u = 0$ on $\partial\Omega\})$ for $0 < \alpha < 1 - \frac{n}{p}$, the solution u is in $C_{2\pi/\kappa}^\beta(\mathbb{R}, C^{1,\alpha}(\overline{\Omega}))$. We insert $u = u(t, x)$ into the quasilinear operator G of (III.4.13), and we define

(III.4.15)
$$\tilde{L}(t)h \equiv \sum_{i,j=1}^n a_{ij}(\nabla u, u, x, \lambda) h_{x_i x_j},$$

$$f(t) \equiv g(\nabla u, u, x, \lambda), \ t \in \mathbb{R},$$

where we suppress the dependence on λ. By the continuous differentiability of a_{ij} and g, the property $u \in C_{2\pi/\kappa}^\beta(\mathbb{R}, C^{1,\alpha}(\overline{\Omega}))$ implies

(III.4.16) $$a_{ij}, f \in C_{2\pi/\kappa}^\beta(\mathbb{R}, C^\alpha(\overline{\Omega})), \quad i, j = 1, \ldots, n.$$

The assumed ellipticity of $\tilde{L}_0 \equiv \tilde{L}(t)$ for $(u, \lambda) = (0, \lambda_0)$ entails the uniform ellipticity of $\tilde{L}(t)$ (uniform with respect to t) for small amplitudes of u in $C^{1,\alpha}(\overline{\Omega})$ and for λ near λ_0. Therefore, (III.4.10) holds for each $\tilde{L}(t)$ and

(III.4.17) $\tilde{L}(t) : C^{2,\alpha}(\overline{\Omega}) \cap \{u | u = 0 \text{ on } \partial\Omega\} \to C^\alpha(\overline{\Omega})$
               is bijective for all $t \in \mathbb{R}$.

In view of (III.4.11), the properties (III.4.16), (III.4.17) imply that

for each $\tau \in \mathbb{R}$ there exists a semigroup

$e^{\tilde{L}(\tau)t}$ on $C^\alpha(\overline{\Omega})$ with estimates (III.4.12), and

(III.4.18)

$$\|(\tilde{L}(t) - \tilde{L}(\tau))\tilde{L}^{-1}(s)u\|_{0,\alpha} \leq C|t - \tau|^\beta \|u\|_{0,\alpha}$$

for all $s, t, \tau \in \mathbb{R}$, $u \in C^\alpha(\overline{\Omega})$, with a uniform constant $C > 0$.

With the aid of (III.4.18), the construction of a so-called *fundamental solution* is accomplished in [162], [52], provided that the semigroups $e^{\tilde{L}(\tau)t}$ are holomorphic without singularity at $t = 0$. In the singular case (III.4.12), however, this construction still provides a fundamental solution, but having weaker properties. We summarize the results of [90]:

Let $Z = C^\alpha(\overline{\Omega})$ and $X = D(\tilde{L}(t)) = C^{2,\alpha}(\overline{\Omega}) \cap \{u | u = 0 \text{ on } \partial\Omega\}$ and $0 < \alpha < \beta \leq 1$ with a gap $\beta - \alpha$ that is sufficiently large (conditions on $\beta - \alpha$ are found in [90]). Then for $T > 0$,

there is a fundamental solution $\tilde{U}(t, \tau) \in L(Z, Z)$, which is continuous for $0 \leq \tau < t \leq T$, such that

$\tilde{U}(t, \tau)\tilde{U}(\tau, s) = \tilde{U}(t, s)$ for $0 \leq s < \tau < t \leq T$,

$\tilde{U}(t, \tau)u \in X$ for $0 \leq \tau < t \leq T$ and $u \in Z$,

(III.4.19) $\|\tilde{U}(t, \tau)u\|_{0,\alpha} \leq C_1(t - \tau)^{-\alpha/2}\|u\|_{0,\alpha}$,

$\|\tilde{L}(s)\tilde{U}(t, \tau)u\|_{0,\alpha} \leq C_2(t - \tau)^{-1-(\alpha/2)}\|u\|_{0,\alpha}$,

for $0 \leq \tau < t \leq T$, $s \in [0, T]$, and $u \in Z$,

$$\frac{\partial}{\partial t}\tilde{U}(t, \tau) = \tilde{L}(t)\tilde{U}(t, \tau) \text{ in } L(Z, Z) \text{ for } 0 \leq \tau < t \leq T.$$

For our purposes, the following *Variation of Constants Formula* is useful:

For $f \in C^\beta([0, T], Z)$ and $u(0) \in Z$
the function obtained by integration in Z,

$$u(t) = \tilde{U}(t, 0)u(0) + \int_0^t \tilde{U}(t, s)f(s)ds,$$

(III.4.20)

is in $C^1((0, T], Z) \cap C((0, T], X)$, and

$$\frac{du}{dt} = \tilde{L}(t)u + f \text{ in } (0, T].$$

(Actually, one can prove $u \in C^{\tilde{\beta}}([\varepsilon, T], X)$ for some $\tilde{\beta} < \beta - (\alpha/2)$, but this is not done in [90].) Using (III.4.20), the regularity of bifurcating $2\pi/\kappa$-periodic solutions of (III.4.14) is immediate: By definitions (III.4.15), every periodic solution of (III.4.14) satisfies

$$(\text{III.4.21}) \qquad u(t) = \tilde{U}(t, 0)u(0) + \int_0^t \tilde{U}(t, s)f(s)ds \text{ in } L^p(\Omega)$$

$$\text{with } u(0) \in W^{2,p}(\Omega) \cap W_0^{1,p}(\Omega) \subset C^{1,\alpha}(\overline{\Omega}).$$

Since $f \in C^{\beta}([0, T], Z)$ for $Z = C^{\alpha}(\overline{\Omega})$, cf. (III.4.16), the function $u \in C_{2\pi/\kappa}^{\beta}(\mathbb{R}, C^{1,\alpha}(\overline{\Omega}))$ satisfies the same integral equation (III.4.21) in $C^{\alpha}(\overline{\Omega})$, so that the regularity (III.4.20) and periodicity of u imply

$$u \in C_{2\pi/\kappa}^1(\mathbb{R}, C^{\alpha}(\overline{\Omega})) \cap C_{2\pi/\kappa}(\mathbb{R}, C^{2,\alpha}(\overline{\Omega})),$$

$$(\text{III.4.22}) \qquad u(t, x) = 0 \text{ for } t \in \mathbb{R} \text{ and } x \in \partial\Omega, \text{ and}$$

$$\frac{du}{dt} = F(u, \lambda) \text{ holds in } C^{\alpha}(\overline{\Omega}).$$

Properties (III.4.22) imply clearly that $u = u(t, x)$ is a classical $2\pi/\kappa$-periodic solution of (III.4.1). As a matter of fact, the notion of a classical solution requires less than (III.4.22). Note that d/dt means the derivative with respect to the norm in $C^{\alpha}(\overline{\Omega})$. We summarize:

Theorem III.4.1 *Consider the evolution equation (III.4.14), where the right-hand side given by (III.4.2) is quasilinear in the sense of (III.4.13). Assume the ellipticity of $D_u F(0, \lambda_0) = L(\lambda_0)$, the spectral conditions (III.4.5), and the nonresonance condition (I.8.14). Then Theorem I.8.2 on Hopf Bifurcation is applicable to the evolution equation (III.4.14) in $L^p(\Omega)$.*

The regularity (III.4.22) implies that the bifurcating $2\pi/\kappa$-periodic solutions are classical solutions of the parabolic problem (III.4.1).

Finally, we mention that the bifurcating curve $\{(u(r), \lambda(r))\}$ of $2\pi/\kappa(r)$-periodic solutions through $(0, \lambda_0)$ with $2\pi/\kappa(0) = 2\pi/\kappa_0$ is continuously differentiable (with respect to r) in $(C_{2\pi/\kappa(r)}^{1+\beta}(\mathbb{R}, L^p(\Omega)) \cap C_{2\pi/\kappa(r)}^{\beta}(\mathbb{R}, W^{2,p}(\Omega))) \times \mathbb{R}$ and therefore, by continuous embedding for $p > n$, it is also continuously differentiable in $(C_{2\pi/\kappa(r)}^{\beta}(\mathbb{R}, C^{1,\alpha}(\overline{\Omega}) \cap \{u|u = 0 \text{ on } \partial\Omega\}) \times \mathbb{R}$ for $0 < \alpha < 1 - \frac{n}{p}$.

Remark III.4.2 *The elegant treatment of Hopf Bifurcation using semigroups – an approach that mimics the finite-dimensional case by considering the parabolic problem (III.4.1) as an ODE (III.4.14) in a Banach space – fails for fully nonlinear parabolic problems, since the Fredholm property (III.4.7) is not true for $Z = C^{\alpha}(\overline{\Omega})$ and $X = C^{2,\alpha}(\overline{\Omega}) \cap \{u|u = 0 \text{ on } \partial\Omega\}$. The reason for this failure is that the space $W = C_{2\pi}^{\beta}(\mathbb{R}, C^{\alpha}(\overline{\Omega}))$ is not adequate: the Hölder* *continuity in space is superimposed by the Hölder continuity in time, which is more than just Hölder continuity in the time–space cylinder $[0,T] \times \overline{\Omega} \equiv \overline{Q}_T$.*

The adequate spaces are given in [53], [121]. Using the notation of [121], we see that the space $H^{\ell+2,\ell/2+1}(\overline{Q}_T)$ for some $0 < \ell < 1$ consists of functions whose time derivatives up to order 1 are uniformly Hölder continuous in time with exponent $\ell/2$ and whose space derivatives up to order 2 are uniformly Hölder continuous in space with exponent ℓ. These functions are also uniformly Hölder continuous with exponent ℓ in the time–space cylinder with weighted distance $|(x,t) - (x',t')| = (|x-x'|^2 + |t-t'|)^{1/2}$; cf. [53]. Note that no mixed derivatives are involved.

We sketch how to prove Hopf Bifurcation for fully nonlinear parabolic problems using the parabolic regularity in those Hölder spaces over the time–space cylinder proved in [53], [121].

We know from Proposition I.8.1 that for $J_0 : Y \cap E \to W$ with $Y = C_{2\pi}^{1+\beta}(\mathbb{R}, L^p(\Omega))$, $E = C_{2\pi}^{\beta}(\mathbb{R}, W^{2,p}(\Omega) \cap W_0^{1,p}(\Omega))$, and $W = C_{2\pi}^{\beta}(\mathbb{R}, L^p(\Omega))$, the decomposition

(III.4.23)
$$W = R(J_0) \oplus N(J_0)$$

holds; cf. (I.8.28). For $\beta = \ell/2$ there is a continuous embedding $H_{2\pi}^{\ell,\ell/2}(\mathbb{R} \times \overline{\Omega}) \subset W$, where the subscript 2π denotes 2π-periodicity in time. The parabolic regularity proved in [53], [121] applies then as follows:

$$J_0 u = f \text{ for } u \in Y \cap E, f \in H_{2\pi}^{\ell,\ell/2}(\mathbb{R} \times \overline{\Omega})$$

(III.4.24)
$$\Rightarrow u \in H_{2\pi}^{\ell+2,\ell/2+1}(\mathbb{R} \times \overline{\Omega}) \cap \{u|u(t,x) = 0 \text{ on } \mathbb{R} \times \partial\Omega\}$$

$$\text{and } \kappa_0 \frac{\partial u}{\partial t} - L_0 u = f \text{ in } \mathbb{R} \times \Omega$$

holds in the classical sense.

In contrast to the "infinite-dimensional ordinary differential operator" $J_0 = \kappa_0 \frac{d}{dt} - L_0$, the operator $\mathcal{J}_0 \equiv \kappa \frac{\partial}{\partial t} - L_0$ is a genuine partial differential operator. A proof of (III.4.24) is rather involved. Since $N(J_0) = N(\mathcal{J}_0) \subset X_\ell \equiv H_{2\pi}^{\ell+2,\ell/2+1}(\mathbb{R} \times \overline{\Omega}) \cap \{u|u(t,x) = 0 \text{ on } \mathbb{R} \times \partial\Omega\}$, analogous arguments as in (III.1.22)–(III.1.28) prove that for $Z_\ell \equiv H_{2\pi}^{\ell,\ell/2}(\mathbb{R} \times \overline{\Omega})$, the parabolic operator

$$\mathcal{J}_0 = \kappa_0 \frac{\partial}{\partial t} - L_0 : X_\ell \to Z_\ell$$

is a (continuous) Fredholm operator of index zero,

(III.4.25)
and in particular, $Z_\ell = R(\mathcal{J}_0) \oplus N(\mathcal{J}_0)$ with projection $Q|_{Z_\ell} : Z_\ell \to N(\mathcal{J}_0)$ along $R(\mathcal{J}_0)$, where Q is defined by (I.8.21) with duality (III.4.4).

(In view of (III.4.23) an adjoint operator is not needed.)

Let F be a fully nonlinear operator defined by (III.4.2). We rewrite (III.4.1) by

(III.4.26)
$$\frac{\partial u}{\partial t} = F(u, \lambda) \text{ in } \mathbb{R} \times \Omega,$$
$$u = 0 \text{ on } \mathbb{R} \times \partial\Omega,$$

and the usual rescaling in time yields $2\pi/\kappa$-periodic solutions of (III.4.26), provided that

(III.4.27)
$$\mathcal{G}(u, \kappa, \lambda) \equiv \kappa\frac{\partial u}{\partial t} - F(u, \lambda) = 0$$
$$\text{for } (u, \kappa, \lambda) \in X_\ell \times \mathbb{R} \times \mathbb{R}.$$

If $G \in C^{k+1}(\mathbb{R}^{n\times m}_{sym} \times \mathbb{R}^n \times \mathbb{R} \times \overline{\Omega} \times \mathbb{R}, \mathbb{R})$, then $\mathcal{G} \in C^k(X_\ell \times \mathbb{R} \times \mathbb{R}, Z_\ell)$.

By $D_u\mathcal{G}(0, \kappa_0, \lambda_0) = \mathcal{J}_0$ and the Fredholm property (III.4.25), the proof for bifurcating nontrivial solutions of $\mathcal{G}(u, \kappa, \lambda) = 0$ near $(0, \kappa_0, \lambda_0)$ via the method of Lyapunov–Schmidt follows precisely the lines of Section I.8. In particular, assumptions (III.4.5) and the nonresonance condition (I.8.14) are sufficient to solve the bifurcation equation nontrivially; that bifurcation equation is of the form (I.8.34), or (I.8.38), since the projection Q has the same structure as in (I.8.21); cf. (III.4.25).

In this way we can prove the existence of a nontrivial curve $\{(u(r), \lambda(r))|r \in (-\delta, \delta)\}$ of $2\pi/\kappa(r)$-periodic solutions of (III.4.26) through $(u(0), \lambda(0)) = (0, \lambda_0)$ with $2\pi/\kappa(0) = 2\pi/\kappa_0$ in $(H^{\ell+2,\ell/2+1}_{2\pi/\kappa(r)}(\mathbb{R} \times \overline{\Omega}) \cap \{u|u(t, x) = 0 \text{ on } \mathbb{R} \times \partial\Omega\}) \times \mathbb{R}$.

The **Bifurcation Formulas** derived in Section I.9 hold accordingly. The evaluation of (I.9.11) (with duality (II.4.4)) is certainly nontrivial, since an inversion of elliptic operators $2i\kappa_0 - L_0$ and L_0 is involved. If, however, $D^2_{uu}F(0, \lambda_0) = 0$, i.e., if the nonlinear part of F is at least of third order, then the bifurcation formulas of Theorem I.9.1 are much simpler.

Finally, the **Principle of Exchange of Stability** holds for nondegenerate as well as for degenerate Hopf Bifurcation for parabolic problems. **Degenerate Hopf Bifurcation** is expounded in Section I.17: In this case, the condition $\text{Re}\mu'(\lambda_0) \neq 0$, cf. (III.4.5)$_5$, is replaced by $\text{Re}\mu^{(m)}(\lambda_0) \neq 0$ for some odd $m \in \mathbb{N}$. If in (III.4.13) the coefficients a_{ij} and the function g are analytic with respect to their real variables $(\nabla u, u, \lambda)$, the quasilinear operator F defined by (III.4.2) is analytic in the sense of (I.16.2) for $X = W^{2,p}(\Omega) \cap W^{1,p}_0(\Omega)$ and $Z = L^p(\Omega)$. Therefore, all results of Section I.17 are applicable to the evolution equation (III.4.14) in $L^p(\Omega)$. The Principle of Exchange of Stability then says that consecutive curves of periodic solutions ordered in the (r, λ)-plane have opposite stability properties (in a possibly weakened sense); cf. Theorem I.17.6. Stability means, by definition, linear stability determined by the second nontrivial Floquet exponent. If that Floquet exponent vanishes, then the Principle of Linearized Stability fails; cf. Section I.12. In case of a nondegenerate Hopf Bifurcation, however, the second Floquet exponent is nonzero along a nondegenerate pitchfork, and the Principle of Exchange of Stability applies; cf. Theorem I.12.2, Corollary I.12.3, and Figure I.12.1.

If the parabolic problem (III.4.1) is semilinear (defined in (III.2.10)), then as proved in [76], linear stability indeed implies nonlinear orbital stability.

III.5 Global Bifurcation and Continuation for Elliptic Problems

Whereas in Section III.2 we study local bifurcations for fully nonlinear elliptic problems, we prove global results only for quasilinear elliptic problems, namely,

$$
\begin{aligned}
& G(\nabla^2 u, \nabla u, u, x, \lambda) \equiv \\
\text{(III.5.1)} \quad & \sum_{i,j=1}^n a_{ij}(\nabla u, u, x, \lambda) u_{x_i x_j} + g(\nabla u, u, x, \lambda) = 0 \text{ in } \Omega, \\
& \hspace{6cm} u = 0 \text{ on } \partial\Omega.
\end{aligned}
$$

Again $\Omega \subset \mathbb{R}^n$ is a bounded domain with a boundary $\partial\Omega$ such that linear elliptic operators have the properties provided in Section III.1. As usual, we assume $G(\mathbf{0}, 0, 0, x, \lambda) = 0$ for all $(x, \lambda) \in \Omega \times \mathbb{R}$, so that we have the trivial solution $(0, \lambda)$ for all $\lambda \in \mathbb{R}$.

The ellipticity of problem (III.5.1) is defined as follows:

$$
\sum_{i,j=1}^n a_{ij}(\mathbf{v}, u, x, \lambda)\xi_i\xi_j \geq d(\mathbf{v}, u, \lambda)\|\xi\|^2
$$

(III.5.2)   for all $x \in \Omega$, $\xi \in \mathbb{R}^n$, with some positive
and continuous function $d : \mathbb{R}^n \times \mathbb{R} \times \mathbb{R} \to \mathbb{R}$.

Then $d(\mathbf{v}, u, \lambda) \geq d_R > 0$ for all $(\mathbf{v}, u, \lambda) \in \mathbb{R}^n \times \mathbb{R} \times \mathbb{R}$ such that $\|\mathbf{v}\| + |u| + |\lambda| \leq R$.

We define $X = C^{2,\alpha}(\overline{\Omega}) \cap \{u | u = 0 \text{ on } \partial\Omega\}, Y = C^{1,\alpha}(\overline{\Omega})$, and $Z = C^\alpha(\overline{\Omega})$.

If a_{ij}, g and their partial derivatives with respect to $(\mathbf{v}, u, x) \in \mathbb{R}^n \times \mathbb{R} \times \overline{\Omega}$ are in $C(\mathbb{R}^n \times \mathbb{R} \times \overline{\Omega} \times \mathbb{R}, \mathbb{R})$, then

$$
L(u, \lambda)h \equiv \sum_{i,j=1}^n a_{ij}(\nabla u, u, x, \lambda)h_{x_i x_j} \text{ satisfies}
$$

$$
L \in C(Y \times \mathbb{R}, L(X, Z)),
$$

(III.5.3)

and $\hat{g} \in C(Y \times \mathbb{R}, Z)$, where

$$
\hat{g}(u, \lambda)(x) = g(\nabla u(x), u(x), x, \lambda).
$$

In case $a_{ij}, g \in C^3(\mathbb{R}^n \times \mathbb{R} \times \overline{\Omega} \times \mathbb{R}, \mathbb{R})$, the mapping

$$F(u, \lambda) \equiv L(u, \lambda)u + \hat{g}(u, \lambda) \text{ satisfies}$$

$$F \in C^2(X \times \mathbb{R}, Z) \text{ and}$$

(III.5.4) $D_u F(u, \lambda)h$

$$= \sum_{i,j=1}^{n} a_{ij}(\nabla u, u, x, \lambda)h_{x_i x_j} + \sum_{i=1}^{n} b_i(\nabla^2 u, \nabla u, u, x, \lambda)h_{x_i}$$

$$+ c(\nabla^2 u, \nabla u, u, x, \lambda)h \quad \text{for } (u, \lambda) \in X \times \mathbb{R}, \ h \in X.$$

Before we apply Theorem II.5.8 to (III.5.1) in its form $F(u, \lambda) = 0$, i.e., before verifying the conditions (II.5.32) on F, we discuss the notion of an odd crossing number of $D_u F(0, \lambda)$ at $\lambda = \lambda_0$. An odd crossing number is equivalent to a change of index $i(F(\cdot, \lambda), 0)$ at $\lambda = \lambda_0$, cf. (II.5.34), and its Definition II.4.1 is here applied to the family

$$D_u F(0, \lambda) = L(0, \lambda) + D_u \hat{g}(0, \lambda), \text{ where}$$

(III.5.5) $L(0, \lambda)$ is given in (III.5.3) and

$$D_u \hat{g}(0, \lambda)h = \sum_{i=1}^{n} g_{v_i}(\mathbf{0}, 0, x, \lambda)h_{x_i} + g_u(\mathbf{0}, 0, x, \lambda)h.$$

For special families $D_u F(0, \lambda)$, an odd crossing number is easily verified; cf. (III.2.8), (III.2.17), (III.2.36). It is the crucial condition for local bifurcation (Theorem II.4.4) as well as for global bifurcation (Theorem II.5.8).

If F is not of class C^2, then Theorem II.3.3 offers a way to obtain the same global alternative as given by Theorem II.5.8, provided that the problem $F(u, \lambda) = 0$ is converted into a problem for a compact perturbation of the identity, that is, $u + f(u, \lambda) = 0$; cf. Definition II.2.1. If this has been done, then the crucial assumption is again an odd crossing number of the family $I + D_u f(0, \lambda)$ at $\lambda = \lambda_0$.

The conversion of $F(u, \lambda) = 0$ to $u + f(u, \lambda) = 0$ is not unique. Nonetheless, one expects that the spectral properties of the families $D_u F(0, \lambda)$ and $I + D_u f(0, \lambda)$ near $\lambda = \lambda_0$ imply simultaneously an odd crossing number as long as the problems $F(u, \lambda) = 0$ and $u + f(u, \lambda) = 0$ are equivalent. That expectation, however, is not at all obvious, as a simple semilinear example shows:

Assume that $F(u, \lambda) = Lu + \hat{g}(u, \lambda)$, where L is some elliptic operator (independent of (u, λ)) that is invertible and where $\hat{g}$ is given in $(III.5.3)_4$. Then the following equivalence is obvious:

(III.5.6) $$F(u, \lambda) \equiv Lu + \hat{g}(u, \lambda) = 0 \text{ for } (u, \lambda) \in X \times \mathbb{R} \Leftrightarrow$$

$$u + f(u, \lambda) \equiv u + L^{-1}\hat{g}(u, \lambda) = 0 \text{ for } (u, \lambda) \in Y \times \mathbb{R}.$$

(Below, we prove that $f : Y \times \mathbb{R} \to Y$ is completely continuous.) It is simple to verify that 0 is an eigenvalue of $D_u F(0, \lambda_0) = L + D_u \hat{g}(0, \lambda_0)$ if and only if 0 is an eigenvalue of $I + D_u f(0, \lambda_0) = I + L^{-1} D_u \hat{g}(0, \lambda_0)$, and that the geometric multiplicities are the same. However, since L^{-1} and $D_u \hat{g}(0, \lambda_0)$ (given by (III.5.5)$_3$) do not commute in general, it is not clear whether the algebraic multiplicities are equal, too. Consequently, it is open whether the two families $D_u F(0, \lambda)$ and $I + D_u f(0, \lambda)$ have simultaneously odd crossing numbers at $\lambda = \lambda_0$. In particular, the computation of the crossing number of the compact perturbation of the identity might cause problems if it is not proved that it has the same parity as the crossing number of the original elliptic family.

Having these difficulties in mind, we prefer to stay first with the original formulation $F(u, \lambda) = 0$ of problem (III.5.1); cf. (III.5.3), (III.5.4).

We verify the conditions (II.5.32) on F.

First of all, $F \in C^2(X \times \mathbb{R}, Z)$ if $a_{ij}, g \in C^3(\mathbb{R}^n \times \mathbb{R} \times \overline{\Omega} \times \mathbb{R}, \mathbb{R})$. Then the coefficients a_{ij}, b_i, and c of $D_u F(u, \lambda)$ as given by (III.5.4) are C^2-functions of their arguments, and by (III.5.2) the linear operator $D_u F(u, \lambda) \in L(X, Z)$ is elliptic for all $(u, \lambda) \in X \times \mathbb{R}$. In view of Section III.1, this proves (i) and (ii) of Definition II.5.1. In order to verify condition (iii) of that definition we have to replace F by $-F$ and accordingly $D_u F(u, \lambda)$ by $-D_u F(u, \lambda)$. By the sectorial property of elliptic operators as stated in (III.4.9), (III.4.10), the spectrum of $-D_u F(u, \lambda)$ is in a sector $\{z \in \mathbb{C} | \ |\arg z| \le \frac{\pi}{2} - \delta\} \cup \{z \in \mathbb{C} | \ |z| \le R\}$ for some $\delta > 0, R > 0$ depending only on the ellipticity $d(\mathbf{v}, u, \lambda) > 0$ in (III.5.2), on the norms $\|a_{ij}\|_{0,\alpha}, \|b_i\|_{0,\alpha}, \|c\|_{0,\alpha}$ of the coefficients of (III.5.4), and on Ω. This proves that a strip $(-\infty, c) \times (-i\varepsilon, i\varepsilon) \subset \mathbb{C}$ for some $c > 0, \varepsilon > 0$ contains at most finitely many eigenvalues of finite algebraic multiplicity; cf. (III.1.38). Finally, a small perturbation of $-D_u F(u, \lambda)$ by a small perturbation of (u, λ) in $X \times \mathbb{R}$ entails a small perturbation of the eigenvalues of $-D_u F(u, \lambda)$ in a sector such that their total number in that strip is stable under small perturbations.

The last property to be verified is *properness*. According to (III.5.4), we decompose $F(u, \lambda) = L(u, \lambda)u + \hat{g}(u, \lambda)$. Let

(III.5.7)
$$F(u_n, \lambda_n) = f_n, \text{ where } f_n \to f \text{ in } Z$$
$$\text{and } \{(u_n, \lambda_n)\} \text{ is bounded in } X \times \mathbb{R}.$$

Without loss of generality, $(\lambda_n)_{n \in \mathbb{N}}$ converges to λ in $\mathbb{R}$, and by compact embedding, $(u_n)_{n \in \mathbb{N}}$ converges to u in Y. This, in turn, implies, by (III.5.3),

(III.5.8)
$$\hat{g}(u_n, \lambda_n) \to \hat{g}(u, \lambda) \text{ in } Z \text{ and}$$
$$\|(L(u_n, \lambda_n) - L(u, \lambda))u_n\|_{0,\alpha} \le \varepsilon \|u_n\|_{2,\alpha} \le \varepsilon M$$

for all $n \ge n_0(\varepsilon)$. By (III.5.7), (III.5.8),

$$L(u, \lambda)u_n =$$

(III.5.9)
$$(L(u, \lambda) - (L(u_n, \lambda_n)))u_n - \hat{g}(u_n, \lambda_n) + f_n$$

$$\text{converges to } -\hat{g}(u, \lambda) + f \text{ in } Z.$$

Since $L(u, \lambda)$ is elliptic, cf. (III.5.2), injective, and therefore bijective by its Fredholm property, the elliptic a priori estimate $\|u_n\|_{2,\alpha} \leq C_R \|L(u, \lambda)u_n\|_{0,\alpha}$, with some constant $C_R > 0$ depending on $\|u\|_{1,\alpha} + |\lambda| \leq R$, implies that $(u_n)_{n\in\mathbb{N}}$ converges to $u \in X$. This proves properness and the admissibility of F according to (II.5.32). Therefore, Theorem II.5.8 is applicable to the problem $F(u, \lambda) = 0$.

Next, we convert $F(u, \lambda) = 0$ into a compact perturbation of the identity $u + f(u, \lambda) = 0$ in such a way that the crossing numbers of $D_u F(0, \lambda)$ and of $I + D_u f(0, \lambda)$ at some $\lambda = \lambda_0$ have the same parity.

Choose a continuous function $c : \mathbb{R} \to \mathbb{R}$ such that

(III.5.10)
$$g_u(0, 0, x, \lambda) + c(\lambda) \leq 0$$
$$\text{for all } (x, \lambda) \in \overline{\Omega} \times \mathbb{R},$$

and define

(III.5.11)
$$\hat{L}(u, \lambda) \equiv L(u, \lambda) + D_u\hat{g}(0, \lambda) + c(\lambda)I.$$

In view of (III.5.5), the lowest-order coefficient of the elliptic operator $\hat{L}(u, \lambda)$ is nonpositive, so that by the maximum (and minimum) principle, $\hat{L}(u, \lambda) \in L(X, Z)$ is injective. Its Fredholm property (cf. Section III.1) implies that it is bijective. We claim that

(III.5.12)
$$\|h\|_{2,\beta} \leq \hat{C}_R \|\hat{L}(u, \lambda)h\|_{0,\beta} \text{ with some } \hat{C}_R > 0,$$
$$\text{for all } (u, \lambda) \in Y \times \mathbb{R} \text{ such that } \|u\|_{1,\alpha} + |\lambda| \leq R,$$
$$\text{and for all } h \in C^{2,\beta}(\overline{\Omega}) \cap \{u | u = 0 \text{ on } \partial\Omega\},$$
$$\text{where } 0 < \beta < \alpha < 1.$$

Assume the existence of sequences $(u_n)_{n\in\mathbb{N}}$, $(\lambda_n)_{n\in\mathbb{N}}$, and $(h_n)_{n\in\mathbb{N}}$ with $\|u_n\|_{1,\alpha} + |\lambda_n| \leq R$, $\|h_n\|_{2,\beta} = 1$, and

(III.5.13)
$$\hat{L}(u_n, \lambda_n)h_n \to 0 \text{ in } C^\beta(\overline{\Omega}) \text{ as } n \to \infty.$$

By $0 < \beta < \alpha < 1$ and a compact embedding, we can assume w.l.o.g. that $u_n \to u$ in $C^{1,\beta}(\overline{\Omega})$ and $\lambda_n \to \lambda$ in $\mathbb{R}$, whence

(III.5.14)
$$\hat{L}(u_n, \lambda_n)h \to \hat{L}(u, \lambda)h \text{ in } C^\beta(\overline{\Omega})$$
$$\text{uniformly for } h \in C^{2,\beta}(\overline{\Omega}) \text{ with } \|h\|_{2,\beta} \leq 1.$$

Therefore,

$$\hat{L}(u, \lambda) h_n$$

(III.5.15)
$$= (\hat{L}(u, \lambda) - \hat{L}(u_n, \lambda_n)) h_n + \hat{L}(u_n, \lambda_n) h_n$$

converges to 0 in $C^\beta(\overline{\Omega})$,

whence $h_n \to 0$ in $C^{2,\beta}(\overline{\Omega})$ as $n \to \infty$

by the elliptic a priori estimate $(III.5.12)_1$ for the single operator $\hat{L}(u, \lambda)$. This contradicts $\|h_n\|_{2,\beta} = 1$ and proves (III.5.12).

We convert $F(u, \lambda) = 0$ for $(u, \lambda) \in X \times \mathbb{R}$ given by $(III.5.4)_1$ equivalently into

$$u + f(u, \lambda) \equiv$$

(III.5.16) $$u + \hat{L}(u, \lambda)^{-1}(\hat{g}(u, \lambda) - D_u \hat{g}(0, \lambda) u - c(\lambda) u) = 0$$

for $(u, \lambda) \in Y \times \mathbb{R}$.

The mapping $f : Y \times \mathbb{R} \to Y$ is completely continuous: First of all, the mapping $(u, \lambda) \mapsto \hat{g}(u, \lambda) - D_u \hat{g}(0, \lambda) u - c(\lambda) u$ is continuous and bounded from $Y \times \mathbb{R}$ into Z, which, in turn, is continuously embedded into $C^\beta(\overline{\Omega})$ for $0 < \beta < \alpha < 1$. The identity $\hat{L}(u, \lambda)^{-1} - \hat{L}(\tilde{u}, \tilde{\lambda})^{-1} = \hat{L}(\tilde{u}, \tilde{\lambda})^{-1}[\hat{L}(\tilde{u}, \tilde{\lambda}) - \hat{L}(u, \lambda)]\hat{L}(u, \lambda)^{-1}$ together with (III.5.12) and $(III.5.3)_2$, where X is replaced by $C^{2,\beta}(\overline{\Omega}) \cap \{u|u = 0 \text{ on } \partial\Omega\}$ and Z is replaced by $C^\beta(\overline{\Omega})$, proves that f is continuous and bounded as a mapping from $Y \times \mathbb{R}$ into $C^{2,\beta}(\overline{\Omega})$. Finally, bounded sets in $C^{2,\beta}(\overline{\Omega})$ are relatively compact in Y, which proves the complete continuity of f.

For the application of Theorem II.3.3 we need also continuous differentiability of f with respect to u along the trivial solution line $\{(0, \lambda)\}$. This is true if in addition to the conditions for (III.5.3), $g \in C(\mathbb{R}, C^2(\overline{V} \times \overline{U} \times \overline{\Omega}))$ for a neighborhood $V \times U$ of $(\mathbf{0}, 0) \in \mathbb{R}^n \times \mathbb{R}$; cf. (III.5.5) and (III.5.17) below. (This nomenclature means that the mapping $\lambda \mapsto g(\cdot, \cdot, \cdot, \lambda)$ is continuous from $\mathbb{R}$ into the Banach space $C^2(\overline{V} \times \overline{U} \times \overline{\Omega})$.)

The crucial condition for global bifurcation for the compact perturbation of the identity $u + f(u, \lambda) = 0$ is an odd crossing number of the family $I + D_u f(0, \lambda)$ at some $\lambda = \lambda_0$. Since $\hat{g}(0, \lambda) = 0$, we obtain from (III.5.16)

$$I + D_u f(0, \lambda) = I - c(\lambda)\hat{L}(0, \lambda)^{-1}, \text{ where by (III.5.11)},$$

(III.5.17) $$\hat{L}(0, \lambda) = L(0, \lambda) + D_u \hat{g}(0, \lambda) + c(\lambda) I =$$

$$D_u F(0, \lambda) + c(\lambda) I; \text{ cf. (III.5.5)}.$$

If 0 is an eigenvalue of $D_u F(0, \lambda_0)$, then by its definition (III.5.10), $c(\lambda_0) < 0$ and $c(\lambda_0)$ and $1/c(\lambda_0)$ are eigenvalues of $\hat{L}(0, \lambda_0)$ and $\hat{L}(0, \lambda_0)^{-1}$, respectively, of the same algebraic multiplicity; cf. (III.1.38). Furthermore,

$\mu(\lambda)$ is in the 0-group of $D_u F(0, \lambda) \Leftrightarrow$

(III.5.18) $\dfrac{\mu(\lambda)}{\mu(\lambda) + c(\lambda)}$ is in the 0-group of $I + D_u f(0, \lambda)$

for $\lambda \in (\lambda_0 - \delta, \lambda_0 + \delta)$ having the same
algebraic multiplicity in both cases.

In view of $\mu(\lambda) + c(\lambda) < 0$ for real $\mu(\lambda)$ and for $\lambda \in (\lambda_0 - \delta, \lambda_0 + \delta)$, provided
that $\delta > 0$ is small enough, this proves that

(III.5.19)   the family $D_u F(0, \lambda)$ has an odd crossing number
at $\lambda = \lambda_0 \Leftrightarrow$
$I + D_u f(0, \lambda)$ has an odd crossing number at $\lambda = \lambda_0$.

Before we summarize our results, we mention that for solution sets $\{(u, \lambda)\}$ $\subset Y \times \mathbb{R}$ of $u + f(u, \lambda) = 0$, connectedness and unboundedness in $Y \times \mathbb{R}$ are equivalent to connectedness and unboundedness in $X \times \mathbb{R}$. (For a proof use a "bootstrapping argument.") By (III.5.11), (III.5.16) the solution sets of $F(u, \lambda) = 0$ and $u + f(u, \lambda) = 0$ coincide. Theorem II.3.3 then yields the following:

Theorem III.5.1 *Assume that the elliptic family* $L(0, \lambda) + D_u \hat{g}(0, \lambda)$ *defined in (III.5.5) has an odd crossing number at some* $\lambda = \lambda_0$. *If the functions* a_{ij} *and* g *and their partial derivatives with respect to* $(\mathbf{v}, u, x) \in \mathbb{R}^n \times \mathbb{R} \times \Omega$ *are continuous and if* $g \in C(\mathbb{R}, C^2(\overline{V} \times \overline{U} \times \overline{\Omega}))$ *for a neighborhood* $V \times U$ *of* $(\mathbf{0}, 0) \in \mathbb{R}^n \times \mathbb{R}$, *then there exists a component* C *in the closure* S *of nontrivial solutions of the quasilinear elliptic problem* $L(u, \lambda)u + \hat{g}(u, \lambda) = 0$ *in* $X \times \mathbb{R}$ *emanating at* $(0, \lambda_0)$ *and subject to the alternatives*
(i) C *is unbounded in* $X \times \mathbb{R}$, *or*
(ii) C *contains some* $(0, \lambda_1)$, *where* $\lambda_0 \neq \lambda_1$.
Here $X = C^{2,\alpha}(\overline{\Omega}) \cap \{u | u = 0 \text{ on } \partial\Omega\}$.

In case of $\dim N(D_u F(0, \lambda_0)) = 1 = \dim N(I + D_u f(0, \lambda_0))$, the application of Theorem II.5.9 gives the following refinement: $C = C^+ \cup C^- \cup \{(0, \lambda_0)\}$ and each of C^+ and C^- either satisfies one of the alternatives (i) and (ii) or (iii) contains a pair $(u, \lambda), (-u, \lambda)$ where $u \neq 0$.

Under the assumptions (III.2.8), the continua $C^\pm$ are locally given by smooth curves $\{(u(s), \lambda(s)) | s \in (0, \delta)\}$ and $\{(u(s), \lambda(s)) | s \in (-\delta, 0)\}$. Globally, not much is known about $C^\pm$ in general. In Section III.6 we provide an example for the second alternative of Theorem III.5.1. In Section III.7 we prove that $C^\pm$ are globally smooth curves parameterized by the amplitude of u if $C^\pm$ are positive and negative branches emanating from the principal eigenvalue. None of these results hold in general, but they require special assumptions.

Remark III.5.2 *The key for the application of the Leray–Schauder degree instead of the degree of proper Fredholm operators is the equivalence*

(III.5.19), which, in turn, relies on the inversion of $\hat{L}(u, \lambda)$. However, there are elliptic problems with nonlinear boundary conditions that cannot be transformed into a compact perturbation of the identity in an obvious way; cf. the problems in nonlinear elasticity considered in [75] and in fluid mechanics studied in [23]. In these cases the degree for Fredholm operators developed in Section II.5 is applicable, and Theorem II.5.8 yields the same alternative as Theorem II.3.3. (If for physical reasons the mapping F is not everywhere defined, there is a third possibility, that C reaches the boundary of the domain of definition of F.)

III.5.1 An Example (Continued)

The Euler–Lagrange equation (III.2.63) for the variational problem (III.2.62) gives rise to a bifurcation problem $F_\varepsilon(v, m) = 0$; cf. (III.2.66), where the mass m serves as a bifurcation parameter. The local bifurcation diagram for fixed small $\varepsilon > 0$ is sketched in Figure III.2.5.

Here we have a one-dimensional kernel of $D_v F_\varepsilon(0, m_n^i), i = 1, 2, \text{cf.} (\text{III}.2.77)$, and the nondegeneracy (III.2.80) implies an odd crossing number of $D_v F_\varepsilon(0, m)$ at $m = m_n^i, i = 1, 2$; cf. (II.4.27).

Due to the nonlocal term, it is simpler to transform $F_\varepsilon(v, m) = 0$ for $(v, m) \in X_n^{2,\alpha} \times \mathbb{R}$ into a compact perturbation of the identity, namely, $v + f_\varepsilon(v, m) = 0$ for $(v, m) \in X_n^\alpha \times \mathbb{R}$, where

$$(\text{III}.5.20) \quad f_\varepsilon(v, m) = -\Delta^{-1} \frac{1}{\varepsilon} \left(W'(m + v) - \int_\Omega W'(m + v) dx \right).$$

Note that $\Delta : X_n^{2,\alpha} \to X_n^\alpha$ is bijective due to the mean value zero; cf. (III.1.35). The mapping $f_\varepsilon : X_n^\alpha \times \mathbb{R} \to X_n^\alpha$ is completely continuous, and by (III.2.80) the family $I + D_v f_\varepsilon(0, m) = I - \frac{1}{\varepsilon} W''(m) \Delta^{-1}$ has an odd crossing number at $m = m_n^i$, $i = 1, 2$, since $D_v F_\varepsilon(0, m) = -\varepsilon \Delta + W''(m) I$ has an odd crossing number at $m = m_n^i$.

The application of Theorem II.3.3 then provides the existence of components $\mathcal{C}_{n,i} \subset X_n^{2,\alpha} \times \mathbb{R}$ in the closure $\mathcal{S}$ of nontrivial solutions of $v + f_\varepsilon(v, m) = 0$ or of $F_\varepsilon(v, m) = 0$ emanating at $(0, m_n^i)$, $i = 1, 2$. Since $\varepsilon > 0$ is fixed in that analysis, we suppress the dependence on ε in the notation $\mathcal{C}_{n,i}$.

As remarked after (III.2.82), the local bifurcating curve $\{(v(s), m(s)) | s \in (-\delta, \delta)\}$ through $(v(0), m(0)) = (0, m_n^i)$ decomposes into two components $\{(v(s), m(s)) | s \in [0, \delta)\}$ and $\{(v(s), m(s)) | s \in (-\delta, 0]\}$, which are transformed into each other by the reversion (III.2.82). These are the local components defined in (II.5.39), which are globally extended to $\mathcal{C}_{n,i}^+$ and $\mathcal{C}_{n,i}^-$; cf. (II.5.40). Since those global components are transformed into each other by the reversion (III.2.82), too, we do not need Theorem II.5.9 in order to state the following for $i = 1, 2$:

(III.5.21)

$$\mathcal{C}_{n,i} = \mathcal{C}_{n,i}^+ \cup \mathcal{C}_{n,i}^- \cup \{(0, m_n^i)\},$$

and each of $\mathcal{C}_{n,i}^+$ and $\mathcal{C}_{n,i}^-$ is subject
to alternative (i) or (ii) of
Theorem III.5.1.

In Section III.6 we show that $\mathcal{C}_{n,i}$ is bounded for $i = 1, 2$, and that $\mathcal{C}_{n,i}$ each connect $(0, m_n^1)$, $(0, m_n^2)$. Therefore, $\mathcal{C}_{n,1} = \mathcal{C}_{n,2}$, and by definition (II.5.40), $\mathcal{C}_{n,i}^+ = \mathcal{C}_{n,i}^-$ for $i = 1, 2$. (Note that $\mathcal{C}_{n,1}^+$ is connected to $\mathcal{C}_{n,1}^-$ through $(0, m_n^2)$.)

III.5.2 Global Continuation

Solutions (u_0, λ_0) of a quasilinear elliptic problem (III.5.1) are possibly locally continued by the Implicit Function Theorem (Theorem I.1.1) and globally extended via the Global Implicit Function Theorem (Theorem II.6.1). We give the details.

As usual, $X = C^{2,\alpha}(\overline{\Omega}) \cap \{u | u = 0 \text{ on } \partial\Omega\}$, $Y = C^{1,\alpha}(\overline{\Omega})$, and $Z = C^{\alpha}(\overline{\Omega})$.

Let $(u_0, \lambda_0) \in X \times \mathbb{R}$ be a solution of (III.5.1), and assuming $g(\mathbf{0}, 0, x, 0) = 0$ for all $x \in \Omega$, we have w.l.o.g. $(u_0, \lambda_0) = (0, 0)$. Assume that $F : X \times \mathbb{R} \to Z$ as given in (III.5.4)$_1$ is continuous and has the regularity (I.1.3) of the local Implicit Function Theorem. If a_{ij}, g, and their partial derivatives with respect to $(\mathbf{v}, u, x) \in \mathbb{R}^n \times \mathbb{R} \times \overline{\Omega}$ are globally continuous, then $F \in C(X \times \mathbb{R}, Z)$; cf. (III.5.3). If in addition, $g(\cdot, \cdot, \cdot, 0) \in C^2(V \times U \times \overline{\Omega})$ for a neighborhood $V \times U$ of $(\mathbf{0}, 0) \in \mathbb{R}^n \times \mathbb{R}$, then $D_u F(0, 0) \in L(X, Z)$ exists. If, moreover, $a_{ij}, g \in C((-\delta, \delta), C^2(\overline{V} \times \overline{U} \times \overline{\Omega}))$, then F has the regularity (I.1.3) of the local Implicit Function Theorem near $(u_0, \lambda_0) = (0, 0)$. (The latter means that the mappings $\lambda \mapsto a_{ij}(\cdot, \cdot, \cdot, \lambda)$ and $\lambda \mapsto g(\cdot, \cdot, \cdot, \lambda)$ are continuous from $(-\delta, \delta)$ into the Banach space $C^2(\overline{V} \times \overline{U} \times \overline{\Omega})$.) Note, however, that according to Remark II.6.2, assumption (II.6.2) can be reduced to (II.6.2)$_1$. In other words, the last assumptions on a_{ij} and g are redundant.

Convert $F(u, \lambda) = 0$ equivalently into the compact perturbation of the identity $u + f(u, \lambda)$ defined in (III.5.16). Then the crucial assumption for the application of Theorem II.6.1 is the following:

(III.5.22)

$$D_u F(0, 0) \in L(X, Z) \text{ is bijective} \Leftrightarrow$$

$$I + D_u f(0, 0) \in L(Y, Y) \text{ is bijective.}$$

Observe that $D_u f(0, 0) \in L(Y, Y)$ exists if $D_u F(0, 0) \in L(X, Z)$ exists. In view of the ellipticity (III.5.2), conditions (III.5.22) are satisfied if

(III.5.23)

0 is not an eigenvalue of
$$D_u F(0, 0) = L(0, 0) + D_u \hat{g}(0, 0); \text{ cf. (III.5.5).}$$

Apply the Leray–Schauder degree to the compact perturbation of the identity (III.5.16) and observe that connectedness and unboundedness of solution sets $\{(u, \lambda)\} \subset X \times \mathbb{R}$ of $F(u, \lambda) = 0$ are equivalent to connectedness and unboundedness of solution sets $\{(u, \lambda)\} \subset Y \times \mathbb{R}$ of $u + f(u, \lambda) = 0$. If $F \in C^2(X \times \mathbb{R}, Z)$, cf. (III.5.4), then apply the degree for Fredholm operators to F. *In any case, we obtain the conclusion of Theorem II.6.1 for the solution set of $F(u, \lambda) = 0$ in $X \times \mathbb{R}$.* Note that $\mathcal{C} \backslash \{(0, 0)\} \neq \emptyset$ also if the regularity of the local Implicit Function Theorem does not hold; cf. Remark II.6.2.

In order to eliminate the second alternative of Theorem II.6.1, we assume that

(III.5.24) $L(u, 0)u + \hat{g}(u, 0) = 0$ implies $u = 0$.

In this case, the hyperplane $X \times \{0\} \subset X \times \mathbb{R}$ contains only the solution $(u_0, \lambda_0) = (0, 0)$, so that the set $\mathcal{C} \backslash \{(0, 0)\}$ is not connected. Therefore, the components $\mathcal{C}^+, \mathcal{C}^-$ of solutions of $F(u, \lambda) = L(u, \lambda)u + \hat{g}(u, \lambda) = 0$ in $X \times \mathbb{R}$ are each unbounded. An a priori estimate like

(III.5.25)
$$L(u, \lambda)u + \hat{g}(u, \lambda) = 0 \Rightarrow$$
$$\|u\|_{2,\alpha} \leq C(\lambda) \text{ for all } \lambda \in \mathbb{R}$$

implies the existence of solutions $(u, \lambda) \in \mathcal{C}$ for all $\lambda \in \mathbb{R}$. In Section III.6 we give explicit conditions such that (III.5.24) holds; cf. Theorem III.6.6. In Section III.7 we investigate the asymptotic behavior of $\mathcal{C}$ for a special problem: If the nonlinearity has a sublinear growth at infinity, then (III.5.25) holds.

Remark III.5.3 *We consider global bifurcation and global continuation of classical solutions of (III.5.1) under strong assumptions on the regularity of the data, i.e., on the mapping G and on the domain Ω. For general bounded domains Ω and less-smooth functions G defining a semilinear elliptic problem of type (III.2.10), the notion of a weak (or generalized) solution is still possible. That definition, however, is apparently not adequate for a global bifurcation or a global continuation analysis. In [74] we overcome this problem by a convenient formulation. As a matter of fact, a weak solution $(u, \lambda) \in C(\overline{\Omega}) \times \mathbb{R}$ solves some problem $u + f(u, \lambda) = 0$, where $f : C(\overline{\Omega}) \times \mathbb{R} \to C(\overline{\Omega})$ is completely continuous. This allows us to use all techniques for local and global analysis expounded in Chapters I and II. The conditions on the boundary $\partial\Omega$ are very general: It suffices that every point of $\partial\Omega$ satisfy an exterior cone condition. Moreover, it is shown in [74] that a positivity of weak solutions is globally preserved. We return to this in the next section.*

III.6 Preservation of Nodal Structure on Global Branches

The global branches (continua) of quasilinear elliptic problems (III.5.1) emanating at $(0, \lambda_0)$, where the linearized problem has an eigenvalue zero, are unbounded or meet the trivial solution line at a different $(0, \lambda_1)$; cf. Theorem III.5.1. In this section we show how to eliminate one alternative in order to get much more insight into the global solution set of (III.5.1).

The essential step is the separation of global branches via the nodal properties of their solutions. This idea was first carried out for nonlinear Sturm–Liouville eigenvalue problems; cf. [26]: The number of nodes (zeros) of a solution in the underlying interval is preserved along each nontrivial branch. This number serves as a "label" for each branch, so that the branches are separated and unbounded.

The extension to two or more independent variables seems straightforward: One has to prove a preservation of nodal patterns. However, it is not at all clear how this can be done in general. As an illustration, consider the well-known linear eigenvalue problem characterizing the frequencies and normal modes of a square membrane. By simple superposition, the number and arrangement of nodal lines along a solution sheet emanating from a repeated eigenvalue can vary drastically. Consequently, there seems to be little hope of attaining results comparable to those of [26] for general quasilinear elliptic problems. However, when the symmetry of a problem fixes the location of particular nodal lines, then we can show that a minimal frozen nodal pattern is indeed globally preserved.

Pick one such nodal domain: The essential step is to prove that positivity of a solution in that domain is preserved under perturbation. This is usually shown by an application of the elliptic minimum (maximum) principle and the Hopf boundary lemma. This, in turn, requires a smooth boundary (or moderate corners for more refined versions). Since we plan to consider problems on polygonal domains in a lattice (see Figures III.1.1–III.1.3), we present a maximum principle suitable to our purpose but that does not require any regularity of the boundary.

III.6.1 A Maximum Principle

Let $\Omega \subset \mathbb{R}^n$ be any bounded domain and let

$$(III.6.1) \qquad Lu = \sum_{i,j=1}^{n} a_{ij}(x) u_{x_i x_j} + \sum_{i=1}^{n} b_i(x) u_{x_i} + c(x)u$$

be an elliptic operator over Ω in the sense of (III.1.1). The coefficients satisfy $a_{ij} \in C^1(\overline{\Omega})$, $b_i, c \in C(\overline{\Omega})$.

Theorem III.6.1 *Suppose that* $v \in C^2(\Omega) \cap C(\overline{\Omega}) \cap H_0^1(\Omega)$ *satisfies*

$$(III.6.2) \qquad\qquad (Lv)(x) \leq 0 \quad \text{for all} \quad x \in \Omega.$$

Then there exists a number $\gamma > 0$*, depending only on the (maximum) norms of* $a_{ij} \in C^1(\overline{\Omega})$*,* $b_i, c \in C(\overline{\Omega})$*, and on the ellipticity of* L*, such that*

$$(III.6.3) \qquad \begin{array}{c} \text{meas}\{x \in \Omega | v(x) < 0\} < \gamma \\ \text{implies } v \equiv 0 \text{ or } v(x) > 0 \text{ for all } x \in \Omega. \end{array}$$

Proof. Let $\mathcal{D} = \{x \in \Omega | v(x) < 0\}$ and we assume that $\mathcal{D} \neq \emptyset$. Then $v \in H_0^1(\mathcal{D})$, since $v = 0$ on $\partial \mathcal{D}$ (cf. the appendix of [70], for example). By Poincaré's inequality (cf. [56]),

$$(III.6.4) \qquad \begin{array}{c} \|v\|_{L^2(\mathcal{D})} \leq \left(\dfrac{|\mathcal{D}|}{\omega_n}\right)^{1/n} \|\nabla v\|_{L^2(\mathcal{D})}, \\[2mm] \text{where } \omega_n \text{ denotes the volume of the unit} \\ \text{ball in } \mathbb{R}^n \text{ and } |\mathcal{D}| = \text{meas}\mathcal{D}. \end{array}$$

Since $v < 0$ in $\mathcal{D}$, it follows from the hypothesis (III.6.2) that

$$(III.6.5) \qquad\qquad (Lv, v)_{L^2(\mathcal{D})} \geq 0,$$

and after integration by parts,

$$(III.6.6) \qquad \begin{array}{c} \displaystyle\int_{\mathcal{D}} \sum_{i,j=1}^n a_{ij}(x) v_{x_i} v_{x_j} dx \leq \int_{\mathcal{D}} \sum_{i=1}^n \hat{b}_i(x) v_{x_i} v + c(x) v^2 dx, \\[2mm] \text{where } \hat{b}_i = b_i - \sum_{j=1}^n \frac{\partial}{\partial x_j} a_{ij}. \end{array}$$

Then ellipticity, uniform estimates of the coefficients $\hat{b}_i$ and c, and the inequality $ab \leq (\varepsilon/2)a^2 + (1/2\varepsilon)b^2$ deliver

$$(III.6.7) \qquad d\|\nabla v\|_{L^2(\mathcal{D})}^2 \leq \varepsilon c_1 \|\nabla v\|_{L^2(\mathcal{D})}^2 + \left(\frac{c_1}{\varepsilon} + c_2\right) \|v\|_{L^2(\mathcal{D})}^2.$$

Choosing $\varepsilon = d/2c_1$ yields

$$(III.6.8) \qquad\qquad \|\nabla v\|_{L^2(\mathcal{D})} \leq C\|v\|_{L^2(\mathcal{D})}$$

with a constant $C > 0$ depending only on the coefficients and on the ellipticity of L. In view of (III.6.4), we obtain, for $|\mathcal{D}| > 0$ and $v < 0$ on $\mathcal{D}$,

$$(III.6.9) \qquad\qquad \left(\frac{\omega_n}{|\mathcal{D}|}\right)^{1/n} \leq C \text{ or } |\mathcal{D}| \geq \omega_n C^{-n} \equiv \gamma.$$

Hence, if $|\mathcal{D}| < \gamma$, then $\mathcal{D} = \emptyset$ and $v \geq 0$ on Ω. But then for $c^- = \min\{c, 0\} \leq 0, c^- \in C(\overline{\Omega})$,

$$(\text{III.6.10}) \qquad \begin{aligned} &\sum_{i,j=1}^{n} a_{ij}(x) v_{x_i} v_{x_j} + \sum_{i=1}^{n} \hat{b}_i(x) v_{x_i} + c^-(x) v \\ &\le L v \le 0 \text{ on } \Omega \text{ by (III.6.2)}. \end{aligned}$$

By the minimum principle in [56] either $v \equiv 0$ or $v(x) > 0$ for all $x \in \Omega$.   □

III.6.2 Global Branches of Positive Solutions on a Domain or on a Lattice

We consider global branches of quasilinear elliptic problems

$$G(\nabla^2 u, \nabla u, u, x, \lambda) \equiv$$

$$\sum_{i,j=1}^{n} a_{ij}(\nabla u, u, x, \lambda) u_{x_i x_j} + g(\nabla u, u, x, \lambda) = 0 \text{ in } \Omega,$$
$$u = 0 \text{ on } \partial\Omega,$$

(III.6.11)   written as (cf. (III.5.3)–(III.5.4))

$$L(u, \lambda) u + \hat{g}(u, \lambda) \equiv F(u, \lambda) = 0$$

for $(u, \lambda) \in X \times \mathbb{R}$, $X = C^{2,\alpha}(\overline{\Omega}) \cap \{u | u = 0 \text{ on } \partial\Omega\}$.

As usual, we consider $F : X \times \mathbb{R} \to Z$, where $Z = C^{\alpha}(\overline{\Omega})$. Clearly, if $g(0, 0, x, \lambda) = 0$ for all $x \in \Omega$ and $\lambda \in \mathbb{R}$, we have the trivial solution line $\{(0, \lambda)\}$. The conditions on local and global bifurcation are given in Sections III.2 and III.5. Here we do not prove existence, but we focus on the qualitative properties of a global branch emanating at $(0, \lambda_0)$. Necessarily, 0 is an eigenvalue of $D_u F(0, \lambda_0) = L(0, \lambda_0) + D_u \hat{g}(0, \lambda_0)$, and we assume that $N(D_u F(0, \lambda_0))$ is spanned by some positive function $\hat{v}_0$. There are two ways of dealing with this situation: In the *first case*,

(III.6.12)
$$\begin{aligned} &\Omega \subset \mathbb{R}^n \text{ is a bounded domain} \\ &\text{with a smooth boundary } \partial\Omega, \text{ and} \\ &0 \text{ is the } principal\ eigenvalue \\ &\text{of the elliptic operator } D_u F(0, \lambda_0). \end{aligned}$$

The existence of a principal eigenvalue is guaranteed either by the Krein–Rutman Theorem or by variational principles (going back to Courant, Fischer, Hilbert, and Weyl) discussed in Remark III.2.1. In any case, a principal eigenvalue 0 is (algebraically) simple.

The *second case* is described as follows: Let $\mathcal{L} \subset \mathbb{R}^2$ be a rectangular, square, or hexagonal lattice as shown in Figures III.1.1–III.1.3. To stay with the nomenclature of Section III.1, let

$$X_D = \{u : \mathbb{R}^2 \to \mathbb{R} | u \text{ has the double periodicity of } \mathcal{L}$$

(III.6.13)   and u has an inverse reflection symmetry

across each line of the lattice $\mathcal{L}\}.$

For the lattices shown in Figure III.1.1, the function spaces X_D are given in (III.1.31) and (III.1.36). The double periodicity of the hexagonal lattice is $u(x_1 + a, x_2) = u(x_1, x_2) = u(x_1 + \frac{1}{2}a, x_2 + \frac{1}{2}\sqrt{3}a)$ for all $(x_1, x_2) \in \mathbb{R}^2$. The inverse reflection symmetries across the lines shown in Figure III.1.2 and Figure III.1.3 can be written down explicitly, too, but we leave it to the reader. (The reflections are all of the form $(x_1, x_2) \mapsto \mathbf{R}(x_1, x_2) + (a_1, a_2)$, where the reflection $\mathbf{R} \in O(2)$ and the translation by (a_1, a_2) leave the respective lattice $\mathcal{L}$ invariant.) We assume that the quasilinear elliptic operator (III.6.11) has the following "equivariance":

(III.6.14) $F : (C^{2,\alpha}(\mathbb{R}^2) \cap X_D) \times \mathbb{R} \to C^{\alpha}(\mathbb{R}^2) \cap X_D.$

A sufficient condition for this is, for instance, that

$$G(\nabla^2, u, \nabla u, u, x, \lambda) \equiv$$

$$\nabla \cdot \mathbf{q}(\nabla u, u, \lambda) + p(\nabla u, u, \lambda), \text{ where}$$

$$\mathbf{q} : \mathbb{R}^2 \times \mathbb{R} \times \mathbb{R} \to \mathbb{R}^2, \ p : \mathbb{R}^2 \times \mathbb{R} \times \mathbb{R} \to \mathbb{R}$$

are smooth functions such that

(III.6.15)
$$\mathbf{q}(\mathbf{R}\mathbf{v}, u, \lambda) = \mathbf{R}\mathbf{q}(\mathbf{v}, u, \lambda),$$
$$\mathbf{q}(\mathbf{v}, -u, \lambda) = \mathbf{q}(\mathbf{v}, u, \lambda),$$
$$p(\mathbf{R}\mathbf{v}, u, \lambda) = p(\mathbf{v}, u, \lambda),$$
$$p(\mathbf{v}, -u, \lambda) = -p(\mathbf{v}, u, \lambda),$$

for all reflections $\mathbf{R} \in O(2)$ that
leave the lattice $\mathcal{L}$ invariant,
and for all $(\mathbf{v}, u, \lambda) \in \mathbb{R}^2 \times \mathbb{R} \times \mathbb{R}.$

In the more general quasilinear case (III.6.11), sufficient conditions for equivariance are readily obtained by differentiations of (III.6.15) via

(III.6.16)
$$a_{ij}(\mathbf{v}, u, \lambda) = \frac{\partial q_i}{\partial v_j}(\mathbf{v}, u, \lambda), \ i, j = 1, 2,$$

$$g(\mathbf{v}, u, \lambda) = \sum_{i=1}^{2} \frac{\partial q_i}{\partial u}(\mathbf{v}, u, \lambda)v_i + p(\mathbf{v}, u, \lambda).$$

We verify the hypotheses (II.5.32) of Theorem II.5.8; i.e., we verify Definition II.5.2 for F as defined in (III.6.11) and acting as in (III.6.14). If the coefficients a_{ij} and g of (III.6.11) are C^3-functions, the operator F of (III.6.14) is of class C^2. The equivariance (III.6.14) implies that

(III.6.17)

$$D_u F(u, \lambda) : C^{2,\alpha}(\mathbb{R}^2) \cap X_D \to C^\alpha(\mathbb{R}^2) \cap X_D,$$

the representation (III.5.4) shows that it is elliptic, and the results of Section III.1, in particular (III.1.33), prove that it is a Fredholm operator of index zero.

To verify the spectral condition (iii) of Definition II.5.1, it suffices to show that the spectrum of $-D_u F(u, \lambda)$ is bounded from below in the following sense: Let Ω be a tile in the lattice $\mathcal{L}$. If $v \in C^{2,\alpha}(\mathbb{R}^2) \cap X_D$, then $v \in C^{2,\alpha}(\overline{\Omega}) \cap \{u | u = 0$ on $\partial\Omega\}$, and for an eigenvalue $\mu \in \mathbb{C}$, we can conclude that

$$-D_u F(u, \lambda) v = \mu v \ (\text{cf. (III.5.4)});$$

$$d \|\nabla v\|_{L^2(\Omega)}^2 \leq \varepsilon c_1 \|\nabla v\|_{L^2(\Omega)}^2 + \left(\frac{c_1}{\varepsilon} + c_2 + \text{Re}\mu\right) \|v\|_{L^2(\Omega)}^2,$$

(III.6.18)

by ellipticity (III.5.2) and estimates as in (III.6.7),

$$-C\|\nabla v\|_{L^2(\Omega)}^2 \leq \text{Re}\mu \|\nabla v\|_{L^2(\Omega)}^2,$$

by Poincaré's inequality; whence $\text{Re}\mu \geq -C$.

The constant $C > 0$ depends only on the ellipticity $d = d(\mathbf{v}, u, \lambda) > 0$, on the coefficients of (III.5.4), and on the tile Ω. By (III.1.38) (which is valid also in this case), a strip $(-\infty, c) \times (-i\varepsilon, i\varepsilon) \subset \mathbb{C}$ contains at most finitely many eigenvalues of finite algebraic multiplicity, and their total number in that strip is stable under small perturbations of $-D_u F(u, \lambda)$. (The (locally) uniform bound $\text{Re}\mu \geq -C$ is a weaker statement than the sectorial property of $-D_u F(u, \lambda)$ as stated in (III.4.9), (III.4.10). This property was used to prove admissibility of $-D_u F(u, \lambda)$ in Section III.5.) The last property, properness, is verified as in (III.5.7)–(III.5.9), since $C^{2,\alpha}(\mathbb{R}^2) \cap X_D$ is compactly embedded into $C^{1,\alpha}(\mathbb{R}^2)$ (by double periodicity).

To summarize, the equivariant quasilinear elliptic problem $F(u, \lambda) = 0$ on the lattice $\mathcal{L}$, where the operator F as given by (III.6.11) acts as in (III.6.14), allows a global bifurcation analysis as stated in Theorem III.5.1 with $X = C^{2,\alpha}(\mathbb{R}^2) \cap X_D$. The crucial assumption is an odd crossing number of the family $D_u F(0, \lambda) = L(0, \lambda) + D_u \hat{g}(0, \lambda)$ at some $\lambda = \lambda_0$.

The subject matter of this section is the bifurcation of positive solutions. As in (III.6.12), this requires that the eigenspace of the eigenvalue 0 of $D_u F(0, \lambda_0)$ be spanned by some positive function. What does this mean on a lattice?

We consider $D_u F(0, \lambda_0) : Z \to Z$ with domain $X = C^{2,\alpha}(\mathbb{R}^2) \cap X_D$ and $Z = C^\alpha(\mathbb{R}^2) \cap X_D$. By the definition of X_D in (III.6.13),

(III.6.19)

all lines in the lattice $\mathcal{L}$ are in the nodal set of all $u \in C^{2,\alpha}(\mathbb{R}^2) \cap X_D$.

Now we are ready to describe the *second case*:

(III.6.20)
Let Ω be a fixed tile in the lattice $\mathcal{L}$
and let 0 be an eigenvalue of the
equivariant and elliptic operator $D_u F(0, \lambda_0)$
with an eigenfunction $\hat{v}_0 \in C^{2,\alpha}(\mathbb{R}^2) \cap X_D$
that is *positive on* Ω.

By the inverse reflection symmetries defining X_D, the eigenfunction $\hat{v}_0$ is positive or negative on all tiles of the lattice $\mathcal{L}$, and the lines of the lattice $\mathcal{L}$ are the precise nodal set of the eigenfunction $\hat{v}_0$.

For the family

$$(III.6.21) \qquad D_u F(0, \lambda) = \Delta + c(\lambda)I,$$

for instance, the assumption (III.6.20) is satisfied for all lattices $\mathcal{L}$ shown in Figures III.1.1–III.1.3, provided that $\lambda = \lambda_0$ satisfies a characteristic equation $c(\lambda) = \mu_0$, where μ_0 depends explicitly on the lattice $\mathcal{L}$. The precise values of μ_0 and the eigenfunctions $\hat{v}_0$ are given in [70]. In any case,

$$(III.6.22) \qquad \dim N(D_u F(0, \lambda_0)) = 1.$$

The argument for (III.6.22) is the following: The eigenfunction $\hat{v}_0 \in C^{2,\alpha}(\mathbb{R}^2) \cap X_D$ satisfies

$$(III.6.23) \qquad \begin{aligned} D_u F(0, \lambda_0)\hat{v}_0 &= 0 \quad \text{in} \quad \Omega, \\ \hat{v}_0 &= 0 \quad \text{on} \quad \partial\Omega, \\ \hat{v}_0 &> 0 \quad \text{in} \quad \Omega. \end{aligned}$$

Since in all cases the elliptic operator $D_u F(0, \lambda_0)$ is formally self-adjoint (cf. (III.6.21)), the Rayleigh quotient $B(v, v)/\|v\|_0^2$ over Ω is maximized by a function $\hat{v}_1 \in H_0^1(\Omega)$ and therefore also by $|\hat{v}_1| \in H_0^1(\Omega)$, which is positive in Ω; cf. (III.1.2) and Remark III.2.1. The last statement follows from interior regularity and the minimum principle (III.6.10). The maximal value of the quotient is the principal eigenvalue of $D_u F(0, \lambda_0)$ over Ω (in its weak formulation), and if it were not 0, then the positive eigenfunctions $\hat{v}_0$ and $\hat{v}_1$ would be orthogonal in $L^2(\Omega)$, a contradiction. Therefore, $\hat{v}_0 = \hat{v}_1$, and 0 is the principal eigenvalue of $D_u F(0, \lambda_0)$ over Ω, which is simple. This argument is valid regardless of the corners of the boundary $\partial\Omega$.

In view of assumption (III.6.12) in the first case, (III.6.22) holds in both cases, where the eigenfunction $\hat{v}_0$ spanning $N(D_u F(0, \lambda_0))$ satisfies (III.6.23). The conditions for an odd crossing number at $\lambda = \lambda_0$ in case of (III.6.22) are given in Section III.2; for (III.6.21) it simply means that λ_0 solves a characteristic equation $c(\lambda) = \mu_0$ and that $c(\lambda)$ is strictly monotonic near $\lambda = \lambda_0$.

The main result of this section is the following:

Theorem III.6.2 *Assume that 0 is a principal eigenvalue of $D_u F(0, \lambda_0) = L(0, \lambda_0) + D_u \hat{g}(0, \lambda_0)$ in the sense of (III.6.12) or (III.6.20). Let C be the* *global continuum in the closure $\mathcal{S}$ of nontrivial solutions of $F(u, \lambda) = 0$ emanating at $(0, \lambda_0)$. Let $\mathcal{C}_{\lambda_0} \subset \mathcal{C}\backslash(\{0\} \times \mathbb{R})$ denote all components whose closures contain $(0, \lambda_0)$. If $(u, \lambda) \in \mathcal{C}_{\lambda_0}$, then either $u > 0$ or $u < 0$ in Ω, where Ω is a bounded domain in $\mathbb{R}^n$ with a smooth boundary $\partial\Omega$, or $\Omega \subset \mathbb{R}^2$ is a tile in the lattice $\mathcal{L}$. This means that for all $(u, \lambda) \in \mathcal{C}_{\lambda_0}$, the lines of the lattice $\mathcal{L}$ constitute the precise nodal set of the function u. All lattices shown in Figures III.1.1–III.1.3 are admitted; cf. also Figure III.6.1.*

If $\mathcal{C} \cap (\{0\} \times \mathbb{R}) = \{(0, \lambda_0)\}$, then $\mathcal{C} = \mathcal{C}_{\lambda_0} \cup \{(0, \lambda_0)\}$, so that the alternatives (i) and (ii) of Theorem III.5.1 hold for at least one component of $\mathcal{C}_{\lambda_0}$. It might happen, however, that some $(0, \lambda_1)$ is in $\overline{\mathcal{C}}_{\lambda_0}$, where $\lambda_1 \neq \lambda_0$, and that some branch emanates at $(0, \lambda_1)$ that is in $\mathcal{C}\backslash(\{0\} \times \mathbb{R})$ but not in $\mathcal{C}_{\lambda_0}$. As we see in Section III.6.3 below, the assumptions (III.6.22), (III.6.23) are then also satisfied for $\lambda = \lambda_1$, and the claim of Theorem III.6.2 holds for $\mathcal{C}_{\lambda_1}$ as well. In this way, we cover all of $\mathcal{C}\backslash(\{0\} \times \mathbb{R})$, and we do not lose information about $\mathcal{C}$ when studying only $\mathcal{C}_{\lambda_0}$.

Proof of Theorem III.6.2: Suppose that $(u, \lambda) \in \mathcal{C}_{\lambda_0}$. Then $h = u \in X = C^{2,\alpha}(\overline{\Omega}) \cap \{u | u = 0 \text{ on } \partial\Omega\}$ is a solution of the **linear** elliptic problem

$$(\text{III.6.24}) \qquad\qquad \tilde{L}(u, \lambda)h = 0,$$

where we define

$$\tilde{L}(u, \lambda)h \equiv$$
$$\sum_{i,j=1}^{n} a_{ij}(\nabla u, u, x, \lambda)h_{x_i x_j} + \sum_{i=1}^{n} \tilde{b}_i(./.)h_{x_i} + \tilde{c}(./.)h,$$

$$(\text{III.6.25}) \quad \tilde{b}_i(\mathbf{v}, u, x, \lambda) = \int_0^1 g_{v_i}(t\mathbf{v}, tu, x, \lambda)dt,$$

$$\tilde{c}(\mathbf{v}, u, x, \lambda) \quad = \int_0^1 g_u(t\mathbf{v}, tu, x, \lambda)dt,$$

$$\text{for } (\mathbf{v}, u, x, \lambda) \in \mathbb{R}^n \times \mathbb{R} \times \Omega \times \mathbb{R}.$$

We establish the claim of Theorem III.6.2 first for local bifurcating solutions. By construction via the method of Lyapunov–Schmidt, all $(u, \lambda) \in \mathcal{C}_{\lambda_0}$ in a small neighborhood of $(0, \lambda_0)$ are of the form

$$(\text{III.6.26}) \qquad\qquad u = s\hat{v}_0 + o(s) \quad \text{in } X \text{ as } s \to 0;$$

cf. Corollary I.2.4 (and (II.5.37)). Since $\hat{v}_0 > 0$ in Ω,

$$(\text{III.6.27}) \qquad\qquad \text{meas}\{x \in \Omega | (u/s)(x) < 0\} \to 0 \text{ as } s \to 0.$$

Since $u/s \in C^{2,\alpha}(\overline{\Omega}) \cap \{u | u = 0 \text{ on } \partial\Omega\}$ and $\tilde{L}(u, \lambda)(u/s) = 0$ in Ω, Theorem III.6.1 implies that $u/s > 0$ in Ω for small $|s|$. Thus $u > 0$ for $s \in (0, \delta)$ and $u < 0$ for $s \in (-\delta, 0)$. (We do not claim that $\mathcal{C}_{\lambda_0}$ consists locally of two curves or that $\mathcal{C}_{\lambda_0}$ has only two components.)

Defining the cones

(III.6.28)
$$K^+ = \{(u, \lambda) \in X \times \mathbb{R} | u > 0 \text{ in } \Omega\},$$
$$K^- = \{(u, \lambda) \in X \times \mathbb{R} | u < 0 \text{ in } \Omega\},$$

then we see that the preceding analysis shows that

(III.6.29)
$$\mathcal{C}_{\lambda_0} \cap (B_r(0) \times (\lambda_0 - \delta, \lambda_0 + \delta)) \subset K^+ \cup K^- \text{ for } B_r(0) \subset X,$$
provided that $r > 0$ and $\delta > 0$ are sufficiently small.

In particular, $K^\pm \cap \mathcal{C}_{\lambda_0}$ are each nonempty.

We show that $K^\pm \cap \mathcal{C}_{\lambda_0}$ are open relative to $\mathcal{C}_{\lambda_0}$.
Let $(u, \lambda) \in K^+ \cap \mathcal{C}_{\lambda_0}$, i.e., $u > 0$ in Ω, and $(\tilde{u}, \tilde{\lambda}) \in \mathcal{C}_{\lambda_0}$ close to (u, λ) in $X \times \mathbb{R}$. Then

(III.6.30)
$$\text{meas}\{x \in \Omega | \tilde{u}(x) < 0\} \to 0$$
$$\text{as } (\tilde{u}, \tilde{\lambda}) \to (u, \lambda) \text{ in } X \times \mathbb{R}.$$

Since $\tilde{u} \in C^{2,\alpha}(\overline{\Omega}) \cap \{u | u = 0 \text{ on } \partial\Omega\}$ and $\tilde{L}(\tilde{u}, \tilde{\lambda})\tilde{u} = 0$ in Ω, Theorem III.6.1 implies $\tilde{u} > 0$ in Ω or $(\tilde{u}, \tilde{\lambda}) \in K^+ \cap \mathcal{C}_{\lambda_0}$.

We show that $K^\pm \cap \mathcal{C}_{\lambda_0}$ are each closed relative to $\mathcal{C}_{\lambda_0}$.
Let $(u_n, \lambda_n) \in K^+ \cap \mathcal{C}_{\lambda_0}$ be such that $(u_n, \lambda_n) \to (u, \lambda) \in \mathcal{C}_{\lambda_0}$ in $X \times \mathbb{R}$. Since $u_n > 0$ in Ω, $u \geq 0$ in Ω, and $\tilde{L}(u, \lambda)u = 0$ in Ω, the argument of (III.6.10) implies $u > 0$ in Ω. Therefore, the limit (u, λ) is in $K^+ \cap \mathcal{C}_{\lambda_0}$.

Let $\mathcal{C}_{\lambda_0}^\pm$ denote the components of $K^\pm \cap \mathcal{C}_{\lambda_0}$ in $\mathcal{C}_{\lambda_0}$. Then we have shown that $K^\pm \cap \mathcal{C}_{\lambda_0} = \mathcal{C}_{\lambda_0}^\pm$ or

(III.6.31)
$$\mathcal{C}_{\lambda_0}^+ \subset K^+, \ \mathcal{C}_{\lambda_0}^- \subset K^-, \text{ and by (III.6.29)},$$
$$\mathcal{C}_{\lambda_0} = \mathcal{C}_{\lambda_0}^+ \cup \mathcal{C}_{\lambda_0}^-,$$

which proves Theorem III.6.2.   □

We remark that $\mathcal{C}_{\lambda_0}^\pm$ might each consist of more than one component. In contrast to the components $\mathcal{C}^\pm$ of (II.5.40), we have obviously $\mathcal{C}_{\lambda_0}^+ \cap \mathcal{C}_{\lambda_0}^- = \emptyset$.

We apply Theorem II.5.9 to the *first case* (III.6.11), (III.6.12), and the statement after Theorem III.5.1 implies that

(III.6.32)
each of $\mathcal{C}_{\lambda_0}^+$ and $\mathcal{C}_{\lambda_0}^-$ satisfies the
alternatives of Theorem III.5.1; i.e.,
it is unbounded or meets $\{0\} \times \mathbb{R}$ at
some $(0, \lambda_1)$ where $\lambda_0 \neq \lambda_1$.

In the *second case* (III.6.14), (III.6.20), the operator $F(\cdot, \lambda)$ in (III.6.14) is odd according to (III.6.15). Therefore, we can apply Corollary II.5.10 directly without converting the problem to a compact perturbation of the identity. In particular,

(III.6.33)
$$\mathcal{C}_{\lambda_0}^- = \{(u,\lambda)|(-u,\lambda) \in \mathcal{C}_{\lambda_0}^+\}, \text{ and}$$
each of $\mathcal{C}_{\lambda_0}^+$ and $\mathcal{C}_{\lambda_0}^-$ satisfies the alternatives
of Theorem III.5.1.

The nodal set of any solution u of the quasilinear elliptic problem (III.6.11) such that $(u,\lambda) \in \mathcal{C}_{\lambda_0}^+ \cup \mathcal{C}_{\lambda_0}^-$ is precisely the same as the nodal set of the eigenfunction $\hat{v}_0$, namely the lines of the lattice $\mathcal{L}$.

III.6.3 Unbounded Branches of Positive Solutions

The positivity (negativity) on Ω along $\mathcal{C}_{\lambda_0}^+(\mathcal{C}_{\lambda_0}^-)$ serves as a useful criterion for eliminating one alternative in (III.6.32) or (III.6.33). Assume that $\mathcal{C}_{\lambda_0}^+$ meets some $(0,\lambda_1)$ where $\lambda_1 \neq \lambda_0$. Then there exists a sequence $(u_n,\lambda_n) \in \mathcal{C}_{\lambda_0}^+$ such that $u_n \to 0$ in X and $\lambda_n \to \lambda_1$ in $\mathbb{R}$. (Recall that $X = C^{2,\alpha}(\overline{\Omega}) \cap \{u|u = 0$ on $\partial\Omega\}$ or $X = C^{2,\alpha}(\mathbb{R}^2) \cap X_D$ in the first or second case, respectively.) Then

$$v_n = u_n/\|u_n\|_{0,\alpha} \in X \text{ solve}$$

(III.6.34)
$$\tilde{L}(u_n,\lambda_n)v_n = 0, \text{ cf. (III.6.24), (III.6.25),}$$

whence $\|v_n\|_{2,\alpha} \leq \tilde{C}$ for all $n \in \mathbb{N}$,

by elliptic a priori estimates, since the coefficients $a_{ij}, \tilde{b}_i, \tilde{c}$ of the operator $\tilde{L}(u_n,\lambda_n)$ are uniformly bounded in $C^{1,\alpha}(\overline{\Omega})$ and $C^{\alpha}(\overline{\Omega})$ for all $n \in \mathbb{N}$. (Note that the elliptic a priori estimates are also valid for $\tilde{L}(u,\lambda) : C^{2,\alpha}(\mathbb{R}^2) \cap X_D \to C^{\alpha}(\mathbb{R}^2) \cap X_D$ as expounded in Section III.1 for elliptic operators on a lattice.)

By a compact embedding we can assume without loss of generality that

$$v_n \to \hat{v}_1 \text{ in } C^{1,\alpha}(\overline{\Omega}), \text{ whence}$$

(III.6.35)
$$L(u_n,\lambda_n)v_n = -\sum_{i=1}^n \tilde{b}_i v_{nx_i} - \tilde{c}v_n \text{ converges to}$$

$$-D_u\hat{g}(0,\lambda_1)\hat{v}_1 \text{ in } C^{\alpha}(\overline{\Omega}) \text{ as } n \to \infty,$$

by definitions (III.6.11) and (III.6.25). On the other hand,

(III.6.36)
$$\|(L(u_n,\lambda_n) - L(0,\lambda_1))v_n\|_{0,\alpha} \leq \varepsilon\|v_n\|_{2,\alpha} \leq \varepsilon\tilde{C}$$

for all $n \geq n_0(\varepsilon)$. By (III.6.35) and an elliptic a priori estimate, this implies

$$L(0,\lambda_1)v_n \to -D_u\hat{g}(0,\lambda_1)\hat{v}_1 \text{ in } C^{\alpha}(\overline{\Omega}),$$

(III.6.37)
$$v_n \to \hat{v}_1 \text{ in } C^{2,\alpha}(\overline{\Omega}), \ \hat{v}_1 \in X, \ \|\hat{v}_1\|_{0,\alpha} = 1,$$

$$D_uF(0,\lambda_1)\hat{v}_1 = L(0,\lambda_1)\hat{v}_1 + D_u\hat{g}(0,\lambda_1)\hat{v}_1 = 0.$$

Finally, $v_n > 0$ in Ω, since $(u_n, \lambda_n) \in \mathcal{C}^+_{\lambda_0}$, so that $\hat{v}_1 \geq 0$, and by the argument of (III.6.10), $\hat{v}_1 > 0$ in Ω.

To summarize, 0 is an eigenvalue of $D_u F(0, \lambda_1)$ with a positive eigenfunction $\hat{v}_1$. Standard arguments imply $\dim N(D_u F(0, \lambda_1)) = 1$ (cf. our comments after (III.6.23)), and 0 is a principal eigenvalue of $D_u F(0, \lambda_1)$ over the domain $\Omega \subset \mathbb{R}^n$ with smooth boundary $\partial \Omega$, cf. (III.6.12), or 0 is a principal eigenvalue of $D_u F(0, \lambda_1)$ operating on a lattice $\mathcal{L}$; cf. (III.6.20), (III.6.23). In view of (III.6.32) or (III.6.33), this observation implies the following theorem:

Theorem III.6.3 *Assume that* 0 *is a principal eigenvalue of*

$$(\text{III.6.38}) \qquad D_u F(0, \lambda) = L(0, \lambda) + D_u \hat{g}(0, \lambda), \quad \text{cf. (III.6.11)},$$

in the sense of (III.6.12) or (III.6.20) if and only if $\lambda = \lambda_0$. *Then the global continua of positive solutions of (III.6.11),* $\mathcal{C}^+_{\lambda_0}$, *and of negative solutions,* $\mathcal{C}^-_{\lambda_0}$, *are each unbounded in* $C^{2,\alpha}(\overline{\Omega}) \times \mathbb{R}$.

We give an application of Theorem III.6.3. Let $\mathcal{L} \subset \mathbb{R}^2$ be a rectangular, square, or hexagonal lattice, and let the quasilinear operator (III.6.11) be equivariant in the sense of (III.6.14). Let $D_u F(0, \lambda) = \Delta + c(\lambda)I$ with some function

$$(\text{III.6.39}) \qquad \begin{aligned} &c : \mathbb{R} \to (0, \infty) \text{ that is strictly monotonic,} \\ &c(\lambda) \to 0 \text{ as } \lambda \to -\infty, \ c(\lambda) \to \infty \text{ as } \lambda \to \infty. \end{aligned}$$

Then the characteristic equations $c(\lambda) = \mu_0 > 0$, given in [70] for a principal eigenvalue 0 of $D_u F(0, \lambda)$, have a unique solution $\lambda = \lambda_0$ for each lattice $\mathcal{L}$. There is also another peculiarity: Depending on the characteristic periodicities of the lattice $\mathcal{L}$ (which are given by one number $a > 0$ for the square and the hexagonal lattice, and by two numbers $a > 0, b > 0$ for the rectangular lattice), each positive value μ_0 occurs on the right-hand side of the characteristic equations $c(\lambda) = \mu_0$. This means the following:

(III.6.40)
> For a semilinear problem $\Delta u + g(u, \lambda) = 0$,
> where $g(-u, \lambda) = -g(u, \lambda)$ and
> $c(\lambda) = g_u(0, \lambda)$ satisfies (III.6.39),
> each point $(0, \lambda_0)$ gives rise to unbounded
> branches $\mathcal{C}^+_{\lambda_0}$ of solutions having the properties
> that for $(u, \lambda) \in \mathcal{C}^+_{\lambda_0}$ the lines of some
> rectangular or hexagonal lattice $\mathcal{L}$ shown in Figure III.6.1
> are the precise nodal set for u. The periodicities of $\mathcal{L}$
> are in one-to-one correspondence with λ_0.

(Statement (III.6.40) holds for more general quasilinear problems $F(u, \lambda) = 0$ that are equivariant with respect to the lattices $\mathcal{L}$; cf. (III.6.15), (III.6.16). For (III.6.40) only $D_u F(0, \lambda) = \Delta + c(\lambda)I$ is needed.)

We show the nodal sets in Figure III.6.1. There are two more lattices, cf. Figures III.1.2, III.1.3, but these are affine linear images of those shown in Figure III.6.1.

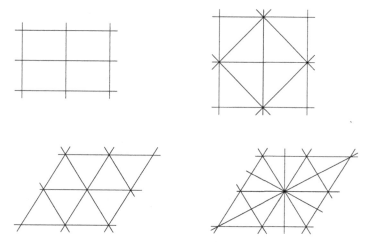

Figure III.6.1

III.6.4 Separation of Branches

The nodal properties labeling each branch are a useful criterion for the separation of branches. This has been carried out in [67] for a quasi-linear elliptic problem over a rectangle $(0, a) \times (0, b)$ subject to homogeneous Dirichlet boundary conditions. Assuming the equivariance of a rectangular lattice, we consider the quasilinear elliptic operator on a rectangular lattice with periodicities $2a$ and $2b$; cf. (III.1.31)$_2$. By (III.6.40), for each $(m, n) \in \mathbb{N} \times \mathbb{N}$ there is a bifurcation point $(0, \lambda_{mn})$ of an unbounded branch $\mathcal{C}^+_{\lambda_{mn}}$, and for $(u, \lambda) \in \mathcal{C}^+_{\lambda_{mn}}$ the lines of a rectangular lattice with periodicities $2a/m$ and $2b/n$ form the precise nodal set of u. Clearly, u satisfies the homogeneous Dirichlet boundary conditions on the boundary of the rectangle $(0, a) \times (0, b)$, so that each unbounded branch $\mathcal{C}^+_{\lambda_{mn}}$ is a solution branch of the quasilinear boundary value problem over the rectangle $(0, a) \times (0, b)$. Having different nodal patterns, all infinitely many branches $\mathcal{C}^+_{\lambda_{mn}}$ are separated. The analogous result holds also for all branches over any triangle in a hexagonal lattice; cf. [70].

The Cahn–Hilliard energy (III.2.62) over a square has many bifurcating global branches of critical points. For that analysis we use the setting $F_\varepsilon(v, m) = 0$, cf. (III.2.66), or equivalently $v + f_\varepsilon(v, m) = 0$, cf. (III.5.20), of the Euler–Lagrange equation (III.2.63). The global branches $\mathcal{C}_{n,i}$ emanate at $(0, m^i_n)$, $i = 1, 2$, and so far, not much more than the Rabinowitz alternative is known about them; cf. (III.5.21). If $(v, m) \in \mathcal{C}_{n,i}$, then v has the double periodicity $2/n$, which, however, would not exclude a union of $\mathcal{C}_{n,i}$ with $\mathcal{C}_{kn,i}$. By a mean value zero, positivity (or negativity) over a tile cannot be used as a label to separate the branches. Below, we continue the analysis of that example, and we introduce a new method for separating the branches $C_{n,i}$ for different n. A side effect of that analysis is that the branches are bounded; i.e., we establish alternative (ii) of Theorem III.5.1.

In Section III.7 we prove that the branches $C_{\lambda_0}^+$ of (III.6.40) for the rectangular and hexagonal lattices on the left of Figure III.6.1 are globally smooth curves parameterized by the amplitude of u, provided that $g(u, \lambda)$ depends linearly on λ, i.e., $g(u, \lambda) = \lambda g(u)$; cf. Remark III.7.18. This statement sharpens considerably all global results of this and the previous sections, since in general, global branches are only continua. For plane symmetric domains Ω with a smooth boundary, Theorem III.7.9 gives conditions for when the global continua $C_{\lambda_0}^{\pm}$ of Theorem III.6.3 are smooth curves parameterized by the amplitude of u.

III.6.5 An Example (Continued)

We continue the analysis of the Euler–Lagrange equation (III.2.63) for the variational problem (III.2.62). In Section III.5 we show that the locally bifurcating curves sketched in Figure III.2.5 are globally extended by $C_{n,i} = C_{n,i}^+ \cup C_{n,i}^- \cup \{(0, m_n^i)\}, i = 1, 2$, cf. (III.5.21), and that each of $C_{n,i}^+$ and $C_{n,i}^-$ is subject to the Rabinowitz alternative. Since both subcontinua are transformed into each other by the reversion (III.2.82), they satisfy the same alternative, but we do not know yet which one. Note that $C_{n,i}^{\pm}$ are not characterized by positivity or negativity, respectively. As a matter of fact, as in seen from their mean value zero, solutions v for $(v, \lambda) \in C_{n,i}^{\pm}$ have no positive or negative sign in the square Ω. Note also that the symmetry class X_n given in (III.2.70) does not separate the branches $C_{n,i}^{\pm}$, since X_n contains the classes X_{kn} for all $k \in \mathbb{N}$.

For the subsequent analysis we adopt the definition of Theorem III.6.2:

(III.6.41) Let $C_{m_n^i} \subset C_{n,i} \backslash (\{0\} \times \mathbb{R})$ denote all components whose closures contain $(0, m_n^i)$, $i = 1, 2$.

Then

(III.6.42) $C_{m_n^i} \cap (B_r(0) \times (m_n^i - \tilde{\delta}, m_n^i + \tilde{\delta})) = C_{m_n^i, loc}^+ \cup C_{m_n^i, loc}^-$,

where $C_{m_n^i, loc}^+ = \{(v(s), m(s)) | s \in (0, \delta)\}$ and $C_{m_n^i, loc}^- = \{(v(s), m(s)) | s \in (-\delta, 0)\}$ are the local solution curves of $F_\varepsilon(v, m) = 0$ through $(v(0), m(0)) = (0, m_n^i)$; cf. (III.2.81). As mentioned previously, since $v(s)$ has no definite sign, the preceding analysis, using positive or negative cones, is not adequate to label the local continua $C_{m_n^i, loc}^{\pm}$ and their global components in $C_{m_n^i}$, denoted by $C_{m_n^i}^{\pm}$. Here is the way out:

The crucial observation is that the partial derivatives of the eigenfunction $\hat{v}_n$, cf. (III.2.69), are negative in the square $\Omega_n = (0, \frac{1}{n}) \times (0, \frac{1}{n})$. We make this more precise now.

If $v \in C^1(\mathbb{R}^2) \cap X_n$, where X_n is defined in (III.2.70), then

$$\frac{\partial}{\partial x_1} v \in X_n^1 = \left\{ w : \mathbb{R}^2 \to \mathbb{R} | w(x_1, x_2) = -w(-x_1, x_2) = \right.$$

$$\left. w(x_1, -x_2) = w\left(x_1 + \frac{2}{n}, x_2\right) = w\left(x_1, x_2 + \frac{2}{n}\right) \right\},$$

(III.6.43)

$$\frac{\partial}{\partial x_2} v \in X_n^2 = \left\{ w : \mathbb{R}^2 \to \mathbb{R} | w(x_1, x_2) = w(-x_1, x_2) = \right.$$

$$\left. - w(x_1, -x_2) = w\left(x_1 + \frac{2}{n}, x_2\right) = w\left(x_1, x_2 + \frac{2}{n}\right) \right\}.$$

In particular,

$$w_j = 0 \text{ for } x_j = \frac{k}{n}, \ k \in \mathbb{Z},$$

(III.6.44)

$$\text{if } w_j \in C(\mathbb{R}^2) \cap X_n^j, \ j = 1, 2.$$

If $(v, m) \in X_n^{2,\alpha} \times \mathbb{R}$ solves $F_\varepsilon(v, m) = 0$, then, by interior elliptic regularity and a "bootstrapping argument," it follows from (III.2.66) for a smooth function W that $v \in C^{3,\alpha}(\mathbb{R}^2)$ and that for $j = 1, 2$,

$$w_j = \frac{\partial}{\partial x_j} v \quad \text{solve}$$

(III.6.45)

$$-\varepsilon \Delta w_j + W''(m + v)w_j = 0 \text{ on } \mathbb{R}^2,$$

$$w_j = 0 \text{ for } x_j = \frac{k}{n}, \ k \in \mathbb{Z}.$$

For $(v, m) \in C_{m_n^i, loc}^{\pm}$ we have by (III.2.81) for $j = 1, 2$,

(III.6.46) $$w_j(s) = s \frac{\partial}{\partial x_j} \hat{v}_n + o(s) \quad \text{in } C^{1,\alpha}(\mathbb{R}^2) \cap X_n^j.$$

From (III.2.69) it follows that

$$\frac{\partial}{\partial x_j} \hat{v}_n < 0 \text{ in the strip } S_n^j = \left\{ 0 < x_j < \frac{1}{n} \right\},$$

(III.6.47) $$\frac{\partial^2}{\partial x_j^2} \hat{v}_n < 0 \text{ for } x_j = 0, \quad \frac{\partial^2}{\partial x_j^2} \hat{v}_n > 0 \text{ for } x_j = \frac{1}{n}, \text{ whence}$$

$$w_j(s) < 0 \text{ in } S_n^j \text{ for } 0 < s < \delta,$$

$$w_j(s) > 0 \text{ in } S_n^j \text{ for } -\delta < s < 0, \ j = 1, 2.$$

We define for $j = 1, 2$,

$$K_{j,n}^- = \left\{ (v, \lambda) \in X_n^{2,\alpha} \times \mathbb{R} \,\middle|\, \frac{\partial}{\partial x_j} v < 0 \text{ in } S_n^j, \right.$$

(III.6.48)
$$\left. \frac{\partial^2}{\partial x_j^2} v < 0 \text{ for } x_j = 0, \ \frac{\partial^2}{\partial x_j^2} v > 0 \text{ for } x_j = \frac{1}{n} \right\},$$

$K_{j,n}^+$ analogously.

Then we have shown in (III.6.47) that

(III.6.49) $\mathcal{C}_{m_n^i, loc}^+ \subset K_{j,n}^-, \quad \mathcal{C}_{m_n^i, loc}^- \subset K_{j,n}^+ \quad$ for $j = 1, 2$.

By the double periodicity of $(v, \lambda) \in K_{j,n}^\pm$,

(III.6.50) $\left| \dfrac{\partial^2}{\partial x_j^2} v \right| \geq d(v) > 0 \text{ for } x_j = 0 \text{ and } x_j = \dfrac{1}{n},$

which implies that $K_{j,n}^\pm \cap \mathcal{C}_{m_n^i}$ are each open relative to $\mathcal{C}_{m_n^i} \subset X_n^{2,\alpha} \times \mathbb{R}$.

We show that $K_{j,n}^\pm \cap \mathcal{C}_{m_n^i}$ are also closed relative to $\mathcal{C}_{m_n^i}$.

Let $(v_k, m_k) \in K_{j,n}^- \cap \mathcal{C}_{m_n^i}$ such that $(v_k, m_k) \to (v, m) \in \mathcal{C}_{m_n^i}$ in $X_n^{2,\alpha} \times \mathbb{R}$. Then $w_j = \frac{\partial}{\partial x_j} v \leq 0$ in S_n^j, and w_j solves (III.6.45); i.e., w_j solves a linear elliptic problem in the strip S_n^j and satisfies homogeneous Dirichlet boundary conditions on ∂S_n^j. The argument of (III.6.10) and the elliptic maximum principle then imply that $w_j < 0$ in S_n^j, since $w \equiv 0$ is excluded. (Note that w_j is periodic along the strip, so that it suffices to consider the elliptic equation in a compact segment of the strip.) The Hopf boundary lemma in [56] finally implies that $|\frac{\partial}{\partial x_j} w_j| > 0$ on ∂S_n^j, so that $(v, m) \in K_{j,n}^- \cap \mathcal{C}_{m_n^i}$.

Let $\mathcal{C}_{m_n^i}^\pm$ denote the components of $K_{j,n}^\mp \cap \mathcal{C}_{m_n^i}$ in $\mathcal{C}_{m_n^i}$. Then, for $j = 1, 2$, $K_{j,n}^\mp \cap \mathcal{C}_{m_n^i} = \mathcal{C}_{m_n^i}^\pm$,

(III.6.51)
$$\mathcal{C}_{m_n^i}^+ \subset K_{j,n}^-, \ \mathcal{C}_{m_n^i}^- \subset K_{j,n}^+, \text{ and by (III.6.49)},$$
$$\mathcal{C}_{m_n^i} = \mathcal{C}_{m_n^i}^+ \cup \mathcal{C}_{m_n^i}^-.$$

(We give here a different argument from that for (III.6.31), since the strips S_n^j are unbounded, but ∂S_n^j are smooth.) Obviously, $\mathcal{C}_{m_n^i}^+ \cap \mathcal{C}_{m_n^i}^- = \emptyset$, and the continua are transformed into each other by the reversion (III.2.82). Note that $\mathcal{C}_{m_n^i}^\pm \subset \mathcal{C}_{n,i}^\pm$, respectively, but as we shall see below, equality does not hold.

Property (III.6.51)$_1$ means, by the definition (III.6.48) of the cones $K_{j,n}^\mp$, that the branches $\mathcal{C}_{m_n^i}^\pm$ are labeled by *nodal properties of the derivatives*. In view of the periodicity (III.6.43), the constant sign of $\frac{\partial}{\partial x_j} v = w_j$ in S_n^j is true for all parallel strips of width $\frac{1}{n}$, with changing sign in adjacent strips. This characteristic nodal pattern implies a separation of branches:

$$(\text{III.6.52}) \qquad \mathcal{C}_{m_n^i} \cap \mathcal{C}_{m_k^j} = \emptyset \text{ for } n \neq k, \ i,j \in \{1,2\}.$$

The knowledge about the precise nodal pattern of $\frac{\partial}{\partial x_j} v = w_j$ on $\mathcal{C}_{m_n^i}$ has more benefits. First of all, it implies a fixed location of all minima and maxima:

If $(v,m) \in \mathcal{C}_{m_n^i}^+ \subset K_{j,n}^-$, then

$$(\text{III.6.53}) \qquad \max_{x \in \mathbb{R}^2} v(x_1, x_2) = v(0,0) = v\left(\frac{2k}{n}, \frac{2\ell}{n}\right),$$

$$\min_{x \in \mathbb{R}^2} v(x_1, x_2) = v\left(\frac{1}{n}, \frac{1}{n}\right) = v\left(\frac{2k+1}{n}, \frac{2\ell+1}{n}\right).$$

This means also that

$$(\text{III.6.54}) \qquad \Delta v(0,0) \leq 0 \quad \text{and} \quad \Delta v\left(\frac{1}{n}, \frac{1}{n}\right) \geq 0.$$

Subtracting the equation $F_\varepsilon(v,m) = 0$, cf. (III.2.66), in $x = (0,0)$ and $x = (\frac{1}{n}, \frac{1}{n})$ results in the nonlocal term droping out, and by (III.6.54) we obtain

$$(\text{III.6.55}) \qquad \begin{aligned} W'(m + v(0,0)) &\leq W'\left(m + v\left(\frac{1}{n}, \frac{1}{n}\right)\right) \\ &\text{for all } (v,m) \in \mathcal{C}_{m_n^i}^+. \end{aligned}$$

On the other hand, by its mean value zero,

$$(\text{III.6.56}) \qquad v\left(\frac{1}{n}, \frac{1}{n}\right) < 0 < v(0,0).$$

By the assumed shape of the graph of W' sketched in Figure III.2.4, the inequalities (III.6.55), (III.6.56) are compatible only if

$$(\text{III.6.57}) \qquad \begin{aligned} &|m| \leq M_1 \text{ and } -M_2 \leq v\left(\frac{1}{n}, \frac{1}{n}\right) < v(0,0) \leq M_2 \\ &\text{for constants } M_1, M_2 > 0 \text{ depending only on } W'. \end{aligned}$$

This gives the uniform a priori estimate

$$(\text{III.6.58}) \qquad \begin{aligned} &|m| + \|v\|_{0,0} = |m| + \max_{x \in \mathbb{R}^2} |v(x)| \leq M_1 + M_2 \\ &\text{for all } (v,m) \in \mathcal{C}_{m_n^i}^+, \text{ independent of } n \in \mathbb{N} \text{ and } \varepsilon > 0. \end{aligned}$$

Using the equation $F_\varepsilon(v,m) = 0$, cf. (III.2.66), a "bootstrapping" delivers the following sequence of estimates via (interior) elliptic a priori estimates

$(\Omega = (0,1) \times (0,1) \subset \tilde{\Omega} = (-1,2) \times (-1,2))$:

$$\varepsilon \|\Delta v\|_{L^2(\tilde{\Omega})} \leq \|W'(m+v)\|_{L^2(\tilde{\Omega})} + 3 \int_{\Omega} |W'(m+v)| dx,$$

whence $\|v\|_{H^2(\Omega)} \leq C_1/\varepsilon$,

(III.6.59) $\|v\|_{0,\alpha} \leq C_2/\varepsilon$ by embedding $H^2(\Omega) \subset C^\alpha(\overline{\Omega})$,

$$\varepsilon \|\Delta v\|_{0,\alpha} \leq \|W'(m+v)\|_{0,\alpha} + \int_{\Omega} |W'(m+v)| dx,$$

whence $\|v\|_{2,\alpha} \leq C_3/\varepsilon^2$,

where for $v \in X_n^{2,\alpha}$ the norms over Ω or $\mathbb{R}^2$ are the same.

To summarize,

(III.6.60) $$|m| + \|v\|_{2,\alpha} \leq C_0/\varepsilon^2 \quad \text{for all} \quad (v,m) \in \mathcal{C}_{m_n^i}^+$$

and for a constant $C_0 > 0$ depending only on W'.

The application of the reversion (III.2.82) gives the same bound also for all $(v,m) \in \mathcal{C}_{m_n^i}^-$. This means that (for fixed $\varepsilon > 0$) $\mathcal{C}_{m_n^i}^\pm$ are bounded in $X_n^{2,\alpha} \times \mathbb{R}$. Since $\mathcal{C}_{m_n^i}^\pm = \mathcal{C}_{n,i}^\pm$, respectively, if $\mathcal{C}_{n,i}^\pm \cap (\{0\} \times \mathbb{R}) = \emptyset$, statement (III.5.21) implies that $\mathcal{C}_{m_n^i}^\pm$ meet the trivial axis $\{0\} \times \mathbb{R}$ apart from $(0, m_n^i)$. By (III.6.42) and the uniqueness of $\mathcal{C}_{m_n^i,loc}^+$ and $\mathcal{C}_{m_n^i,loc}^-$, $\mathcal{C}_{m_n^i}^+$ and $\mathcal{C}_{m_n^i}^-$ also consist of one component.

In order to decide on the target of $\mathcal{C}_{m_n^1}^+$, say, we adapt the arguments (III.6.34)–(III.6.37) to (III.6.45). Let $(v_k, m_k) \in \mathcal{C}_{m_n^1}^+$ be such that $v_k \to 0$ in $X_n^{2,\alpha}$ and $m_k \to \tilde{m}_1 \neq m_n^1$ in $\mathbb{R}$. Then $w_{1k} = \frac{\partial}{\partial x_1} v_k$ satisfies

$$-\varepsilon \Delta w_{1k} + W''(m_k + v_k) w_{1k} = 0 \quad \text{on} \quad \mathbb{R}^2,$$

(III.6.61) $$w_{1k} = 0 \quad \text{for} \quad x_1 = 0, \frac{1}{n},$$

$$\frac{\partial}{\partial x_2} w_{1k} = 0 \quad \text{for} \quad x_2 = 0, \frac{1}{n},$$

by the symmetries and periodicities of X_n^1; cf.(III.6.43). (Interior) elliptic a priori estimates for (III.6.61)$_1$ imply, as in (III.6.34),

$$\|w_{1k}\|_{2,\alpha} / \|w_{1k}\|_{0,\alpha} \leq \tilde{C}, \quad \text{or} \quad (\text{w.l.o.g.}),$$

(III.6.62) $$w_{1k} / \|w_{1k}\|_{0,\alpha} \to w_1 \quad \text{in} \quad C^\alpha(\mathbb{R}^2) \text{ as } k \to \infty \text{ and}$$

$$\|w_1\|_{0,\alpha} = 1.$$

Using again (III.6.61)$_1$, we see that the same "bootstrapping" as in (III.6.35)–(III.6.37) proves that

$$w_{1k}/\|w_{1k}\|_{0,\alpha} \to w_1 \text{ in } C^{2,\alpha}(\mathbb{R}^2) \text{ as } k \to \infty \text{ and}$$

$$-\varepsilon \Delta w_1 + W''(\tilde{m}_1)w_1 = 0 \quad \text{on} \quad \mathbb{R}^2,$$

(III.6.63)

$$w_1 = 0 \quad \text{for} \quad x_1 = 0, \frac{1}{n},$$

$$\frac{\partial}{\partial x_2} w_1 = 0 \quad \text{for} \quad x_2 = 0, \frac{1}{n},$$

since clearly, $w_1 \in X_n^1$. By (III.6.51), $w_{1k} < 0$ in $\Omega_n = \left(0, \frac{1}{n}\right) \times \left(0, \frac{1}{n}\right)$ and therefore $w_1 \leq 0$ in Ω_n, but in view of (III.6.62), $w_1 \not\equiv 0$. Therefore, w_1 is an eigenfunction of the Laplacian over Ω_n with mixed boundary conditions (III.6.63)$_{3,4}$ for the eigenvalue $W''(\tilde{m}_1)/\varepsilon$. On the other hand, $\frac{\partial}{\partial x_1}\hat{v}_n$ is an eigenfunction of the same problem for the eigenvalue $-n^2\pi^2$; cf. (III.2.69).

Since Δ with the mixed boundary conditions is symmetric with respect to the scalar product in $L^2(\Omega_n)$, eigenfunctions for different eigenvalues are orthogonal with respect to that scalar product. But $\frac{\partial}{\partial x_1}\hat{v}_n < 0$ and $w_1 \leq 0$ in Ω_n, so that orthogonality is excluded. Therefore, $W''(\tilde{m}_1)/\varepsilon = -n^2\pi^2$ or $\tilde{m}_1 = m_n^2$; cf. (III.2.77). Since the reversion (III.2.82) transforms $\mathcal{C}_{m_n^1}^+$ into $\mathcal{C}_{m_n^1}^-$, the component $\mathcal{C}_{m_n^1}^-$ meets $(0, m_n^2)$ as well. We summarize:

(III.6.64)

The bounded solution continua $\mathcal{C}_{m_n^1}^\pm$ emanating at $(0, m_n^1)$ meet the trivial solution line at $(0, m_n^2)$, where m_n^1, m_n^2 are the two solutions of the characteristic equation $W''(m) = -\varepsilon n^2\pi^2$. The solution continua $\mathcal{C}_{m_n^2}^\pm$ emanating at $(0, m_n^2)$ meet $(0, m_n^1)$ and therefore $\mathcal{C}_{m_n^1}^\pm = \mathcal{C}_{m_n^2}^\pm$, respectively.

We sketch the global continua $\mathcal{C}_{m_n^i}^\pm$ in Figure III.6.2.

The entire analysis for the solution continua $\mathcal{C}_{m_n^i}^\pm$ of $F_\varepsilon(v, m) = 0$, where F_ε is defined in (III.2.66), is valid for fixed small $\varepsilon > 0$. Therefore, we suppress the dependence on ε in the notation $\mathcal{C}_{m_n^i}^\pm$.

Note that our analysis does not imply that $\mathcal{C}_{m_n^i}^\pm$ are globally smooth curves: The topological argument proving Theorem III.5.1 guarantees only continua. However, a numerical path-following device suggests curves, and these curves have two additional turning points, sketched in Figure III.6.2. In [107] we give the arguments for those turning points:

The turning points are closely related to the singular limits of solutions on $\mathcal{C}^{\pm}_{m^i_n}$ as $\varepsilon \searrow 0$. Recall that in the variational problem (III.2.62) the parameter ε represents the interfacial energy. For $\varepsilon = 0$ that variational problem is simple, and minimizers for m in the so-called spinodal region are piecewise constant. The interface between the two constant concentrations of the binary alloy form a characteristic pattern, which, however, is not determined by the model (III.2.62) for $\varepsilon = 0$. It is believed that patterns that are selected by singular limits of conditionally critical points of $E_\varepsilon(u)$ as ε tends to zero are physically relevant, in particular, if the critical points are minimizers of $E_\varepsilon(u)$. In this case, the "Minimal Interface Criterion" of Modica holds, which means that the patterns with minimal interfaces are selected by singular limits of minimizers. In [106], [107] we show that this pattern formation is not the only possible one: We prove that all conditionally critical points of $E_\varepsilon(u)$ on the global continua $\mathcal{C}^{\pm}_{m^i_n}$ converge to minimizers of $E_0(u)$ as ε tends to zero, for fixed m in the spinodal region. (We give the proof of the convergence to conditionally critical points of $E_0(u)$ in Remark III.6.4.) They form patterns that do not necessarily have minimal interfaces but that are determined by the symmetries and monotonicities of $u = m + v \in X_n$. If the pattern in the limit has no minimal interface, then according to the "Minimal Interface Criterion," the critical points on $\mathcal{C}^{\pm}_{m^i_n}$ are not minimizers of the energy $E_\varepsilon(u)$ (under the constraint of prescribed mass m). This explains the two turning points sketched in Figure III.6.2: The local unstable solution curves shown in Figure III.2.5 gain stability beyond the first turning points according to the "Principle of Exchange of Stability" of Section I.7 and sketched in Figure I.7.1. But that stability is lost when the branch turns back again, so that there is a "middle branch" that is not stable. According to the comments on the stability in Section III.2, cf. (III.2.91), the unstable critical points $u = m + v$ are not minimizers of the energy $E_\varepsilon(u)$. Nonetheless, as ε tends to zero they form characteristic patterns of minimizers with nonminimal interfaces shown in [107]. We recall, however, that we have no proof that the solution branches $\mathcal{C}^{\pm}_{m^i_n}$ are globally curves. Such global smoothness is proved for a special class of problems in Section III.7.

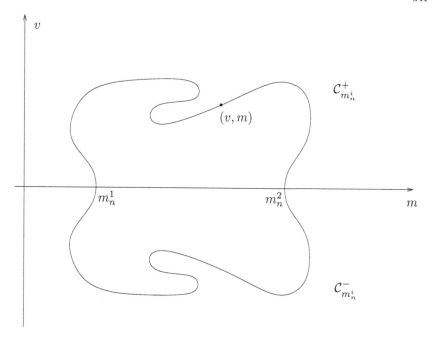

Figure III.6.2

Remark III.6.4 *We give a proof that solutions of the Euler–Lagrange equation (III.2.63) on the continua $\mathcal{C}^{\pm}_{m^i_n}$ converge for a mean value m in the spinodal region to a nontrivial conditionally critical point of the enery $E_0(u)$ (see (III.2.62)) as ε tends to zero. That proof contains some useful arguments. The proof that the limit is a global minimizer, however, is too special and too technical to give here.*

So far we have suppressed the dependence on $\varepsilon > 0$, but apparently this dependence will be crucial now. Therefore we denote the continua by $\mathcal{C}^{\pm}_{m^i_n,\varepsilon}$. By (III.2.77) the bifurcation points $(0, m^1_n)$ and $(0, m^2_n)$ tend to $(0, m_1)$ and $(0, m_2)$, respectively, as ε tends to zero; see Figure III.2.4. Therefore (III.6.64) guarantees that for any $m \in (m_1, m_2)$, which is the so-called spinodal region, a singular limit analysis makes sense.

We fix $m \in (m_1, m_2)$ and we consider the set $\{v_{\varepsilon,m} | (v_{\varepsilon,m}, m) \in \mathcal{C}^{+}_{m^i_n,\varepsilon}$ for $0 < \varepsilon < \varepsilon_0\}$, where ε_0 might depend on m. By the uniform estimate (III.6.58) there is a sequence $(\varepsilon_k)_{k\in\mathbb{N}}$ with $\lim_{k\to\infty} \varepsilon_k = 0$ such that

$$\text{(III.6.65)} \qquad w\text{-}\lim_{k\to\infty} v_{\varepsilon_k,m} = v_{0,m} \in L^2(\Omega),$$

where "w-$\lim$" denotes the weak limit in $L^2(\Omega)$. In particular,

$$\text{(III.6.66)} \qquad 0 = \int_\Omega v_{\varepsilon_k,m} dx \to \int_\Omega v_{0,m} dx = 0, \text{ as } k \to \infty,$$

i.e., the limit function $v_{0,m}$ has also mean value zero. In order to sharpen the convergence (III.6.65) we generalize Helly's theorem in [44] on pointwise convergence of monotonic sequences to two dimensions.

Let, as before, $\Omega_n = (0, \frac{1}{n}) \times (0, \frac{1}{n})$ and let K denote the positive closed cone $K = \{(x_1, x_2) | 0 \leq x_1, 0 \leq x_2\}$ in $\mathbb{R}^2$. We define an order in $\mathbb{R}^2$ by

$$(III.6.67) \qquad\qquad x \leq y \Leftrightarrow y - x \in K,$$

and in view of (III.6.51) (see also (III.6.48)) we have monotonicity with respect to that order:

$$(III.6.68) \qquad x \leq y \Rightarrow v_{\varepsilon_k,m}(x) \geq v_{\varepsilon_k,m}(y) \quad \text{for all } x, y \in \overline{\Omega}_n.$$

Let $\{x^j | j \in \mathbb{N}\}$ be a countable dense set in $\overline{\Omega}_n$. By the uniform boundedness of $(v_{\varepsilon_k,m}(x^j))_{k \in \mathbb{N}}$, cf. (III.6.58), a common procedure of elementary analysis yields a subsequence, again denoted by $(\varepsilon_k)_{k \in \mathbb{N}}$, such that

$$(III.6.69) \qquad \lim_{k \to \infty} v_{\varepsilon_k,m}(x^j) = v_0(x^j) \quad \text{exists in } \mathbb{R} \text{ for all } j \in \mathbb{N}.$$

By (III.6.68) we obtain the monotonicity

$$(III.6.70) \qquad x^i \leq x^j \Rightarrow v_0(x^i) \geq v_0(x^j) \quad \text{for all } i, j \in \mathbb{N}.$$

We extend v_0 on $\overline{\Omega}_n$ as follows:

$$(III.6.71) \qquad v_0(x) = \inf\{v_0(x^i) | x^i \leq x\} \quad \text{for any } x \in \overline{\Omega}_n.$$

Then the monotonicity (III.6.70) is preserved on $\overline{\Omega}_n$:

$$(III.6.72) \qquad x \leq y \Rightarrow v_0(x) \geq v_0(y) \quad \text{for all } x, y \in \overline{\Omega}_n.$$

Furthermore $|v_0(x)| \leq M_2$ for all $x \in \overline{\Omega}_n$; cf. (III.6.57). We say that v_0 has a jump at $x \in \overline{\Omega}_n$ if

$$(III.6.73) \qquad \inf\{v_0(x^i) | x^i \leq x\} > \sup\{v_0(x^j) | x \leq x^j\}.$$

On each line that is parallel to the x_1-axis the order (III.6.67) coincides with the usual order on the real line, and it is well known that the monotonic function v_0 has at most countably many jumps on such a line. Therefore, by Fubini's theorem, the set

$$(III.6.74) \qquad \begin{aligned} &J = \{x \in \overline{\Omega}_n | v_0 \text{ has a jump at } x\} \\ &\text{has measure zero.} \end{aligned}$$

Now let $x \in \overline{\Omega}_n \backslash J$ and $\eta > 0$ be given. In view of equality in (III.6.73) we find two points x^i and x^j satisfying

(III.6.75) $x^i \leq x \leq x^j$ and $0 \leq v_0(x^i) - v_0(x^j) \leq \dfrac{\eta}{3}$.

By (III.6.68), (III.6.69) we obtain

$$|v_{\varepsilon_k,m}(x) - v_{\varepsilon_\ell,m}(x)| = v_{\varepsilon_k,m}(x) - v_{\varepsilon_\ell,m}(x), \quad \text{say},$$

(III.6.76) $\leq v_{\varepsilon_k,m}(x^i) - v_0(x^i) + v_0(x^i) - v_0(x^j) + v_0(x^j) - v_{\varepsilon_\ell,m}(x^j)$

$$\leq \frac{\eta}{3} + \frac{\eta}{3} + \frac{\eta}{3} = \eta \quad \text{for } k, \ell \geq k_0(\eta).$$

(If $|v_{\varepsilon_k,m}(x) - v_{\varepsilon_\ell,m}(x)| = v_{\varepsilon_\ell,m}(x) - v_{\varepsilon_k,m}(x)$ the same arguments hold.)
Therefore $(v_{\varepsilon_k,m}(x))_{k \in \mathbb{N}}$ is a Cauchy sequence for all $x \in \overline{\Omega}_n \backslash J$, whence

(III.6.77) $\lim\limits_{k \to \infty} v_{\varepsilon_k,m}(x) = \tilde{v}_0(x)$ for all $x \in \overline{\Omega}_n \backslash J$.

We show that v_0 defined in (III.6.71) coincides with $\tilde{v}_0$ on $\overline{\Omega}_n \backslash J$: For some
fixed $x \in \overline{\Omega}_n \backslash J$,

(III.6.78)
$$\tilde{v}_0(x) - v_0(x) = \tilde{v}_0(x) - v_{\varepsilon_k,m}(x) + v_{\varepsilon_k,m}(x) - v_{\varepsilon_k,m}(x^i)$$
$$+ v_{\varepsilon_k,m}(x^i) - v_0(x^i) + v_0(x^i) - v_0(x).$$

Choose $x^i \leq x$ such that $v_0(x^i) - v_0(x) \leq \frac{\eta}{3}$ and choose after this $k \in \mathbb{N}$
such that $\tilde{v}_0(x) - v_{\varepsilon_k,m}(x) \leq \frac{\eta}{3}$ and $v_{\varepsilon_k,m}(x^i) - v_0(x^i) \leq \frac{\eta}{3}$; cf. (III.6.77),
(III.6.69). Since by the monotonicity (III.6.68), $v_{\varepsilon_k,m}(x) - v_{\varepsilon_k,m}(x^i) \leq 0$, we
obtain
(III.6.79) $\tilde{v}_0(x) - v_0(x) \leq \eta.$
For the same fixed $x \in \overline{\Omega}_n \backslash J$,

(III.6.80)
$$v_0(x) - \tilde{v}_0(x) = v_0(x) - v_0(x^j) + v_0(x^j) - v_{\varepsilon_k,m}(x^j)$$
$$+ v_{\varepsilon_k,m}(x^j) - v_{\varepsilon_k,m}(x) + v_{\varepsilon_k,m}(x) - \tilde{v}_0(x).$$

Choose $x \leq x^j$ such that $v_0(x) - v_0(x^j) \leq \frac{\eta}{3}$ and choose after this $k \in \mathbb{N}$ such
that $v_0(x^j) - v_{\varepsilon_k,m}(x^j) \leq \frac{\eta}{3}$ and $v_{\varepsilon_k,m}(x) - \tilde{v}_0(x) \leq \frac{\eta}{3}$. Since by monotonicity,
$v_{\varepsilon_k,m}(x^j) - v_{\varepsilon_k,m}(x) \leq 0$, we obtain

(III.6.81) $v_0(x) - \tilde{v}_0(x) \leq \eta.$

Estimates (III.6.79), (III.6.81) prove $v_0 = \tilde{v}_0$ on $\overline{\Omega}_n \backslash J$ and by (III.6.77),

(III.6.82) $\lim\limits_{k \to \infty} v_{\varepsilon_k,m}(x) = v_0(x)$ for all $x \in \overline{\Omega}_n \backslash J$.

Lebesgue's dominated convergence theorem then yields, in view of (III.6.74),

(III.6.83) $\displaystyle\lim_{k\to\infty}\int_{\Omega_n} v_{\varepsilon_k,m}\varphi dx = \int_{\Omega_n} v_0\varphi dx$ *for all* $\varphi \in L^2(\Omega_n)$,

and therefore, by (III.6.65),

(III.6.84) $v_0 = v_{0,m} \in L^2(\Omega_n)$.

Repeating the above arguments for all squares of side length $\frac{1}{n}$ *in* $\Omega = (0,1) \times (0,1)$ *and using the respective monotonicities of* $v_{\varepsilon_k,m}$ *in these squares we obtain a limit function* $v_0 = v_{0,m} \in L^2(\Omega)\cap X_n$; *cf. (III.2.70). We summarize:*

(III.6.85)

> *Let for a sequence* $(\varepsilon_k)_{k\in\mathbb{N}}$ *with* $\displaystyle\lim_{k\to\infty}\varepsilon_k = 0$,
>
> $(v_{\varepsilon_k,m}, m) \in \mathcal{C}^+_{m^i_n,\varepsilon_k}$ *for m in the spinodal region* (m_1, m_2).
>
> *Then a subsequence, again denoted by* $(v_{\varepsilon_k,m})_{k\in\mathbb{N}}$,
>
> *has a limit* $v_{0,m}$ *in the following sense:*
>
> $\displaystyle\lim_{k\to\infty} v_{\varepsilon_k,m}(x) = v_{0,m}(x)$ *almost everywhere in* Ω.
>
> *That limit function* $v_{0,m}$ *is in* $L^\infty(\Omega)$, *it has*
>
> *mean value zero, and it has the same symmetries,*
>
> *periodicities, and monotonicities as* $v_{\varepsilon_k,m}$.

Next we show that $u_{0,m} = m + v_{0,m}$ *is a critical point of* $E_0(u)$ *under the constraint* $\int_\Omega u dx = m$. *Indeed, by* $F_{\varepsilon_k}(v_{\varepsilon_k,m}, m) = 0$, *cf. (III.2.66), we obtain after an integration by parts*

$$\int_\Omega (-\varepsilon_k v_{\varepsilon_k,m}\Delta\varphi + W'(m + v_{\varepsilon_k,m}))\varphi dx = 0$$

(III.6.86) *for all test functions* $\varphi \in C^\infty(\mathbb{R}^2)$ *satisfying*

$$\int_\Omega \varphi dx = 0 \quad \text{and Neumann boundary conditions on } \partial\Omega.$$

Passing in (III.6.86) to the limit yields by Lebesgue's dominated convergence theorem

(III.6.87) $\displaystyle\int_\Omega W'(u_{0,m})\varphi dx = 0, \quad \int_\Omega u_{0,m} dx = m,$

for all test functions characterized in (III.6.86).

Therefore $W'(u_{0,m}) = \lambda$ *and* $u_{0,m} \in L^\infty(\Omega)$ *is a critical point of* $E_0(u)$ *under the constraint* $\int_\Omega u dx = m$.

A trivial critical point is the constant function $u_0 = m$, *which is not a minimizer. Thus the whole procedure makes sense only if we show that* $u_{0,m} \neq m$ *or that* $v_{0,m} \neq 0$.

The function $\frac{\partial}{\partial x_1} v_{\varepsilon,m}$ satisfies the boundary value problem (III.6.61), which can be interpreted as an eigenvalue problem for $-\Delta$ with weight function over Ω_n, namely,

(III.6.88)
$$\Delta w + \lambda r_\varepsilon w = 0 \quad in \quad \Omega_n,$$
with boundary conditions (III.6.61)$_{2,3}$ and
$$r_\varepsilon = -W''(m + v_{\varepsilon,m}).$$

As pointed out in Remark III.2.3, the negative eigenfunction $\frac{\partial}{\partial x_1} v_{\varepsilon,m}$ belongs to the positive principal eigenvalue $\lambda_\varepsilon = \frac{1}{\varepsilon}$, which is characterized as follows:

(III.6.89)
$$\lambda_\varepsilon = \min \left\{ \frac{\|\nabla w\|^2_{L^2(\Omega_n)}}{(r_\varepsilon w, w)_{L^2(\Omega_n)}} \left| \begin{array}{l} 0 \neq w \in W^{1,2}(\Omega_n), \\ (r_\varepsilon w, w)_{L^2(\Omega_n)} > 0, \\ w \text{ satisfies the Dirichlet} \\ \text{boundary conditions (III.6.61)}_2 \end{array} \right. \right\}.$$

Usually the weight function r_ε is positive, but that positivity is not necessary. What is needed is that the set in (III.6.89) not be empty and that Poincaré's inequality hold for the admitted class of functions. That is true for the Dirichlet boundary conditions on two opposite sides of the square. The minimum is attained at an eigenfunction having constant sign in Ω_n. The Neumann boundary conditions on the two other sides of the square are automatically fulfilled by a minimizer of (III.6.89), since they are the natural boundary conditions for the variational problem (III.6.89). For more details see [77].

The convergence in (III.6.85) implies by Lebesgue's dominated convergence theorem

(III.6.90)
$$\lim_{k \to \infty} (r_{\varepsilon_k} w, w)_{L^2(\Omega_n)} = -(W''(m + v_{0,m})w, w)_{L^2(\Omega_n)}$$
for all $w \in W^{1,2}(\Omega_n)$.

Assume that $-W''(m + v_{0,m}) = r_0$ is positive on a set of positive measure in Ω_n. Then the (weak) eigenvalue problem (III.6.88) with weight function r_0 has a positive principal eigenvalue λ_0 with an eigenfunction $w_0 \in W^{1,2}(\Omega_n)$ satisfying the Dirichlet boundary condition (III.6.61)$_2$. Furthermore, that eigenfunction w_0 has a constant sign in Ω_n and minimizes (III.6.89), where r_ε is replaced by r_0. We deduct from (III.6.89) and (III.6.90)

(III.6.91)
$$\frac{1}{\varepsilon_k} = \lambda_{\varepsilon_k} \leq \frac{\|\nabla w_0\|^2_{L^2(\Omega_n)}}{(r_{\varepsilon_k} w_0, w_0)_{L^2(\Omega_n)}} \leq \frac{\|\nabla w_0\|^2_{L^2(\Omega_n)}}{(r_0 w_0, w_0)_{L^2(\Omega_n)}} + 1$$
$$= \lambda_0 + 1 \quad for \ all \ k \geq k_0.$$

The contradiction of (III.6.91) proves that $W''(m + v_{0,m}) \geq 0$ almost everywhere in Ω_n. Looking at Figure III.2.4, we see that this is possible only if $v_{0,m} \neq 0$ for $m \in (m_1, m_2)$.

Since $W'(m + v_{0,m}) = W'(u_{0,m}) = \lambda$ in Ω, the critical point $u_{0,m}$ with mean value m is piecewise constant, having two values that are not in the spinodal region (m_1, m_2). The monotonicity with respect to the order (III.6.67) in Ω_n restricts the distribution of these two values in Ω_n, and by reflections and periodic extension it suggests a pattern shown in [106].

So far, no minimizing property of $u_{0,m}$ for the functional $E_0(u)$ under the constraint $\int_\Omega u\,dx = m$ has been proved. A subtle analysis [107], however, reveals that the singular limit $u_{0,m}$ fulfills a second Weierstrass–Erdmann corner condition that is known for minimizers of one-dimensional variational problems, namely

(III.6.92) $\begin{aligned} &W(u_{0,m}) - u_{0,m}W'(u_{0,m}) \quad \text{is continuous, i.e.,}\\ &\text{is constant in } \Omega_n. \end{aligned}$

The condition (III.6.92) allows for $u_{0,m}$ only the values α and β shown in Figure III.2.4. Consequently, the singular limit $u_{0,m}$ is a global minimizer of $E_0(u) = \int_\Omega W(u)\,dx$ under the constraint $\int_\Omega u\,dx = m$, where the mean value m is in the spinodal region. (The simple arguments for the last statement can be found in [109], e.g..)

Remark III.6.5 *The methods related to the Euler–Lagrange equation of the Cahn–Hilliard energy, expounded in Sections III.2.5, III.5.1, and III.6.5, can also be applied to other problems. In [120] we investigate radially symmetric critical points of nonconvex functionals of the form*

$$\text{(III.6.93)}\qquad E(u) = \int_{B_R(0)} W(\nabla u) + G(u)\,dx$$

over a ball $B_R(0)$ in $\mathbb{R}^n$. Here W depends only on the Euclidean norm of ∇u and is a nonconvex two-well potential as sketched in Figure III.2.4. Consequently, the direct methods of the calculus of variations are not applicable and the Euler–Lagrange equation is not elliptic. Therefore, the existence of critical points, in particular of minimizers, is not at all obvious, even if the functional is coercive and bounded from below.

One possibility of overcoming the difficulties is given by a singular perturbation of higher order,

$$\text{(III.6.94)}\quad E_\varepsilon(u) = \int_{B_R(0)} \frac{1}{2}\varepsilon(\Delta u)^2 + W(\nabla u) + G(u)\,dx \quad \text{for } \varepsilon > 0.$$

Since (III.6.94) has a uniformly elliptic Euler–Lagrange equation of fourth order, the chances of proving the existence of critical points are increased. Moreover, if critical points of (III.6.94) converge as ε tends to zero, then one can hope for critical points of (III.6.93). For the first step, i.e., to prove the existence of a solution of the Euler–Lagrange equation, we embed (III.6.94) *into a family of functionals*

$$(III.6.95) \quad E_{\varepsilon,\lambda}(u) = \int_{B_R(0)} \frac{1}{2}\varepsilon(\Delta u)^2 + W(\nabla u, \lambda) + G(u)dx, \quad \lambda \in \mathbb{R},$$

where $E_{\varepsilon,0}(u) = E_\varepsilon(u)$. Assuming $\nabla W(0, \lambda) = 0$ (where ∇ denotes the gradient with respect to the first variables) and $G'(0) = 0$, we see that the functional (III.6.95) has a trivial critical point $u = 0$ for all $\lambda \in \mathbb{R}$ given by the trivial solution $(0, \lambda)$ of the corresponding Euler–Lagrange equation.

A rough analysis shows that the trivial critical point is not a minimizer for $E_{\varepsilon,0}(u) = E_\varepsilon(u)$. In order to find nontrivial critical points we restrict the functionals to radially symmetric functions. This is possible if $W(\cdot, \lambda)$ is radially symmetric, and it is adequate to the problem, since one can show that under certain additional conditions, a minimizer of $E(u)$ is radially symmetric, if it exists. The mathematical benefits of this restriction, however, are striking: A bifurcation analysis of the one-dimensional Euler–Lagrange equation of fourth order gives us pairs $\pm\lambda_k \in \mathbb{R}$, $k \in \mathbb{N} \cup \{0\}$, where a local bifurcation with a one-dimensional kernel takes place. At each $(0, \pm\lambda_k)$ Theorem I.5.1 is applicable, and the bifurcation formulas of Section I.6 yield pitchfork bifurcations. For a global continuation of the local bifurcating curves we apply Theorem II.5.9, and we obtain global continua $\mathcal{C}^{\pm}_{\pm\lambda_k}$ emanating at $(0, \pm\lambda_k)$ and satisfying the alternatives given in that theorem. In general, however, we cannot decide which of the alternatives is valid. We discuss only $\mathcal{C}^{\pm}_{\pm\lambda_0}$ in more detail. Here $\lambda_0 > 0$ is closely related to the principal eigenvalue of the negative Laplacian $-\Delta$ over the ball $B_R(0)$ satisfying homogeneous Dirichlet boundary conditions. The corresponding eigenfunction spanning the one-dimensional kernel is not only positive but also radially symmetric. These geometric properties are preserved for all functions of the global continuum $\mathcal{C}^{+}_{\lambda_0}$. In order to prove this as we do in Section III.6.2 we have to ensure the validity of a maximum principle. However, such a maximum principle applies only to special elliptic equations of fourth order. The Euler–Lagrange equation of (III.6.95) is not admissible in general, unless it is restricted to radially symmetric functions.

An a priori estimate of $\mathcal{C}^{+}_{\lambda_0}$ rules out unboundedness, and by its positivity it is not connected to $\mathcal{C}^{-}_{\lambda_0}$ or to any other $\mathcal{C}^{\pm}_{\pm\lambda_k}$ for $k \geq 1$. The only possibility left is $\mathcal{C}^{+}_{\lambda_0} = \mathcal{C}^{+}_{-\lambda_0}$, in other words, the continuum $\mathcal{C}^{+}_{\lambda_0}$ connects $(0, +\lambda_0)$ and $(0, -\lambda_0)$. In particular, there is a radially symmetric nontrivial critical point of (III.6.95) for any $\lambda \in (-\lambda_0, \lambda_0)$. Note that this is true for any fixed $\varepsilon > 0$.

Making use of the positivity and further geometric properties such as monotonicity of the critical points on $\mathcal{C}^{+}_{\lambda_0}$, we can prove the compactness of the set of critical points on $\mathcal{C}^{+}_{\lambda_0}$ for fixed $\lambda \in (-\lambda_0, \lambda_0)$ as ε tends to zero. In particular, we obtain a singular limit for $\lambda = 0$. This singular limit is a nonnegative radially symmetric critical point of $E(u) = E_0(u)$ given in (III.6.93).

However, that singular limit could be trivial, and therefore it would not be of any interest. Another application of all geometric properties of the nontrivial critical points together with a subtle analysis yields finally that the singular limit is positive in $B_R(0)$.

Under additional assumptions we show that this method gives indeed a minimizer of $E(u)$, which is unique. All details of the proofs can be found in [120].

III.6.6 Global Branches of Positive Solutions via Continuation

We consider (III.6.11) with $G(\mathbf{0}, 0, 0, x, 0) = 0$ for all $x \in \Omega$ such that we have the solution $(u_0, \lambda_0) = (0, 0) \in X \times \mathbb{R}$. We prove a global continuation of that solution by unbounded branches of positive and negative solutions. To that end we assume that $\Omega \subset \mathbb{R}^n$ is a bounded domain with a smooth boundary $\partial \Omega$ and that the functions a_{ij} and g in (III.6.11) satisfy the regularity assumptions of Section III.5. Moreover,

(III.6.96)
$$
\begin{aligned}
&g(\mathbf{0}, 0, x, \lambda) \geq 0 \quad \text{for } \lambda \geq 0, \\
&g(\mathbf{0}, 0, x, \lambda) \leq 0 \quad \text{for } \lambda \leq 0, \\
&\text{and for all } x \in \Omega, \text{ but} \\
&g(\mathbf{0}, 0, x, \lambda) \neq 0 \quad \text{for some } x \in \Omega \text{ if } \lambda \neq 0.
\end{aligned}
$$

Apart from assumption (III.6.96) for all $(x, \lambda) \in \Omega \times \mathbb{R}$ we need also the following global and local properties of the function g:

(III.6.97)
$$
\begin{aligned}
&g_u(\mathbf{v}, u, x, 0) \leq 0 \quad \text{for all} \quad (\mathbf{v}, u, x) \in \mathbb{R}^n \times \mathbb{R} \times \Omega, \\
&g(\mathbf{v}, u, x, \lambda) \geq 0 \quad \text{for } \lambda \geq 0, \\
&g(\mathbf{v}, u, x, \lambda) \leq 0 \quad \text{for } \lambda \leq 0 \text{ or} \\
&g_u(\mathbf{v}, u, x, \lambda) \leq 0 \quad \text{or} \\[4pt]
&g_u(\mathbf{v}, u, x, \lambda)u \geq 0 \quad \text{for } \lambda \geq 0, \\
&g_u(\mathbf{v}, u, x, \lambda)u \leq 0 \quad \text{for } \lambda \leq 0, \\
&\text{and for } (\mathbf{v}, u, x, \lambda) \in \mathbb{R}^n \times \mathbb{R} \times \Omega \times \mathbb{R} \\
&\text{such that } \|\mathbf{v}\| + |u| + |\lambda| \leq \delta.
\end{aligned}
$$

As in (III.6.24), every solution $(u, \lambda) \in X \times \mathbb{R}$ of $F(u, \lambda) = 0$ yields a solution $h = u$ of the *linear* elliptic problem

(III.6.98)
$$
\begin{aligned}
\tilde{L}(u, \lambda)h &= -g \quad \text{in} \quad \Omega, \\
h &= 0 \quad \text{on} \quad \partial\Omega,
\end{aligned}
$$

where $\tilde{L}(u,\lambda)$ is defined in (III.6.25) and $g = g(\mathbf{0},0,x,\lambda)$. By (III.6.96)–(III.6.98),

$$D_u F(0,0)h = 0 \quad \text{for} \quad h \in X \Leftrightarrow$$

(III.6.99)
$$\tilde{L}(0,0)h = 0 \quad \text{in} \quad \Omega,$$
$$h = 0 \quad \text{on} \quad \partial\Omega,$$

where $\tilde{c}(\mathbf{0},0,x,0) \leq 0$; cf.(III.6.25). Therefore, the maximum (minimum) principle implies that $h = 0$ and 0 is not an eigenvalue of $D_u F(0,0)$. In view of the ellipticity of $D_u F(0,0)$, the point 0 is in the resolvent set of $D_u F(0,0)$, and $D_u F(0,0) \in L(X,Z)$ is bijective. By the Implicit Function Theorem, Theorem I.1.1, the solution $(u_0, \lambda_0) = (0,0)$ is locally continued by a curve $\{(u(\lambda),\lambda)|\lambda \in (-\delta,\delta)\}$ and globally extended by a continuum $\mathcal{C}$ subject to the alternative described in Theorem II.6.1; see Section III.5.2, (III.5.22)–(III.5.24).

We rule out alternative (ii) of Theorem II.6.1. By (III.6.97)$_1$,

$$F(u,0) = 0 \quad \text{for} \quad u \in X \Leftrightarrow$$

(III.6.100)
$$\tilde{L}(u,0)u = 0 \quad \text{in} \quad \Omega,$$
$$u = 0 \quad \text{on} \quad \partial\Omega,$$

where $\tilde{c}(\nabla u, u, x, 0) \leq 0$, cf. (III.6.25), whence $u = 0$ by the maximum (minimum) principle. Therefore, $\mathcal{C}\backslash\{(0,0)\}$ is not connected, since for

(III.6.101)
$$\mathcal{C} = \{(0,0)\} \cup \mathcal{C}^+ \cup \mathcal{C}^-,$$

the components $\mathcal{C}^+$ and $\mathcal{C}^-$ are separated by the hyperplane $X \times \{0\}$. As in Theorem II.6.1, let

$$\mathcal{C}^+ \text{ denote the component of } \{(x(\lambda),\lambda)|\lambda \in (0,\delta)\},$$
(III.6.102) $\quad \mathcal{C}^- \text{ denote the component of } \{(x(\lambda),\lambda)|\lambda \in (-\delta,0)\},$
$$\text{each of which is unbounded in } X \times \mathbb{R}.$$

By (III.6.100),

(III.6.103)
$$\mathcal{C}^+ \subset X \times (0,\infty), \ \mathcal{C}^- \subset X \times (-\infty,0),$$

and we claim that $u > 0$ in Ω for $(u,\lambda) \in \mathcal{C}^+$ and $u < 0$ in Ω for $(u,\lambda) \in \mathcal{C}^-$. We start with the local curve $\{(u(\lambda),\lambda)|\lambda \in (0,\delta)\}$. By (III.6.96)–(III.6.98),

$$L(u(\lambda),\lambda)u(\lambda) \leq 0 \text{ in } \Omega \text{ or}$$

(III.6.104)
$$\tilde{L}(u(\lambda),\lambda)u(\lambda) = 0 \text{ in } \Omega,$$
$$u(\lambda) = 0 \text{ on } \partial\Omega,$$

where $\tilde{c}(\nabla u(\lambda), u(\lambda), x, \lambda) \leq 0$ when (III.6.97)$_4$ is valid for $(\mathbf{v}, u, x, \lambda) \in \mathbb{R}^n \times \mathbb{R} \times \Omega \times \mathbb{R}$ with $\|\mathbf{v}\| + |u| + |\lambda| \leq \delta$. In case of (III.6.97)$_5$ we have

$$(\text{III.6.105}) \quad L(u(\lambda), \lambda)u(\lambda) + \sum_{i=1}^{n} \tilde{b}_i(\nabla u(\lambda), u(\lambda), x, \lambda)u(\lambda)_{x_i} \leq 0 \text{ in } \Omega,$$
$$u(\lambda) = 0 \text{ on } \partial\Omega,$$

since $\tilde{c}(\nabla u(\lambda), u(\lambda)x, \lambda)u(\lambda) \geq 0$. Recall that $L(u, \lambda)$ is defined as in (III.6.11). In all cases (III.6.104), (III.6.105) the minimum principle implies $u(\lambda) > 0$ in Ω or $u(\lambda) \equiv 0$, which is excluded for $\lambda \in (0, \delta)$ by (III.6.96). For $\lambda \in (-\delta, 0)$ the assumptions (III.6.96), (III.6.97) imply by the maximum principle that $u(\lambda) < 0$ in Ω. Thus

$$(\text{III.6.106}) \quad \begin{aligned} \{(u(\lambda), \lambda)|\lambda \in (0, \delta)\} &\subset K^+, \\ \{(u(\lambda), \lambda)|\lambda \in (-\delta, 0)\} &\subset K^-, \end{aligned}$$

where the cones $K^{\pm}$ are defined in (III.6.28). Once we get started in the cones, the further extension of $K^{\pm} \cap C^{\pm} \neq \emptyset$ in $C^{\pm}$ follows precisely the lines of (III.6.28)–(III.6.31). We can show that $K^{\pm} \cap C^{\pm}$ are each open and closed relative to $C^{\pm}$, since by (III.6.103), we can use (III.6.96)$_1$ in (III.6.98) for $(u, \lambda) \in C^+$, and (III.6.96)$_2$ for $(u, \lambda) \in C^-$. Note that assumptions (III.6.97) are no longer used when (III.6.106) is known. Thus, by connectedness of $C^{\pm}$, we obtain $K^{\pm} \cap C^{\pm} = C^{\pm}$ or

$$(\text{III.6.107}) \quad C^+ \subset K^+ \quad \text{and} \quad C^- \subset K^-.$$

We summarize:

Theorem III.6.6 *The quasilinear elliptic problem (III.6.11) satisfying (III.6.96), (III.6.97) possesses an unbounded positive solution continuum $C^+ \subset X \times (0, \infty)$ and an unbounded negative solution continuum $C^- \subset X \times (-\infty, 0)$, both of which emanate from the solution $(u_0, \lambda_0) = (0, 0)$.*

In the next section we show how positivity (negativity) is used to obtain more information about qualitative properties of global branches. In particular, for a special problem (III.7.1), global positive or negative solution continua are smooth curves parameterized by the amplitude of u; cf. Theorems III.7.9 and III.7.10. Moreover, under suitable growth conditions on the nonlinearity, their asymptotic behavior can be determined; cf. Theorem III.7.17.

Remark III.6.7 *In Remark III.5.3 we mention the existence of global branches of weak solutions of a semilinear elliptic problem over a bounded domain with a nonsmooth boundary. Since these weak solutions are in $C(\overline{\Omega})$, pointwise positivity (or negativity) in Ω is defined. Since the Maximum Principle of Theorem III.6.1 can be generalized to weak solutions, the existence of global positive and negative branches of weak solutions can be proved under much more general assumptions on the data of the problem; cf. [74].*

III.7 Smoothness and Uniqueness of Global Positive Solution Branches

In this section we study the model problem

(III.7.1)
$$\Delta u + \lambda g(u) = 0 \quad \text{in } \Omega,$$
$$u = 0 \quad \text{on } \partial\Omega,$$

where Ω is a bounded domain in $\mathbb{R}^n$ having a smooth boundary $\partial\Omega$. The function $g : \mathbb{R} \to \mathbb{R}$ is smooth (C^2 is enough), and first we assume $g(0) = 0$, so that we have the trivial solution $(0, \lambda)$ for all $\lambda \in \mathbb{R}$.

Let $g'(0) > 0$. We know from Section III.2.4 that every eigenvalue $\mu_n > 0$ of $-\Delta$ subject to the homogeneous boundary conditions (III.7.1)$_2$ provides a bifurcation point $(0, \lambda_n) = (0, \mu_n/g'(0))$ for (III.7.1). In this section we study the branch emanating at $(0, \lambda_0)$. To this purpose we define two positive values of λ:

(III.7.2)
$\mu_0 = \lambda_0 g'(0)$ is the principal eigenvalue of $-\Delta$ on Ω subject to homogeneous Dirichlet boundary conditions; i.e., μ_0 is the smallest positive and simple eigenvalue of $-\Delta$ with positive eigenfunction $\hat{v}_0$, and $\mu_1 = \lambda_1 g'(0)$ is the second eigenvalue of $-\Delta$; i.e., $0 < \lambda_0 < \lambda_1$.

Then, by the results of Section III.6, in particular of Theorem III.6.3,

(III.7.3)
there exist an unbounded continuum $C_{\lambda_0}^+$ of positive solutions and an unbounded continuum $C_{\lambda_0}^-$ of negative solutions of (III.7.1) in $X \times \mathbb{R}$ emanating at $(0, \lambda_0)$.

Recall that $X = C^{2,\alpha}(\overline{\Omega}) \cap \{u|u = 0 \text{ on } \partial\Omega\}$.

(The conditions for local and global bifurcation are easy to verify: Define $F(u, \lambda) = \Delta u + \lambda g(u)$ for $(u, \lambda) \in X \times \mathbb{R}$. Then $D_{u\lambda}^2 F(0, \lambda)h = L_\lambda(\lambda)h = g'(0)h$; cf. (III.2.8), and the simplicity of the principal eigenvalue and the (formal) self-adjointness of $L(\lambda_0)$ imply condition (I.5.3) or (III.2.8). Moreover, this condition is equivalent to an odd crossing number of $D_u F(0, \lambda) = L(\lambda) = \Delta + \lambda g'(0)I$; cf. (I.7.36). To be more precise, the greatest eigenvalue of $L(\lambda)$ is $\mu(\lambda) = (\lambda - \lambda_0)g'(0)$, and it crosses the imaginary axis at $\mu(\lambda_0) = 0$ with "nonvanishing speed" $\mu'(\lambda_0) = g'(0) > 0$. This describes a loss of stability of the trivial solution $(0, \lambda)$: it is stable for $\lambda < \lambda_0$ and unstable for $\lambda > \lambda_0$.)

The topological methods, however, for proving the existence of $C = \{(0, \lambda_0)\} \cup C_{\lambda_0}^+ \cup C_{\lambda_0}^-$ do not give more information than (III.7.3). In particular, the following questions have no answer yet:

- Are $\mathcal{C}^{\pm}_{\lambda_0}$ smooth curves in $X \times \mathbb{R}$?
- What is the asymptotic behavior of $\mathcal{C}^{\pm}_{\lambda_0}$ as $\|u\|_{2,\alpha} + |\lambda| \to \infty$?
- Are there other positive (negative) solutions of (III.7.1) that are not on $\mathcal{C}^{+}_{\lambda_0}(\mathcal{C}^{-}_{\lambda_0})$?

We give some answers using first growth conditions on g. Assume that

$$
\text{(III.7.4)} \qquad
\begin{aligned}
& 0 \le g(u)u \le g'(0)u^2 \text{ for } u \in \mathbb{R}, \\
& \text{i.e., } g \text{ grows at most linearly.}
\end{aligned}
$$

Then, for $(u, \lambda) \in \mathcal{C}^{\pm}_{\lambda_0}$,

$$
\text{(III.7.5)} \qquad (-\Delta u, u)_0 = \|\nabla u\|_0^2 = \lambda(g(u), u)_0 \le \lambda g'(0)\|u\|_0^2.
$$

Since the principal eigenvalue $\lambda_0 g'(0)$ is the minimal value of the Rayleigh quotient, i.e.,

$$
\text{(III.7.6)} \qquad \lambda_0 g'(0) = \min\left\{ \frac{\|\nabla u\|_0^2}{\|u\|_0^2} \middle| u \in H_0^1(\Omega), u \ne 0 \right\},
$$

cf. Remark III.2.1, we obtain from (III.7.5)

$$
\text{(III.7.7)} \qquad \lambda \ge \lambda_0 \quad \text{for all} \quad (u, \lambda) \in \mathcal{C}^{\pm}_{\lambda_0}.
$$

For a *sublinear growth* of g the bifurcation is supercritical.

For a superlinear growth, i.e.,

$$
\text{(III.7.8)} \qquad g(u)u > g'(0)u^2 \quad \text{for} \quad u \in \mathbb{R}, u \ne 0,
$$

we obtain for $(u, \lambda) \in \mathcal{C}^{\pm}_{\lambda_0}$,

$$
\text{(III.7.9)} \qquad
\begin{aligned}
-\Delta u - \lambda g'(0)u &= \lambda(g(u) - g'(0)u) \text{ in } \Omega, \\
u &= 0 \text{ on } \partial\Omega, \text{ and} \\
g(u) - g'(0)u &\gtrless 0 \text{ in } \Omega.
\end{aligned}
$$

Since (III.7.1) has no nontrivial solution for $\lambda = 0$ and since $\mathcal{C}^{\pm}_{\lambda_0}$ are connected, we have $\lambda > 0$ for all $(u, \lambda) \in \mathcal{C}^{\pm}_{\lambda_0}$.

Let $\hat{v}_0$ denote the positive and normalized eigenfunction of $-\Delta$ for the principal eigenvalue $\mu_0 = \lambda_0 g'(0)$. Then

$$
\text{(III.7.10)} \qquad
\begin{aligned}
(\lambda_0 - \lambda)g'(0)(u, \hat{v}_0)_0 &= (-\Delta u - \lambda g'(0)u, \hat{v}_0)_0 \\
&= \lambda(g(u) - g'(0)u, \hat{v}_0)_0 \gtrless 0 \text{ by (III.7.9)}, \\
(u, \hat{v}_0)_0 &\gtrless 0 \text{ for } (u, \lambda) \in \mathcal{C}^{\pm}_{\lambda_0}, \text{ whence} \\
0 &< \lambda < \lambda_0 \text{ for all } (u, \lambda) \in \mathcal{C}^{\pm}_{\lambda_0}.
\end{aligned}
$$

For a *superlinear growth* of g the bifurcation is subcritical.

By unboundedness of $\mathcal{C}_{\lambda_0}^{\pm}$, this means that $\|u\|_{2,\alpha}$ is unbounded for $(u, \lambda) \in \mathcal{C}_{\lambda_0}^{\pm}$, but it does not imply any specific asymptotic behavior. In order to sharpen the result of (III.7.10), we assume that

$$\lim_{u \to \infty} g'(u) = g'(\infty), \ \lim_{u \to -\infty} g'(u) = g'(-\infty) \text{ exist,}$$

(III.7.11) $\quad 0 < g'(0) < g'(\pm\infty) \leq \dfrac{\lambda_1}{\lambda_0} g'(0), \text{ cf. (III.7.2),}$

$$g''(u) > 0 \text{ for } u > 0,$$
$$g''(u) < 0 \text{ for } u < 0.$$

By $g(0) = 0$, the assumptions (III.7.11) imply (III.7.8) and

(III.7.12) $\qquad 0 < \dfrac{g(u)}{u} < g'(u) < g'(\pm\infty) \text{ for } u \neq 0.$

In the following we let (cf. (III.2.47))

(III.7.13)
$\mu_0(\rho)$ denote the principal eigenvalue with positive continuous weight function ρ of $-\Delta$ on Ω subject to homogeneous Dirichlet boundary conditions and $\mu_1(\rho)$ the second eigenvalue with weight function ρ.

According to (III.7.2), $\mu_0(1) = \mu_0$. The eigenvalue $\mu_0(\rho)$ is given as in (III.7.6), where the denominator of the quotient is replaced by $(\rho u, u)_0$, cf. Remark 2.3, and by the minimax principle (see [24]), the second eigenvalue is described as

(III.7.14) $\quad \mu_1(\rho) = \max_{0 \neq v \in H_0^1(\Omega)} \min \left\{ \dfrac{\|\nabla u\|_0^2}{(\rho u, u)_0} \,\Big|\, 0 \neq u \in H_0^1(\Omega), (\rho u, v)_0 = 0 \right\}.$

Both variational characterizations imply the monotonicity of the eigenvalues with respect to the weight function:

(III.7.15)
$$0 < \rho_1(x) \leq \rho_2(x) \text{ for all } x \in \Omega$$
$$\Rightarrow \mu_0(\rho_1) \geq \mu_0(\rho_2),$$
$$\mu_0(\rho_1) > \mu_0(\rho_2) \text{ if } \rho_1 \neq \rho_2,$$
$$\mu_1(\rho_1) \geq \mu_1(\rho_2).$$

For $(u, \lambda) \in \mathcal{C}_{\lambda_0}^{\pm}$,

(III.7.16) $\qquad \Delta u + \lambda \dfrac{g(u)}{u} u = 0 \quad \text{in } \Omega,$

$$u = 0 \quad \text{on } \partial\Omega,$$

so that $\lambda = \mu_0(g(u)/u)$, which is the principal eigenvalue of $-\Delta$ with weight function $g(u)/u$. (Note that $g(u)/u \in C(\overline{\Omega})$.) By (III.7.12), (III.7.15),

(III.7.17)
$$\lambda = \mu_0(g(u)/u) > \mu_0(g'(u)) > \mu_0(g'(\pm\infty)) = \lambda_0 \frac{g'(0)}{g'(\pm\infty)},$$
$$\mu_1(g(u)/u) \geq \mu_1(g'(u)) \geq \mu_1(g'(\pm\infty)) = \lambda_1 \frac{g'(0)}{g'(\pm\infty)} \geq \lambda_0,$$

where we use (III.7.11)$_2$. This restricts the interval of values of λ to

(III.7.18)
$$0 < \lambda_0 \frac{g'(0)}{g'(\pm\infty)} < \lambda < \lambda_0 \quad \text{for all} \ (u,\lambda) \in \mathcal{C}^{\pm}_{\lambda_0};$$

cf. (III.7.10). For $F(u,\lambda) = \Delta u + \lambda g(u)$ and $(u,\lambda) \in \mathcal{C}^{\pm}_{\lambda_0}$,

(III.7.19)
$$D_u F(u,\lambda)h = 0 \ \text{for} \ h \in X \Leftrightarrow$$
$$\Delta h + \lambda g'(u)h = 0 \quad \text{in} \ \Omega,$$
$$h = 0 \quad \text{on} \ \partial\Omega,$$

which means that λ is an eigenvalue with positive weight $g'(u)$ of $-\Delta$. By (III.7.17), (III.7.18),

(III.7.20)
$$\mu_0(g'(u)) < \lambda < \lambda_0 \leq \mu_1(g'(u)),$$

so that λ is between the first (principal) and the second eigenvalues with weight $g'(u)$ of $-\Delta$. Therefore,

(III.7.21)
$$D_u F(u,\lambda) \in L(X,Z) \ \text{is bijective}$$
$$\text{for all} \ (u,\lambda) \in \mathcal{C}^{\pm}_{\lambda_0}.$$

Recall that $Z = C^{\alpha}(\overline{\Omega})$. By the Implicit Function Theorem (Theorem I.1.1) every solution $(u,\lambda) \in \mathcal{C}^{\pm}_{\lambda_0}$ of $F(u,\lambda) = 0$ is locally continued by a smooth curve parameterized by λ. Therefore, by connectedness,

(III.7.22)
$$\mathcal{C}^{\pm}_{\lambda_0} \ \text{are globally smooth curves}$$
$$\text{parameterized by} \ \lambda \in (\lambda^{\pm}_{\infty}, \lambda_0) \subset \left(\lambda_0 \frac{g'(0)}{g'(\pm\infty)}, \lambda_0\right);$$

cf. (III.7.18). Let $(\lambda_n)_{n \in \mathbb{N}}$ be any sequence such that $\lambda_n \searrow \lambda^+_{\infty}$, say, and $\|u(\lambda_n)\|_{2,\alpha} \to \infty$ by unboundedness of $\mathcal{C}^+_{\lambda_0} = \{(u(\lambda),\lambda)|\lambda \in (\lambda^+_{\infty}, \lambda_0)\}$ in $X \times \mathbb{R}$. Then $(u(\lambda_n))_{n \in \mathbb{N}}$ is unbounded in $C(\overline{\Omega})$, too, since its boundedness implies boundedness in $C^{2,\alpha}(\overline{\Omega})$ by the following "bootstrapping": Boundedness of $(u(\lambda_n))_{n \in \mathbb{N}}$, and therefore of $(\lambda_n g(u(\lambda_n)))_{n \in \mathbb{N}}$, in $C(\overline{\Omega})$ implies by (III.7.1) and an elliptic a priori estimate boundedness of $(u(\lambda_n))_{n \in \mathbb{N}}$ in $W^{2,p}(\overline{\Omega})$ for every $1 < p < \infty$. A continuous embedding into $C^{\alpha}(\overline{\Omega})$ for $p > n$ yields boundedness of $(u(\lambda_n))_{n \in \mathbb{N}}$, and therefore of $(\lambda_n g(u(\lambda_n)))_{n \in \mathbb{N}}$, in $C^{\alpha}(\overline{\Omega})$, and (III.7.1) together with another elliptic a priori estimate implies finally boundedness of $(u(\lambda_n))_{n\in\mathbb{N}}$ in $C^{2,\alpha}(\overline{\Omega})$.

Passing to a subsequence, assume that $\|u(\lambda_n)\|_\infty \to \infty$ as $\lambda_n \searrow \lambda_\infty^+$. (Here $\|\quad\|_\infty$ is the maximum norm in $C(\overline{\Omega})$, also denoted by $\|\quad\|_{0,0}$.) Setting $v_n = u(\lambda_n)/\|u(\lambda_n)\|_\infty$ and $s_n = 1/\|u(\lambda_n)\|_\infty$ yields that $\|v_n\|_\infty = 1$, $s_n \to 0$, and v_n solves

$$\Delta v_n + \lambda_n s_n g\left(\frac{v_n}{s_n}\right) = 0 \quad \text{in } \Omega,$$

(III.7.23)

$$v_n = 0 \quad \text{on } \partial\Omega,$$
$$v_n > 0 \quad \text{in } \Omega,$$

By (III.7.11), $g(u)/u \to g'(\infty)$ as $u \to \infty$, whence $|g(u) - g'(\infty)u| \leq \varepsilon u$ for $u \geq M(\varepsilon)$. Therefore, for $M = M(\varepsilon)$ and positive u,

$$\max_{x\in\overline{\Omega}} |g(u(x)) - g'(\infty)u(x)|$$

(III.7.24)

$$\leq \max_{0\leq u(x)\leq M} |./.| + \max_{u(x)\geq M} |./.| \leq C_g(M) + \varepsilon\|u\|_\infty$$

$$\leq 2\varepsilon\|u\|_\infty \quad \text{if } \|u\|_\infty \geq C_g(M)/\varepsilon.$$

Since $\|v_n\|_\infty = 1$, the estimate (III.7.24) implies

$$\left\|s_n g\left(\frac{v_n}{s_n}\right) - g'(\infty)v_n\right\|_\infty \leq 2\varepsilon$$

(III.7.25)

$$\text{for all } 0 < s_n \leq \varepsilon/C_g(M) \text{ or}$$

$$s_n g\left(\frac{v_n}{s_n}\right) - g'(\infty)v_n \to 0 \text{ in } C(\overline{\Omega}) \text{ as } n \to \infty.$$

By boundedness of $\left(\lambda_n s_n g\left(\frac{v_n}{s_n}\right)\right)_{n\in\mathbb{N}}$ in $C(\overline{\Omega})$, cf. (III.7.25), the same bootstrapping as described before yields via (III.7.23) boundedness of $(v_n)_{n\in\mathbb{N}}$ in $W^{2,p}(\Omega)$ for $p > n$ and, by a compact embedding into $C^\alpha(\overline{\Omega})$, the convergence of a subsequence $(v_{n_k})_{k\in\mathbb{N}}$ to some $v \in C^\alpha(\overline{\Omega})$ with $\|v\|_\infty = 1$ and $v \geq 0$ in Ω. By (III.7.25) this implies

(III.7.26)

$$\lambda_{n_k} s_{n_k} g\left(\frac{v_{n_k}}{s_{n_k}}\right) \to \lambda_\infty^+ g'(\infty)v \text{ in } C(\overline{\Omega})$$

$$\text{as } k \to \infty.$$

Finally, the elliptic a priori estimate for (III.7.23) in $L^p(\Omega)$ gives by (III.7.26) the convergence of $(v_{n_k})_{k\in\mathbb{N}}$ to v in $W^{2,p}(\Omega)$, and we obtain by (III.7.23) in the limit

$$\Delta v + \lambda_\infty^+ g'(\infty)v = 0 \quad \text{in } \Omega,$$

(III.7.27)

$$v = 0 \quad \text{on } \partial\Omega,$$
$$v \geq 0 \quad \text{in } \Omega.$$

Since $v \in C^\alpha(\overline{\Omega})$, elliptic regularity clearly provides $v \in C^{2,\alpha}(\overline{\Omega})$. The elliptic maximum principle implies $v > 0$ in Ω, since $v \equiv 0$ is excluded by $\|v\|_\infty =$

1. Therefore, $\lambda_\infty^+ g'(\infty) = \mu_0$, which is the principal eigenvalue of $-\Delta$; cf. (III.7.2). By $\mu_0 = \lambda_0 g'(0)$, we obtain

(III.7.28) $\qquad \lambda_\infty^+ = \lambda_0 \dfrac{g'(0)}{g'(\infty)} \quad$ and $\quad \lambda_\infty^- = \lambda_0 \dfrac{g'(0)}{g'(-\infty)}$

by analogous arguments. In view of (III.7.22), this proves that the global branches $\mathcal{C}_{\lambda_0}^\pm$ are smooth curves that are parameterized by λ over the intervals $(\lambda_0 g'(0)/g'(\pm,\infty), \lambda_0)$, respectively.

We claim the asymptotic behavior

(III.7.29) $\qquad\qquad\qquad \lim\limits_{\lambda \searrow \lambda_\infty^\pm} \|u(\lambda)\|_\infty = \infty.$

If (III.7.29) is not true, there exists a sequence $(\lambda_n)_{n\in\mathbb{N}}$ such that $\lambda_n \searrow \lambda_\infty^+$, say, and $\|u(\lambda_n)\|_\infty \le C$ for $(u(\lambda_n), \lambda_n) \in \mathcal{C}_{\lambda_0}^+$. By the "bootstrapping" expounded before, the boundedness of $(u(\lambda_n))_{n\in\mathbb{N}}$ in $C(\overline{\Omega})$ implies via (III.7.1) the boundedness of $(u(\lambda_n))_{n\in\mathbb{N}}$ in $C^{2,\alpha}(\overline{\Omega})$ and, by compact embedding, the convergence of a subsequence $(u(\lambda_{n_k}))_{k\in\mathbb{N}}$ to some u in $C^\alpha(\overline{\Omega})$ with $u \ge 0$ in Ω. By the convergence of $(\lambda_{n_k} g(u(\lambda_{n_k})))_{k\in\mathbb{N}}$ to $\lambda_\infty^+ g(u)$ in $C^\alpha(\overline{\Omega})$, an elliptic a priori estimate for (III.7.1) in $C^\alpha(\overline{\Omega})$ implies the convergence of $(u(\lambda_{n_k}))_{k\in\mathbb{N}}$ to u in $C^{2,\alpha}(\overline{\Omega})$, and (u, λ_∞^+) solves (III.7.1). (This property that boundedness of solutions implies relative compactness is called "properness" of the mapping $\Delta u + \lambda g(u)$; cf. (III.5.7)–(III.5.9).) By its construction, $(u, \lambda_\infty^+) \in \overline{\mathcal{C}}_{\lambda_0}^+$, and by Theorem III.6.2, either $u > 0$ or $u \equiv 0$ in Ω. In the second case, $(0, \lambda_\infty^+)$ would be a bifurcation point for (III.7.1) from the trivial solution line $\{(0, \lambda)\}$, which is excluded, since $\lambda_\infty^+ g'(0) < \lambda_0 g'(0) = \mu_0$, which is the principal (= smallest) eigenvalue of $-\Delta$. Therefore, $u > 0$ in Ω and $(u, \lambda_\infty^+) \in \mathcal{C}_{\lambda_0}^+$, contradicting (III.7.18) by (III.7.28). This proves (III.7.29).

By the asymptotic behavior (III.7.29) we can choose a sequence $\lambda_n \searrow \lambda_\infty^+$ such that the analysis (III.7.23)–(III.7.27) is valid. Since the limit satisfying (III.7.27) is unique by $\|v\|_\infty = 1$, we can conclude that the entire sequence (v_n) converges to v in $W^{2,p}(\Omega)$ (for any $1 < p < \infty$). Accordingly, (III.7.26) is true for any sequence $v_n = u(\lambda_n)/\|u(\lambda_n)\|_\infty = u(\lambda_n)s_n$ where $\lambda_n \searrow \lambda_\infty^+$.

We claim that the convergence (III.7.26) also takes place in $C^\alpha(\overline{\Omega})$. For a proof we give some well-known facts about convergence in Hölder spaces. By the simple estimate

(III.7.30) $\qquad \dfrac{|f(x) - f(y)|}{|x - y|^\alpha} \le 2 \left(\dfrac{|f(x) - f(y)|}{|x - y|^\beta} \right)^{\alpha/\beta} \|f\|_\infty^{1-(\alpha/\beta)}$

$\qquad\qquad$ for $0 < \alpha < \beta \le 1$, $x, y \in \Omega$, $x \ne y$,

the following property is immediate:

Let $(f_n)_{n \in \mathbb{N}} \subset C^\beta(\overline{\Omega})$ be such that

(III.7.31) $\quad f_n \to 0$ in $C(\overline{\Omega})$ and $\|f_n\|_{0,\beta} \leq C$ as $n \to \infty$.

Then $f_n \to 0$ in $C^\alpha(\overline{\Omega})$ for $0 < \alpha < \beta \leq 1$ as $n \to \infty$.

If $(f_n)_{n \in \mathbb{N}} \subset C^1(\overline{\Omega})$ (not to be confused with the Hölder space $C^\beta(\overline{\Omega})$ with $\beta = 1$), then the mean value theorem implies $\|f_n\|_{0,1} \leq C$, provided that $\|\nabla f_n\|_\infty \leq \tilde{C}$ as $n \to \infty$. Using this observation we apply (III.7.31) with $\beta = 1$ to

(III.7.32)
$$f_n = s_n g\left(\frac{v_n}{s_n}\right) - g'(\infty)v \text{ for which}$$
$$\nabla f_n = g'\left(\frac{v_n}{s_n}\right)\nabla v_n - g'(\infty)\nabla v \text{ and}$$
$$\|\nabla f_n\|_\infty \leq C_{g'}(\|\nabla v_n\|_\infty + \|\nabla v\|) \leq \tilde{C} \text{ as } n \to \infty.$$

Here assumption $(\text{III.7.11})_1$ gives the constant $C_{g'}$, and the convergence $v_n \to v$ in $W^{2,p}(\Omega)$ (which implies the convergence $v_n \to v$ in $C^1(\overline{\Omega})$ by embedding $W^{2,p}(\Omega) \subset C^1(\overline{\Omega})$ for $p > n$) gives the boundedness of $\|\nabla v_n\|_\infty$. This proves that

(III.7.33) $\quad \lambda_n s_n g\left(\dfrac{v_n}{s_n}\right) \to \lambda_\infty^+ g'(\infty)v \quad$ in $C^\alpha(\overline{\Omega})$,

and the elliptic a priori estimate for (III.7.23) in $C^\alpha(\overline{\Omega})$ implies the convergence $v_n \to v$ in $C^{2,\alpha}(\overline{\Omega})$.

We summarize:

Theorem III.7.1 *Let* $\mathcal{C}_{\lambda_0}^\pm$ *be the unbounded continua of positive (negative) solutions of (III.7.1) emanating at* $(0, \lambda_0)$, *where the function* g *satisfies* $g(0) = 0$, $g'(0) > 0$, *and (III.7.11). Then* $\mathcal{C}_{\lambda_0}^\pm$ *are smooth curves parameterized by* $\lambda \in (\lambda_\infty^\pm, \lambda_0) = (\lambda_0 g'(0)/g'(\pm\infty), \lambda_0)$, *respectively. Furthermore,*

(III.7.34)
$$\lim_{\lambda \searrow \lambda_\infty^\pm} \|u(\lambda)\|_\infty = \infty \text{ for } (u(\lambda), \lambda) \in \mathcal{C}_{\lambda_0}^\pm,$$
$$\lim_{\lambda \searrow \lambda_\infty^\pm} \frac{u(\lambda)}{\|u(\lambda)\|_\infty} = \hat{v}_0 \text{ in } C^{2,\alpha}(\overline{\Omega}),$$
where $\hat{v}_0$ *is the positive (negative)*
eigenfunction of $-\Delta$ *to the principal*
eigenvalue $\mu_0 = \lambda_0 g'(0) = \lambda_\infty^\pm g'(\pm\infty)$.

Finally, the trivial solution line $\{(0, \lambda)\}$ *is stable for* $\lambda < \lambda_0$ *and unstable for* $\lambda > \lambda_0$, *whereas* $\mathcal{C}_{\lambda_0}^\pm$ *are unstable.*

The instability of the global curves $\mathcal{C}_{\lambda_0}^\pm$ is not a consequence of the local Principle of Exchange of Stability expounded in Section I.7. It is a consequence of (III.7.20) as follows:

Let $\tilde{v}_0$ (depending on u) be the normalized positive eigenfunction for the principal eigenvalue $\mu_0(g'(u))$ with weight function $g'(u)$ of $-\Delta$ on Ω subject to homogeneous Dirichlet boundary conditions. Then the greatest eigenvalue of $D_u F(u, \lambda)$ for $(u, \lambda) \in C_{\lambda_0}^{\pm}$ is

$\max\{-\|\nabla v\|_0^2 + \lambda(g'(u)v, v)_0 | v \in H_0^1(\Omega), \|v\|_0 = 1\}$

$\geq -\|\nabla \tilde{v}_0\|_0^2 + \lambda(g'(u)\tilde{v}_0, \tilde{v}_0)_0 > -\|\nabla \tilde{v}_0\|_0^2 + \mu_0(g'(u))(g'(u)\tilde{v}_0, \tilde{v}_0)_0 = 0.$

The positivity of the greatest eigenvalue of $D_u F(u, \lambda)$ entails the instability of u.

We sketch $C_{\lambda_0}^+$ in Figure III.7.4. The monotonicity with respect to the amplitude is true for symmetric domains in $\mathbb{R}^2$; cf. Theorems III.7.9, III.7.17.

Properties (III.7.34) suggest a Bifurcation from Infinity at $\lambda = \lambda_\infty^{\pm}$. We prove "bifurcation" in the sense of Theorem I.20.1 under more restrictive assumptions on g at $u = \infty$ that give $\lambda_\infty^{\pm} = \lambda_\infty$. On the other hand, we drop the assumption $g(0) = 0$.

III.7.1 Bifurcation from Infinity

Problem (III.7.1) provides an example of Bifurcation from Infinity in the sense of Theorem I.20.1 under the assumptions

(III.7.35)
$$\lim_{|u| \to \infty} g'(u) = g'(\infty) > 0, \text{ i.e., } g'(\infty) = g'(-\infty),$$
$$|ug''(u)| \leq C_1 \text{ for all } u \in \mathbb{R}, \text{ with some } C_1 > 0.$$

Let $\mu_0 = \lambda_\infty g'(\infty)$ be the principal eigenvalue of $-\Delta$ and let $\hat{v}_0$ be a positive eigenfunction spanning $N(\Delta + \mu_0 I) \subset X = C^{2,\alpha}(\overline{\Omega}) \cap \{u | u = 0 \text{ on } \partial\Omega\}$. Setting

(III.7.36)
$$F(u, \lambda) = L(\lambda)u + \lambda R(u),$$
$$L(\lambda)u = \Delta u + \lambda g'(\infty)u,$$
$$R(u) = g(u) - g'(\infty)u,$$

we see that the assumptions (I.20.2), (I.20.3) with $A(\lambda) = L(\lambda)$, $\lambda_0 = \lambda_\infty$, and (I.20.5) are satisfied. (Recall that $\hat{v}_0 \notin R(\Delta + \mu_0 I) = \text{span}[\hat{v}_0]^{\perp} \cap Z$, where $Z = C^{\alpha}(\overline{\Omega})$ and the orthogonal complement is taken with respect to the scalar product in $L^2(\Omega)$.) The crucial assumption for Theorem I.20.1 is (I.20.4): Let U be a neighborhood of $\hat{v}_0$ in X. Then it has to be shown that

$$sg\left(\frac{v}{s}\right) - g'(\infty)v \to 0 \text{ in } Z = C^\alpha(\overline{\Omega}),$$

(III.7.37) $$\left(g'\left(\frac{v}{s}\right) - g'(\infty)\right)h \to 0 \text{ in } Z \text{ as } s \to 0,$$

for all $v \in U \subset X$ and

uniformly for $h \in X$ with $\|h\|_{2,\alpha} \le 1$.

By (III.7.35)$_1$, $g(u)/u \to g'(\infty)$ as $|u| \to \infty$, whence $|g(u) - g'(\infty)u| \le \varepsilon|u|$ for $|u| \ge M(\varepsilon)$. Therefore, for $M = M(\varepsilon)$,

(III.7.38)
$$\max_{x \in \overline{\Omega}} |g(u(x)) - g'(\infty)u(x)|$$
$$\le \max_{|u(x)| \le M} |./.| + \max_{|u(x)| \ge M} |./.| \le C_g(M) + \varepsilon\|u\|_\infty$$
$$\le 2\varepsilon\|u\|_\infty \text{ if } \|u\|_\infty \ge C_g(M)/\varepsilon.$$

Using $\|v\|_\infty \le c_1$ for all $v \in U$, we see that the estimate (III.7.38) implies

$$\left\|sg\left(\frac{v}{s}\right) - g'(\infty)v\right\|_\infty \le 2c_1\varepsilon$$

(III.7.39) for all $|s| \le \varepsilon c_1/C_g(M)$ or

$$sg\left(\frac{v}{s}\right) - g'(\infty)v \to 0 \text{ in } C(\overline{\Omega}) \text{ as } s \to 0.$$

This argument does not apply to (III.7.37)$_2$, where we need the following properties:

$$v \in U \subset X \text{ implies } v(x) > 0 \text{ for all } x \in \Omega,$$

$$0 < v(x) \le \varepsilon \text{ if } \text{dist}(x, \partial\Omega) \le \delta_1(\varepsilon),$$

(III.7.40) $$v(x) \ge d(\delta_1) > 0 \text{ if } \text{dist}(x, \partial\Omega) \ge \delta_1 > 0,$$

$$h \in X \text{ with } \|h\|_{2,\alpha} \le 1 \text{ implies}$$

$$|h(x)| \le \varepsilon \text{ if } \text{dist}(x, \partial\Omega) \le \delta_2(\varepsilon).$$

Note that $U \subset X$ is a neighborhood of $\hat{v}_0$ that is positive in Ω and for which Hopf's boundary lemma is valid on $\partial\Omega$.

By (III.7.35)$_1$, $|g'(u) - g'(\infty)| \le C_{g'}$ for all $u \in \mathbb{R}$, whence

$$\left|(g'\left(\frac{v(x)}{s}\right) - g'(\infty))h(x)\right| \le C_{g'}\varepsilon$$

if $\text{dist}(x, \partial\Omega) \le \delta_2(\varepsilon)$,

(III.7.41)
$$\left|\left(g'\left(\frac{v(x)}{s}\right) - g'(\infty)\right)h(x)\right| \le \varepsilon$$

if $\text{dist}(x, \partial\Omega) \ge \delta_2(\varepsilon)$ and $|s| \le \delta(\varepsilon)$, or

$$\left(g'\left(\frac{v}{s}\right) - g'(\infty)\right)h \to 0 \text{ in } C(\overline{\Omega}) \text{ as } s \to 0,$$

uniformly for $h \in X$ with $\|h\|_{2,\alpha} \le 1$.

We use the property (III.7.31) in order to prove (III.7.37). Let $(s_n)_{n\in\mathbb{N}} \subset \mathbb{R}$ be such that $s_n \to 0$. Then for

$$f_n = s_n g\left(\frac{v_n}{s_n}\right) - g'(\infty)v \text{ we obtain}$$

(III.7.42)
$$\nabla f_n = \left(g'\left(\frac{v}{s_n}\right) - g'(\infty)\right)\nabla v, \text{ whence}$$

$$\|\nabla f_n\|_\infty \le C_{g'}\|\nabla v\|_\infty \le C_{g'}c_2 = \tilde{C},$$

where $C_{g'}$ depends only on the uniform bound of $|g'(u)|$, cf. (III.7.35)$_1$, and c_2 is a uniform bound of $\|\nabla v\|_\infty$ for all $v \in U$. The application of (III.7.31) with $\beta = 1$ then implies (III.7.37)$_1$. For

$$f_n = \left(g'\left(\frac{v}{s_n}\right) - g'(\infty)\right)h \text{ we obtain}$$

(III.7.43)
$$\nabla f_n = \frac{1}{s_n}g''\left(\frac{v}{s_n}\right)(\nabla v)h + \left(g'\left(\frac{v}{s_n}\right) - g'(\infty)\right)\nabla h.$$

The second term of (III.7.43)$_2$ is estimated as in (III.7.42) when c_2 is a uniform bound of $\|\nabla h\|_\infty$ for all $h \in X$ with $\|h\|_{2,\alpha} \le 1$. For the first term we use the uniform estimate

$$|h(x)| \le c_3 v(x) \text{ for all } x \in \Omega,$$

(III.7.44)
$$\text{for all } h \in X \text{ with } \|h\|_{2,\alpha} \le 1,$$

$$\text{and for all } v \in U \subset X.$$

Indeed, the positive eigenfunction $\hat{v}_0$ with nonzero normal derivatives on $\partial\Omega$ (Hopf's boundary lemma) bounds all functions $h \in X = C^{2,\alpha}(\overline{\Omega}) \cap \{u|u = 0$ on $\partial\Omega\}$ with $\|h\|_\infty + \|\nabla h\|_\infty \le \|h\|_{2,\alpha} \le 1$ pointwise in Ω by $|h(x)| \le \tilde{c}_3\hat{v}_0(x)$ for some $\tilde{c}_3 > 0$. This estimate persists for all $v \in U \subset X$ with a uniform constant c_3, since U is a neighborhood of $\hat{v}_0$ in X. Using (III.7.44) we obtain for the first term (III.7.43)$_2$,

$$\left| \frac{1}{s_n} g'' \left(\frac{v(x)}{s_n} \right) \nabla v(x) h(x) \right|$$

(III.7.45)
$$\leq c_3 \left| \frac{v(x)}{s_n} g'' \left(\frac{v(x)}{s_n} \right) \right| \|\nabla v(x)\|$$

$$\leq c_3 C_1 \|\nabla v\|_\infty \leq \tilde{C} \text{ by (III.7.35)}_2,$$

uniformly for $v \in U \subset X$, $\|h\|_{2,\alpha} \leq 1$.

By (III.7.31) this proves (III.7.37)$_2$, and all assumptions of Theorem I.20.1 are verified.

Theorem III.7.2 *Under the assumptions (III.7.35) there is a unique continuous curve $\{(v(s), \lambda(s)) | s \in (-\delta, \delta)\}$ in $(C^{2,\alpha}(\overline{\Omega}) \cap \{u | u = 0 \text{ on } \partial\Omega\}) \times \mathbb{R}$ through $(v(0), \lambda(0)) = (\hat{v}_0, \lambda_\infty)$ such that the functions*

(III.7.46)
$$u(s) = \frac{v(s)}{s} \text{ for } s \in (-\delta, \delta) \backslash \{0\}$$

solve (III.7.1) with $\lambda = \lambda(s)$.

Here $\mu_0 = \lambda_\infty g'(\infty)$ is the principal eigenvalue of $-\Delta$ with positive eigenfunction $\hat{v}_0$.

The property

(III.7.47)
$$\lim_{s \to 0} \|u(s)\|_{2,\alpha} = \infty, \quad \lim_{s \to 0} \lambda(s) = \lambda_\infty$$

explains the nomenclature "Bifurcation from Infinity" at $\lambda = \lambda_\infty$. Since $\hat{v}_0 > 0$ in Ω and $\hat{v}_0$ has nonzero normal derivatives on $\partial\Omega$ (Hopf's boundary lemma), we have $v(s) > 0$ in Ω for $s \in (-\delta, \delta)$ such that

(III.7.48)
$$u(s) > 0 \text{ in } \Omega \quad \text{for } s \in (0, \delta),$$
$$u(s) < 0 \text{ in } \Omega \quad \text{for } s \in (-\delta, 0).$$

If the global continua $\mathcal{C}^\pm_{\lambda_0}$ exist (cf. Theorem III.7.1), one expects that the ends of $\mathcal{C}^\pm_{\lambda_0}$ near (∞, λ_∞) are on the curve (III.7.46) bifurcating from infinity. We show more:

Let $(u_n, \lambda_n) \in X \times \mathbb{R}$ be positive (negative)

solutions of (III.7.1) such that $\|u_n\|_\infty \to \infty$

and $\lambda_n \to \hat{\lambda}_\infty^+ (\hat{\lambda}_\infty^-)$ as $n \to \infty$. Then

(III.7.49) $u_n = \dfrac{v(s_n)}{s_n} = u(s_n)$, $\lambda_n = \lambda(s_n)$,

for some $s_n \in (0, \delta)((-\delta, 0))$ with $s_n \to 0$ and where

$\{(v(s), \lambda(s))\}$ is the unique curve of Theorem III.7.2,

and therefore $\hat{\lambda}_\infty^+ = \hat{\lambda}_\infty^- = \lambda_\infty$.

For a proof we confine ourselves to the case of positive u_n. The same arguments that prove Theorem III.7.1 yield

$$\frac{u_n}{\|u_n\|_\infty} \to v \text{ in } C^{2,\alpha}(\overline{\Omega}) \text{ as } n \to \infty,$$

where $v > 0$ solves

(III.7.50) $\Delta v + \hat{\lambda}_\infty^+ g'(\infty)v = 0$ in Ω,

$$v = 0 \text{ on } \partial\Omega,$$

whence $v = \hat{v}_0$ and $\hat{\lambda}_\infty^+ = \lambda_\infty$.

Normalizing $\|\hat{v}_0\|_0 = 1$ (where $\| \quad \|_0$ and $(\quad , \quad)_0$ denote the norm and the scalar product in $L^2(\Omega)$), we obtain for $s_n = 1/(u_n, \hat{v}_0)_0$ that

$v_n = s_n u_n \in X$ satisfies

$v_n \to \hat{v}_0$ in X, $\lambda_n \to \lambda_\infty$, $s_n \to 0$ as $n \to \infty$,

(III.7.51) $\tilde{F}(v_n, \lambda_n, s_n) = (0, 0)$, where

$$\tilde{F}(v, \lambda, s) = \left(L(\lambda)v + \lambda s R\left(\frac{v}{s}\right), (v, \hat{v}_0)_0 - 1 \right);$$

cf. (III.7.36). On the other hand, the triplet from Theorem III.7.2, $(v(s), \lambda(s), s)$ $\in X \times \mathbb{R} \times \mathbb{R}$, solves (III.7.51)$_3$, too, so that uniqueness of the solutions near $(\hat{v}_0, \lambda_\infty, 0)$ implies (III.7.49). (In the proof of Theorem I.20.1 the duality $\langle v, \hat{v}_0' \rangle$ can be replaced by $(v, \hat{v}_0)_0$.) Statement (III.7.49) implies in particular that

(III.7.52)   under the assumptions (III.7.11) and (III.7.35), the global curves $\mathcal{C}_{\lambda_0}^\pm$ described in Theorem III.7.1 connect the local curves bifurcating from 0 at $\lambda = \lambda_0$ and bifurcating from infinity at $\lambda = \lambda_\infty$.

Finally, the asymptotic behaviors (III.7.33) and (III.7.47) are equivalent:

(III.7.53) $\lim\limits_{\lambda \searrow \lambda_\infty} \|u(\lambda)\|_{2,\alpha} = \infty \Leftrightarrow \lim\limits_{\lambda \searrow \lambda_\infty} \|u(\lambda)\|_\infty = \infty.$

This complete description of the global positive (negative) solution branches $\mathcal{C}^{\pm}_{\lambda_0}$ requires rather restrictive assumptions on the nonlinearity g. In the sequel we prove smoothness of positive solution branches under quite general assumptions on g, but the domain Ω is symmetric and partially convex in $\mathbb{R}^2$.

Remark III.7.3 *Bifurcation from Infinity for (III.7.1) can be proved under weaker assumptions on g than stated in (III.7.35). To be more precise, assumption (III.7.35)$_2$ is not necessary, and (III.7.35)$_1$ can be weakened to* $\lim_{|u| \to \infty} g(u)/u = K$. *Following [149], we consider* $\Delta : X \to Z$ *as a continuous bijection for* $X = C^{2,\alpha}(\overline{\Omega}) \cap \{u | u = 0 \text{ on } \partial\Omega\}$, $Z = C^\alpha(\overline{\Omega})$, *and also for* $X = W^{2,p}(\Omega) \cap W^{1,p}_0(\Omega)$, $Z = L^p(\Omega)$, *cf. (III.1.8), (III.1.40). Then (III.7.1) is equivalent to the fixed-point problem*

$$\text{(III.7.54)} \qquad u = -\lambda\Delta^{-1}g(u), \ u \in C(\overline{\Omega}).$$

Setting $A(\lambda) = I + \lambda K\Delta^{-1}$ *and* $R(u) = \Delta^{-1}(g(u) - Ku)$, *we see that the inversion* $w = u/\|u\|^2_\infty$ *transforms*

$$\text{(III.7.55)} \qquad \begin{aligned} &A(\lambda)u + \lambda R(u) = 0, \ (u, \lambda) \in C(\overline{\Omega}) \times \mathbb{R}, \ into \\ &A(\lambda)w + \lambda\|w\|^2_\infty R(w/\|w\|^2_\infty) = 0, \ (w, \lambda) \in C(\overline{\Omega}) \times \mathbb{R}. \end{aligned}$$

Assumption $|g(u) - Ku| = o(|u|)$ *at* $|u| = \infty$ *implies for*

$$\tilde{R}(w) \equiv \|w\|^2_\infty R(w/\|w\|^2_\infty) \ that$$

$$\text{(III.7.56)} \qquad \tilde{R}(w) = o(\|w\|_\infty) \ at \ w = 0 \ in \ C(\overline{\Omega})$$

and $\tilde{R} : C(\overline{\Omega}) \to C(\overline{\Omega})$ *is compact; see (III.7.38) and [149].*

Since $A(\lambda) = I + \lambda K\Delta^{-1}$ *is a compact perturbation of the identity, cf. Definition II.2.1, the analysis of bifurcation from the trivial solution line for* $A(\lambda)w + \lambda\tilde{R}(w) = 0$ *allows the application of the Leray–Schauder degree; cf. Sections II.2, II.3. In particular, if* $\mu_k = \lambda_k K$ *is an eigenvalue of* $-\Delta$ *on* Ω *(subject to homogeneous Dirichlet boundary conditions) of odd multiplicity, then* $(0, \lambda_k)$ *is a bifurcation point for* $A(\lambda)w + \lambda\tilde{R}(w) = 0$; *cf. Theorems II.3.2 and II.3.3. The inversion* $u = w/\|w\|^2_\infty$ *then provides bifurcation for* $A(\lambda)u + \lambda R(u) = 0$ *at* (∞, λ_k), *which, in turn, proves Bifurcation from Infinity for (III.7.1) at* $\lambda = \lambda_k$. *Furthermore, the alternative of the Global Rabinowitz Bifurcation Theorem is valid for* $A(\lambda)w + \lambda\tilde{R}(w) = 0$. *However, the translation of that alternative to an alternative for Bifurcation from Infinity after inversion is somewhat awkward, and it is found in [149]. Note that the statements of Theorems I.20.1 and III.7.2 are much sharper. The bifurcating continuum obtained by the application of the Leray–Schauder degree does not necessarily consist of a unique curve as stated in Theorem III.7.2. Since we need that uniqueness in the sequel, the stronger assumptions (III.7.35) are* *justified. Our analysis, however, applies only if $\mu_0 = \lambda_\infty g'(\infty)$ is the principal eigenvalue of $-\Delta$.*

III.7.2 Local Parameterization of Positive Solution Branches over Symmetric Domains

We consider (III.7.1) over a domain having the following properties:

(III.7.57)
$$\begin{aligned} &\Omega \subset \mathbb{R}^2 \text{ is bounded with a smooth boundary } \partial\Omega, \\ &\Omega \text{ is symmetric with respect to the } x\text{- and } y\text{-axes}, \\ &\text{and } \Omega \text{ is partially convex; i.e.,} \\ &\text{if } (x_1, y), (x_2, y) \in \Omega \text{ then } (tx_1 + (1-t)x_2, y) \in \Omega, \\ &\text{if } (x, y_1), (x, y_2) \in \Omega \text{ then } (x, ty_1 + (1-t)y_2) \in \Omega, \\ &\text{for all } t \in [0, 1]. \end{aligned}$$

Remark III.7.4 *For the sake of convenience we restrict the presentation to domains (III.7.57). There are, however, other cases to which the same or modified arguments as given in this sequel can be applied: For instance,*

(III.7.58)   *$\Omega \subset \mathbb{R}^2$ is a rectangle*
                        or an equilateral triangle.

Consider $\Delta u + \lambda g(u) = 0$ with $g(-u) = -g(u)$ on a rectangular or hexagonal lattice $\mathcal{L}$ as described in Section III.6. As summarized in (III.6.40), each point $(0, \lambda_0)$ with $\lambda_0 > 0$ gives rise to unbounded branches of solutions whose nodal set consists precisely of the lines of a particular rectangular or hexagonal lattice whose periodicities are in one-to-one correspondence with λ_0. On each tile Ω of the lattice, the solution u is positive or negative. Whereas a rectangle Ω has the symmetries of (III.7.57), an equilateral triangle Ω has three symmetry axes. What is needed is for each half of Ω on one side of a symmetry axis to be an optimal cap in the sense of [55]. Then, according to a celebrated result of [55], positive solutions on Ω have the symmetries of Ω; cf. Proposition III.7.5 below. Having this in mind, an extension to smooth and convex domains having the symmetries of any regular polygon is possible, too. In Remark III.7.18 we summarize our results for domains other than (III.7.57).

We quote from [55] the following result:

Proposition III.7.5 *Assume that $u \in C^2(\overline{\Omega})$ is a positive solution of (III.7.1), where $\lambda > 0$ and $g(0) \geq 0$. Then u has the same symmetries as Ω; i.e., for $(x, y) \in \Omega$,*

(III.7.59)
$$\begin{aligned} u(-x, y) &= u(x, y), \\ u(x, -y) &= u(x, y), \end{aligned}$$

and moreover,

(III.7.60)
$$u_x < 0 \text{ on } \{(x,y) \in \Omega | x > 0\},$$
$$u_y < 0 \text{ on } \{(x,y) \in \Omega | y > 0\}.$$

Finally,

(III.7.61)
$$u_\nu < 0 \quad \text{on} \quad \partial\Omega$$

for the derivative in the direction of the exterior normal unit vector ν on $\partial\Omega$.

Note that properties (III.7.59), (III.7.60) are valid for every g and $\lambda \neq 0$.

The last property (III.7.61) follows from Hopf's boundary lemma using the trick (III.6.10) and the assumptions $\lambda > 0, g(0) \geq 0$: If $g(0) > 0$, then for x near $\partial\Omega$, (III.7.1) implies $\Delta u(x) = -\lambda g(u(x)) < 0$, and the usual boundary lemma applies. If $g(0) = 0$, then (III.7.1) implies $\Delta u + \lambda c u = 0$, where $c = g(u)/u \in C(\overline{\Omega})$. Since $u > 0$ in Ω, we obtain, as in (III.6.10), $\Delta u + \lambda c^- u \leq 0$, and again the usual Hopf lemma applies.

As stated in Remark III.7.8 below, property (III.7.61) is not necessarily true if $g(0) < 0$.

Another consequence of Proposition III.7.5 is that for positive solutions of (III.7.1) (for all $\lambda \neq 0$ and for all g),

(III.7.62)
$$\|u\|_\infty = \max_{(x,y)\in\Omega} u(x,y) = u(0,0).$$

Lemma III.7.6 *Let $u \in C^2(\overline{\Omega})$ be a positive solution of (III.7.1) where $g(0) \geq 0$. Then the linear problem*

(III.7.63)
$$\Delta v + \lambda g'(u)v = 0 \quad \text{in } \Omega,$$
$$v = 0 \quad \text{on } \partial\Omega,$$

has no nontrivial solution $v \in C^2(\overline{\Omega})$ having the symmetries

(III.7.64)
$$v(-x,y) = v(x,y),$$
$$v(x,-y) = v(x,y),$$

and satisfying

(III.7.65)
$$v(0,0) = 0.$$

If the homogeneous boundary conditions (III.7.63)$_2$ are replaced by

(III.7.66)
$$v > 0 \quad \text{or} \quad v < 0 \quad \text{on} \quad \partial\Omega,$$

then problems (III.7.63)$_1$, (III.7.64), (III.7.65), (III.7.66) also have no solution.

Proof. We assume the existence of some nontrival $v \in C^2(\overline{\Omega})$ having the properties (III.7.64), (III.7.65), solving (III.7.63)$_1$, and satisfying $v = 0$ or $v > 0$ or $v < 0$ on $\partial\Omega$. By (III.7.65), the origin $(0,0)$ is in the nodal set

N of v. The complement $\Omega \backslash N$ decomposes into nodal domains, and by the maximum principle, v has opposite signs in two adjacent domains. (The sign of $\lambda g'(u) = c$ plays no role for this argument; see the trick of (III.6.10).)

Since $(0,0) \in N$, the symmetry of v implies that there is a nodal domain $\mathcal{D}$ of v in $\Omega \cap \{x > 0\}$ or in $\Omega \cap \{y > 0\}$ and $v = 0$ on $\partial \mathcal{D}$, $v > 0$ in $\mathcal{D}$. (In case of (III.7.66) we have $\overline{\mathcal{D}} \subset \Omega$.) Therefore, depending on its sign, the number $\lambda \neq 0$ is the largest negative or smallest positive eigenvalue of the weak eigenvalue problem for $-\Delta$ with weight function $g'(u)$ over $\mathcal{D}$ subject to homogeneous Dirichlet boundary conditions; cf. Remark III.2.3, in particular (III.2.46), (III.2.47). It is characterized by an extremal property of

$$(\text{III.7.67}) \qquad \frac{\|\nabla v\|^2_{L^2(\mathcal{D})}}{(g'(u)v, v)_{L^2(\mathcal{D})}} \quad \text{among} \quad v \in H_0^1(\mathcal{D})$$
$$\text{such that } (g'(u)v, v)_{L^2(\mathcal{D})} > 0 \text{ or } < 0.$$

Since we need a piecewise smoothness of $\partial \mathcal{D}$ (which is true but which is not easy to prove), we proceed as follows: Let $\mathcal{D} \subset D \subset \Omega \cap \{x > 0\}$, say, where ∂D is piecewise smooth. Then the largest negative (or smallest positive) eigenvalue μ_0 of the same weak eigenvalue problem over D is not smaller (or not larger) than λ. This follows by the extremal property of the "principal" eigenvalue μ_0. If $\mu_0 = \lambda$, we replace v by the positive eigenfunction over D that satisfies (III.7.63), where Ω is replaced by D. If $\lambda < \mu_0 < 0$ or $0 < \mu_0 < \lambda$, then we shrink D in $\Omega \cap \{x > 0\}$, preserving its piecewise smooth boundary, so that the "principal" eigenvalue μ_0 becomes λ. This is possible, since by Poincaré's inequality (III.6.4),

$$(\text{III.7.68}) \qquad \begin{aligned} &|\mu_0| \geq c|D|^{-1} \text{ for a uniform } c > 0 \\ &\text{and } |D| = \text{meas} D. \end{aligned}$$

Therefore, replacing $\mathcal{D}$ by D if necessary, we can assume that v satisfies (III.7.63), where Ω is replaced by $\mathcal{D}$, that $v > 0$ in $\mathcal{D}$, that $\mathcal{D}$ is in $\Omega \cap \{x > 0\}$, say, and that $\partial \mathcal{D}$ is piecewise smooth.

Using (III.7.1) for u (which is certainly in $C^3(\overline{\Omega})$ by elliptic regularity), we obtain by differentiation

$$(\text{III.7.69}) \qquad \Delta u_x + \lambda g'(u)u_x = 0 \quad \text{in } \overline{\mathcal{D}},$$

and Green's formula then gives, in view of $v = 0$ on $\partial \mathcal{D}$,

$$(\text{III.7.70}) \qquad \int_{\partial \mathcal{D}} \frac{\partial v}{\partial n} u_x = 0,$$

where n is the exterior normal unit vector on $\partial \mathcal{D}$. However,

$$\frac{\partial v}{\partial n} < 0 \text{ on } \partial\mathcal{D}\backslash\{ \text{ singular points}\},$$

by Hopf's boundary lemma using $v > 0$ as in (III.6.10),

(III.7.71) $u_x < 0$ on $\partial\mathcal{D} \cap \{x > 0\}$,

$$u_x = 0 \text{ on } \partial\mathcal{D} \cap \{x = 0\},$$

by Proposition III.7.5,

contradicting (III.7.70).   □

Our **Main Result on Local Parameterization of Positive Solutions** reads as follows:

Theorem III.7.7 *Let $(u_0, \lambda_0) \in C^{2,\alpha}(\overline{\Omega}) \times \mathbb{R}$ be a solution of (III.7.1) such that $u_0 > 0$ in Ω and $\lambda_0 > 0$. We assume $g(0) \geq 0$ and that the domain Ω satisfies (III.7.57). According to (III.7.62), we denote its amplitude by $\|u_0\|_\infty = u_0(0,0) = p_0$. Then there are a neighborhood $U \times V$ of $(u_0, \lambda_0) \in C^{2,\alpha}(\overline{\Omega}) \times \mathbb{R}$ and a curve of class C^1 if g is of class C^2,*

(III.7.72)
$$\mathcal{C} = \{(u(p), \lambda(p))|p \in (p_0 - \delta, p_0 + \delta)\}$$

$$\text{through } (u(p_0), \lambda(p_0)) = (u_0, \lambda_0),$$

such that all solutions of (III.7.1) in $U \times V$ are on the curve $\mathcal{C}$, $u(p) > 0$ in Ω, and

(III.7.73) $\|u(p)\|_\infty = u(p)(0,0) = p$ for $p \in (p_0 - \delta, p_0 + \delta)$.

In other words, the curve $\mathcal{C}$ is parameterized by the amplitude p of $u(p)$.

Proof. The result follows from the Implicit Function Theorem of Section I.1 in a suitable setting. As usual, we set $X = C^{2,\alpha}(\overline{\Omega}) \cap \{u|u = 0 \text{ on } \partial\Omega\}$, $Z = C^\alpha(\overline{\Omega})$, and from (III.1.26), (III.1.29), we know that

$$L : X \to Z, \text{ defined by}$$

(III.7.74) $Lv = \Delta v + \lambda_0 g'(u_0)v,$

is a (bounded) Fredholm operator of index zero.

For the symbols $\sigma, \tau \in \{+, -\}$ we introduce four symmetry classes,

(III.7.75) $\Sigma^{(\sigma,\tau)} = \{u : \overline{\Omega} \to \mathbb{R}|u(-x, y) = \sigma u(x, y), u(x, -y) = \tau u(x, y)\},$

and by the symmetry $u_0 \in \Sigma^{(+,+)}$ stated in Proposition III.7.5,

$$L : X^{(\sigma,\tau)} \to Z^{(\sigma,\tau)} \text{ with}$$

(III.7.76)
$$X^{(\sigma,\tau)} = X \cap \Sigma^{(\sigma,\tau)}, \ Z^{(\sigma,\tau)} = Z \cap \Sigma^{(\sigma,\tau)},$$

is a Fredholm operator of index zero

for all $\sigma, \tau \in \{+, -\}$.

For a proof of (III.7.76) we refer to the arguments given in Section III.1 proving (III.1.33): Choose a constant $c \geq 0$ such that $L - cI : X \to Z$ becomes bijective. Since the operator $L - cI$ commutes with every (inverse) reflection across the x- or y-axis, the unique solution u of $Lu - cu = f$ is necessarily in $X^{(\sigma,\tau)}$ if $f \in Z^{(\sigma,\tau)}$. Therefore, $L - cI : X^{(\sigma,\tau)} \to Z^{(\sigma,\tau)}$ is bijective, which implies (III.7.76) by the same arguments that (III.1.8) implies (III.1.12).

Next we introduce a mapping

$$G : X^{(+,+)} \times \mathbb{R} \times (0, \infty) \to Z^{(+,+)} \times \mathbb{R},$$

(III.7.77)
$$G(w, \lambda, t) \equiv \left(\Delta w + \frac{\lambda p_0}{t} g\left(\frac{t}{p_0} w \right), w(0,0) - p_0 \right),$$

which is of class C^1 and $G(u_0, \lambda_0, p_0) = (0,0)$. We prove that its derivative

(III.7.78)
$$D_{(w,\lambda)} G(u_0, \lambda_0, p_0)[v, \lambda] = (Lv + \lambda g(u_0), v(0,0)),$$

$$D_{(w,\lambda)} G(u_0, \lambda_0, p_0) : X^{(+,+)} \times \mathbb{R} \to Z^{(+,+)} \times \mathbb{R},$$

is bijective.

  Case I. The kernel $N(L)$ is nontrivial in $X^{(+,+)}$.

  By Lemma III.7.6 we know that $v(0,0) \neq 0$ for all nontrivial $v \in N(L)$. Therefore,

(III.7.79)
$$\dim N(L) = 1,$$

since otherwise, there would be some nontrivial $v \in N(L) \subset X^{(+,+)}$ with $v(0,0) = 0$. We claim that

(III.7.80)
$$g(u_0) \notin R(L) \subset Z^{(+,+)}.$$

The rescaled function $u_0^s(x,y) = u_0(sx, sy)$ solves $\Delta u_0^s + s^2 \lambda_0 g(u_0^s) = 0$ for all $s > 0$. Differentiating this equation with respect to s at $s = 1$ yields

(III.7.81)
$$v_1(x,y) = x u_{0x}(x,y) + y u_{0x}(x,y), \text{ solving}$$

$$\Delta v_1 + \lambda_0 g'(u_0) v_1 = -2\lambda_0 g(u_0).$$

Furthermore,

(III.7.82)
$$v_1 \in \Sigma^{(+,+)} \quad \text{and} \quad v_1 < 0 \text{ on} \quad \partial\Omega$$

by assumption (III.7.57) on Ω and by (III.7.61) of Proposition III.7.5. Assume that $g(u_0) \in R(L)$. Then

$$Lv_0 = 2\lambda_0 g(u_0) \text{ for some } v_0 \in X^{(+,+)},$$

(III.7.83)
$$L(v + v_0 + v_1) = 0,$$

$$(v + v_0 + v_1)(0,0) = 0, \text{ and}$$

$$v + v_0 + v_1 < 0 \text{ on } \partial\Omega \text{ for some } v \in N(L),$$

contradicting Lemma III.7.6. This proves (III.7.80).

Since $v(0,0) \neq 0$ for nontrivial $v \in N(L)$, we derive from (III.7.76), (III.7.79), (III.7.80) that $D_{(w,\lambda)}G(u_0, \lambda_0, p_0)$ as given by (III.7.78) is surjective. To prove its injectivity, let

(III.7.84)
$$D_{(w,\lambda)}G(u_0, \lambda_0, p_0)[v, \lambda] = (Lv + \lambda g(u_0), v(0,0)) = (0,0),$$

for some $(v, \lambda) \in X^{(+,+)} \times \mathbb{R}.$

By (III.7.80) we obtain $\lambda = 0$ and $v \in N(L)$. Lemma III.7.6 then implies that $v = 0$.

   Case II. The kernel $N(L)$ is trivial in $X^{(+,+)}$.

   In this case, (III.7.76) implies that $L : X^{(+,+)} \to Z^{(+,+)}$ is bijective and therefore

(III.7.85)
$$Lv_2 = g(u_0) \quad \text{for some} \quad v_2 \in X^{(+,+)}.$$

We claim that $v_2(0,0) \neq 0$. If $v_2(0,0) = 0$, then the function v_1 from (III.7.81) yields

$$L(v_1 + 2\lambda_0 v_2) = 0, \; v_1 + 2\lambda_0 v_2 \in \Sigma^{(+,+)},$$

(III.7.86)
$$(v_1 + 2\lambda_0 v_2)(0,0) = 0, \text{ and}$$

$$v_1 + 2\lambda_0 v_2 < 0 \text{ on } \partial\Omega,$$

contradicting Lemma III.7.6. For given $(f, \mu) \in Z^{(+,+)} \times \mathbb{R}$,

$$D_{(w,\lambda)}G(u_0, \lambda_0, p_0)[v, \lambda] = (f, \mu) \text{ for}$$

(III.7.87)
$$v = L^{-1}f - \lambda v_2 \in X^{(+,+)},$$

$$\lambda = ((L^{-1}f)(0,0) - \mu)/v_2(0,0) \in \mathbb{R},$$

which proves the surjectivity of $D_{(w,\lambda)}G(u_0, \lambda_0, p_0)$. Assume (III.7.84). If $\lambda = 0$, then $v = 0$. If $\lambda \neq 0$, using the function v_1 from (III.7.81),

$$L(\lambda v_1 - 2\lambda_0 v) = 0, \; \lambda v_1 - 2\lambda_0 v \in \Sigma^{(+,+)},$$

(III.7.88)
$$(\lambda v_1 - 2\lambda_0 v)(0,0) = 0, \text{ and}$$

$$\text{sign}(\lambda v_1 - 2\lambda_0 v) = -\text{sign}\lambda \text{ on } \partial\Omega,$$

contradicting Lemma III.7.6. This proves the bijectivity of $D_{(w,\lambda)}G(u_0, \lambda_0, p_0)$ in all cases.

   By the Implicit Function Theorem, there exists a C^1-curve

$$\{(w(t), \lambda(t)) | t \in (p_0 - \delta, p_0 + \delta)\} \text{ in } X^{(+,+)} \times \mathbb{R} \text{ through}$$

(III.7.89) $(w(p_0), \lambda(p_0)) = (u_0, \lambda_0)$ and

$$G(w(t), \lambda(t), t) = (0, 0) \text{ for all } t \in (p_0 - \delta, p_0 + \delta),$$

and all solutions of $G(w, \lambda, t) = (0, 0)$ in a neighborhood of (u_0, λ_0, p_0) are on that curve. Obviously,

(III.7.90)
$$(u(t), \lambda(t)) = \left(\frac{t}{p_0} w(t), \lambda(t) \right)$$

solves (III.7.1) for all $t \in (p_0 - \delta, p_0 + \delta)$.

Since $u(p_0) = u_0 > 0$ in Ω and $u_{0\nu} < 0$ on $\partial\Omega$, cf. (III.7.61), we obtain, for sufficiently small $\delta > 0$,

(III.7.91) $u(t) > 0 \quad \text{in } \Omega \quad \text{for all} \quad t \in (p_0 - \delta, p_0 + \delta)$.

Therefore, (III.7.62) implies

(III.7.92) $\|u(t)\|_\infty = u(t)(0, 0) = \dfrac{t}{p_0} w(t)(0, 0) = t$

by the second equation of $G(w, \lambda, t) = (0, 0)$. This proves that the parameter t is indeed the amplitude p.

Finally, every solution (u, λ) of (III.7.1) in a neighborhood $U \times V$ of (u_0, λ_0) in $X \times \mathbb{R}$ is positive, and therefore, by Proposition III.7.5, $u \in X^{(+,+)}$, and $(w, \lambda, t) = \left(\frac{p_0}{p} u, \lambda, p \right)$ solves $G(w, \lambda, t) = (0, 0)$ for $p = u(0, 0)$. Since (w, λ, t) is in a neighborhood of (u_0, λ_0, p_0) in $X^{(+,+)} \times \mathbb{R} \times (0, \infty)$, the solution (w, λ, t) is on the curve (III.7.89), or (u, λ) is on the curve $\mathcal{C}$ given by (III.7.90).   □

Remark III.7.8 *Note that Theorem III.7.7 applies to every positive solution of (III.7.1); i.e., (u_0, λ_0) is not necessarily obtained by a bifurcation from the trivial solutions. Indeed, if $g(0) > 0$, such a trivial solution line does not exist. The condition $g(0) \geq 0$, however, cannot be dropped in Theorem III.7.7. In [78], Remark 4.2, we give a counterexample for (III.7.1) over a ball $\Omega = B_R(0) \subset \mathbb{R}^2$ with $g(0) = -\varepsilon < 0$ having a solution*

(III.7.93)
*(u_0, λ_0) such that $u_0 > 0$ in Ω and $\lambda_0 > 0$,
$u_{0\nu} = 0$ on $\partial\Omega$, cf. (III.7.61),
but there exists no positive solution for $p < p_0$,
where p_0 is the amplitude of u_0,
and there is only a curve of positive solutions
$\{(u(\lambda), \lambda) | 0 < \lambda < \lambda_0\}$ with amplitudes
$p > p_0$ that ends in (u_0, λ_0).*

The continuation of that curve for $\lambda > \lambda_0$ consists of radially symmetric solutions that are no longer positive. As mentioned in [78], at (u_0, λ_0) a symmetry-breaking bifurcation of axially symmetric solutions (which are nonpositive) takes place, too.

We can clearly apply Theorem III.7.7 to any positive (negative) solution (u, λ) of (III.7.1) on a global continuum $\mathcal{C}_{\lambda_0}^+ (\mathcal{C}_{\lambda_0}^-)$ bifurcating from the trivial solution line at $(0, \lambda_0)$; cf. Theorem III.6.3 or (III.7.3). This yields the following smoothness:

Theorem III.7.9 *Assume $g(0) = 0$, $g'(0) > 0$, and that the domain Ω satisfies (III.7.57). Then the unbounded continua $\mathcal{C}_{\lambda_0}^+$ of positive and $\mathcal{C}_{\lambda_0}^-$ of negative solutions of (III.7.1) emanating at $(0, \lambda_0)$ are each smooth curves parameterized by the amplitude of u.*

For a proof, observe that $\lambda_0 = \mu_0/g'(0) > 0$ and $\lambda > 0$ for all $(u, \lambda) \in \mathcal{C}_{\lambda_0}^{\pm}$. Finally, if $u < 0$ solves (III.7.1), then $\hat{u} = -u > 0$ solves $\Delta\hat{u} + \lambda\hat{g}(\hat{u}) = 0$, where $\hat{g}(\hat{u}) = -g(-\hat{u})$. Therefore, Theorem III.7.7 applies also to negative solutions of (III.7.1).

We can also apply Theorem III.7.7 to any positive (negative) solution (u, λ) of (III.7.1) on a global continuum $\mathcal{C}^+(\mathcal{C}^-)$ emanating from the solution $(0, 0)$ and obtained by the Global Implicit Function Theorem; cf. Theorem III.6.6. This yields the following result:

Theorem III.7.10 *Assume $g(0) > 0$ and that the domain Ω satisfies the condition (III.7.57). Then the unbounded continua $\mathcal{C}^+$ of positive and $\mathcal{C}^-$ of negative solutions of (III.7.1) emanating from $(0, 0)$ are each smooth curves parameterized by the amplitude of u.*

For a proof, observe that $\lambda > 0$ for $(u, \lambda) \in \mathcal{C}^+$ and $\lambda < 0$ for $(u, \lambda) \in \mathcal{C}^-$. If $(u, \lambda) \in \mathcal{C}^-$, then $\hat{u} = -u > 0$ solves $\Delta\hat{u} + (-\lambda)\hat{g}(\hat{u}) = 0$, where $\hat{g}(\hat{u}) = g(-\hat{u})$ and $-\lambda > 0$, $\hat{g}(0) > 0$.

We recall that all continua mentioned in Theorems III.7.9 and III.7.10 are in the symmetry class $X^{(+,+)} \times \mathbb{R}$.

In Theorem III.7.17 the asymptotic behavior of the continua $\mathcal{C}_{\lambda_0}^{\pm}$ or $\mathcal{C}^{\pm}$ at $\|u\|_\infty = \infty$ is investigated.

III.7.3 Global Parameterization of Positive Solution Branches over Symmetric Domains and Uniqueness

We consider (III.7.1) over a domain having the properties (III.7.57). We prove an a priori bound for the parameter λ if u is positive.

Lemma III.7.11 *Let* $(u, \lambda) \in X \times \mathbb{R}$ $(X = C^{2,\alpha}(\overline{\Omega}) \cap \{u|u = 0 \text{ on } \partial\Omega\})$ *be a positive solution of (III.7.1) where g satisfies*

$$\begin{aligned} & g(u) \geq 0 \text{ for } u \geq 0, \\ (\text{III.7.94}) \quad & g(u) > 0 \text{ for } u \in (a, \infty), \ a \geq 0, \\ & \lim_{u \to \infty} g(u)/u = K > 0. \end{aligned}$$

If $\|u\|_\infty = u(0,0) \geq a + \varepsilon$, *then*

$$(\text{III.7.95}) \qquad\qquad \lambda \in (0, \Lambda)$$

for some constant $\Lambda > 0$ *depending only on* Ω, *g*, *and* $\varepsilon > 0$.

Proof. A positive solution of (III.7.1) with a function g satisfying (III.7.94) can exist only for $\lambda > 0$: Since for $\lambda = 0$ problem (III.7.1) has only the trivial solution $u = 0$, we assume $\lambda < 0$. Then $\Delta u = -\lambda g(u) \geq 0$ by (III.7.94)$_1$, so that by the maximum principle, u is constant if it attains a nonnegative maximum in Ω. This contradiction proves $\lambda > 0$.

We assume that there is a sequence of positive solutions (u_n, λ_n) of (III.7.1) such that
$$(\text{III.7.96}) \qquad \|u_n\|_\infty \geq a + \varepsilon, \ \lim_{n \to \infty} \lambda_n = \infty.$$

Case I. The amplitudes $\|u_n\|_\infty$ are unbounded.

Without loss of generality, $\|u_n\|_\infty \to \infty$ as $n \to \infty$. Then the rescaled functions
$$\tilde{u}_n(x, y) = u_n(x/\sqrt{\lambda_n}, y/\sqrt{\lambda_n}),$$

$$v_n = \tilde{u}_n/\|\tilde{u}_n\|_\infty \text{ with } s_n = 1/\|\tilde{u}_n\|_\infty \text{ solve}$$

$$(\text{III.7.97}) \qquad \Delta v_n + s_n g\left(\frac{v_n}{s_n}\right) = 0 \text{ in } \Omega_n = \sqrt{\lambda_n}\Omega,$$
$$v_n = 0 \text{ on } \partial\Omega_n,$$

$$1 \geq v_n > 0 \text{ in } \Omega_n, \ v_n(0,0) = 1,$$

where we use also (III.7.62). By the assumption (III.7.94) on g, the arguments for (III.7.24), (III.7.25) prove that

$$(\text{III.7.98}) \qquad \max_{(x,y)\in\Omega_n} \left| s_n g\left(\frac{v_n}{s_n}\right) - K v_n \right| \to 0 \text{ as } n \to \infty,$$
$$\text{whence } 0 \leq s_n g\left(\frac{v_n}{s_n}\right) \leq C \text{ on } \Omega_n \text{ for all } n \in \mathbb{N}.$$

For every ball $B_R(0)$ there is an $n_R \in \mathbb{N}$ such that $B_R(0) \subset \Omega_n$ for all $n \geq n_R$. Following the arguments after (III.7.25), elliptic a priori estimates imply

$$(\text{III.7.99}) \qquad \|v_n\|_{W^{2,2}(B_R(0))} \leq C_R \quad \text{for all} \quad n \geq n_R,$$

where C_R depends only on the radius R and on g. By a compact embedding $W^{2,2}(B_R(0)) \subset C(\overline{B}_R(0))$, a diagonal process for $n \to \infty$ and $R = N \to \infty$ yields a subsequence $(v_{n_k})_{k \in \mathbb{N}}$ of $(v_n)_{n \in \mathbb{N}}$ and some

(III.7.100)
$$v \in C(\mathbb{R}^2) \text{ such that}$$
$$v_{n_k} \to v \text{ in } C(\overline{B}_R(0)) \text{ for each } R > 0.$$

By (III.7.98),

(III.7.101)
$$s_{n_k} g\left(\frac{v_{n_k}}{s_{n_k}}\right) \to Kv \text{ in } C(\overline{B}_R(0)), \text{ whence}$$
$$v_{n_k} \to v \text{ in } W^{2,2}(B_R(0)) \text{ for each } R > 0,$$

where we use also the elliptic equation (III.7.97)$_3$ and an elliptic a priori estimate. Therefore, $\Delta v + Kv = 0$ in $B_R(0)$ for each $R > 0$, and by the embedding $W^{2,2}(B_R(0)) \subset C^\alpha(\overline{B}_R(0))$ we obtain finally

(III.7.102)
$$\Delta v + Kv = 0 \text{ in } \mathbb{R}^2, \ v \in C^{2,\alpha}(\mathbb{R}^2),$$
$$0 \le v \le 1, \ v(0,0) = 1,$$

contradicting a result on bounded superharmonic functions on $\mathbb{R}^2$ (Liouville's Theorem; see [56]).

Case II. The amplitudes $\|u_n\|_\infty$ are bounded.

Without loss of generality, $\lim_{n \to \infty} \|u_n\|_\infty = p \in (a, \infty)$. We define v_n and Ω_n as in (III.7.97), and we obtain as before (III.7.99), (III.7.100). Instead of (III.7.101), we get this time

(III.7.103)
$$s_{n_k} g\left(\frac{v_{n_k}}{s_{n_k}}\right) \to \frac{g(pv)}{p} \quad \text{in} \quad C(\overline{B}_R(0))$$

for each $R > 0$, which, in turn, yields

(III.7.104)
$$\Delta v + g(pv)/p = 0 \text{ in } \mathbb{R}^2, \ v \in C^{2,\alpha}(\mathbb{R}^2),$$
$$0 \le v \le 1, \ v(0,0) = 1.$$

In this case, the assumption (III.7.94)$_1$ implies that v is a bounded superharmonic function on $\mathbb{R}^2$. Therefore, Liouville's Theorem implies $v(x,y) \equiv v(0,0) = 1$ for all $(x,y) \in \mathbb{R}^2$. This gives $g(p) = 0$, contradicting (III.7.94)$_2$ for $p \in (a, \infty)$. $\qquad\square$

We continue to consider the model problem (III.7.1) over a domain Ω having the properties (III.7.57).

Theorem III.7.12 *We assume for $g : \mathbb{R} \to \mathbb{R}$ that g is of class C^2 and satisfies*

$$g(u) \geq 0 \ for \ u \geq 0,$$

(III.7.105)
$$g(u) > 0 \ for \ u \in (a, \infty), \ a \geq 0,$$

$$\lim_{|u| \to \infty} g'(u) = g'(\infty) > 0,$$

$$|ug''(u)| \leq C_1 \ for \ all \ u \in \mathbb{R} \ and \ some \ C_1 > 0.$$

Then for each $p \in (a, \infty)$ *there is a unique solution* $(u, \lambda) = (u(p), \lambda(p)) \in (C^{2,\alpha}(\overline{\Omega}) \cap \{u | u = 0 \ on \ \partial\Omega\}) \times \mathbb{R}$ *of (III.7.1) having the properties*

(III.7.106)
$$u(p) > 0 \ in \ \Omega, \ \lambda(p) > 0,$$

$$\|u(p)\|_\infty = p.$$

Furthermore,

(III.7.107)
$$\{(u(p), \lambda(p)) | p \in (a, \infty)\} \ forms \ a \ curve$$

$$of \ class \ C^1 \ in \ C^{2,\alpha}(\overline{\Omega}) \times \mathbb{R}.$$

Proof. By Theorem III.7.2 there is a unique curve $\{(u(s), \lambda(s)) | s \in (0, \delta$ and $\lambda(s) \to \lambda_\infty > 0$ as $s \searrow 0$; cf. (III.7.48) and (III.7.53). Therefore,

(III.7.108)
$$\hat{p} = \inf\{r \in (a, \infty)| \ there \ is \ a \ positive \ solution \ (u, \lambda)$$

$$of \ (III.7.1) \ with \ \|u(p)\|_\infty = p \ for \ all \ p \in (r, \infty)\}$$

exists in $[a, \infty)$. We claim that $\hat{p} = a$.

Assume $\hat{p} > a$. Let $(u_n, \lambda_n) \in X \times \mathbb{R}$, $X = C^{2,\alpha}(\overline{\Omega}) \cap \{u | u = 0 \ on \ \partial\Omega\}$, be a sequence of positive solutions of (III.7.1) with $\|u_n\|_\infty = p_n \searrow \hat{p} \geq a + \varepsilon$. By Lemma III.7.11, the sequence $(\lambda_n)_{n \in \mathbb{N}}$ is bounded in $\mathbb{R}$, and therefore $((u_n, \lambda_n))_{n \in \mathbb{N}}$ is bounded in $X \times \mathbb{R}$; cf. the arguments after (III.7.22). By properness of the mapping $F : X \times \mathbb{R} \to Z = C^\alpha(\overline{\Omega})$ given by $F(u, \lambda) = \Delta u + \lambda g(u)$, the sequence $((u_n, \lambda_n))_{n \in \mathbb{N}}$ is relatively compact in $X \times \mathbb{R}$; cf. (III.5.7)–(III.5.9).

Choose a subsequence of $((u_n, \lambda_n))_{n \in \mathbb{N}}$ converging to $(\hat{u}, \hat{\lambda}) \in X \times \mathbb{R}$ solving (III.7.1). Then $\hat{u} \geq 0$, $\|\hat{u}\|_\infty = \hat{p} > 0$, and the maximum principle gives $\hat{u} > 0$ in Ω and also $\hat{\lambda} > 0$. Therefore, Theorem III.7.7 implies the existence of a curve $\{(u(p), \lambda(p)) | p \in (\hat{p} - \delta, \hat{p} + \delta)\}$ through $(\hat{u}, \hat{\lambda})$ with $u(p) > 0$ in Ω and $\|u(p)\|_\infty = p$ for $p \in (\hat{p} - \delta, \hat{p} + \delta)$. This contradicts the definition (III.7.108) of $\hat{p}$, proving $\hat{p} = a$.

Next we prove uniqueness. Assume that

(III.7.109)
$$B = \{p \in (a, \infty)| \ there \ are \ at \ least \ two \ positive$$

$$solutions \ (u, \lambda) \ of \ (III.7.1) \ with \ \|u\|_\infty = p\}$$

is not empty. By (III.7.49) and Lemma III.7.11, the set B is bounded in (a, ∞). Therefore, $\hat{b} = \sup B$ exists in (a, ∞). If $\hat{b} \in B$, then Theorem III.7.7 provides a contradiction to the definition of $\hat{b}$. If $\hat{b} \notin B$, then by properness of $F(u, \lambda) = \Delta u + \lambda g(u)$, there is a unique solution $(\hat{u}, \hat{\lambda})$ of (III.7.1) with

$\hat{u} > 0$ in Ω and $\|\hat{u}\|_\infty = \hat{b}$. By definition of $\hat{b}$, at least two different sequences (u_n^i, λ_n^i), $i = 1, 2$, with $\|u_n^1\|_\infty = \|u_n^2\|_\infty < \hat{b}$ converge to $(\hat{u}, \hat{\lambda})$ in $X \times \mathbb{R}$, contradicting again Theorem III.7.7. Therefore, the set B has to be empty, which proves the uniqueness.

By the local result of Theorem III.7.7, the curve (III.7.107) is clearly globally of class C^1 in $X \times \mathbb{R}$.   □

Proposition III.7.13 *Assume $g(0) \geq 0$ and $g(p_0) = 0$ for some $p_0 > 0$. Then:*

(i) *There is no positive solution $(u, \lambda) \in X \times \mathbb{R}$ of (III.7.1) with $\|u\|_\infty = p_0$.*

(ii) *If $(u_n, \lambda_n) \in X \times \mathbb{R}$ is a sequence of positive solutions of (III.7.1) such that $\|u_n\|_\infty \to p_0$, then $(\mathrm{sign}(g(\|u_n\|_\infty)))\lambda_n \to \infty$ as $n \to \infty$.*

Proof. Assume the existence of a positive solution with $\|u\|_\infty = p_0$. By Proposition III.7.5 and (III.7.62),

$$w = u - p_0 \text{ satisfies}$$

$$w(0, 0) = 0, \; w < 0 \text{ in } \Omega \backslash \{(0, 0)\},$$

(III.7.110) $\Delta w + \lambda c w = 0$ in Ω, where

$$c(x, y) = \begin{cases} g(u(x, y))/w(x, y)) & \text{for } (x, y) \in \overline{\Omega} \backslash \{(0, 0)\}, \\ g'(p_0) & \text{for } (x, y) = (0, 0). \end{cases}$$

Then the function c is continuous on $\overline{\Omega}$, but the properties of w contradict the maximum principle (by the trick (III.6.10), no sign condition on λc is necessary for $w \leq 0$). This proves (i).

Case (ii) can occur only if $g(\|u_n\|_\infty) \neq 0$. Since $\|u_n\|_\infty = u_n(0, 0)$ (cf. (III.7.62)), the relations $\Delta u_n(0, 0) = -\lambda_n g(u_n(0, 0)) \leq 0$ imply $\mathrm{sign} g(\|u_n\|_\infty) = \mathrm{sign} \lambda_n \neq 0$. If the sequence $(|\lambda_n|)_{n \in \mathbb{N}}$ were bounded, by properness of the mapping $F : X \times \mathbb{R} \to Z$ given by $F(u, \lambda) = \Delta u + \lambda g(u)$, the sequence $((u_n, \lambda_n))_{n \in \mathbb{N}}$ would be relatively compact in $X \times \mathbb{R}$; cf. (III.5.7)–(III.5.9). The limit of a subsequence would give a positive solution $(\hat{u}, \hat{\lambda}) \in X \times \mathbb{R}$ of (III.7.1) with $\|\hat{u}\|_\infty = p_0$, contradicting (i). Therefore, no subsequence of $(|\lambda_n|)_{n \in \mathbb{N}}$ can be bounded, which proves (ii).   □

Next we give our **Main Result on Positive Solutions** of (III.7.1) over symmetric domains Ω satisfying (III.7.57).

Theorem III.7.14 *We assume for $g : \mathbb{R} \to \mathbb{R}$ that g is of class C^2 and satisfies*

(III.7.111) $g(u) \geq 0$ *for $u \in [0, b)$, $b \leq \infty$.*

Then for each $p \in (0, b)$ where $g(p) > 0$ there is a unique solution $(u, \lambda) = (u(p), \lambda(p)) \in (C^{2,\alpha}(\overline{\Omega}) \cap \{u | u = 0 \text{ on } \partial\Omega\}) \times \mathbb{R}$ of (III.7.1) having the properties

$$u(p) > 0 \text{ in } \Omega, \ \lambda(p) > 0,$$

(III.7.112)

$$\|u(p)\|_\infty = p.$$

For $p \in (0, b)$ where $g(p) = 0$, no positive solution u of (III.7.1) with $\|u\|_\infty = p$ exists.

To each maximal interval $(p_1, p_2) \subset (0, b)$ where $g(p) > 0$ there corresponds precisely one maximal curve $\{(u(p), \lambda(p)) | p \in (p_1, p_2)\} \subset X \times \mathbb{R}$ of positive solutions of (III.7.1) that is of class C^1.

If for $i \in \{1, 2\}$ the point p_i is in $(0, b)$, then $g(p_i) = 0$ and $\lambda(p) \to \infty$ as $p \in (p_1, p_2)$ and $p \to p_i$.

Proof. Let $g(p) > 0$ for $p \in (0, b)$. We consider problem (III.7.1) with a modified function $\tilde{g} : \mathbb{R} \to \mathbb{R}$ of class C^2 such that

(III.7.113)

$$\tilde{g}(u) = g(u) \text{ for } u \in [0, p],$$

$$\tilde{g} \text{ satisfies (III.7.105) for some } a \in [0, p).$$

Such a function $\tilde{g}$ certainly exists. Theorem III.7.12 then guarantees a unique solution $(u(p), \lambda(p)) \in X \times \mathbb{R}$ of the modified problem having the properties (III.7.106). By (III.7.113)$_1$, this solution is also a unique solution of the original problem satisfying (III.7.112). (Nonuniqueness would imply nonuniqueness of the modified problem as well, contradicting Theorem III.7.12.) By Theorem III.7.7 on local parameterization, each such solution $(u(p), \lambda(p))$ is on a curve of class C^1 parameterized by the amplitude p of $u(p)$. Proposition III.7.13 finally excludes positive solutions with amplitudes p where $g(p) = 0$, and it gives the asymptotic behavior $\lambda(p) \to \infty$ as the parameter p approaches a zero of the function g.   □

Remark III.7.15 *In [78], Remark 6.4, we show by a counterexample that the assumption (III.7.111) cannot be dropped in Theorem III.7.14. We give an example for (III.7.1) over a ball $\Omega = B_R(0) \subset \mathbb{R}^2$ with $g(0) = 0$, $g(u) > 0$ for $u > p_0$, but $g(u) < 0$ for $u \in (0, p_0)$ having no positive solution with amplitude $\|u\|_\infty = p \in (p_0, \frac{3}{2}p_0)$. In view of $g(0) = 0$, Theorem III.7.7 is valid.*

The assumptions (III.7.57) on the domain Ω cannot be dropped either. This is shown in [33] by various counterexamples.

III.7.4 Asymptotic Behavior at $\|u\|_\infty = 0$ and $\|u\|_\infty = \infty$

We keep the assumptions of Theorem III.7.14 for problem (III.7.1) over a domain Ω satisfying (III.7.57). If $g(u) > 0$ for $u \in (0, \delta)$ or for $u \in (a, \infty)$, then (III.7.1) possesses unique positive solutions $(u, \lambda) \in (C^{2,\alpha}(\overline{\Omega}) \cap \{u | u = 0$

on $\partial\Omega\}) \times \mathbb{R}$ with amplitudes $\|u\|_\infty = p \in (0, \delta)$ or $\|u\|_\infty = p \in (a, \infty)$ and with some $\lambda = \lambda(p) > 0$. We investigate the asymptotic behavior of $\lambda(p)$ as $p \to 0$ and $p \to \infty$.

Theorem III.7.16 *Under the assumptions of Theorem III.7.14 and $g(u) > 0$ for $u \in (0, \delta)$ the following hold:*

(i) If $g(0) > 0$, then $\lambda(p) \to 0$ as $p \to 0$.

(ii) If $g(0) = 0$, $g'(0) > 0$, then $\lambda(p) \to \lambda_0 = \mu_0/g'(0)$ as $p \to 0$, where μ_0 is the principal eigenvalue of $-\Delta$; cf. (III.7.2).

(iii) If $g(0) = 0$, $g'(0) = 0$, then $\lambda(p) \to \infty$ as $p \to 0$.

Proof. Let $\gamma = \lim_{u \to 0} g(u)/u \in [0, \infty]$ be the limit in each of the three cases. Then, for given $\varepsilon > 0$ or $M > 0$,

$$\gamma - \varepsilon < g(u)/u < \gamma + \varepsilon \text{ or}$$

(III.7.114) $$\qquad M < g(u)/u \text{ for } \gamma = \infty,$$

$$\text{for all } 0 < u < \delta(\varepsilon) \text{ or } 0 < u < \delta(M).$$

We define for a positive solution $(u, \lambda) \in X \times \mathbb{R}$ of (III.7.1) with amplitude $\|u\|_\infty = p < \delta(\varepsilon)$ or $\|u\|_\infty = p < \delta(M)$ the function $\rho = g(u)/u$, which is continuous and positive in Ω. As in (III.7.13), we denote by $\mu_0(\rho)$ the principal eigenvalue with weight function ρ of $-\Delta$ on Ω subject to homogeneous Dirichlet boundary conditions. Then $\mu_0(\rho)$ minimizes the quotient $\|\nabla u\|_0^2/(\rho u, u)_0$ in $H_0^1(\Omega)\backslash\{0\}$; cf. (III.7.6). Therefore, the estimates (III.7.114) for ρ imply, in view of the monotonicity (III.7.15),

$$\mu_0(\rho) < \mu_0(M) \text{ in case (i), where } \gamma = \infty,$$

$$\mu_0(\gamma + \varepsilon) < \mu_0(\rho) < \mu_0(\gamma - \varepsilon) \text{ in case (ii),}$$

(III.7.115) $$\qquad \text{where } \gamma = g'(0) > 0,$$

$$\mu_0(\varepsilon) < \mu_0(\rho) \text{ in case (iii), where } \gamma = 0.$$

On the other hand, if $\rho = \text{const}$, then clearly, $\mu_0(\rho) = \mu_0/\rho$, where μ_0 is the principal eigenvalue of $-\Delta$ with weight $\rho = 1$. Finally, if (u, λ) is a positive solution of (III.7.1), then clearly, $\lambda = \mu_0(\rho)$; cf. (III.7.16). Combining these observations with (III.7.115) proves Theorem III.7.16.   □

In case (ii), the curve of positive solutions emanating at $(0, \lambda_0)$ bifurcates from the trivial solution line $\{(0, \lambda)\}$ (note that $g(0) = 0$). The maximal curve provided by Theorem III.7.14 is the unbounded continuum $\mathcal{C}_{\lambda_0}^+$ introduced in (III.7.3). The curve $\mathcal{C}_{\lambda_0}^+$ is parameterized by the amplitude $p = \|u\|_\infty$, and it exists as long as $g(p) > 0$. If $g(p_0) = 0$ for some $p_0 > 0$, then $\lambda(p) \to \infty$ as $p \nearrow p_0$. If $g(u) > 0$ for all $u \in (0, \infty)$, then the asymptotic behavior at $\|u\|_\infty = \infty$ is partially determined by the following theorem.

Theorem III.7.17 *Under the assumptions of Theorem III.7.14 for* $b = \infty$
and $g(u) > 0$ *for* $u \in (a, \infty)$, *the following hold:*
(i) If $\lim_{u \to \infty} g(u)/u = K > 0$, *then* $\lambda(p) \to \lambda_\infty = \mu_0/K$ *as* $p \to \infty$, *where*
μ_0 *is the principal eigenvalue of* $-\Delta$; *cf. (III.7.2).*
(ii) If $\lim_{u \to \infty} g(u)/u = 0$, *then* $\lambda(p) \to \infty$ *as* $p \to \infty$.

Proof. (i) Let $(u(p_n), \lambda(p_n)) \in X \times \mathbb{R}$ be a sequence of positive solutions of
(III.7.1) such that $\|u(p_n)\|_\infty = p_n \to \infty$ as $n \to \infty$. By the proof of Lemma
III.7.11, Case I, the sequence $(\lambda(p_n))_{n \in \mathbb{N}}$ is bounded, and the arguments for
(III.7.23)–(III.7.27), (III.7.98)–(III.7.102) prove that the only possible clus-
ter point for $(\lambda(p_n))_{n \in \mathbb{N}}$ is λ_∞ with $\lambda_\infty K = \mu_0$, where μ_0 is the principal
eigenvalue of $-\Delta$.
(ii) Assume that a sequence of positive solutions $(u(p_n), \lambda(p_n)) \in X \times \mathbb{R}$
satisfies $\|u(p_n)\|_\infty = p_n \to \infty$ and $0 < \lambda(p_n) < \Lambda$. Without loss of generality,
we assume that $\lambda(p_n) \to \lambda \geq 0$. Again the arguments for (III.7.23)–(III.7.27),
(III.7.98)–(III.7.102) prove that (for a subsequence) $u(p_{n_k})/\|u(p_{n_k})\|_\infty \to v$
in $W^{2,2}(\Omega)$, $\lambda(p_{n_k})g(u(p_{n_k}))/p_{n_k} \to 0$ in $C(\overline{\Omega})$ as $k \to \infty$, and $\Delta v = 0$ in
Ω, $v = 0$ on $\partial\Omega$, $\|v\|_\infty = v(0,0) = 1$, which is a contradiction.   □

The case $\lim_{u \to \infty} g(u)/u = \infty$ is obviously more involved. We refer to
[103], for example, where we discuss the exponential growth of g and prove
that $\lambda(p) \to 0$ as $p \to \infty$.

Remark III.7.18 *In Section III.6 we prove the existence of global solution*
branches of $\Delta u + \lambda g(u) = 0$ *with* $g(-u) = -g(u)$ *that are characterized by a*
fixed rectangular or hexagonal nodal pattern of their solutions; cf. (III.6.40).
As mentioned in Remark III.7.4, these continua can also be considered as
positive solution branches over a tile Ω *of the respective lattice* $\mathcal{L}$ *which is a*
rectangle or an equilateral triangle. These tiles Ω *are admitted for Proposition*
III.7.5, and accordingly, positive solutions have the symmetries of Ω, *and*
the directional derivatives orthogonal to a symmetry axis are negative; cf.
(III.7.60). Relation (III.7.61), however, does not hold at the corners of $\partial\Omega$.
Nonetheless, a modified Lemma III.7.6 holds when $(0,0)$ *is replaced by the*
center of Ω, *i.e., the intersection of the symmetry axes. Therefore, Theorem*
III.7.7 is valid for every solution $(u, \lambda) \in (C^{2,\alpha}(\mathbb{R}^2) \cap X_D) \times \mathbb{R}$ *such that*
$u > 0$ *in a tile* Ω *and* $\lambda > 0$. *Since the unbounded branches* $\mathcal{C}^+_{\lambda_0}$ *of (III.6.40)*
emanating at $(0, \lambda_0)$ *for some* $\lambda_0 > 0$ *stay in the half-space* $(C^{2,\alpha}(\mathbb{R}^2) \times X_D) \times$
$(0, \infty)$, *they are globally smooth curves parameterized by the amplitude of* u;
cf. Theorem III.7.9. This sharpens considerably the result of (III.6.40).

All other results apart from uniqueness in this section about global parame-
terization and the asymptotic behavior are also valid for $\Delta u + \lambda g(u) = 0$ *with*
$g(-u) = -g(u)$ *on a rectangular or hexagonal lattice. The reason for that*
exception is that we have not proved the uniqueness of the curve bifurcating
from infinity; cf. Theorem III.7.2 and (III.7.49). The uniqueness claimed in
Theorem III.7.12 relies on the uniqueness near infinity. (We believe that The-
orem III.7.2 holds also on a lattice, but its proof does not apply to polygonal
domains, since we need a smooth boundary for Hopf's boundary lemma.)

For smooth and convex domains having the symmetries of a regular poly-gon, all results on local and global parameterization, uniqueness, and asymp-totic behavior of this section hold. Needless to say, they hold also for an interval, i.e., if (III.7.1) is an ODE with two-point boundary conditions.

We sketch some typical cases in Figure III.7.1–Figure III.7.3.

(III.7.116)

In Figure III.7.1,
$g(u) > 0$ for $u \in [0, p_1) \cup (p_1, \infty), g(p_1) = 0,$

$\lim_{u \to \infty} g(u)/u = K.$

In Figure III.7.2,
$g(u) > 0$ for $u \in (0, p_1) \cup (p_1, \infty), g(p_1) = 0,$

$g(0) = 0, g'(0) > 0, \lim_{u \to \infty} g(u)/u = 0.$

In Figure III.7.3,
$g(u) > 0$ for $u \in (0, p_1) \cup (p_1, p_2) \cup (p_2, \infty), g(p_i) = 0, i = 1, 2,$

$g(0) = 0, \ g'(0) = 0, \ g(u) \sim e^u$ as $u \to \infty.$

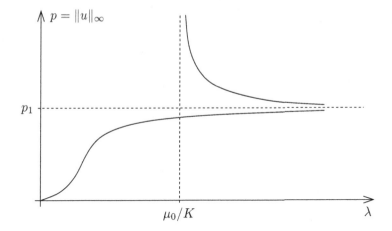

Figure III.7.1

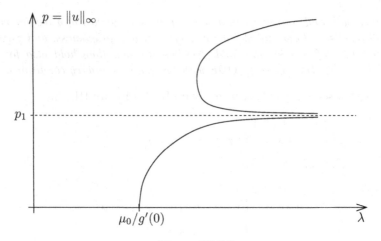

Figure III.7.2

If $g(u) > 0$ for all $u \in (0, \infty)$, then the asymptotic behavior at the zeros $g(p_i) = 0$ is eliminated, and the asymptotic behavior at $\|u\|_\infty = 0$ and at $\|u\|_\infty = \infty$ is connected by a smooth curve. Under the conditions (III.7.11), a parameterization by the amplitude p as well as by λ is possible, and we sketch $\mathcal{C}_{\lambda_0}^+$ in Figure III.7.4. The trivial solution line is stable for $\lambda < \mu_0/g'(0)$ and unstable for $\lambda > \mu_0/g'(0)$, whereas $\mathcal{C}_{\lambda_0}^+$ is unstable; cf. Theorem III.7.1.

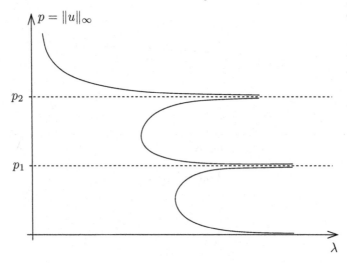

Figure III.7.3

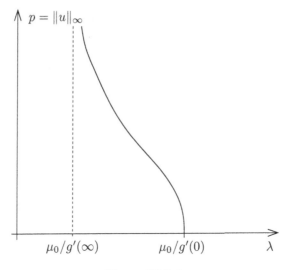

Figure III.7.4

III.7.5 Stability of Positive Solution Branches

We discuss the stability properties of the global curves of positive solutions given by Theorems III.7.9 and III.7.10. These curves are parameterized by the amplitude p of u. For analytical $g : \mathbb{R} \to \mathbb{R}$ we apply the results of Section I.16, starting with the local Principle of Exchange of Stability given by formula (I.16.51). Since the greatest eigenvalue of the linearization along the trivial solution line is $\mu(\lambda) = (\lambda - \lambda_0)g'(0)$ (see the comments after (III.7.3)), the trivial solution is stable for $\lambda < \lambda_0$ and unstable for $\lambda > \lambda_0$. Excluding vertical bifurcation, a supercritically bifurcating curve is stable and a subcritically bifurcating curve is unstable. The local curve given by Theorem III.7.10 is stable, since all eigenvalues of $D_u F(u, \lambda) = \Delta + \lambda g'(u)I$ are negative for (u, λ) near $(0, 0) \in X^{(+,+)} \times \mathbb{R}$. Recall that $F(u, \lambda) = \Delta u + \lambda g(u)$ defines a mapping $F : X^{(+,+)} \times \mathbb{R} \mapsto Z^{(+,+)}$; cf. (III.7.75), (III.7.76). Accordingly, the subsequent stability analysis admits only perturbations in the symmetry class $\Sigma^{(+,+)}$. Whereas this weakens the notion of stability, it sharpens that of instability. For a discussion of the critical eigenvalue perturbations along global curves we need the following lemma.

Lemma III.7.19 *Let* $(u, \lambda) \in C_{\lambda_0}^+$ *or* C^+ *given by Theorem III.7.9 or III.7.10. Both global curves are parameterized by the amplitude p of u, i.e., $(u, \lambda) = (u(p), \lambda(p))$ for $p > 0$. Then the following holds ($\dot{} = \frac{d}{dp}$):*

$$\dot\lambda(p) \neq 0 \quad \Rightarrow \quad D_u F(u(p), \lambda(p)) : X^{(+,+)} \to Z^{(+,+)}$$

is bijective or 0 is not an eigenvalue

(III.7.117)   of $D_u F(u(p), \lambda(p))$,

$$\dot\lambda(p) = 0 \quad \Rightarrow \quad 0 \neq \dot u(p) \in N(D_u F(u(p), \lambda(p))$$

and $(D_\lambda F(u(p), \lambda(p)), \dot u(p))_0 \neq 0.$

Proof. Assume in case $\dot\lambda(p) \neq 0$ that $0 \neq v \in N(D_u F(u(p), \lambda(p)))$. Differentiating $F(u(p), \lambda(p)) = 0$ with respect to p we obtain

(III.7.118) $D_u F(u(p), \lambda(p))\dot u(p) + D_\lambda F(./.)\dot\lambda(p) = 0,$

and using the symmetry of $D_u F(u, \lambda) = \Delta + \lambda g'(u)I$ with respect to the scalar product $(\ ,\ )_0$ in $L^2(\Omega)$, (III.7.118) yields after scalar multiplication by v

(III.7.119) $(D_\lambda F(u(p), \lambda(p)), v)_0 = 0,$

where we also used $\dot\lambda(p) \neq 0$. By the surjectivity of $D_{(u,\lambda)}F(u(p), \lambda(p)) :$ $X^{(+,+)} \times \mathbb{R} \to Z^{(+,+)}$ proved for (III.7.78), there is a $(w, \lambda) \in X^{(+,+)} \times \mathbb{R}$ such that $D_u F(u(p), \lambda(p))w + D_\lambda F(./.)\lambda = v$. Scalar multiplication by v, using $D_u F(u(p), \lambda(p))v = 0$ and $(D_\lambda F(u(p), \lambda(p)), v)_0 = 0$, gives $\|v\|_0^2 = 0$ or $v = 0$. This contradiction proves that $D_u F(u(p), \lambda(p))$ is injective and, by the Fredholm property, that it is bijective.

If $\dot\lambda(p) = 0$ then (III.7.118) proves $\dot u(p) \in N(D_u F(u(p), \lambda(p)))$. We show that $\dot u(p) \neq 0$. By (III.7.90) we have

(III.7.120)
$$u(t) = \frac{t}{p}w(t), \quad \text{where}$$
$$w(p) = u(p) \quad \text{and} \quad w(t)(0,0) = p \text{ for } t \in (p - \delta, p + \delta).$$

This implies $\dot u(p) = \frac{1}{p}u(p) + \dot w(p)$ and $\dot w(p)(0,0) = 0$. Therefore $\dot u(p)(0,0) = 1$, cf. (III.7.92), and $\dot u(p) \neq 0$. Again by the surjectivity of $D_{(u,\lambda)}F(u(p), \lambda(p))$ there is a $(w, \lambda) \in X^{(+,+)} \times \mathbb{R}$ such that $D_u F(u(p), \lambda(p))w + D_\lambda F(./.)\lambda = \dot u(p)$. Using $\dot u(p) \in N(D_u F(u(p), \lambda(p))$ and the symmetry of $D_u F(u(p), \lambda(p))$, we obtain after scalar multiplication by $\dot u(p)$,

(III.7.121) $\lambda(D_\lambda F(u(p), \lambda(p)), \dot u(p))_0 = \|\dot u(p)\|_0^2 > 0,$

which proves the last statement of (III.7.117).   □

A consequence of (III.7.117)$_1$ is that

(III.7.122)
the stability property of the curve $\{(u(p), \lambda(p))\}$

does not change as long as $\dot\lambda(p) \neq 0.$

In case $\dot\lambda(p_0) = 0$,

$$\hat{v}_0 = c_0 \dot{u}(p_0), \quad \text{normalized to } \|\hat{v}_0\|_0 = 1 \text{ by } c_0 > 0,$$

(III.7.123)

$$\text{spans } N(D_u F(u(p_0), \lambda(p_0))); \quad \text{cf. (III.7.79).}$$

For a Lyapunov–Schmidt reduction we decompose

(III.7.124)
$$X^{(+,+)} = N(D_u F(u(p_0), \lambda(p_0))) \oplus (R(D_u F(./.)) \cap X^{(+,+)}),$$
$$Z^{(+,+)} = R(D_u F(u(p_0), \lambda(p_0))) \oplus N(D_u F(./.)),$$

with orthogonal (with respect to $(\ ,\ )_0$) projections $Qu = (u, \hat{v}_0)_0 \hat{v}_0$ and $P = Q|_{X^{(+,+)}}$.

Setting $\hat{F}(u,p) = F(u(p_0 + p) + u, \lambda(p_0 + p))$ we have $\hat{F}(0,p) = 0$ for all p near 0 and $D_u \hat{F}(0,0) = D_u F(u(p_0), \lambda(p_0))$. Let $\hat{\Phi}(s,p)$ be the scalar bifurcation function for $\hat{F}(u,p) = 0$ obtained by the method of Lyapunov–Schmidt near $(u,p) = (0,0)$ according to the decomposition (III.7.124) and let $\Phi(s,\lambda)$ be the scalar bifurcation function for $F(u,\lambda) = 0$ near $(u,\lambda) = (u(p_0), \lambda(p_0))$ according to the same decomposition. (The "scalar bifurcation function" is derived from the bifurcation function (I.2.9) when $v \in N$ and $\Phi \in N$ are identified with their coordinates $s = (v, \hat{v}_0)_0$ and $(\Phi, \hat{v}_0)_0$ with respect to the basis $\{\hat{v}_0\} \subset N$, respectively; cf. also (I.19.5). Moreover, since $v = Qu$ and $\Phi = QF$, these coordinates coincide with $(u, \hat{v}_0)_0$ and $(F, \hat{v}_0)_0$, respectively.) Then, in view of $F(u(p_0 + p), \lambda(p_0 + p)) = 0$, we have $\Phi(s(p_0 + p), \lambda(p_0 + p)) = 0$ for all p near 0, where $Qu(p_0 + p) = (u(p_0 + p), \hat{v}_0)_0 \hat{v}_0 = s(p_0 + p) \hat{v}_0$. Furthermore,

(III.7.125) $\hat{\Phi}(s,p) = \Phi((u(p_0 + p), \hat{v}_0)_0 + s, \lambda(p_0 + p)); \quad$ cf. (I.16.39).

By $u(p_0 + p) = u(p_0) + p\dot{u}(p_0) +$ h.o.t. we obtain with (III.7.123)

$$(u(p_0 + p), \hat{v}_0)_0 = (u(p_0), \hat{v}_0)_0 + p(\dot{u}(p_0), \hat{v}_0)_0 + \text{ h.o.t.}$$

(III.7.126)

$$= s(p_0 + p) = s(p_0) + \dot{s}(p_0)p + \text{ h.o.t. with } \dot{s}(p_0) = \frac{1}{c_0}.$$

Assume now that

(III.7.127) $\dot{\lambda}(p_0) = \cdots = \lambda^{(k-1)}(p_0) = 0, \ \lambda^{(k)}(p_0) \neq 0 \quad \text{for } k \geq 2.$

Differentiating $\hat{\Phi}(0,p) = \Phi(s(p_0 + p), \lambda(p_0 + p)) = 0$ k times with respect to p at $p = 0$, (III.7.127) implies

$$D_s \Phi(s(p_0), \lambda(p_0)) = \cdots = D_s^{k-1} \Phi(s(p_0), \lambda(p_0)) = 0,$$

(III.7.128)
$$\frac{d}{dp} D_s \Phi(s(p_0 + p), \lambda(p_0 + p))|_{p=0} = \cdots = \frac{d^{k-2}}{dp^{k-2}} D_s \Phi(./.)|_{p=0} = 0,$$

$$\frac{d^{k-1}}{dp^{k-1}} D_s \Phi(s(p_0 + p), \lambda(p_0 + p))|_{p=0} = -c_0 D_\lambda \Phi(./.)|_{p=0} \lambda^{(k)}(p_0).$$

On the other hand, Theorem I.16.6 links (III.7.128) to the simple eigenvalue perturbation $D_u F(u(p), \lambda(p))v(p) = \mu(p)v(p)$ with $0 \neq v(p) \in X^{(+,+)}$ for p near p_0 as follows:

$$\mu(p_0) = \dot{\mu}(p_0) = \cdots = \mu^{(k-2)}(p_0) = 0,$$

$$(III.7.129) \qquad \mu^{(k-1)}(p_0) = -c_0 D_\lambda \Phi(s(p_0), \lambda(p_0))\lambda^{(k)}(p_0)$$

$$= -c_0(D_\lambda F(u(p_0), \lambda(p_0)), \hat{v}_0)_0 \lambda^{(k)}(p_0) \neq 0,$$

where we use the last statement of (III.7.117). Formula (III.7.129) generalizes formula (I.7.31). (For $k = 2$, (III.7.129) and (I.7.31) differ by the factor c_0 which has the following reason: For (I.7.31) we choose a parameterization such that $\|\dot{x}(0)\| = \|\hat{v}_0\| = 1$, whereas here $\|\dot{u}(p_0)\|_0 = \|\frac{1}{c_0}\hat{v}_0\|_0 = \frac{1}{c_0}$. If we parameterize the solution curve as $(\hat{u}(t), \hat{\lambda}(t)) = (u(p_0 + c_0 t), \lambda(p_0 + c_0 t))$, then $\|\dot{\hat{u}}(0)\|_0 = 1$ and there is no longer a factor c_0 in (III.7.129).)

Formula (III.7.129) has the following consequences for the global curves $\mathcal{C}_{\lambda_0}^+$ and $\mathcal{C}^+$ parameterized by the amplitude p; recall that the number $k \geq 2$ is defined in (III.7.127) and that $\mu(p)$ is the simple eigenvalue perturbation $D_u F(u(p), \lambda(p))v(p) = \mu(p)v(p)$ for p near p_0.

(III.7.130)

> If k is odd then $(u(p_0), \lambda(p_0))$
> is not a turning point and
> sign$\mu(p)$ does not change near $p = p_0$.
>
> If k is even then $(u(p_0), \lambda(p_0))$
> is a turning point and
> sign$\mu(p)$ changes at $p = p_0$.

Supercritically bifurcating curves $\mathcal{C}_{\lambda_0}^+$ and $\mathcal{C}^+$ are stable for amplitudes $p > 0$ up to the first point $(u(p_0), \lambda(p_0))$ where $\dot{\lambda}(p_0) = 0$; cf. (III.7.122). If (III.1.127) holds with an odd k, then the stability is preserved for $p > p_0$ (and p near p_0). Let $(u(p_0), \lambda(p_0))$ be a first point on $\mathcal{C}_{\lambda_0}^+$ or $\mathcal{C}^+$ where (III.1.127) holds with an even k. By (III.7.130) the stability is lost at this turning point.

To be more precise, in this case $\mu(p_0) = 0$ is the greatest eigenvalue, i.e., it is the principal eigenvalue of $D_u F(u(p_0), \lambda(p_0)) = \Delta + \lambda(p_0)g'(u(p_0))I$, and the eigenfunction $\dot{u}(p_0)$ is positive in Ω. Furthermore, $\lambda^{(k)}(p_0) < 0$, since $\lambda(p_0)$ is a local maximum at the first turning point. Since $\mu(p) < 0$ for $p < p_0$ and $\mu(p) > 0$ for $p > p_0$ (and p near p_0), $\mu^{(k-1)}(p_0) > 0$. Formula (III.7.129) then implies that $(D_\lambda F(u(p_0), \lambda(p_0)), \dot{u}(p_0))_0 = \int_\Omega g(u(p_0))\dot{u}(p_0)dx > 0$.

Beyond the first turning point or on a subcritically bifurcating curve $\mathcal{C}_{\lambda_0}^+$ the solution $u(p)$ is unstable. Its instability is preserved as long as $\dot{\lambda}(p) \neq 0$ or if (III.7.127) holds with an odd k. Let $(u(p_0), \lambda(p_0))$ be the next turning point when the amplitude p is increased. By (III.7.130) the critical eigenvalue $\mu(p)$

changes sign at $p = p_0$. However, $\mu(p_0) = 0$ is not necessarily the principal eigenvalue. It could also be the second eigenvalue.

In the first case the principal eigenvalue becomes negative, which means that $\mu^{(k-1)}(p_0) < 0$. In view of $\lambda^{(k)}(p_0) > 0$ $(\lambda(p_0)$ is a local minimum), formula (III.7.129) is satisfied only if $\int_\Omega g(u(p_0))\dot{u}(p_0)dx > 0$. In this case the curve regains stability for $p > p_0$ (and p near p_0).

In the second case the second eigenvalue becomes positive. By the same arguments as before, this is possible only if $\int_\Omega g(u(p_0))\dot{u}(p_0)dx < 0$. In this case the curve remains unstable for $p > p_0$ (and p near p_0).

In the first case $\dot{u}(p_0)$ is positive in Ω; in the second case the eigenfunction $\dot{u}(p_0)$ has precisely two nodal domains in Ω; cf. [24]. Since $\dot{u}(p_0) \in X^{(+,+)}$, the nodal line is a symmetric closed curve in Ω with the origin in its interior. This is not a priori excluded; note that the function $g'(u(p_0))$ does not necessarily have a constant sign.

To summarize, in contrast to a loss of stability, a gain of stability at a turning point is not guaranteed. We sketch some possibilities in Figure III.7.5, where the first two curves represent $\mathcal{C}^+_{\lambda_0}$ and the third curve sketches $\mathcal{C}^+$; cf. Theorems III.7.9 and III.7.10.

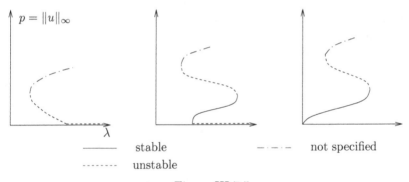

Figure III.7.5

We mention that a curve $\mathcal{C}^+_{\lambda_0}$ can have infinitely many turning points as the amplitude tends to infinity. A specific example is given in [112].

III.8 Notes and Remarks to Chapter III

The Fredholm property of elliptic operators over bounded domains is clearly well known. Since a single suitable reference for all our purposes is probably not available in closed form, we include a proof for convenience. The Fredholm property of elliptic operators on a lattice, however, is not so well established. A proof for Neumann boundary conditions is given in [49]. In that paper we prove also how a fixed rectangular nodal structure preserved on global branches breaks up under a small perturbation.

Local and global bifurcation for elliptic problems is a subject of countless papers that have appeared in the last thirty years. We try to give a synopsis from what we have learned from those and in particular from what we have learned from our teacher, K. Kirchgässner, and also from M. Crandall and P. Rabinowitz during a visit in Madison in 1977/78. The same holds for Hopf bifurcation for parabolic problems, and we mention our starting point [115]. (For references about Hopf bifurcation see the Notes and Remarks to Chapter I, Section I.22.)

The discussion of the stationary Cahn–Hilliard model, serving as a paradigm in Sections III.2, III.5, III.6, is taken from [106], [107].

In contrast to elliptic and parabolic problems, hyperbolic problems are very rare in bifurcation theory. Most contributions to the one-dimensional wave equation are devoted to forced nonlinear vibrations. We mention [147], [148], [138], and the references cited there. By its nature, this is not a bifurcation problem, and its approach uses a broad selection of nonlinear functional analysis. Free vibrations "in the large" are found in [152] via global methods of the calculus of variations. Bifurcation of free vibrations in the spirit of a Lyapunov Center Theorem is proved in [25] by methods of KAM Theory; cf. Remark III.3.1. Our approach to free nonlinear vibrations goes exclusively back to [92], and it was resumed in [110], [101], [72], [73], and [61].

Preservation of nodal structure on global branches was first proved in [26], whose ideas were extended to higher dimensions in [66], [67], [68], [70]. In [71] we took the challenge to prove it for elliptic systems and in [69] for fully nonlinear elliptic problems (whose results are not included in this book).

Global positive (and negative) solution branches of elliptic problems have been investigated by many people. It started with the pioneering paper [147], which established the Rabinowitz alternative for each of the branches of positive and negative solutions. Later, more qualitative properties of these branches, depending on the data of the problem, became an issue; cf. the reviews [5], [127], and also [143], for example. The question about smoothness and uniqueness has been answered only for a class of ODEs [156] and for problems over a ball: Due to the radial symmetry of positive solutions, they satisfy an ODE; cf. [161], [141], for example. To the best of our knowledge, smoothness and uniqueness of positive solution branches over more general but still symmetric domains has been investigated only in [103], [104], [78]. The proof of Lemma III.7.11 is essentially due to [32]. An iteration scheme that selects stable positive solution branches in a neighborhood of the bifurcation point is presented in [116].

References

1. R.A. Adams. *Sobolev Spaces*. Academic Press, San Diego–London, 1978.
2. S. Agmon. On the eigenfunctions and on the eigenvalues of general elliptic boundary value problems. *Communications on Pure and Applied Mathematics*, 15:119–147, 1962.
3. J.C. Alexander and J.A. Yorke. Global bifurcation of periodic orbits. *American Journal of Mathematics*, 100:263–292, 1978.
4. E.L. Allgower and K. Georg. *Numerical Continuation Methods. An Introduction*. Springer-Verlag, Berlin–Heidelberg–New York, 1990.
5. H. Amann. Fixed point equations and nonlinear eigenvalue problems in ordered Banach spaces. *SIAM Review*, 18:620–709, 1976.
6. H. Amann. *Ordinary Differential Equations*. de Gruyter, Berlin, 1990.
7. H. Amann. *Linear and Quasilinear Parabolic Problems. Volume I: Abstract Linear Theory*. Birkhäuser, Basel–Boston–Berlin, 1995.
8. H. Amann and S. Weiss. On the uniqueness of the topological degree. *Mathematische Zeitschrift*, 130:39–54, 1973.
9. A. Ambrosetti. Branching points for a class of variational operators. *Journal d'Analyse Mathématique*, 76:321–335, 1998.
10. A. Ambrosetti and G. Prodi. *A Primer of Nonlinear Analysis*. Cambridge University Press, Cambridge, 1993.
11. S. Antman. *Nonlinear Problems of Elasticity*. Springer-Verlag, New York–Berlin–Heidelberg, 1995.
12. P. Bachmann. *Zahlentheorie. Vierter Teil. Die Arithmetik der quadratischen Formen*. Johnson Reprint Corp. 1968, Leipzig, 1898.
13. P. Benevieri and M. Furi. A simple notion of orientability for Fredholm maps of index zero between Banach manifolds and degree theory. *Annales des Sciences Mathématiques du Québec*, 22:131–148, 1998.
14. R. Böhme. Die Lösungen der Verzweigungsgleichungen für nichtlineare Eigenwertprobleme. *Mathematische Zeitschrift*, 127:105–126, 1970.
15. B. Buffoni and J. Toland. *Analytic Theory of Global Bifurcation*. Princeton University Press, Princeton, NJ, 2003.
16. G. Buttazzo, M. Giaquinta, and S. Hildebrandt. *One–dimensional Variational Problems*. Clarendon Press, Oxford, 1998.
17. P. Chossat and G. Iooss. *The Couette–Taylor Problem*. Springer-Verlag, New York–Berlin–Heidelberg, 1994.
18. P. Chossat and R. Lauterbach. *Methods in Equivariant Bifurcations and Dynamical Systems*. World Scientific Publ. Co., Singapore, 2000.
19. S.-N. Chow and J.K. Hale. *Methods of Bifurcation Theory*. Springer-Verlag, New York–Berlin–Heidelberg, 1982.

20. S.-N. Chow and R. Lauterbach. A bifurcation theorem for critical points of variational problems. *Nonlinear Analysis. Theory, Methods & Applications*, 12:51–61, 1988.

21. S.-N. Chow, J. Mallet-Paret, and J.A. Yorke. Global Hopf bifurcation from a multiple eigenvalue. *Nonlinear Analysis. Theory, Methods & Applications*, 2:753–763, 1978.

22. C. Conley. *Isolated invariant sets and the Morse index*. Regional Conference Series in Mathematics, No. 38. American Mathematical Society, Providence, R.I., 1978.

23. A. Constantin and W. Strauss. Exact periodic traveling water waves with vorticity. *Comptes Rendus de l'Académie des Sciences, Paris, Series I*, 335:797–800, 2002.

24. R. Courant and D. Hilbert. *Methods of Mathematical Physics*. Interscience, New York, 1953.

25. W. Craig and C.E. Wayne. Newton's method and periodic solutions of nonlinear wave equations. *Communications on Pure and Applied Mathematics*, 46:1409–1498, 1993.

26. M.G. Crandall and P.H. Rabinowitz. Nonlinear Sturm–Liouville eigenvalue problems and topological degree. *Journal of Mathematical Mechanics*, 19:1083–1102, 1970.

27. M.G. Crandall and P.H. Rabinowitz. Bifurcation from simple eigenvalues. *Journal of Functional Analysis*, 8:321–340, 1971.

28. M.G. Crandall and P.H. Rabinowitz. Bifurcation, perturbation of simple eigenvalues, and linearized stability. *Archive for Rational Mechanics and Analysis*, 52:161–180, 1973.

29. M.G. Crandall and P.H. Rabinowitz. The Hopf Bifurcation Theorem in Infinite Dimensions. *Archive for Rational Mechanics and Analysis*, 67:53–72, 1977.

30. E.N. Dancer. On the structure of solutions of non-linear eigenvalue problems. *Indiana University Mathematics Journal*, 23:1069–1076, 1974.

31. E.N. Dancer. Boundary-value problems for ordinary differential equations on infinite intervals. *Proceedings of the London Mathematical Society*, 30:76–94, 1975.

32. E.N. Dancer. On the number of positive solutions of weakly nonlinear elliptic equations when a parameter is large. *Proceedings of the London Mathematical Society*, 53:429–452, 1986.

33. E.N. Dancer. The effect of domain shape on the number of positive solutions of certain nonlinear equations II. *Journal of Differential Equations*, 87:316–339, 1990.

34. K. Deimling. *Nonlinear Functional Analysis*. Springer-Verlag, Berlin–New York–Heidelberg, 1985.

35. M. Demazure. *Bifurcations and Catastrophes. Geometry of Solutions to Nonlinear Problems*. Springer-Verlag, Berlin–Heidelberg–New York, 2000.

36. O. Diekmann, S.A. van Gils, S.M. Verduyn Lunel, and H.-O. Walther. *Delay Equations*. Springer-Verlag, New York–Berlin–Heidelberg, 1995.

37. J. Dieudonné. Sur le polygone de Newton. *Archiv der Mathematik*, 2:49–55, 1950.

38. J. Dieudonné. *Foundations of Modern Analysis*. Academic Press, New York, 1964.

39. N. Dunford and J. Schwartz. *Linear Operators, Part I: General Theory*. Wiley-Interscience, New York, 1964.

40. G. Eisenack and C. Fenske. *Fixpunkttheorie*. Bibliographisches Institut, Mannheim, 1978.

41. K.D. Elworthy and A.J. Tromba. Degree theory on Banach manifolds. *Proceedings of Symposia in Pure Mathematics, Volume 18, Part 1*. American Mathematical Society, Providence, R.I., pages 86–94, 1970.

42. J. Esquinas. Optimal multiplicity in local bifurcation theory. II: General case. *Journal of Differential Equations*, 75:206–215, 1988.

43. J. Esquinas and J. Lopez-Gomez. Optimal multiplicity in local bifurcation theory I. Generalized generic eigenvalues. *Journal of Differential Equations*, 71:72–92, 1988.

44. G.M. Ewing. *Calculus of Variations with Applications*. Courier Dover Publications, New York, 1985.

45. C. Fenske. Analytische Theorie des Abbildungsgrades für Abbildungen in Banachräumen. *Mathematische Nachrichten*, 48:279–290, 1971.

46. C. Fenske. Extensio gradus ad quasdam applicationes Fredholmii. *Mitteilungen aus dem Mathematischen Seminar Giessen*, 121:65–70, 1976.

47. M. Fečkan. *Topological Degree Approach to Bifurcation Problems*. Springer Science + Business Media B.V., Dordrecht, NL, 2008.

48. B. Fiedler. An index for global Hopf bifurcation in parabolic systems. *Journal für die reine und angewandte Mathematik*, 359:1–36, 1985.

49. P.C. Fife, H. Kielhöfer, S. Maier-Paape, and T. Wanner. Perturbation of doubly periodic solution branches with applications to the Cahn–Hilliard equation. *Physica D*, 100:257–278, 1997.

50. P.M. Fitzpatrick and J. Pejsachowicz. Parity and generalized multiplicity. *Transactions of the American Mathematical Society*, 326:281–305, 1991.

51. P.M. Fitzpatrick, J. Pejsachowicz, and P.J. Rabier. Orientability of Fredholm Families and Topological Degree for Orientable Nonlinear Fredholm Mappings. *Journal of Functional Analysis*, 124:1–39, 1994.

52. A. Friedman. *Partial Differential Equations*. Holt, Rinehart and Winston, New York, 1969.

53. A. Friedman. *Partial Differential Equations of Parabolic Type*. Robert E. Krieger Publ. Comp., Malabar, Florida, 1983.

54. R.E. Gaines and J. Mawhin. *Coincidence Degree, and Nonlinear Differential Equations*. Lectures Notes in Mathematics, Volume 568. Springer-Verlag, Berlin–Heidelberg–New York, 1977.

55. B. Gidas, W.-M. Ni, and L. Nirenberg. Symmetry and related properties via the maximum principle. *Communications in Mathematical Physics*, 68:209–243, 1979.

56. D. Gilbarg and N.S. Trudinger. *Elliptic Partial Differential Equations of Second Order*. Springer-Verlag, Berlin–Heidelberg–New York, 1983.

57. M. Golubitsky, J.E. Marsden, I. Stewart, and M. Dellnitz. The Constrained Liapunov–Schmidt Procedure and Periodic Orbits. *Fields Institute Communications*, 4:81–127, 1995.

58. M. Golubitsky and D.G. Schaeffer. *Singularities and Groups in Bifurcation Theory, Volume I*. Springer-Verlag, Berlin–Heidelberg–New York, 1985.

59. M. Golubitsky, I. Stewart, and D.G. Schaeffer. *Singularities and Groups in Bifurcation Theory, Volume II*. Springer-Verlag, Berlin–Heidelberg–New York, 1988.

60. J. Guckenheimer and P. Holmes. *Nonlinear Oscillations, Dynamical Systems, and Bifurcations of Vector Fields*. Springer-Verlag, Berlin–Heidelberg–New York, 1983.

61. C. Gugg, T.J. Healey, H. Kielhöfer, and S. Maier-Paape. Nonlinear Standing and Rotating Waves on the Sphere. *Journal of Differential Equations*, 166:402–442, 2000.

62. J.K. Hale. *Ordinary Differential Equations*. John Wiley & Sons, New York, London, Sydney, 1969.

63. J.K. Hale. *Theory of Functional Differential Equations*. Springer-Verlag, New York–Berlin–Heidelberg, 1977.

64. J.K. Hale and H. Koçak. *Dynamics and Bifurcations*. Springer-Verlag, New York–Berlin–Heidelberg, 1991.

65. M. Haragus and G. Iooss. *Local Bifurcations, Center Manifolds, and Normal Forms in Infinite-Dimensional Dynamical Systems*. Springer-Verlag, New York–Berlin–Heidelberg, 2011.

66. T.J. Healey. Global bifurcation and continuation in the presence of symmetry with an application to solid mechanics. *SIAM Journal of Mathematical Analysis*, 19:824–840, 1988.

67. T.J. Healey and H. Kielhöfer. Symmetry and nodal properties in the global bifurcation analysis of quasi-linear elliptic equations. *Archive for Rational Mechanics and Analysis*, 113:299–311, 1991.

68. T.J. Healey and H. Kielhöfer. Hidden Symmetry of Fully Nonlinear Boundary Conditions in Elliptic Equations: Global Bifurcation and Nodal Structure. *Results in Mathematics*, 21:83–92, 1992.

69. T.J. Healey and H. Kielhöfer. Positivity of global branches of fully nonlinear elliptic boundary value problems. *Proceedings of the American Mathematical Society*, 115:1031–1036, 1992.

70. T.J. Healey and H. Kielhöfer. Preservation of nodal structure on global bifurcating solutions branches of elliptic equations with symmetry. *Journal of Differential Equations*, 106:70–89, 1993.

71. T.J. Healey and H. Kielhöfer. Separation of global solution branches of elliptic systems with symmetry via nodal properties. *Nonlinear Analysis. Theory, Methods & Applications*, 21:665–684, 1993.

72. T.J. Healey and H. Kielhöfer. Free nonlinear vibrations for a class of two-dimensional plate equations: Standing and discrete-rotating waves. *Nonlinear Analysis. Theory, Methods & Applications*, 29:501–531, 1997.

73. T.J. Healey, H. Kielhöfer, and E.L. Montes-Pizarro. Free nonlinear vibrations for plate equations on the equilateral triangle. *Nonlinear Analysis. Theory, Methods & Applications*, 44:575–599, 2001.

74. T.J. Healey, H. Kielhöfer, and C.A. Stuart. Global branches of positive weak solutions of semilinear elliptic problems over non-smooth domains. *Proceedings of the Royal Society of Edinburgh*, 124 A:371–388, 1994.

75. T.J. Healey and H.C. Simpson. Global continuation in nonlinear elasticity. *Archive for Rational Mechanics and Analysis*, 143:1–28, 1998.

76. D. Henry. *Geometric Theory of Semilinear Parabolic Equations*. Lecture Notes in Mathematics, Volume 840. Springer-Verlag, Berlin–Heidelberg–New York, 1981.

77. P. Hess and T. Kato. On some linear and nonlinear eigenvalue problems with an indefinite weight function. *Communications in Partial Differential Equations*, 5:999–1030, 1980.

78. M. Holzmann and H. Kielhöfer. Uniqueness of global positive solution branches of nonlinear elliptic problems. *Mathematische Annalen*, 300:221–241, 1994.

79. E. Hopf. Abzweigung einer periodischen Lösung von einer stationären Lösung eines Differentialsystems. *Berichte der Sächsischen Akademie der Wissenschaften*, 94:1–22, 1942.

80. G. Iooss and M. Adelmeyer. *Topics in Bifurcation Theory and Applications*. Advanced Series in Nonlinear Dynamics. Vol.3, 2nd edition, World Scientific, Singapore, 1999.

81. G. Iooss and D. Joseph. *Elementary Stability and Bifurcation Theory*. Springer-Verlag, New York–Berlin–Heidelberg, 1980.

82. J. Ize. *Bifurcation Theory for Fredholm Operators*. Memoirs of the American Mathematical Society 174, Providence, 1976.

83. J. Ize. Obstruction theory and multiparameter Hopf bifurcation. *Transactions of the American Mathematical Society*, 289:757–792, 1985.

84. J. Ize. Topological bifurcation. *Topological nonlinear analysis: degree, singularity, and variations*. Birkhäuser-Verlag, Boston, MA, pages 341–463, 1995.

85. J. Ize, J. Massabo, and A. Vignoli. Degree theory for equivariant maps. I. *Transactions of the American Mathematical Society*, 315:433–510, 1989.

86. T. Kato. *Perturbation Theory for Linear Operators*. Springer-Verlag, Berlin–Heidelberg–New York, 1984.

87. H.B. Keller. *Numerical Methods in Bifurcation Problems*. Springer-Verlag, Berlin–Heidelberg–New York, 1987.

88. H. Kielhöfer. Halbgruppen und semilineare Anfangs-Randwertprobleme. *Manuscripta Mathematica*, 12:121–152, 1974.

89. H. Kielhöfer. Stability and Semilinear Evolution Equations in Hilbert Space. *Archive for Rational Mechanics and Analysis*, 57:150–165, 1974.

90. H. Kielhöfer. Existenz und Regularität von Lösungen semilinearer parabolischer Anfangs- Randwertprobleme. *Mathematische Zeitschrift*, 142:131–160, 1975.

91. H. Kielhöfer. On the Lyapunov Stability of Stationary Solutions of Semilinear Parabolic Differential Equations. *Journal of Differential Equations*, 22:193–208, 1976.

92. H. Kielhöfer. Bifurcation of Periodic Solutions for a Semilinear Wave Equation. *Journal of Mathematical Analysis and Applications*, 68:408–420, 1979.

93. H. Kielhöfer. Generalized Hopf Bifurcation in Hilbert Space. *Mathematical Methods in the Applied Sciences*, 1:498–513, 1979.

94. H. Kielhöfer. Hopf Bifurcation at Multiple Eigenvalues. *Archive for Rational Mechanics and Analysis*, 69:53–83, 1979.

95. H. Kielhöfer. Degenerate Bifurcation at Simple Eigenvalues and Stability of Bifurcating Solutions. *Journal of Functional Analysis*, 38:416–441, 1980.

96. H. Kielhöfer. A Bunch of Stationary or Periodic Solutions Near an Equilibrium by a Slow Exchange of Stability. *Nonlinear Differential Equations: Invariance, Stability, and Bifurcation*, Academic Press, New-York, pages 207–219, 1981.

97. H. Kielhöfer. Floquet Exponents of Bifurcating Periodic Orbits. *Nonlinear Analysis. Theory, Methods & Applications*, 6:571–583, 1982.

98. H. Kielhöfer. Multiple eigenvalue bifurcation for Fredholm operators. *Journal für die reine und angewandte Mathematik*, 358:104–124, 1985.

99. H. Kielhöfer. Interaction of periodic and stationary bifurcation from multiple eigenvalues. *Mathematische Zeitschrift*, 192:159–166, 1986.

100. H. Kielhöfer. A Bifurcation Theorem for Potential Operators. *Journal of Functional Analysis*, 77:1–8, 1988.

101. H. Kielhöfer. A Bifurcation Theorem for Potential Operators and an Application to Wave Equations. *Proceedings Int. Conf. on Bifurcation Theory and Its Num. Anal. in Xi'an, China*, Xi'an Jiaotong University Press, pages 270–276, 1989.

102. H. Kielhöfer. Hopf Bifurcation from a Differentiable Viewpoint. *Journal of Differential Equations*, 97:189–232, 1992.

103. H. Kielhöfer. Smoothness and asymptotics of global positive branches of $\Delta u + \lambda f(u) = 0$. *ZAMP. Zeitschrift für Angewandte Mathematik und Physik*, 43:139–153, 1992.

104. H. Kielhöfer. Smoothness of global positive branches of nonlinear elliptic problems over symmetric domains. *Mathematische Zeitschrift*, 211:41–48, 1992.

105. H. Kielhöfer. Generic S^1-Equivariant Vector Fields. *Journal of Dynamics and Differential Equations*, 6:277–300, 1994.

106. H. Kielhöfer. Pattern formation of the stationary Cahn–Hilliard model. *Proceedings of the Royal Society of Edinburgh*, 127 A:1219–1243, 1997.

107. H. Kielhöfer. Minimizing sequences selected via singular perturbations, and their pattern formation. *Archive for Rational Mechanics and Analysis*, 155:261–276, 2000.

108. H. Kielhöfer. Critical points of nonconvex and noncoercive functionals. *Corrigenda. Calculus of Variations*, 21:429–436, 2004.

109. H. Kielhöfer. *Variationsrechnung*. Vieweg + Teubner Verlag, Wiesbaden, 2010.

110. H. Kielhöfer and P. Kötzner. Stable periods of a semilinear wave equation and bifurcation of periodic solutions. *ZAMP. Zeitschrift für Angewandte Mathematik und Physik*, 38:204–212, 1987.

111. H. Kielhöfer and R. Lauterbach. On the Principle of Reduced Stability. *Journal of Functional Analysis*, 53:99–111, 1983.

112. H. Kielhöfer and S. Maier. Infinitely many positive solutions of semilinear elliptic problems via sub- and supersolutions. *Communications in Partial Differential Equations*, 18:1219–1229, 1993.

113. K. Kirchgässner. Multiple Eigenvalue Bifurcation for Holomorphic Mappings. *Contributions to Nonlinear Functional Analysis*. Academic Press, New York, pages 69–99, 1977.

114. K. Kirchgässner. Wave solutions of reversible systems and applications. *Journal of Differential Equations*, 45:113–127, 1982.

115. K. Kirchgässner and H. Kielhöfer. Stability and Bifurcation in Fluid Dynamics. *Rocky Mountain Journal of Mathematics*, 3:275–318, 1973.

116. K. Kirchgässner and J. Scheurle. Verzweigung und Stabilität von Lösungen semilinearer elliptischer Randwertprobleme. *Jahresbericht der Deutschen Mathematiker-Vereinigung*, 77:39–54, 1975.

117. K. Kirchgässner and J. Scheurle. Global branches of periodic solutions of reversible systems. *Recent Contributions to Nonlinear Partial Differential Equations*. Research Notes in Mathematics, 50, Pitman, Boston–London–Melbourne, pages 103–130, 1981.

118. M.A. Krasnosel'skii. *Topological Methods in the Theory of Nonlinear Integral Equations*. Pergamon Press, Oxford, 1964.

119. S. Krömer, T.J. Healey, and H. Kielhöfer. Bifurcation with a two-dimensional kernel. *Journal of Differential Equations*, 220:234–258, 2006.

120. S. Krömer and H. Kielhöfer. Radially symmetric critical points of non-convex functionals. *Proceedings of the Royal Society of Edinburgh*, 138A:1261–1280, 2008.

121. O.A. Ladyzhenskaja, V.A. Solonnikov, and N.N. Ural'ceva. *Linear and Quasilinear Equations of Parabolic Type*. AMS Transl. Math. Monographs, Vol. 23, Providence, Rhode-Island, 1968.

122. O.A. Ladyzhenskaja and N.N. Ural'ceva. *Linear and Quasilinear Elliptic Equations*. Academic Press, New York-London, 1968.

123. B. Laloux and J. Mawhin. Multiplicity, Leray–Schauder Formula and Bifurcation. *Journal of Differential Equations*, 24:301–322, 1977.

124. R. Lauterbach. Hopf bifurcation from a turning point. *Journal für die reine und angewandte Mathematik*, 360:136–152, 1985.

125. J. Leray and J. Schauder. Topologie et équations fonctionelles. *Annales Scientifiques de l'École Normale Supérieure*, 51:45–78, 1934.

126. J.L. Lions and E. Magenes. *Problèmes aux limites non homogènes et applications. Non-homogeneous boundary value problems and applications*. Dunod, Paris, Springer-Verlag, Berlin–Heidelberg–New York, 1972.

127. P.L. Lions. On the existence of positive solutions of semilinear elliptic equations. *SIAM Review*, 24:441–467, 1982.

128. N.G. Lloyd. *Degree Theory*. Cambridge University Press, Cambridge, London, 1978.

129. J. Lopez-Gomez. Multiparameter Local Bifurcation Based on the Linear Part. *Journal of Mathematical Analysis and Applications*, 138:358–370, 1989.

130. T. Ma and S. Wang. Bifurcation of Nonlinear Equations: I. Steady State Bifurcation. *Methods and Applications of Analysis*, 11:155–178, 2004.

131. T. Ma and S. Wang. *Bifurcation Theory and Applications*. World Scientific, Series on Nonlinear Science, Series A, Vol.53, Singapore, 2005.

132. K.W. MacEwen and T.J. Healey. A Simple Approach to the 1:1 Resonance Bifurcation in Follower-Load Problems. *Preprint*, 2001.

133. R.J. Magnus. A Generalization of Multiplicity and the Problem of Bifurcation. *Proceedings of the London Mathematical Society*, 32:251–278, 1976.

134. J. Mallet-Paret and J.A. Yorke. Snakes: Oriented families of periodic orbits, their sources, sinks, and continuation. *Journal of Differential Equations*, 43:419–450, 1982.

135. A. Marino. La biforcazione nel caso variazionale. Confer. Sem. Mat. Univ. Bari, 132, Bari, 1973.

136. J. Mawhin. Equivalence theorems for nonlinear operator equations and coincidence degree theory for some mappings in locally convex topological vector spaces. *Journal of Differential Equations*, 12:610–636, 1972.

137. J. Mawhin. *Topological Degree Methods in Nonlinear Boundary Value Problems*. Regional conference series in mathematics, Volume 40. American Mathematical Society, Providence, R.I., 1977.

138. J. Mawhin. Nonlinear functional analysis and periodic solutions of semilinear wave equations. *Nonlinear Phenomena in Mathematical Sciences*. Academic Press, New York, pages 671–681, 1982.

139. A. Mielke. Reduction of quasilinear elliptic equations in cylindrical domains with applications. *Mathematical Methods in the Applied Sciences*, 10:51–66, 1988.

140. A. Mielke. The Ginzburg–Landau Equation in Its Role as a Modulation Equation. *Handbook of Dynamical Systems*, 2:759–834, 2002. Edited by B. Fiedler, Elsevier, Amsterdam.

141. T. Ouyang and J. Shi. Exact multiplicity of positive solutions for a class of semilinear problems. II. *Journal of Differential Equations*, 158:94–151, 1999.

142. A. Pazy. *Semigroups of Linear Operators and Applications to Partial Differential Equations*. Springer-Verlag, New York–Berlin–Heidelberg, 1983.

143. H.O. Peitgen and K. Schmitt. Global topological perturbations of nonlinear elliptic eigenvalue problems. *Mathematical Methods in the Applied Sciences*, 5:376–388, 1983.

144. G.H. Pimbley. *Eigenfunction Branches of Nonlinear Operators, and Their Bifurcations*. Lecture Notes in Mathematics, Volume 104. Springer-Verlag, Berlin–Heidelberg–New York, 1969.

145. F. Quinn and A. Sard. Hausdorff conullity of critical images of Fredholm maps. *American Journal of Mathematics*, 94:1101–1110, 1972.

146. P.J. Rabier. Generalized Jordan chains and two bifurcation theorems of Krasnoselskii. *Nonlinear Analysis. Theory, Methods & Applications*, 13:903–934, 1989.

147. P.H. Rabinowitz. Some global results for nonlinear eigenvalue problems. *Journal of Functional Analysis*, 7:487–513, 1971.

148. P.H. Rabinowitz. Time periodic solutions of nonlinear wave equations. *Manuscripta Mathematica*, 5:165–194, 1971.

149. P.H. Rabinowitz. On Bifurcation From Infinity. *Journal of Differential Equations*, 14:462–475, 1973.

150. P.H. Rabinowitz. Variational methods for nonlinear eigenvalue problems. *C.I.M.E. III. Ciclo, Varenna, 1974*, Edizione Cremonese, Roma, pages 139–195, 1974.

151. P.H. Rabinowitz. A Bifurcation Theorem for Potential Operators. *Journal of Functional Analysis*, 25:412–424, 1977.

152. P.H. Rabinowitz. Free vibrations for a semilinear wave equation. *Communications on Pure and Applied Mathematics*, 31:31–68, 1978.

153. E.L. Reiss. Column buckling – an elementary example of bifurcation. In J.B. Keller and S. Antman, editors, *Bifurcation Theory and Nonlinear Eigenvalue Problems*. Benjamin, New York, 1969.

154. P. Sarreither. Transformationseigenschaften endlicher Ketten und allgemeine Verzweigungsaussagen. *Mathematica Scandinavica*, 35:115–128, 1974.

155. D.H. Sattinger. *Topics in Stability and Bifurcation Theory*. Lecture Notes in Mathematics, Volume 309. Springer-Verlag, Berlin–Heidelberg–New York, 1972.

156. R. Schaaf. *Global Solution Branches of Two Point Boundary Value Problems*. Lecture Notes in Mathematics, Volume 1458. Springer-Verlag, Berlin–Heidelberg–New York, 1990.

157. D. S. Schmidt. Hopf's bifurcation theorem and the center theorem of Liapunov. *Celestical Mechanics*, 9:81–103, 1976.

158. M. Sevryuk. *Reversible Systems*. Lecture Notes in Mathematics, Volume 1211. Springer-Verlag, Berlin–Heidelberg–New York, 1986.

159. R. Seydel. *Practical Bifurcation and Stability Analysis: From Equilibrium to Chaos*. 2nd ed. Springer-Verlag, New York, 1994.

160. S. Smale. An Infinite Dimensional Version of Sard's Theorem. *American Journal of Mathematics*, 87:861–866, 1965.

161. J.A. Smoller and A.G. Wasserman. Existence, uniqueness, and nondegeneracy of positive solutions of semilinear elliptic equations. *Communications in Mathematical Physics*, 95:129–159, 1984.

162. P.E. Sobolevskii. Equations of Parabolic Type in Banach Space. *American Mathematical Society, Translation Series II*, 49:1–62, 1966.

163. M.M. Vainberg and V.A. Trenogin. *Theory of Branching of Solutions of Non-linear Equations*. Noordhoff International Publishers, Leyden, 1974.

164. A. Vanderbauwhede. *Local Bifurcation and Symmetry*. Research Notes in Mathematics, 75. Pitman, Boston–London–Melbourne, 1982.

165. A. Vanderbauwhede. Hopf bifurcation for equivariant conservative and time-reversible systems. *Proceedings of the Royal Society of Edinburgh*, 116 A:103–128, 1990.

166. W. von Wahl. Gebrochene Potenzen eines elliptischen Operators und parabolische Differentialgleichungen in Räumen hölderstetiger Funktionen. *Nachrichten der Akademie der Wissenschaften Göttingen, II. Math. Physik. Klasse*, 11:231–258, 1972.

167. H.F. Weinberger. On the stability of bifurcating solutions. *Nonlinear Analysis. A Collection of Papers in Honor of Erich Rothe*. Academic Press, New York, pages 219–233, 1978.

168. D. Westreich. Bifurcation at eigenvalues of odd multiplicity. *Proceedings of the American Mathematical Society*, 41:609–614, 1973.

169. G. T. Whyburn. *Topological Analysis*. Princeton University Press, New Jersey, 1958.

170. K. Yosida. *Functional Analysis*. Springer-Verlag, Berlin–Heidelberg–New York, 1980.

171. E. Zeidler. *Nonlinear Functional Analysis and Its Applications. I: Fixed-Point Theorems*. Springer-Verlag, New York–Berlin–Heidelberg, 1986.

Index